Methods in Molecular Biology

Series Editor
John M. Walker
School of Life and Medical Sciences
University of Hertfordshire
Hatfield, Hertfordshire, UK

For further volumes:
http://www.springer.com/series/7651

For over 35 years, biological scientists have come to rely on the research protocols and methodologies in the critically acclaimed *Methods in Molecular Biology* series. The series was the first to introduce the step-by-step protocols approach that has become the standard in all biomedical protocol publishing. Each protocol is provided in readily-reproducible step-by-step fashion, opening with an introductory overview, a list of the materials and reagents needed to complete the experiment, and followed by a detailed procedure that is supported with a helpful notes section offering tips and tricks of the trade as well as troubleshooting advice. These hallmark features were introduced by series editor Dr. John Walker and constitute the key ingredient in each and every volume of the *Methods in Molecular Biology* series. Tested and trusted, comprehensive and reliable, all protocols from the series are indexed in PubMed.

Stem Cell Assays

Methods and Protocols

Edited by

Nagarajan Kannan

Stem Cell and Cancer Biology Laboratory, Division of Experimental Medicine and Pathology, Department of Laboratory Medicine and Pathology, Mayo Clinic, Rochester, MN, USA

Philip Beer

Hull York Medical School, York, UK

Humana Press

Editors
Nagarajan Kannan
Stem Cell and Cancer Biology
Laboratory, Division of Experimental
Medicine and Pathology, Department
of Laboratory Medicine and Pathology
Mayo Clinic
Rochester, MN, USA

Philip Beer
Hull York Medical School
York, UK

ISSN 1064-3745 ISSN 1940-6029 (electronic)
Methods in Molecular Biology
ISBN 978-1-0716-1978-0 ISBN 978-1-0716-1979-7 (eBook)
https://doi.org/10.1007/978-1-0716-1979-7

This Humana imprint is published by the registered company Springer Science+Business Media, LLC, part of Springer Nature.
The registered company address is: 1 New York Plaza, New York, NY 10004, U.S.A.

Preface

Stem cell science is a rapidly growing field with complex research methodologies applied across different laboratories. It is important to have common practices for routinely used methods in order to collectively advance the research enterprise and to avoid getting bogged down by issues relating to reproducibility, which has come to become the bane of science in recent years.

As a humble contribution to the ongoing greater community effort to bring more reproducibility in stem cell research, we bring you this *Methods in Molecular Biology* series book titled *Stem Cell Assays*. The book is a collection of 39 protocols applicable to fields of stem cell biology, regenerative medicine, and cancer biology. In the first edition of *Stem Cell Assays*, our intention is to bring to you a broad collection of protocols used in stem cell investigation, rather than focus solely on protocols related to assaying stem/progenitor activity. Some protocols capture innate behaviors of the primitive cells while others describe manipulation of such cell systems. We have not attempted to address shortcomings of methods used in stem cell science, and the book is not meant to provide guidelines for the conduct of stem cell investigation. Rather, the book is intended to be used as a laboratory manual, to encourage researchers to explore new techniques and approaches.

We hope this handy book will be a guide for researchers in both academia and industry. For academics, we provide an overview of the different types of assays available to help answer the myriad questions still vexing the stem field. For industry, we provide a snapshot of the current state of the art in model systems that have potential utility in many fields, particularly the field of drug discovery and development where these advanced assays can be leveraged to ascertain therapeutic efficacy and screen for unwanted toxicities.

The protocols described here are split thematically into 4 areas: Chapters 1–8 are focused on embryonic stem cells (ES) and induced pluripotent cells (iPS), Chapters 9–17 on ES/iPSC-derived tissue stem cells, Chapters 18–31 on tissue resident stem cells, and Chapters 32–39 on assays for primitive malignant cells. We have also included chapters on topics such as CRISPR/Cas9 engineering and transgenic model development due to their wide application in stem cell research.

We would like to highlight that the field of cancer stem cells has evolved uncontrollably in recent years, much like the tissues they represent. There are currently no formal guidelines or standardized methods in stem cell research, and we have not significantly influenced the chapters based on our own personal viewpoints.

We would like to thank all our authors and the publisher for their collegiality, cooperation, and understanding during the uncertainty of the pandemic. We hope our readers will find the chapters as interesting as we did while reviewing them.

Rochester, MN, USA — *Nagarajan Kannan*
York, UK — *Philip Beer*

Contents

Contributors

SATISH KUMAR ADIGA • *Division of Clinical Embryology, Department of Reproductive Science, Kasturba Medical College, Manipal, Manipal Academy of Higher Education, Manipal, India*

YOKO AMAGASE • *Department of Pathophysiology, Faculty of Pharmaceutical Sciences, Doshisha Women's College of Liberal Arts, Kodo, Kyotanabe, Kyoto, Japan*

REZA ARDEHALI • *Division of Cardiology, Department of Internal Medicine, David Geffen School of Medicine, University of California, Los Angeles, Los Angeles, CA, USA; Eli and Edythe Broad Stem Cell Research Center, University of California, Los Angeles, Los Angeles, CA, USA; Molecular, Cellular and Integrative Physiology Graduate Program, University of California, Los Angeles, Los Angeles, CA, USA; Molecular Biology Institute, University of California, Los Angeles, Los Angeles, CA, USA*

PAIGE C. ARNESON-WISSINK • *Department of Biochemistry and Molecular Biology, Mayo Clinic, Rochester, MN, USA*

MARTA BAIOCCHI • *Department of Oncology and Molecular Medicine, Istituto Superiore di Sanità, Rome, Italy*

JAMIE BAKKUM-GAMEZ • *Department of Gynecological Surgery, Mayo Clinic, Rochester, MN, USA; Mayo Clinic Cancer Center, Mayo Clinic, Rochester, MN, USA*

GANG BAO • *Department of Bioengineering, Rice University, Houston, TX, USA*

SUSAN C. BARNETT • *College of Medical, Veterinary and Life Sciences, Institute of Infection, Immunity and Inflammation, University of Glasgow, Glasgow, UK*

CHEK MEI BOK • *Stem Cells and Diabetes Laboratory, Institute of Molecular and Cell Biology, A*STAR, Singapore, Singapore*

RUMELA CHAKRABARTI • *Department of Biomedical Sciences, School of Veterinary Medicine, University of Pennsylvania, Philadelphia, PA, USA*

CHI-WA CHENG • *The Cardiology Division, Department of Medicine, Li Ka Shing Faculty of Medicine, Queen Mary Hospital, The University of Hong Kong, Hong Kong, SAR, China; Hong Kong-Guangdong Stem Cell and Regenerative Medicine Research Centre, The University of Hong Kong and Guangzhou Institutes of Biomedicine and Health, Hong Kong, SAR, China*

MITRA COWAN • *McGill Integrated Core for Animal Modeling (MICAM), McGill University, Montreal, QC, Canada*

DAPHNE NORMA CRASTA • *Division of Reproductive Biology, Department of Reproductive Science, Kasturba Medical College, Manipal, Manipal Academy of Higher Education, Manipal, India; Stem Cell and Cancer Biology Laboratory, Division of Experimental Pathology and Laboratory Medicine, Department of Laboratory Medicine and Pathology, Mayo Clinic, Rochester, MN, USA*

CHIARA D'ANTONI • *Department of Biology and Biotechnologies "Charles Darwin", Sapienza University of Rome, Rome, Italy; Laboratory Affiliated to Istituto Pasteur Italia-Fondazione Cenci Bolognetti, Department of Biology and Biotechnologies "Charles Darwin", Sapienza University of Rome, Rome, Italy*

MARIA LAURA DE ANGELIS • *Department of Oncology and Molecular Medicine, Istituto Superiore di Sanità, Rome, Italy*

PETRA DEN HOLLANDER • *Department of Translational Molecular Pathology, The University of Texas MD Anderson Cancer Center, Houston, TX, USA*
ABHIJEET P. DESHMUKH • *Department of Translational Molecular Pathology, The University of Texas MD Anderson Cancer Center, Houston, TX, USA*
JADE DESJARDINS • *McGill Integrated Core for Animal Modeling (MICAM), McGill University, Montreal, QC, Canada*
NIVEDHITHA DEVARAJU • *Centre for Stem Cell Research (a unit of inStem, Bangalore), Christian Medical College Campus Bagayam, Vellore, Tamil Nadu, India; Manipal Academy of Higher Education, Mangalore, Karnataka, India*
JASON D. DOLES • *Department of Biochemistry and Molecular Biology, Mayo Clinic, Rochester, MN, USA*
ESTHER EBERHARDT • *Department of Anesthesiology, Uniklinik RWTH Aachen, Aachen, Germany*
ALLISON D. EBERT • *Department of Cell Biology, Neurobiology and Anatomy, Medical College of Wisconsin, Milwaukee, WI, USA*
MIGUEL A. ESTEBAN • *Hong Kong-Guangdong Stem Cell and Regenerative Medicine Research Centre, The University of Hong Kong and Guangzhou Institutes of Biomedicine and Health, Hong Kong, SAR, China; Laboratory of Integrative Biology, Guangzhou Institutes of Biomedicine and Health, Chinese Academy of Sciences, Guangzhou, China; Joint School of Life Sciences, Guangzhou Medical University and Guangzhou Institutes of Biomedicine and Health, Guangzhou, China*
LIANG FENG • *Stem Cell and Cancer Biology Laboratory, Division of Experimental Pathology and Laboratory Medicine, Department of Laboratory Medicine and Pathology, Mayo Clinic, Rochester, MN, USA*
CLIFFORD D. L. FOLMES • *Stem Cell and Regenerative Metabolism Laboratory, Departments of Cardiovascular Medicine, Biochemistry and Molecular Biology, and Center for Regenerative Medicine, Mayo Clinic, Scottsdale, AZ, USA*
MATTHEW J. FORD • *Department of Human Genetics, Rosalind and Morris Goodman Cancer Institute, McGill University, Montreal, QC, Canada*
FEDERICA FRANCESCANGELI • *Department of Oncology and Molecular Medicine, Istituto Superiore di Sanità, Rome, Italy*
XIUGONG GAO • *Division of Toxicology, Office of Applied Research and Safety Assessment, Center for Food Safety and Applied Nutrition, U.S. Food and Drug Administration, Laurel, MD, USA*
MARIA GIOVANNA GARONE • *Department of Biology and Biotechnologies "Charles Darwin", Sapienza University of Rome, Rome, Italy*
THORARINN GUDJONSSON • *Stem Cell Research Unit, Department of Anatomy, Faculty of Medicine, Biomedical Center, School of Health Sciences, University of Iceland, Reykjavik, Iceland; Department of Laboratory Hematology, Landspitali—University Hospital, Reykjavik, Iceland*
STEPHEN J. HAGGARTY • *Chemical Neurobiology Laboratory, Center for Genomic Medicine, Department of Neurology, Massachusetts General Hospital and Harvard Medical School, Boston, MA, USA*
BYLGJA HILMARSDOTTIR • *Department of Pathology, Landspitali—University Hospital, Reykjavik, Iceland*

WAI-IN HO • *The Cardiology Division, Department of Medicine, Li Ka Shing Faculty of Medicine, Queen Mary Hospital, The University of Hong Kong, Hong Kong, SAR, China; Hong Kong-Guangdong Stem Cell and Regenerative Medicine Research Centre, The University of Hong Kong and Guangzhou Institutes of Biomedicine and Health, Hong Kong, SAR, China*

ANAN A. ISHTIAH • *Stem Cell and Gene Therapy Group, Regenerative Medicine Cluster, Advanced Medical and Dental Institute (IPPT), Universiti Sains Malaysia, Sains@Bertam, Penang, Malaysia*

HIROKO IZUMI-NAKASEKO • *Department of Pharmacology, Faculty of Medicine, Toho University, Ota-ku, Tokyo, Japan*

XUANMAO JIAO • *Pennsylvania Cancer and Regenerative Medicine Research Center, Baruch S. Blumberg Institute, Wynnewood, PA, USA; Xavier University School of Medicine, Woodbury, NY, USA*

ROBIYA JOSEPH • *Department of Translational Molecular Pathology, The University of Texas MD Anderson Cancer Center, Houston, TX, USA*

HO SUN JUNG • *Department of Cell and Regenerative Biology, University of Wisconsin School of Medicine and Public Health, Madison, WI, USA*

GURUPRASAD KALTHUR • *Division of Reproductive Biology, Department of Reproductive Science, Kasturba Medical College, Manipal, Manipal Academy of Higher Education, Manipal, India; Stem Cell and Cancer Biology Laboratory, Division of Experimental Pathology and Laboratory Medicine, Department of Laboratory Medicine and Pathology, Mayo Clinic, Rochester, MN, USA*

NAGARAJAN KANNAN • *Stem Cell and Cancer Biology Laboratory, Division of Experimental Medicine and Pathology, Department of Laboratory Medicine and Pathology, Mayo Clinic, Rochester, MN, USA*

PRITINDER KAUR • *Curtin Medical School, Curtin University, Bentley, WA, Australia; Curtin Health Innovation Research Institute, Bentley, WA, Australia*

ALEXANDER KELLER • *Research Group Reproduction and Genetics, Vrije Universiteit Brussel, Jette, Belgium*

LISA M. KIM • *Department of Pharmacology, University of Minnesota Medical School, Minneapolis, MN, USA*

PAUL Y. KIM • *Department of Pharmacology, University of Minnesota Medical School, Minneapolis, MN, USA*

DOREEN KLOOS • *Institute of Experimental Hematology, Hannover Medical School, Hannover, Germany; REBIRTH, Research Center for Translational and Regenerative Medicine, Hannover Medical School, Hannover, Germany; Department for Pediatric Pulmonology, Allergology and Neonatology, Hannover Medical School, Hannover, Germany*

NUŠA KRIVEC • *Research Group Reproduction and Genetics, Vrije Universiteit Brussel, Jette, Belgium*

NICK A. KUBURICH • *Department of Translational Molecular Pathology, The University of Texas MD Anderson Cancer Center, Houston, TX, USA*

AGATA KURZYK • *Department of Cancer Biology, Maria Sklodowska-Curie National Research Institute of Oncology, Warsaw, Poland*

NICO LACHMANN • *REBIRTH, Research Center for Translational and Regenerative Medicine, Hannover Medical School, Hannover, Germany; Department for Pediatric*

Pulmonology, Allergology and Neonatology, Hannover Medical School, Hannover, Germany; RESIST Cluster of Excellence, Hannover Medical School, Hannover, Germany; Biomedical Research in Endstage and Obstructive Lung Disease Hannover (BREATH), German Center for Lung Research (DZL), Hannover, Germany

ANGELIKA LAMPERT • *Institute of Physiology, Uniklinik RWTH Aachen, Aachen, Germany*

CIARAN M. LEE • *APC Microbiome Ireland, University College Cork, Cork, Ireland*

CHEUK T. LEUNG • *Department of Pharmacology, University of Minnesota Medical School, Minneapolis, MN, USA; Masonic Cancer Center, University of Minnesota Medical School, Minneapolis, MN, USA*

RONG LI • *Division of Toxicology, Office of Applied Research and Safety Assessment, Center for Food Safety and Applied Nutrition, U.S. Food and Drug Administration, Laurel, MD, USA*

ZHIPING LI • *Pennsylvania Cancer and Regenerative Medicine Research Center, Baruch S. Blumberg Institute, Wynnewood, PA, USA; Xavier University School of Medicine, Woodbury, NY, USA*

SUSAN L. LINDSAY • *College of Medical, Veterinary and Life Sciences, Institute of Infection, Immunity and Inflammation, University of Glasgow, Glasgow, UK*

CHUN LIU • *Stanford Cardiovascular Institute, Stanford University School of Medicine, Stanford, CA, USA*

IAN LUECKE • *Department of Cell Biology, Neurobiology and Anatomy, Medical College of Wisconsin, Milwaukee, WI, USA*

SENDURAI A. MANI • *Department of Translational Molecular Pathology, The University of Texas MD Anderson Cancer Center, Houston, TX, USA*

CHRISTINA MARKOULI • *Research Group Reproduction and Genetics, Vrije Universiteit Brussel, Jette, Belgium*

ANGELICA MEDINA • *Department of Biological Science, Florida State University, Tallahassee, FL, USA*

KATHERINE MINTER-DYKHOUSE • *Stem Cell and Regenerative Metabolism Laboratory, Departments of Cardiovascular Medicine, Biochemistry and Molecular Biology, and Center for Regenerative Medicine, Mayo Clinic, Scottsdale, AZ, USA*

KENJI MISHIMA • *Division of Pathology, Department of Oral Diagnostic Sciences, School of Dentistry, Showa University, Tokyo, Japan*

A. MOAURO • *Graduate Program in Physiology and Osteopathic Medicine, Michigan State University, East Lansing, MI, USA*

KUMARASAMYPET M. MOHANKUMAR • *Centre for Stem Cell Research (a unit of inStem, Bangalore), Christian Medical College Campus Bagayam, Vellore, Tamil Nadu, India*

AJEYA NANDI • *Department of Biomedical Sciences, School of Veterinary Medicine, University of Pennsylvania, Philadelphia, PA, USA*

GHATA NANDI • *Chemical Neurobiology Laboratory, Center for Genomic Medicine, Department of Neurology, Massachusetts General Hospital and Harvard Medical School, Boston, MA, USA*

ANIKA NEUREITER • *Institute of Physiology, Uniklinik RWTH Aachen, Aachen, Germany*

KWONG-MAN NG • *The Cardiology Division, Department of Medicine, Li Ka Shing Faculty of Medicine, Queen Mary Hospital, The University of Hong Kong, Hong Kong, SAR, China; Hong Kong-Guangdong Stem Cell and Regenerative Medicine Research Centre, The University of Hong Kong and Guangzhou Institutes of Biomedicine and Health, Hong Kong, SAR, China*

NATASHA HUI JIN NG • *Stem Cells and Diabetes Laboratory, Institute of Molecular and Cell Biology, A*STAR, Singapore, Singapore*

NGOC B. NGUYEN • *Division of Cardiology, Department of Internal Medicine, David Geffen School of Medicine, University of California, Los Angeles, Los Angeles, CA, USA; Eli and Edythe Broad Stem Cell Research Center, University of California, Los Angeles, Los Angeles, CA, USA; Molecular, Cellular and Integrative Physiology Graduate Program, University of California, Los Angeles, Los Angeles, CA, USA*

SO HYUN PARK • *Department of Bioengineering, Rice University, Houston, TX, USA*

VIVEK I. PATEL • *Xavier University School of Medicine, Woodbury, NY, USA*

RICHARD G. PESTELL • *Pennsylvania Cancer and Regenerative Medicine Research Center, Baruch S. Blumberg Institute, Wynnewood, PA, USA; Xavier University School of Medicine, Woodbury, NY, USA; The Wistar Cancer Center, Wistar Institute, Philadelphia, PA, USA*

ARASH PEZHOUMAN • *Division of Cardiology, Department of Internal Medicine, David Geffen School of Medicine, University of California, Los Angeles, Los Angeles, CA, USA; Eli and Edythe Broad Stem Cell Research Center, University of California, Los Angeles, Los Angeles, CA, USA*

ZALITHA PIETERSE • *Curtin Medical School, Curtin University, Bentley, WA, Australia; Curtin Health Innovation Research Institute, Bentley, WA, Australia*

RONG QIAO • *Division of Cardiology, Department of Internal Medicine, David Geffen School of Medicine, University of California, Los Angeles, Los Angeles, CA, USA*

VIGNESH RAJENDIRAN • *Centre for Stem Cell Research (a unit of inStem, Bangalore), Christian Medical College Campus Bagayam, Vellore, Tamil Nadu, India*

A. RALSTON • *Department of Biochemistry and Molecular Biology, Michigan State University, East Lansing, MI, USA*

NITHIN SAM RAVI • *Centre for Stem Cell Research (a unit of inStem, Bangalore), Christian Medical College Campus Bagayam, Vellore, Tamil Nadu, India*

NORIDZZAIDA RIDZUAN • *Lung Stem Cell and Gene Therapy Group, Regenerative Medicine Cluster, Advanced Medical and Dental Institute (IPPT), Universiti Sains Malaysia, Kepala Batas, Penang, Malaysia*

ALESSANDRO ROSA • *Department of Biology and Biotechnologies "Charles Darwin", Sapienza University of Rome, Rome, Italy; Laboratory Affiliated to Istituto Pasteur Italia-Fondazione Cenci Bolognetti, Department of Biology and Biotechnologies "Charles Darwin", Sapienza University of Rome, Rome, Italy; Center for Life Nano- & Neuro-Science, Fondazione Istituto Italiano di Tecnologia, Rome, Italy*

ALICIA SAARINEN • *Stem Cell and Regenerative Metabolism Laboratory, Departments of Cardiovascular Medicine, Biochemistry and Molecular Biology, and Center for Regenerative Medicine, Mayo Clinic, Scottsdale, AZ, USA*

STEPHANIE SANTARRIAGA • *Department of Cell Biology, Neurobiology and Anatomy, Medical College of Wisconsin, Milwaukee, WI, USA*

MENGCHENG SHEN • *Stanford Cardiovascular Institute, Stanford University School of Medicine, Stanford, CA, USA*

MARK E. SHERMAN • *Mayo Clinic Cancer Center, Mayo Clinic, Rochester, MN, USA; Department of Quantitative Health Sciences, Mayo Clinic, Jacksonville, FL, USA*

ALLISON SHEVTSOV • *Division of Cardiology, Department of Internal Medicine, David Geffen School of Medicine, University of California, Los Angeles, Los Angeles, CA, USA*

ANNA KAREN SIGURDARDOTTIR • *Stem Cell Research Unit, Department of Anatomy, Faculty of Medicine, Biomedical Center, School of Health Sciences, University of Iceland, Reykjavik, Iceland*

M. Catarina Silva • *Chemical Neurobiology Laboratory, Center for Genomic Medicine, Department of Neurology, Massachusetts General Hospital and Harvard Medical School, Boston, MA, USA*

Igor I. Slukvin • *Wisconsin National Primate Research Center, University of Wisconsin Graduate School, Madison, WI, USA; Department of Cell and Regenerative Biology, University of Wisconsin School of Medicine and Public Health, Madison, WI, USA; Department of Pathology and Laboratory Medicine, University of Wisconsin Medical School, Madison, WI, USA*

Rebecca Sherrard Smith • *College of Medical, Veterinary and Life Sciences, Institute of Infection, Immunity and Inflammation, University of Glasgow, Glasgow, UK*

Claudia Spits • *Research Group Reproduction and Genetics, Vrije Universiteit Brussel, Jette, Belgium*

Robert L. Sprando • *Division of Toxicology, Office of Applied Research and Safety Assessment, Center for Food Safety and Applied Nutrition, U.S. Food and Drug Administration, Laurel, MD, USA*

Jeffrey Stulberg • *Department of Cell & Systems Biology, University of Toronto, Toronto, ON, Canada*

Atsushi Sugiyama • *Department of Pharmacology, Faculty of Medicine, Toho University, Ota-ku, Tokyo, Japan; Department of Translational Research and Cellular Therapeutics, Faculty of Medicine, Toho University, Ota-ku, Tokyo, Japan*

Yoshinori Takei • *Department of Translational Research and Cellular Therapeutics, Faculty of Medicine, Toho University, Ota-ku, Tokyo, Japan*

Wei Xuan Tan • *Stem Cells and Diabetes Laboratory, Institute of Molecular and Cell Biology, A*STAR, Singapore, Singapore; Department of Medicine, Yong Loo Lin School of Medicine, National University of Singapore, Singapore, Singapore*

Junichi Tanaka • *Division of Pathology, Department of Oral Diagnostic Sciences, School of Dentistry, Showa University, Tokyo, Japan*

Hengli Tang • *Department of Biological Science, Florida State University, Tallahassee, FL, USA*

Adrian Kee Keong Teo • *Stem Cells and Diabetes Laboratory, Institute of Molecular and Cell Biology, A*STAR, Singapore, Singapore; Department of Medicine, Yong Loo Lin School of Medicine, National University of Singapore, Singapore, Singapore; Department of Biochemistry, Yong Loo Lin School of Medicine, National University of Singapore, Singapore, Singapore*

Gunnhildur Asta Traustadottir • *Stem Cell Research Unit, Department of Anatomy, Faculty of Medicine, Biomedical Center, School of Health Sciences, University of Iceland, Reykjavik, Iceland*

Vincent Tropepe • *Department of Cell & Systems Biology, University of Toronto, Toronto, ON, Canada*

Hung-Fat Tse • *The Cardiology Division, Department of Medicine, Li Ka Shing Faculty of Medicine, Queen Mary Hospital, The University of Hong Kong, Hong Kong, SAR, China; Hong Kong-Guangdong Stem Cell and Regenerative Medicine Research Centre, The University of Hong Kong and Guangzhou Institutes of Biomedicine and Health, Hong Kong, SAR, China; Department of Medicine, The University of Hong Kong-Shenzhen Hospital, Shenzhen, China*

Enkhtuul Tsogtbaatar • *Stem Cell and Regenerative Metabolism Laboratory, Departments of Cardiovascular Medicine, Biochemistry and Molecular Biology, and Center for Regenerative Medicine, Mayo Clinic, Scottsdale, AZ, USA*

GENE I. UENISHI • *Wisconsin National Primate Research Center, University of Wisconsin Graduate School, Madison, WI, USA*
CHANDAN UPADHYAYA • *Xavier University School of Medicine, Woodbury, NY, USA*
SUHAS VASAIKAR • *Department of Translational Molecular Pathology, The University of Texas MD Anderson Cancer Center, Houston, TX, USA*
RUI WEI • *The Cardiology Division, Department of Medicine, Li Ka Shing Faculty of Medicine, Queen Mary Hospital, The University of Hong Kong, Hong Kong, SAR, China; Hong Kong-Guangdong Stem Cell and Regenerative Medicine Research Centre, The University of Hong Kong and Guangzhou Institutes of Biomedicine and Health, Hong Kong, SAR, China*
MAX S. WICHA • *Forbes Institute for Cancer Discovery, University of Michigan Medical School, Ann Arbor, MI, USA*
DARIUS WIDERA • *Stem Cell Biology and Regenerative Medicine Group, School of Pharmacy, University of Reading, Reading, UK*
JOSEPH C. WU • *Stanford Cardiovascular Institute, Stanford University School of Medicine, Stanford, CA, USA; Division of Cardiology, Department of Medicine, Stanford University School of Medicine, Stanford, CA, USA; Department of Radiology, Stanford University School of Medicine, Stanford, CA, USA*
BADRUL HISHAM YAHAYA • *Lung Stem Cell and Gene Therapy Group, Regenerative Medicine Cluster, Advanced Medical and Dental Institute (IPPT), Universiti Sains Malaysia, Kepala Batas, Penang, Malaysia*
YOJIRO YAMANAKA • *Department of Human Genetics, Rosalind and Morris Goodman Cancer Institute, McGill University, Montreal, QC, Canada; McGill Integrated Core for Animal Modeling (MICAM), McGill University, Montreal, QC, Canada*
WENMEI YANG • *Stem Cell and Cancer Biology Laboratory, Division of Experimental Pathology and Laboratory Medicine, Department of Laboratory Medicine and Pathology, Mayo Clinic, Rochester, MN, USA*
JEFFREY J. YOURICK • *Division of Toxicology, Office of Applied Research and Safety Assessment, Center for Food Safety and Applied Nutrition, U.S. Food and Drug Administration, Laurel, MD, USA*
ANN ZEUNER • *Department of Oncology and Molecular Medicine, Istituto Superiore di Sanità, Rome, Italy*
ZHAO ZHANG • *Pennsylvania Cancer and Regenerative Medicine Research Center, Baruch S. Blumberg Institute, Wynnewood, PA, USA*
HUI ZHAO • *Stem Cell and Cancer Biology Laboratory, Division of Experimental Pathology and Laboratory Medicine, Department of Laboratory Medicine and Pathology, Mayo Clinic, Rochester, MN, USA*
JUN ZHAO • *Pennsylvania Cancer and Regenerative Medicine Research Center, Baruch S. Blumberg Institute, Wynnewood, PA, USA*
YANG ZHAO • *Division of Toxicology, Office of Applied Research and Safety Assessment, Center for Food Safety and Applied Nutrition, U.S. Food and Drug Administration, Laurel, MD, USA*

Part I

ES/iPS

Chapter 1

Rapid and Highly Efficient Isolation and Purification of Human Induced Pluripotent Stem Cells

Xiugong Gao, Robert L. Sprando, and Jeffrey J. Yourick

Abstract

Human induced pluripotent stem cells (iPSCs) hold great promise for biomedical applications. However, establishment of new iPSC lines still presents many challenges. Here we describe a simple yet highly efficient two-step protocol for the isolation and purification of human iPSC lines. The first step adapts iPSCs to single cell culture and passaging, promoting survival and self-renewal; the second step enables the isolation and purification of bona fide iPSCs from a mixed population using column-based positive selection of cells expressing pluripotency markers such as TRA-1-60. Both steps utilize commercially available reagents. Using this protocol, iPSCs can be purified from cell preparations containing differentiated or unreprogrammed cells, or even be isolated directly from reprogramming vessels. The protocol could be adopted for high throughput isolation and expansion of iPSC lines and facilitate the widespread use of iPSCs in future applications.

Key words Induced pluripotent stem cell (iPSC), Reprogramming, Isolation, Purification, Single cell passaging, Magnetic-activated cell sorting (MACS)

1 Introduction

The advent of human induced pluripotent stem cells (iPSCs) has revolutionized the stem cell field [1, 2]. These cells show remarkable similarities to, but overcome the ethical hurdles associated with, human embryonic stem cells (ESCs); in addition, these cells could be derived from virtually any individual with specific disease or genetic background. Therefore, iPSCs hold great promise for cell therapy development, disease modeling, drug discovery, and toxicological applications [3].

Currently one of the major bottlenecks in the generation of new iPSC lines is the identification and isolation of bona fide iPSC colonies from derivation vessels containing a mixture of cells, usually dominated by unreprogrammed source cells and partially reprogrammed intermediates; this presents a considerable obstacle and often requires training and expertise in advanced cell culture

Nagarajan Kannan and Philip Beer (eds.), *Stem Cell Assays: Methods and Protocols*, Methods in Molecular Biology, vol. 2429, https://doi.org/10.1007/978-1-0716-1979-7_1,

techniques. Current methods of identification include visual inspection of colony morphology, live alkaline phosphatase (AP) staining [4], or live immunostaining of pluripotency surface markers such as SSEA4 and TRA-1-60 [5]. Nevertheless, the manual isolation process remains highly subjective, and colonies being isolated are oftentimes contaminated with unreprogrammed or partially reprogrammed cells. To enable applications that require de novo iPSC derivation, more efficient methods for identifying, isolating, and purifying reprogrammed cells are needed.

Magnetic-activated cell sorting (MACS) was first described more than two decades ago [6], and since then has been extensively used to isolate a wide variety of cell types [7]. More recently, it has been integrated into improved protocols for automated iPSC selection and passaging [8, 9]. Here we describe a protocol that uses MACS, coupled with single cell culture and passaging, for rapid and highly efficient isolation and expansion of iPSCs. It can be used to purify iPSCs from cell preparations heavily "contaminated" by differentiated or unreprogrammed cells, and partially reprogrammed cells (Fig. 1). The protocol can even isolate iPSC lines directly from derivation plates (*see* **Note 1**). It addresses the aforementioned issue by simplifying the isolation and passaging process, eliminating subjectivity in colony selection, and allowing many lines to be derived at once, thereby drastically reducing the time burden in iPSC generation. Moreover, this protocol can potentially be applied to fully automated high throughput iPSC derivation and maintenance.

2 Materials

1. Phosphate buffered saline (PBS) without Ca^{2+} and Mg^{2+} (PBS−/−).
2. PBS with 0.90 mM Ca^{2+} and 0.50 mM Mg^{2+} (PBS+/+).
3. TrypLE Select 1× without phenol red (Gibco) (*see* **Note 2**).
4. Cellartis DEF-CS 500 Culture System (including Basal Medium, COAT-1, GF-1, GF-2, and GF-3) (Takara) (*see* **Note 3**).
5. 6-well cell culture plates (with tissue culture-treated polystyrene surface).
6. Hemocytometer or automated cell counter.
7. Pre-separation Filters (30 μm) (Miltenyi Biotec).
8. Anti-TRA-1-60 MicroBeads (Miltenyi Biotec) (*see* **Note 4**).
9. MACS Device (Fig. 2) including MultiStand, MiniMACS or OctoMACS Separator, and MS Columns (all from Miltenyi Biotec).

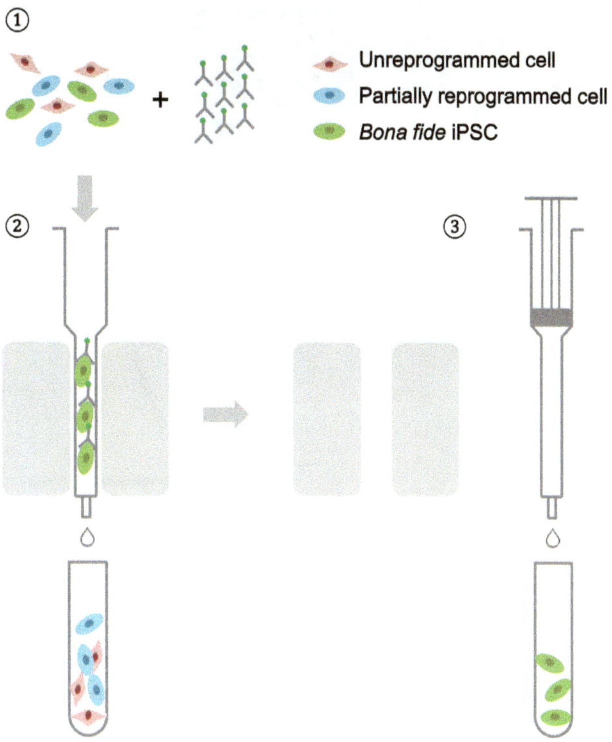

Fig. 1 Schematic representation of iPSC purification by MACS. ① Cell mixture containing unreprogrammed somatic cells, partially reprogrammed cells, and completely reprogrammed cells is mixed with MACS MicroBeads coupled to antibodies targeting iPSC surface makers (such as TRA-1-60). ② Cells are then loaded onto a MACS column placed in a MACS separator (a strong permanent magnet), and a high-gradient magnetic field is induced on the column matrix, which is strong enough to retain cells labeled with minimal amounts of MACS MicroBeads. Unlabeled cells pass through or are washed away from the column. ③ Labeled cells are eluted from the column after removal of the column from the magnet. The entire procedure takes less than 30 min

10. Rho-associated protein kinase (ROCK) inhibitor (*see* **Note 5**).
11. Benchtop centrifuge.
12. Cell culture incubator.

3 Methods

3.1 Adaption of iPSCs to Single Cell Culture and Passaging

1. Grow the iPSC culture to be purified (*see* **Note 6**) in 6-well plates in the original culture system according to the protocol of the previous system until ready for passaging (*see* **Notes 7–9**).
2. Coat the cell culture vessel (*see* **Note 10**) with COAT-1. Dilute Cellartis DEF-CS COAT-1 in PBS+/+ at 1:20 or 1:10 ratio (*see*

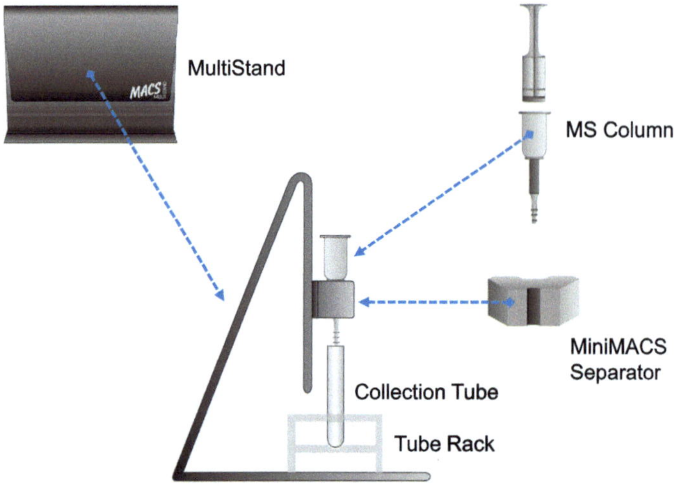

Fig. 2 Illustration of the MACS device assembly. A MiniMACS (or OctoMACS) Separator is attached to the MultiStand and an MS Column is placed in the separator. A collection tube is placed under the column

Note 11). Mix the diluted solution gently and thoroughly by pipetting up and down a few times. Add 1 ml of the diluted solution to each well of a 6-well plate (0.1 ml/cm^2), and tilt back and forth to make sure that the entire surface is covered. Place the plate in an incubator at 37 °C for ≥20 min, or at room temperature for ≥30 min. Aspirate the coating solution from the wells right before use. There is no need to rinse the wells with PBS or culture medium.

3. Prepare appropriate volume of supplemented Cellartis DEF-CS thaw/passage medium by adding GF-1 (1:333), GF-2 (1:1000), and GF-3 (1:1000) to the Cellartis DEF-CS Basal Medium. Prepare the medium freshly and warm to 37 °C before use (*see* **Note 12**). Discard any leftover warm medium.
4. For single cell dissociation and passaging, aspirate old medium from the wells and wash cells once with PBS−/−. Add 200 μl TrypLE Select to each well and incubate at 37 °C for 5–10 min (*see* **Note 13**) to detach the cells. Resuspend the cells in 2–3 ml supplemented Cellartis DEF-CS thaw/passage medium and pipet up and down gently for several times to ensure a single cell suspension (*see* **Note 14**).
5. Count cells using a hemocytometer or an automated cell counter. Cells may be subjected to MACS purification (Subheading 3.2) from this point. Alternatively, for further passaging, adjust the cell density using fresh medium to 2.0–2.5 × 10^5 cells/ml, and seed 2.0 ml (*see* **Note 15**) cell suspension per well of a 6-well plate coated with COAT-1.

6. Gently shake the plate on an even surface back and forth then left and right several times to ensure that cells are dispersed evenly over the surface. Culture the cells in incubator at 37 °C, 5% CO_2, and ≥90% humidity.
7. Change medium daily with 2.5–4.5 ml/well fresh maintenance medium (*see* **Note 16**) until cells reach confluence (1.5–3.0 × 10^5 cells/cm^2), which normally occurs 3–4 days post passage (*see* **Note 17**).
8. Dissociate cells into single cell suspension and split at 1:3 to 1:6 into wells of a new tissue culture plate coated with COAT-1.
9. Repeat **steps 4–8** if further passaging of the iPSC culture is needed (*see* **Note 9**).

3.2 iPSC Purification by Magnetic-Activated Cell Sorting (MACS)

1. Prepare a homogeneous single cell suspension (*see* **Note 18**) from the iPSC culture to be purified as described in Subheading 3.1, **step 4**. Pass the cell suspension through a 30 μm pre-separation filter (*see* **Note 19**) to remove cell clumps which may clog the column. Count cells using a hemocytometer or an automated cell counter, and examine cell purity using flow cytometry analysis (Fig. 3) (*see* **Note 20**).
2. Centrifuge the cell suspension (≤2 × 10^8 total cells) (*see* **Note 21**) at 300 × *g* for 5 min. Aspirate supernatant completely.
3. Per 2 × 10^6 total cells (*see* **Note 22**), resuspend the cell pellet in 80 μl of Cellartis DEF-CS medium supplemented with GF-3/ROCK inhibitor (*see* **Note 5**) and add 20 μl of Anti-TRA-1-60 MicroBeads. Mix well and incubate for 5 min in a refrigerator (2–8 °C) (*see* **Note 23**).
4. Adjust volume to 1.0 ml total using Cellartis DEF-CS medium supplemented with GF-3/ROCK inhibitor, and proceed to magnetic separation in the next steps.
5. Place an MS Column in the magnetic field of a MiniMACS™ or OctoMACS™ Separator. Prepare the MS Column by rinsing with 0.5 ml Cellartis DEF-CS medium supplemented with GF-3/ROCK inhibitor.
6. Apply 1.0 ml of cell suspension labeled with Anti-TRA-1-60 MicroBeads onto the column. Collect flow-through containing unlabeled cells.
7. Wash column with 0.5 ml of Cellartis DEF-CS medium supplemented with GF-3/ROCK inhibitor. Repeat the wash for two additional times (*see* **Note 24**). Collect unlabeled cells that pass through and combine with the flow-through from **step 6**.
8. Remove the column from the separator and place it on a collection tube. Pipette 1.0 ml of Cellartis DEF-CS medium supplemented with GF-3/ROCK inhibitor onto the column.

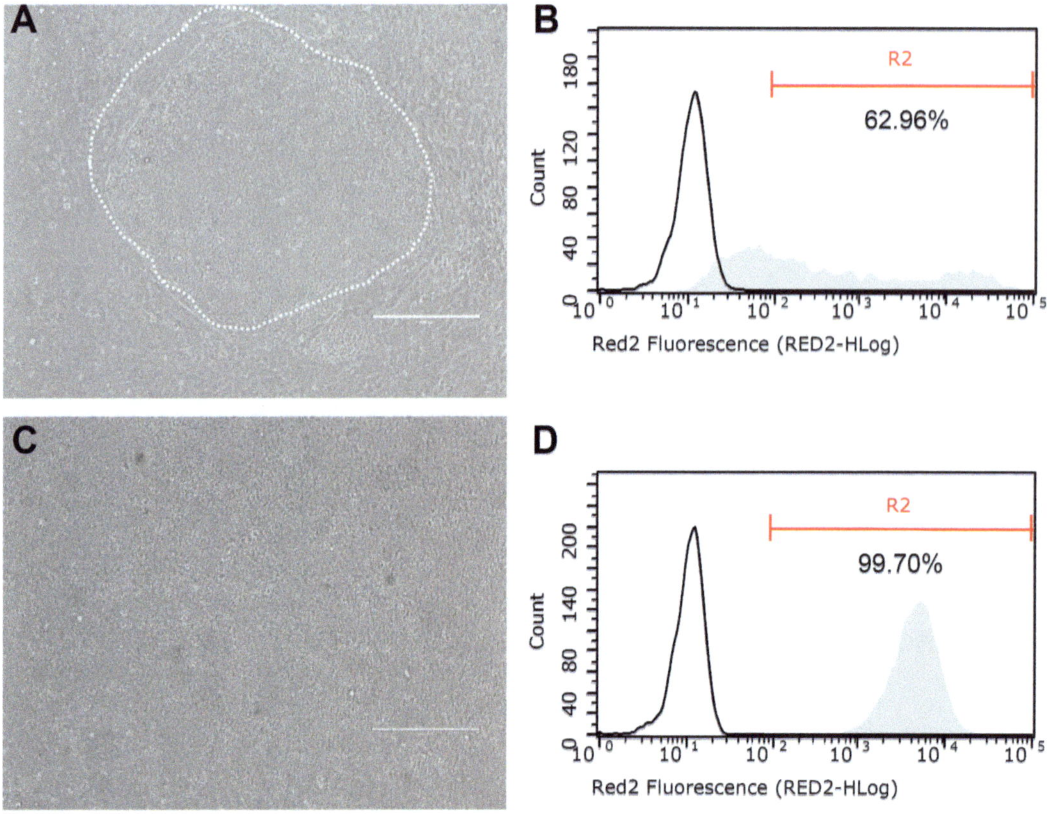

Fig. 3 An exemplary iPSC purification by MACS. (**a**) Representative cell culture micrograph after cells being transitioned to Cellartis DEF-CS Culture System but before MACS purification, where cells typically show a patch of densely packed iPSCs (inside dotted lines) surrounded by loosely packed contaminating somatic cells. (**b**) Representative flow cytometry histogram on pluripotency marker SSEA4 of the cell population before purification. (**c**) Representative cell culture micrograph after cells are purified by MACS and reach confluency after plating. (**d**) Representative flow cytometry histogram on pluripotency marker SSEA4 of the cell population after purification. Scale bars, 400 μm

Immediately flush out the magnetically labeled cells by firmly but slowly (*see* **Note 25**) pushing the plunger into the column.

9. Count the eluted cells using a hemocytometer or an automated cell counter, and examine cell purity using flow cytometry analysis (Fig. 3) (*see* **Note 20**).
10. Plate adequate number of cells in a well of a 6-well plate for expansion (*see* **Note 26**).
11. Repeat Subheading 3.1, **steps 5–8** until reaching desired passage number (*see* **Notes 27–29**).
12. Bank cells and check pluripotency and karyotype of the iPSC line at designated frequency (*see* **Note 30**).

4 Notes

1. Direct isolation from the reprogramming culture will produce a polyclonal population of iPSCs, as individual clones are not maintained independently but are instead pooled together. Therefore, this protocol may have the drawback that individual clones cannot be analyzed or compared with one another. On the other hand, the polyclonal system has the advantages of simplifying the derivation process, averaging all clones, and making the workload more manageable. If individual clones are desired, they can be derived by single cell cloning [10] from the polyclonal population.
2. TrypLE Select contains a recombinant protease that cleaves peptide bonds on the C-terminal sides of lysine and arginine, and can be substituted for trypsin in existing protocols for the dissociation of a wide range of adherent mammalian cells. We recommend using TrypLE Select instead of trypsin/EDTA for the dissociation of iPSCs as it is gentle on cells and can be inactivated simply by dilution with basal medium, avoiding the need for trypsin inhibitors such as FBS.
3. The Cellartis DEF-CS Culture System, recognized for its suitability for genome engineering [11] and single cell cloning [10], promotes reliable growth of iPSCs in a feeder-free and defined environment. Cells are grown in this medium as a homogeneous monolayer and are enzymatically passaged as single cells that maintain pluripotency with a stable karyotype [12]. The formula of the medium composition is proprietary. We recommend adapting the iPSC line to the Cellartis DEF-CS Culture System before MACS purification, as we have found that iPSCs being transitioned to this medium retain high viability (>80%) after MACS purification [13]. However, other culture systems supporting single cell passage and culture can also be tested.
4. Undifferentiated pluripotent stem cells express a number of surface antigens such as TRA-1-60 and TRA-1-81, which are downregulated upon differentiation [14]. It is possible to substitute the TRA-1-60 antibody conjugated to the magnetic MicroBeads with any other antibody directed toward a cell-surface epitope specific to pluripotent stem cells. Miltenyi Biotec offers MicroBeads conjugated with a number of other antibodies, e.g., CD326 (EpCAM), SSEA1 (CD15), and SSEA4, thus providing flexibility to the system. However, we prefer the Anti-TRA-1-60 MicroBeads as they allow a 10 min isolation procedure that minimizes cell handling time.
5. In order to improve survival of iPSCs during and after MACS separation, the magnetic labeling, separation, and subculturing

should be performed in standard culture medium supplemented with ROCK inhibitor. We recommend using Cellartis DEF-CS medium supplemented with GF-1, GF-2, and GF-3 (ROCK inhibitor). GF-3 can be replaced with 10 μM Y-27632 or 2 μM Thiazovivin.

6. iPSCs can be purified from cell preparations contaminated with differentiated or unreprogrammed cells, and partially reprogrammed cells. It is also feasible to isolate iPSCs directly from reprogramming vessels. We have been able to isolate iPSC lines from cell preparations that contain <5% iPSCs.
7. The steps under Subheading 3.1 are needed for iPSCs maintained in medium other than the Cellartis DEF-CS Culture System. These steps can be omitted if the iPSC line is already cultured in this medium.
8. iPSC lines maintained in other culture systems can be readily transitioned to the DEF-CS Culture System either as fresh cultures or from cryopreserved vials. Fresh cultures can be transferred when passaging and cryopreserved cells can be thawed directly into the Cellartis DEF-CS Culture System. We recommend using fresh cultures, i.e., first thaw and culture the cells in the original medium until passaging.
9. It takes 2–5 passages to adapt a cell line to the Cellartis DEF-CS Culture System. At the first few passages, some cell characteristics might be different from those observed in the original culture system. As the Cellartis DEF-CS system utilizes single cell passaging, the morphology of cells differs from that in other culture systems usually using aggregate passaging methods. In addition, newly passaged cells tend to spread out. However, the cells get denser with proliferation, and the typical undifferentiated stem cell morphology (e.g., high nucleus to cytoplasm ratio, and prominent nucleoli) reappears. We also recommend further passaging be conducted after the MACS purification (Subheading 3.2, **steps 10–12**).
10. We recommend using 6-well plate of cell culture grade. The number of cells in one well of a 6-well plate (~2.0×10^6) is enough as starting material for subsequent MACS purification.
11. Ensure that PBS+/+ (not PBS−/−) is used for the dilution of COAT-1. Cells will not attach if COAT-1 is diluted in PBS−/−. For cells transitioning from other culture systems, a higher concentration of COAT-1 may increase the success of transition. Use a dilution of 1:10 or 1:5 during the first few passages to provide extra support during the adaptation process.
12. Warm all other reagents to room temperature before use.
13. The best time should be determined empirically. Check cells under microscope periodically to monitor the dissociation

process. Detachment can be facilitated by tapping the side of the plate firmly but gently. Avoid incubating the cells with TrypLE Select for too long (>10 min), otherwise cells will aggregate and cannot be dissociated into single cell suspension.

14. Since TrypLE Select is diluted ≥10× fold after adding ≥2 ml medium, there is no need to wash the cells after dissociation. Otherwise, if TrypLE Select is diluted 10× fold, centrifuge the cells at 200 × *g* for 2–5 min and resuspend in fresh medium.

15. The volume of the seeding cell suspension can vary between 1.5 and 4.5 ml depending on cell density of the suspension. However, the final seeding density should be in the range of 4.0–5.0 × 10^4 cells/cm^2. Use 4.0 × 10^4 cells/cm^2 if leaving the cells 4 days between passages and 5.0 × 10^4 cells/cm^2 if leaving 3 days between passages. Adjust the density to suit your particular cell line as appropriate.

16. The Cellartis DEF-CS maintenance medium only contains GF-1 (1:333) and GF-2 (1:1000). Do not add GF-3 (a ROCK inhibitor) to the maintenance medium. Use larger volume (≥4.0 ml/well) of medium if the old medium turns yellow at the time of medium exchange due to high metabolic activity from increased cell number.

17. Cell preparations with low iPSC purity will grow slower than pure iPSCs. Also, cells just transitioned from other culture systems might initially grow at a slower rate. Sometimes it may need 5–7 days for cells to reach the density for passaging. If cells remain sparse after 7 days in culture, a passage is still recommended.

18. Purification by MACS requires a homogeneous single cell suspension. A prerequisite and key component of the current protocol is single cell culture and passaging of iPSCs. For optimal performance, it is crucial to obtain a homogeneous single cell suspension before magnetic labeling.

19. Moisten the filter with 1–2 ml of PBS or cell culture medium before use.

20. We recommend evaluating cell purity before and after MACS purification using flow cytometry analysis of surface marker expression (e.g., TRA-1-60, SSEA4) so that purification efficiency could be assessed [13].

21. The MS Column has a capacity of 1 × 10^7 magnetically labeled cells from up to 2 × 10^8 total cells. We recommend using one column for each well of a 6-well plate, which yield ~2 × 10^6 total cells at the time of passaging.

22. Volumes given for magnetic labeling are for up to 2 × 10^6 total cells. When working with 2 × 10^6 cells, use the same volumes as indicated. When working with higher cell numbers, scale up all reagent volumes and total volumes accordingly.

23. Pre-cool all solutions in a refrigerator, keep cells cold all the time, and work quickly to prevent capping of antibodies on the cell surface and nonspecific cell labeling. The recommended incubation temperature is 2–8 °C. Higher temperatures and/or longer incubation times may lead to nonspecific cell labeling. However, working on ice is not recommended.
24. Wait until the column reservoir is just turning empty before starting the next washing step.
25. Bubbles will form in the flow-through if the plunger moves too quickly due to sudden release of pressure.
26. A critical technical detail for the success of the protocol is that after MACS purification and during subsequent single cell culture, iPSCs should be plated at high density ($4.0–5.0 \times 10^4$ cells/cm^2) to maintain cell–cell contact in order to maintain genomic stability. It has been shown that iPSCs are susceptible to genomic abnormalities [15], especially during enzymatic single cell passaging [16]. Maintaining cell–cell contact during passaging improves cell survival [17], thus reducing cell stress and maintaining genetic stability of the cells. Use 12-well or 24-well plate if the total number of cells is too less for a well of a 6-well plate.
27. After purification, cells appear as a homogeneous population with undifferentiated morphology. And the cells expand rapidly, usually reaching the desired density for further passaging in 3 days (Fig. 4).
28. In some cases, even though iPSC purity immediately after MACS may be lower than 90%, the purity can be improved to >99% after continuous passaging in the Cellartis DEF-CS medium. This is because the Cellartis DEF-CS medium favors iPSC growth over other cell types, therefore iPSCs eventually outgrow other cell types and become dominant in the population.

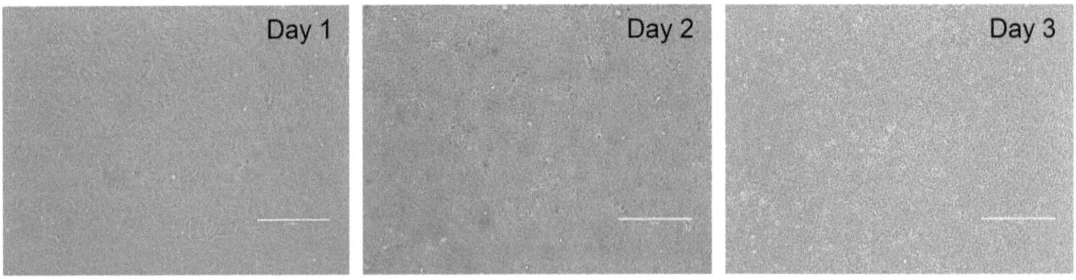

Fig. 4 Single cell passage and expansion of iPSCs. Representative cell culture micrographs of cells one day, two days, and three days after passaging are shown. On day 3, cells reach confluence (with cell density $\geq 2.0 \times 10^5$ cells/cm^2 as shown) and are ready for passaging. Scale bars, 400 μm

29. We recommend passaging iPSC lines until at least passage 15 before using them for downstream applications such as directed differentiation in order for the cells to lose the episomes transfected during reprogramming [18], and for the cells to erase epigenetic memories from the source cell [19].
30. We recommend banking the cells every other passage and testing pluripotency and karyotype every 5 passages. Methods for pluripotency and karyotype testing as well as other iPSC characterization methods could be found in [5].

Acknowledgments

We thank Drs. Marianna D. Solomotis and Mary E. Torrence for critically reading the manuscript as well as for their support of the work. The statements made in this article are those of the authors and do not necessarily represent views, opinions, or policies of the U.S. Food and Drug Administration.

References

1. Takahashi K, Tanabe K, Ohnuki M, Narita M, Ichisaka T, Tomoda K, Yamanaka S (2007) Induction of pluripotent stem cells from adult human fibroblasts by defined factors. Cell 131(5):861–872. https://doi.org/10.1016/j.cell.2007.11.019
2. Yu J, Vodyanik MA, Smuga-Otto K, Antosiewicz-Bourget J, Frane JL, Tian S, Nie J, Jonsdottir GA, Ruotti V, Stewart R, Slukvin II, Thomson JA (2007) Induced pluripotent stem cell lines derived from human somatic cells. Science (New York, NY) 318(5858):1917–1920. https://doi.org/10.1126/science.1151526
3. Shi Y, Inoue H, Wu JC, Yamanaka S (2017) Induced pluripotent stem cell technology: a decade of progress. Nat Rev Drug Discov 16(2):115–130. https://doi.org/10.1038/nrd.2016.245
4. Singh U, Quintanilla RH, Grecian S, Gee KR, Rao MS, Lakshmipathy U (2012) Novel live alkaline phosphatase substrate for identification of pluripotent stem cells. Stem Cell Rev 8(3):1021–1029. https://doi.org/10.1007/s12015-012-9359-6
5. Asprer JS, Lakshmipathy U (2015) Current methods and challenges in the comprehensive characterization of human pluripotent stem cells. Stem Cell Rev 11(2):357–372. https://doi.org/10.1007/s12015-014-9580-6
6. Miltenyi S, Muller W, Weichel W, Radbruch A (1990) High gradient magnetic cell separation with MACS. Cytometry 11(2):231–238. https://doi.org/10.1002/cyto.990110203
7. Grutzkau A, Radbruch A (2010) Small but mighty: how the MACS-technology based on nanosized superparamagnetic particles has helped to analyze the immune system within the last 20 years. Cytometry A 77(7):643–647. https://doi.org/10.1002/cyto.a.20918
8. Dick E, Matsa E, Young LE, Darling D, Denning C (2011) Faster generation of hiPSCs by coupling high-titer lentivirus and column-based positive selection. Nat Protoc 6(6):701–714. https://doi.org/10.1038/nprot.2011.320
9. Valamehr B, Abujarour R, Robinson M, Le T, Robbins D, Shoemaker D, Flynn P (2012) A novel platform to enable the high-throughput derivation and characterization of feeder-free human iPSCs. Sci Rep 2:213. https://doi.org/10.1038/srep00213
10. Feng Q, Shabrani N, Thon JN, Huo H, Thiel A, Machlus KR, Kim K, Brooks J, Li F, Luo C, Kimbrel EA, Wang J, Kim KS, Italiano J, Cho J, Lu SJ, Lanza R (2014) Scalable generation of universal platelets from human induced pluripotent stem cells. Stem Cell Rep 3(5):817–831. https://doi.org/10.1016/j.stemcr.2014.09.010
11. Valton J, Cabaniols JP, Galetto R, Delacote F, Duhamel M, Paris S, Blanchard DA, Lebuhotel C, Thomas S, Moriceau S, Demirdjian R, Letort G, Jacquet A,

Gariboldi A, Rolland S, Daboussi F, Juillerat A, Bertonati C, Duclert A, Duchateau P (2014) Efficient strategies for TALEN-mediated genome editing in mammalian cell lines. Methods 69(2):151–170. https://doi.org/10.1016/j.ymeth.2014.06.013

12. Asplund A, Pradip A, van Giezen M, Aspegren A, Choukair H, Rehnstrom M, Jacobsson S, Ghosheh N, El Hajjam D, Holmgren S, Larsson S, Benecke J, Butron M, Wigander A, Noaksson K, Sartipy P, Bjorquist P, Edsbagge J, Kuppers-Munther B (2016) One standardized differentiation procedure robustly generates homogenous hepatocyte cultures displaying metabolic diversity from a large panel of human pluripotent stem cells. Stem Cell Rev 12(1):90–104. https://doi.org/10.1007/s12015-015-9621-9
13. Gao X, Sprando RL, Yourick JJ (2018) A rapid and highly efficient method for the isolation, purification, and passaging of human-induced pluripotent stem cells. Cell Reprogram 20(5): 282–288. https://doi.org/10.1089/cell.2018.0022
14. Andrews PW, Banting G, Damjanov I, Arnaud D, Avner P (1984) Three monoclonal antibodies defining distinct differentiation antigens associated with different high molecular weight polypeptides on the surface of human embryonal carcinoma cells. Hybridoma 3(4):347–361. https://doi.org/10.1089/hyb.1984.3.347
15. Mayshar Y, Ben-David U, Lavon N, Biancotti JC, Yakir B, Clark AT, Plath K, Lowry WE, Benvenisty N (2010) Identification and classification of chromosomal aberrations in human induced pluripotent stem cells. Cell Stem Cell 7(4):521–531. https://doi.org/10.1016/j.stem.2010.07.017
16. Bai Q, Ramirez JM, Becker F, Pantesco V, Lavabre-Bertrand T, Hovatta O, Lemaitre JM, Pellestor F, De Vos J (2015) Temporal analysis of genome alterations induced by single-cell passaging in human embryonic stem cells. Stem Cells Dev 24(5):653–662. https://doi.org/10.1089/scd.2014.0292
17. Li L, Bennett SA, Wang L (2012) Role of E-cadherin and other cell adhesion molecules in survival and differentiation of human pluripotent stem cells. Cell Adhes Migr 6(1):59–70. https://doi.org/10.4161/cam.19583
18. Okita K, Matsumura Y, Sato Y, Okada A, Morizane A, Okamoto S, Hong H, Nakagawa M, Tanabe K, Tezuka K, Shibata T, Kunisada T, Takahashi M, Takahashi J, Saji H, Yamanaka S (2011) A more efficient method to generate integration-free human iPS cells. Nat Methods 8(5):409–412. https://doi.org/10.1038/nmeth.1591
19. Liang G, Zhang Y (2013) Genetic and epigenetic variations in iPSCs: potential causes and implications for application. Cell Stem Cell 13(2):149–159. https://doi.org/10.1016/j.stem.2013.07.001

Chapter 2

Artificial Activation of Murine Oocytes Using Strontium to Derive Haploid and Diploid Parthenotes

Daphne Norma Crasta, Satish Kumar Adiga, Nagarajan Kannan, and Guruprasad Kalthur

Abstract

Parthenogenesis is a common reproductive strategy among lower animals that involves the development of an embryo from an oocyte, without any contribution from spermatozoon. This phenomenon does not occur naturally in placental mammals. However, the mammalian oocytes can be artificially activated in vitro using mechanical, electrical, and chemical stimuli which can develop up to the blastocyst stage. In this chapter, we describe the protocol for generating haploid and diploid parthenotes from mouse oocytes using strontium as the activating agent under in vitro conditions.

Key words Artificial oocyte activation, Murine oocytes, Haploid, Diploid, Strontium, Cytochalasin D

1 Introduction

Parthenogenesis is defined as the development of an embryo from oocyte without any contribution from spermatozoa. It is a common reproductive strategy among the lower animals such as bees, flies, aphids, some lizards and birds [1–3]. Parthenogenesis does not occur naturally in placental mammals except in the Lt/Sv strain of mice, in which it is known to progress up to the somite stage after implantation [4]. Jacques Loeb was the first to demonstrate artificial parthenogenesis in vitro (1899) in unfertilized sea urchin eggs using hypertonic solutions [5]. Since then, due to its wide application in the developmental biology field, various protocols involving mechanical, electrical, and chemical stimuli have been developed for the successful artificial activation of mammalian oocytes in vitro [6, 7].

Parthenotes have several applications in the field of reproductive and developmental biology. Due to their uniparental origin, they serve as an excellent model that can help in understanding the contribution of maternal and paternal factors during early embryo

Nagarajan Kannan and Philip Beer (eds.), *Stem Cell Assays: Methods and Protocols*, Methods in Molecular Biology, vol. 2429, https://doi.org/10.1007/978-1-0716-1979-7_2, © Springer Science+Business Media, LLC, part of Springer Nature 2022

development [8]. Their major application is in assisted reproductive techniques (ART), where they can be used in assessing the developmental competence of oocytes exposed to various experimental conditions to understand their possible influence on early embryo development. Since dogma commonly holds that life begins only after the union of male and female gametes, parthenotes serve as alternate model to assess the developmental potential of oocytes due to fewer ethical restrictions. Further, in the field of regenerative medicine, they may be used to generate embryonic stem cells, which have less chance of rejection by the host [9].

The most common method of activation is by using chemical agents such as ethanol, calcium ionophore, strontium, and phorbol esters. Strontium (Sr^{2+}) activates oocytes by inducing multiple calcium elevations that closely mimic the sperm-induced oscillations of Ca^{2+} during fertilization. Strontium exposure leads to the extrusion of the second polar body from the oocytes (*see* Fig. 1), resulting in haploid parthenotes [10, 11]. To prevent the extrusion of second polar body, cytoskeletal inhibitors such as Cytochalasin D can be effectively used along with strontium which can generate a diploid parthenote [11] (*see* Fig. 1).

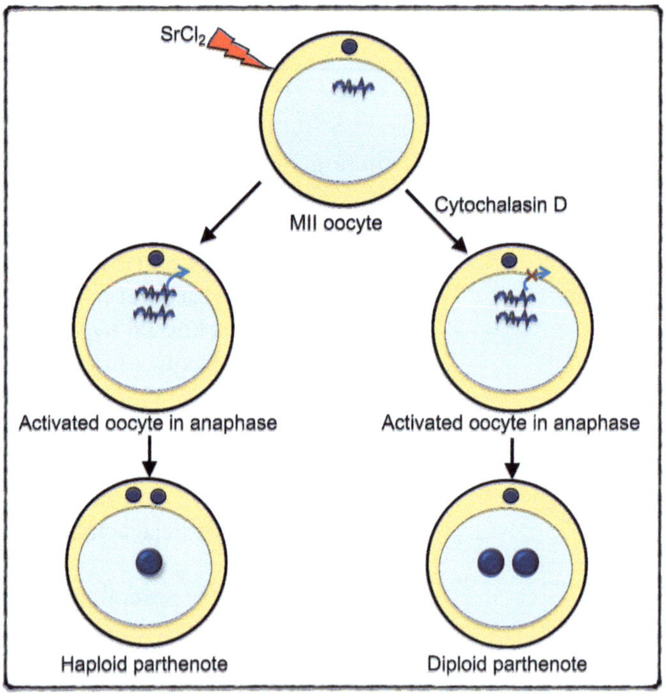

Fig. 1 Schematic representation showing generation of haploid and diploid parthenogenetic embryos

The procedure described here involves multiple steps starting from female mice injections, dissection, and oviduct collection. The oocyte cumulus complexes retrieved from the oviduct are subsequently subjected to parthenogenetic activation in vitro using strontium chloride as activating agent in presence and absence of cytochalasin D to obtain diploid and haploid parthenotes, respectively.

2 Materials

All the solutions used for the experiments are prepared using sterile Milli-Q water and analytical grade reagents. The solutions are filter-sterilized using a 0.2 μ Millex GV syringe filter and prewarmed to 37 °C before use. The tools used for the experiments are heat sterilized. All the work surfaces are cleaned using 70% ethanol at least 1 h prior to use (*see* **Note 1**).

2.1 Superovulation of Mice

1. 8–12-week-old female Swiss albino mice.
2. Pregnant Mare Serum Gonadotropin (PMSG).
3. Human Chorionic Gonadotropin (hCG).
4. Normal Saline (0.9% NaCl).
5. Sterile 1 mL syringes with hypodermic needle.
6. Sterile rubber gloves.

2.2 Dissection of Mice

1. Dry bath to maintain 37 °C.
2. 1.5 mL microcentrifuge tubes for organ collection.
3. M2 medium: $NaHCO_3$ 0.35 g/L, $CaCl_2 \cdot 2H_2O$ 0.24 g/L, $MgSO_4$ 0.1649 g/L, KCl 0.3563 g/L, KH_2PO_4 0.162 g/L, NaCl 0.5531 g/L, Bovine Serum Albumin 1.0 g/L, D-(+)-Glucose 1.0 g/L, HEPES × Na 0.0106 g/L, Phenol Red·Na 0.016 g/L, Sodium Pyruvate 0.0363 g/L, Sodium L-lactate 2.95 g/L, Penicillin–Streptomycin 50,000 U/L. Dissolve all components in milli-Q water and filter-sterilize. Store at 4 °C.
4. Sterile dissection board.
5. Forceps of different types (*see* **Note 2**).
6. Dissection scissors of different types (*see* **Note 3**).
7. Milli-Q water.
8. 50 mL beaker.
9. Laboratory tissue paper.
10. Sterile rubber gloves.
11. Paper pins.

2.3 Oocyte Cumulus Complex (OCC) Collection

1. Laminar air flow with a heat stage set at 37 °C fitted with a stereo microscope.
2. Sterile 35 mm tissue culture grade Petri dishes.
3. M2 medium.
4. M16 medium (without Ca^{2+} and Mg^{2+}): $NaHCO_3$ 2.101 g/L, KCl 0.3563 g/L, KH_2PO_4 0.1619 g/L, NaCl 5.5319 g/L, Bovine Serum Albumin 4.0 g/L, D-(+)-Glucose 1.0019 g/L, Phenol Red·Na 0.01 g/L, Sodium Pyruvate 0.0363 g/L, Sodium L-lactate 2.61 g/L, Penicillin–Streptomycin 50,000 U/L. Dissolve all components in milli-Q water and filter-sterilize. Store at 4 °C.
5. Sterile paraffin oil, prewarmed.
6. Sterile 1 mL syringes.
7. 2–20 μL pipette and sterile pipette tips.
8. 1 mL pipette and sterile pipette tips.

2.4 Parthenogenetic Activation of Oocytes

1. Haploid activation medium: M16 medium (without Ca^{2+} and Mg^{2+}), containing 10 mM Strontium chloride (*see* **Note 4**).
2. Diploid activation medium: M16 medium (without Ca^{2+} and Mg^{2+}), containing 10 mM Strontium chloride and 1 μg/mL of Cytochalasin D.
3. Sterile paraffin oil, prewarmed.
4. Sterile 35 mm tissue culture grade Petri dishes.
5. Incubator maintained at 37 °C and 5% CO_2.
6. 2–20 μL pipette and sterile pipette tips.
7. 1 mL pipette and sterile pipette tips.

2.5 Denuding of OCC

1. M16 medium: $NaHCO_3$ 2.101 g/L, $CaCl_2{\cdot}2H_2O$ 0.25137 g/L, $MgSO_4$ 0.143276 g/L, KCl 0.356349 g/L, KH_2PO_4 0.161959 g/L, NaCl 5.5319304 g/L, Bovine Serum Albumin 4.0 g/L, D-(+)-Glucose 1.001912 g/L, Phenol Red·Na 0.01 g/L, Sodium Pyruvate 0.0363 g/L, Sodium L-lactate 2.61 g/L, Penicillin–Streptomycin 50,000 U/L. Dissolve all components in milli-Q water and filter-sterilize. Store at 4 °C.
2. Hyaluronidase (0.75 mg/mL) in M16 medium.
3. Sterile paraffin oil, prewarmed.
4. Sterile 35 mm tissue culture grade Petri dishes.
5. 2–20 μL pipette and sterile pipette tips.
6. Stripper pipette.
7. Sterile stripper tips (135 μm and 175 μm).

2.6 Embryo Culture

1. M16 medium.
2. Sterile paraffin oil, prewarmed.
3. Sterile 35 mm tissue culture grade Petri dishes.
4. Incubator set at 37 °C and 5% CO_2.

2.7 Pronuclear (PN) Stage Assessment and Embryo Culture

1. Inverted microscope with heat stage maintained at 37 °C.

3 Methods

3.1 Superovulation of Mice

1. Dilute 50 IU of PMSG using normal saline (0.9% NaCl) to make a final concentration of 5 IU in every 0.1 mL.
2. Load a sterile 1 mL syringe with 0.1 mL of 5 IU PMSG. Make sure to avoid air bubbles.
3. Restrain an 8–12-week-old adult female Swiss albino mouse in one hand (*see* **Note 5**) and inject 5 IU of PMSG intra-peritoneally (i.p.) between the fourth and fifth mammary gland (*see* **Note 6**).
4. 48 h after the PMSG injection, inject 10 IU of hCG (100 IU diluted using normal saline to make 10 IU in every 0.1 mL) intra-peritoneally between the fourth and fifth mammary gland.

3.2 Dissection of Mice

1. Post 13.5 h of hCG injection, humanely sacrifice the mouse using a locally approved procedure (*see* **Note 7**).
2. Place the mouse on the dissection board with the abdominal region facing upwards and pin the limbs firmly to the board.
3. Clean the skin around the abdomen region of the mouse using a tissue dipped in sterile milli-Q water.
4. Pinch up the skin of the lower abdomen region using a toothed-forceps and give a cut on the skin and peritoneum using a straight dissection scissors.
5. Make a V-shaped cut onto the skin and peritoneum on either sides of the abdomen and pull the skin upwards exposing the fat pad.
6. Move the fat pad downwards using sterile pointed forceps to expose the internal organs. Move upwards the digestive organs of the mouse and locate the female reproductive organs present under it (*see* Fig. 2).
7. Hold the uterine horn pinched up using forceps and separate the fat from underneath it, along its length and from around the ovary using bent sharp scissors.

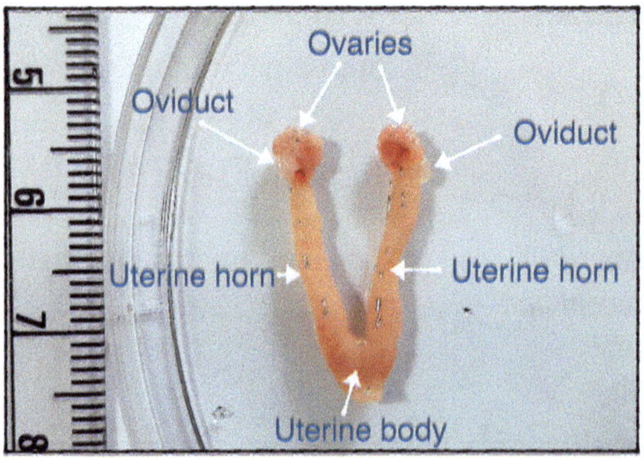

Fig. 2 Female reproductive system in mice

8. Collect the oviduct along with the ovary and a small part of the uterine horn (*see* **Note 8**).
9. Immediately transfer the collected organs into prewarmed M2 medium in a microcentrifuge tube and place in a dry bath maintained at 37 °C.
10. Repeat **steps 7–9** to collect the oviduct on the other side.

3.3 Oocyte Cumulus Complex (OCC) Collection

1. Transfer the collected organs using forceps into a sterile tissue culture grade petri dish containing 2 mL of prewarmed M2 medium placed on the heat stage of the laminar air flow (LAF) maintained at 37 °C.
2. Using two sterile 1 mL syringes (*see* **Note 9**), position the oviduct region under the stereo microscope fitted to the laminar air flow unit maintained at 37 °C, and observe the bulged portion of the oviduct (*see* Fig. 3a).
3. Using the syringe needle, make a cut at one end of the bulged oviduct (*see* **Note 10**) to release out the OCCs.
4. Pipette out the OCCs that are released into the medium (*see* Fig. 3b) using a 2–20 μL pipette and transfer them into sterile tissue culture grade petri dish containing droplets of 30 μL medium (M16 medium without Ca^{2+} and Mg^{2+}) covered with oil (*see* Fig. 4a).
5. Collect the OCC bunch from the other oviduct in the same manner (*see* **Note 11**).

3.4 Parthenogenetic Activation of Oocytes

1. Wash the OCCs 2–3 times in medium (M16 medium without Ca^{2+} and Mg^{2+}) by transferring them from one medium droplet to the next, using a 2–20 μL pipette while observing under the stereo microscope (*see* Fig. 4b).
2. After the quick washes, transfer the OCCs into either the haploid or the diploid activation medium.

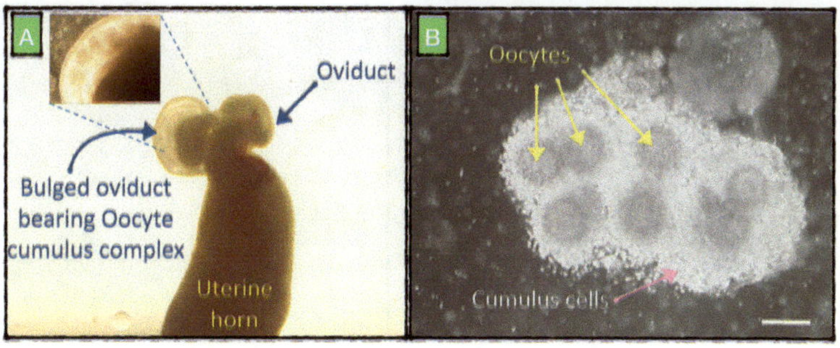

Fig. 3 (**a**) Oviduct bearing the oocyte cumulus complexes (OCCs). (**b**) Clutch of OCCs released from oviduct (scale bar = 200 μm)

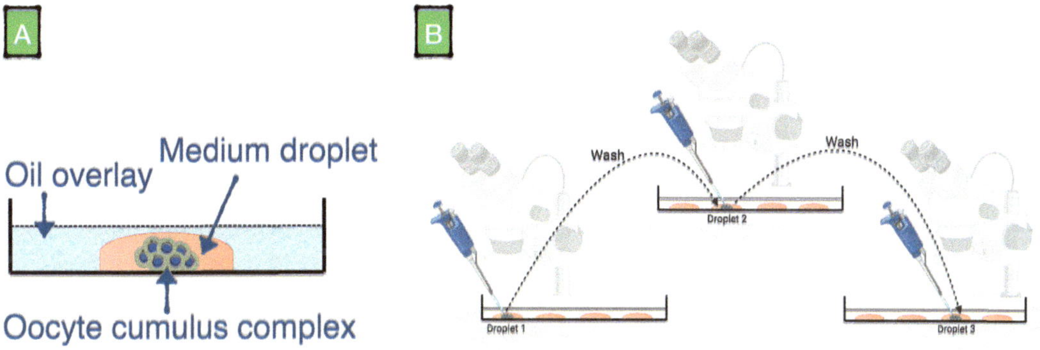

Fig. 4 (**a**) Diagrammatic representation of oocyte cumulus complexes (OCCs) released into medium droplet overlaid with oil. (**b**) Diagrammatic representation of washing of OCCs in medium droplets using micropipette. Figure 4b has been created with Biorender.com

3. To obtain haploid parthenogenetic embryos, transfer the OCC bunches into sterile tissue culture grade petri dish containing 30 μL droplets of the haploid activation medium covered with oil.
4. To obtain diploid parthenogenetic embryos, transfer the OCC bunches into sterile tissue culture grade petri dish containing 30 μL droplets of the diploid activation medium covered with oil.
5. Place the petri dishes in the incubator at 37 °C and 5% CO_2 for 3 h to allow oocyte activation.

3.5 Denuding of Oocyte Cumulus Complexes

1. After 3 h of incubation in activation medium, pick the OCCs (*see* Fig. 5) by observing under the stereo microscope using a 2–20 μL pipette, and transfer them to a 40 μL droplet of hyaluronidase (0.75 mg/mL in M16 medium) covered with oil for 1 min.
2. Focus the edge of the droplet under the stereo microscope and adjust the light of the microscope to view the denuded oocytes.

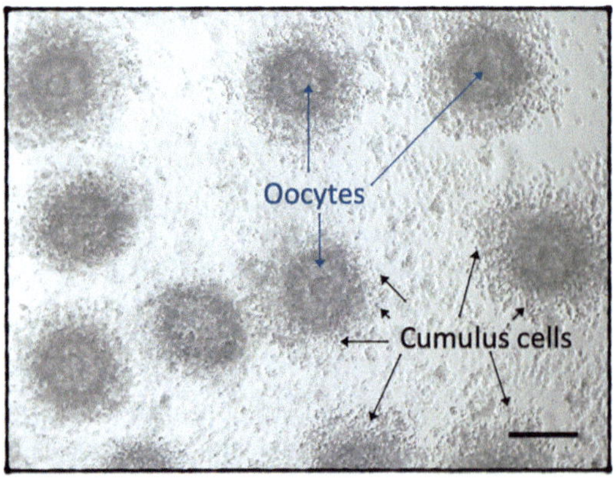

Fig. 5 Oocyte cumulus complex post 3 h of parthenogenetic activation (scale bar = 200 μm)

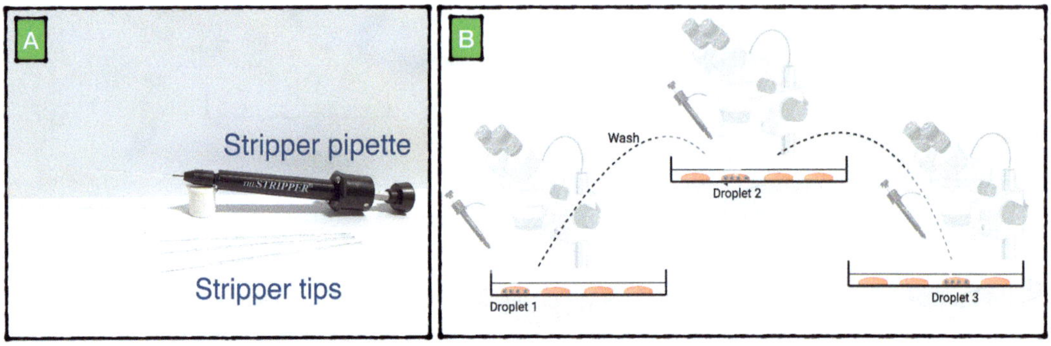

Fig. 6 (**a**) Stripper pipette with a sterile stripper tip. (**b**) Diagrammatic representation of washing of embryos in medium droplets overlaid with oil, using stripper pipette. Figure 6b has been created with Biorender.com

3. Pick the denuded oocytes using sterile stripper tips (135 μm or 175 μm) attached to a stripper pipette (*see* Fig. 6a) under the stereo microscope and transfer them to 20 μL M16 wash droplets covered with oil (*see* **Note 12**).

3.6 Embryo Culture

1. Gently rinse the activated oocytes in 3–4 droplets of medium using stripper pipette (*see* Fig. 6b) and then transfer them to the embryo culture dish containing 20 μL droplets of M16 medium covered with oil (*see* **Note 13**).
2. Place the culture dish in the incubator maintained at 37 °C and 5% CO_2.

3.7 Pronuclear (PN) Stage Assessment and Embryo Culture

1. Post 2 h of completion of activation, observe the embryo culture dish under an inverted microscope (200× magnification) with a heat stage maintained at 37 °C, and count the number of pronuclei in the activated oocytes (*see* Fig. 7).

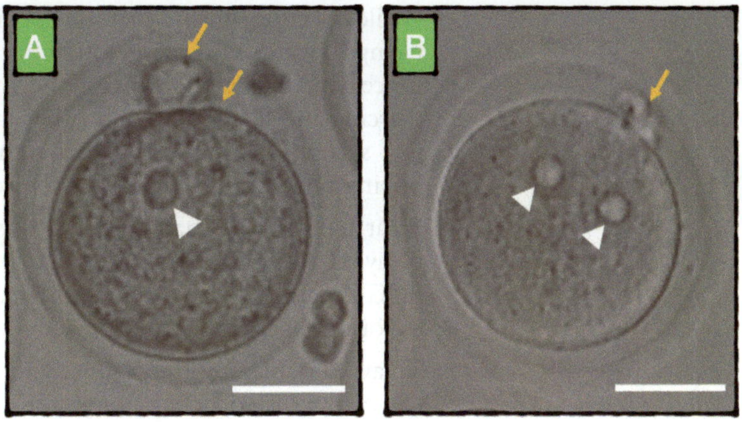

Fig. 7 (**a**) Haploid parthenogenetic embryo having one pronucleus. (**b**) Diploid parthenogenetic embryo having two pronuclei. White arrowhead indicates pronucleus; yellow arrow indicates polar body (scale bar = 25 μm)

2. The embryos having 1 pronucleus (1 PN) and 2 polar bodies (2 PB) are haploid parthenotes.
3. The embryos having 2 pronuclei (2 PN) and 1 polar body (1 PB) are diploid parthenotes (*see* **Note 14**).
4. Place the embryo culture dish back in the incubator and observe the parthenotes once every 24 h post activation until they reach the blastocyst stage, to assess the developmental potential (*see* **Note 15**).

4 Notes

1. In vitro exposure of oocytes to ethanol exerts a toxic effect on their developmental potential.
2. Toothed-forceps can be used for pinching and lifting the skin of the mice; thin pointed forceps can be used for organ collection.
3. Straight pointed dissection scissors can be used for cutting open the skin and peritoneal cavity; curved-sharp scissors or curved micro-scissors can be used for organ collection.
4. Sr^{2+} fails to induce changes in Ca^{2+} levels in oocytes when incubated in the Ca^{2+}-containing medium [11].
5. Pick up the mouse by holding the scruff of its neck as close to the ears as possible. Make sure that you hold enough skin, so that the mouse cannot turn its head. Hold the tail by twisting it around the little finger.
6. Use a hypodermic needle for the injection and pierce the skin and abdominal muscles at an angle of 45° to the body of the mouse to inject the solution into the intra-peritoneal cavity.

While injecting, ensure that the opening of the syringe needle is facing upwards and the graduations on the syringe are visible. Once you pierce the skin of the mouse, take care to avoid piercing the diaphragm and other internal organs. Wait for a few seconds before withdrawing the needle, so as to avoid oozing out of the solution from injected region.

7. Wear gloves while handling and dissecting the mouse. Avoid delaying the dissection and OCC collection beyond 14 h of hCG injection, as it can yield post-mature oocytes and can lead to lot of degeneration during activation.
8. To avoid damaging the oviduct and losing the OCC bunches, the oviduct is collected along with the ovary and a small portion of the uterine horn. The bulged region can later be observed and dissected under a stereo microscope in sterile conditions.
9. Use one straight needle and one bent needle (bent to 90° using a forceps). Keep the organs in place under the stereomicroscope using the straight needle pricked into the uterine horn region, and rotate the organs using the bent needle to locate the bulged region of the oviduct.
10. Make a cut close to the narrow end of the bulged oviduct and carefully push out the OCC bunches from the other end using the bent needle, so that the entire OCC bunch is released out at once without any damage to the oocytes.
11. The paraffin oil should be prewarmed along with a small amount of medium at the bottom of the tube (1:10 ratio) at least for 1 h, to allow equilibration of reactive substances from the oil to the medium. The sterile petri dishes containing droplets of prewarmed medium covered with paraffin oil should be prepared and placed in the incubator at least 30 min prior to the procedure. The overlaid oil must be just enough to cover the medium droplets. Too much volume of oil can hinder the gaseous exchange into medium.
12. Before picking the denuded activated oocytes from the droplets of medium, pipette in some plain medium into the stripper tip to reduce the capillary action that may suddenly pull in the oocytes and lead to losing them. This also helps to avoid expelling of air bubbles while releasing the oocytes. Always keep the oocytes at the tip of the stripper tip while transferring from one droplet of medium to the next.
13. Around 12–15 activated oocytes can be placed in each droplet of 20 μL M16 medium. Group culture of embryos is found to be beneficial due to the concentration of autocrine growth factors [12].

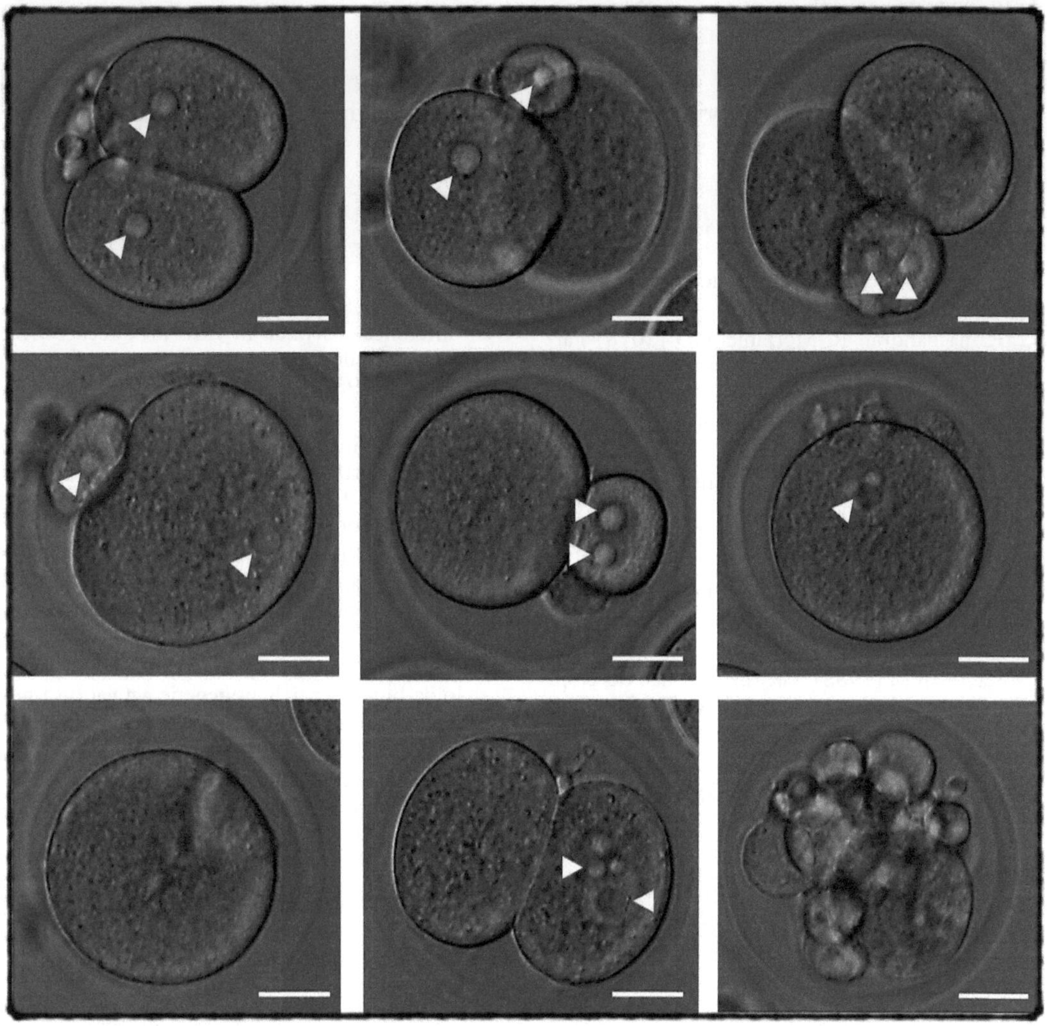

Fig. 8 Variations in embryos apart from 2 pronuclei, observed upon parthenogenetic activation using strontium chloride and cytochalasin D; white arrowhead indicates pronucleus (scale bar = 25 μm)

14. In addition to the 2 PN and 1 PB containing embryo, we find a small proportion of embryos having 2 cell with 1 PN in each blastomere (early dividing); or 1 cell with 1 PN in the cell and the other PN in an extruded cytoplasmic fragment (*see* Fig. 8). These irregular embryos may be eliminated from the culture.

15. The parthenogenetic embryos have a delayed timeline compared to the normally fertilized embryos. The haploid parthenotes take 5–6 days post activation to reach the blastocyst stage, the diploid parthenotes take 4–5 days post activation to reach the blastocyst stage, while the normally fertilized embryos reach blastocyst stage by 3.5–4.5 days post insemination.

Acknowledgments

The author DNC acknowledges DST-Inspire fellowship and Fulbright-Nehru Doctoral Research Fellowship to conduct doctoral studies in NK's laboratory.

References

1. Mittwoch U (1978) Parthenogenesis. J Med Genet 15(3):165–181
2. Ramachandran R, McDaniel CD (2018) Parthenogenesis in birds: a review. Reproduction 155(6):R245–R257
3. Suomalainen E (1950) Pathenogenesis in animals. Adv Genet 3:193–253
4. Eppig JJ (1978) Developmental potential of LT/Sv parthenotes derived from oocytes matured in vivo and in vitro. Dev biol 65(1): 244–249
5. Loeb J (1913) Artificial parthenogenesis and fertilization. The University of Chicago Press, Chicago, Ill
6. Kaufman M (1983) Early mammalian development: parthenogenetic studies. Development and cell biology series. Cambridge University Press, Cambridge
7. Kharche SD, Birade HS (2013) Parthenogenesis and activation of mammalian oocytes for *in vitro* embryo production: a review. Adv Biosci Biotechnol 4:170–182
8. Gómez E, Gutiérrez-Adán A, Díez C, Bermejo-Alvarez P, Muñoz M, Rodriguez A, Otero J, Alvarez-Viejo M, Martín D, Carrocera S, Caamaño JN (2009) Biological differences between in vitro produced bovine embryos and parthenotes. Reproduction 137 (2):285–295
9. Espejel S, Eckardt S, Harbell J, Roll GR, McLaughlin KJ, Willenbring H (2014) Brief report: parthenogenetic embryonic stem cells are an effective cell source for therapeutic liver repopulation. Stem Cells 32(7):1983–1988
10. Kline D, Kline JT (1992) Repetitive calcium transients and the role of calcium in exocytosis and cell cycle activation in the mouse egg. Dev Biol 149(1):80–89
11. Liu L, Trimarchi JR, Keefe DL (2002) Haploidy but not parthenogenetic activation leads to increased incidence of apoptosis in mouse embryos. Biol Reprod 66(1):204–210
12. Lane M, Gardner DK (1992) Effect of incubation volume and embryo density on the development and viability of mouse embryos in vitro. Hum Reprod 7(4):558–562

Chapter 3

Generation of Human iPSC from Small Volume Peripheral Blood Samples

Doreen Kloos and Nico Lachmann

Abstract

The generation of induced pluripotent stem cells (iPSCs) from patients has opened new doors to gain insights into disease pathophysiology and treatment. In particular, the generation of iPSCs from patients who suffer from inherited diseases is of great interest, as most of these diseases are rare and not well studied. As most affected patients are diagnosed during infancy, the derivation of somatic cells for the generation of iPSCs is very much limited. Here we describe a protocol for the generation of human iPSCs from non-mobilized peripheral blood. This protocol can be adapted to all volumes of blood, starting from 1 ml of peripheral blood. Isolated and processed cells can be used for the generation of iPSCs by both, lentiviral and Sendai virus mediated reprogramming, allowing for the rapid generation of patient-specific iPSCs.

Key words iPSC, Reprogramming, Peripheral blood, Lentivirus

1 Introduction

Induced pluripotent stem cells (iPSCs) can be derived from various somatic cells by a process referred to as "*reprogramming*" [1, 2]. Different cell sources and techniques have been described over the past years, highlighting reprogramming as an easy-to-use technique in modern labs [3]. Given the individual laboratory equipment and expertise, the most common cell types used for reprogramming of human cells are hair follicle cells, keratinocytes, fibroblasts, and cells from peripheral blood [4]. The latter especially is of great interest as blood can be obtained by a non-surgical procedure from nearly all patients. Generation of patient-specific iPSCs is of particular interest for patients who suffer from inherited diseases [5]. While most if not all of these genetic disorders are rare and patient material is limited, iPSC technology allows for the unlimited cultivation and differentiation of these cells into the affected cell type, providing a unique cell source for biomedical research. In fact, congenital diseases are typically diagnosed during

Nagarajan Kannan and Philip Beer (eds.), *Stem Cell Assays: Methods and Protocols*, Methods in Molecular Biology, vol. 2429, https://doi.org/10.1007/978-1-0716-1979-7_3,

infancy or early childhood, limiting the derivation of suitable somatic cell types for the generation of iPSCs. While blood is usually considered the tissue of choice for reprogramming, the age, body weight, and overall health status of the patient can limit the volume which can be obtained from the young patient. As an example, Adenosine Deaminase (ADA)-deficient severe combined immunodeficiency (ADA-SCID) is generally diagnosed within the first 12 months of age, with most of the patients dying before the age of two without treatment. Given the nature of the disease, only a few milliliters of blood can be obtained from these patients [6].

We here demonstrate the generation of human iPSCs from small volumes of peripheral blood, using a protocol which can be adapted to the individual volume obtained from the patient. After the isolation of peripheral blood mononuclear cells (PBMCs) by Ficoll gradient, a threshold of approximately 8×10^6 cells should be considered for the isolation of $CD34^+$ cells. Subsequent cultivation of either $CD34^+$ cells and/or PBMCs in small volumes of hematopoietic medium 24 h (h) prior to lentiviral mediated reprogramming allows for the proper generation of human iPSCs for further expansion and freezing.

In our lab, this protocol is used regularly for the generation of patient-specific iPSCs from different immunodeficient patients [7–10]. Moreover, this protocol is also used by other collaborative investigators to generate human iPSC from small volume blood samples [11–13].

2 Materials

Buffers and reagents are prepared in either deionized water or in 1× phosphate-buffered saline (PBS), which can be commercially acquired or manufactured in-house.

2.1 Isolation of Blood Cells

1. 1× PBS: Dissolve 8 g NaCl (final concentration: 137 mM), 200 mg of KCl (final concentration: 2.7 mM), 1.44 g of Na_2HPO_4 (final concentration 10 mM), and 245 mg of KH_2PO_4 (final concentration: 1.8 mM) in 800 ml deionized water. Adjust solution to pH ≈ 7.4. Add deionized water until volume is 1 l and sterile-filter through a 0.22 μm Corning filter.
2. PBS/2 mM EDTA (dilution buffer): Mix 2 ml of a 0.5 M Na_2EDTA solution with 248 ml of 1× PBS in a glass flask. Warm up an aliquot of the buffer to room temperature (*see* **Note 1**).
3. Ficoll (1.077 g/ml) separating solution (*see* **Note 2**).
4. Red cell lysis buffer (RCL): Weigh 4.15 g NH_4Cl (final concentration 155 mM) and 2.3 g $KHCO_3$ (final concentration 46 mM) and transfer them into a 500 ml glass flask. Fill up to

400 ml with deionized water. Add 0.5 ml 0.5 M Na_2EDTA-solution (final concentration 0.5 mM) and stir until salts are dissolved. Fill up to 500 ml with deionized water and sterile-filter through a 0.22 μm filter (*see* **Note 3**).

5. PBS/2 mM EDTA/FCS 2%: Add 2 ml of a 0.5 M Na_2EDTA solution and 10 ml of fetal calf serum (FCS) in a glass flask with 238 ml of 1× PBS. Store and use buffer at 4 °C (*see* **Note 4**).

2.2 Isolation of CD34⁺ Cells

1. Magnetic antibody labeling kit for $CD34^+$ cells (*see* **Note 5**).

2.3 Freezing of Donor Cells

1. Freezing medium: Prepare a sufficient amount of freezing medium by mixing 90% FCS and 10% Dimethyl sulfoxide (DMSO) in a 15 ml tube.
2. Freezing container (with, e.g., isopropanol).

2.4 Culture of PBMCs or $CD34^+$ Cells

1. Complete hematopoietic medium: Use an appropriate volume of basic hematopoietic medium (e.g., StemSpan™ SFEM, *see* **Note 6**) and freshly add 100 ng/ml hSCF, 50 ng/ml hTPO, 100 ng/ml hFLT3L, and 100 U/ml Penicillin/Streptomycin (Pen/Strep). Warm up medium to 37 °C shortly before use.
2. 24-well cell culture plate, suspension.

2.5 Transduction of PBMCs or $CD34^+$ Cells

1. PBS/2 mM EDTA/FCS 2% as described in Subheading 2.1, **item 5**.
2. High titer reprogramming virus (*see* **Note 7**).
3. Polybrene solution: Weight 8 mg/ml polybrene, dissolve it in 1× PBS, and sterilize through a 0.22 μm filter. Before use warm up to 37 °C (*see* **Note 8**).
4. Complete hematopoietic medium, as described in Subheading 2.4, **item 1** (*see* **Note 6**).
5. Mini-Rotator, which fits into an incubator.
6. 24-well cell culture plate, suspension.

2.6 Cultivation of Transduced Cells

1. 0.7 M valproic acid: Dissolve 100.94 mg/ml valproic acid in 1× PBS and sterilize through a 0.22 μm filter (*see* **Note 9**).
2. Complete hematopoietic medium, as described in Subheading 2.4, **item 1** (*see* **Note 6**).
3. 12-well cell culture plate, suspension.

2.7 Transfer to Feeder Cells

1. Complete hematopoietic medium, as described in Subheading 2.4, **item 1** (*see* **Note 6**).
2. Complete human embryonic stem cell medium (hESC-medium): Use KnockOut (KO) DMEM supplemented with 20% KnockOut (KO) serum replacement, 100 U/ml

Pen/Strep, 1% L-Glutamine 200 mM, 1% non-essential amino acids 100×, 0.2% ß-Mercaptoethanol 50 mM (*see* **Note 10**). Prewarm an aliquot and add human basic-fibroblast-growth-factor (bFGF) (final concentration 20 ng/ml) freshly before use.

3. Rock inhibitor (Y-27632 dihydrochloride/RI): Prepare a 10 μM stock in 1× PBS, sterile-filter, make aliquots, and use 1:1000 (*see* **Note 11**).
4. Irradiated CF1 feeder cells (mouse embryonic fibroblasts): Seed 170,000 cells per well on a gelatine-coated 6-well cell culture plate (adherent) 1 day before usage (*see* **Note 12**).

2.8 Subcloning and Expansion of iPSCs

1. Complete human embryonic stem cell medium (hESC-medium), as described in Subheading 2.7, **item 2** (*see* **Note 10**).
2. Irradiated CF1 feeder cells: Seed 80,000 cells per well on a gelatine-coated 12-well cell culture plate (adherent) 1 day before usage (*see* **Note 12**).
3. Collagenase IV: To generate a 25,000 U collagenase IV stock, dissolve collagenase IV powder with 1× Hank's Balanced Salt Solution (HBSS) (with calcium and magnesium) and sterile-filter. Dilute stock to 250 U/ml in KO DMEM, prepare aliquots, and warm up the amount you need (*see* **Note 13**).
4. Irradiated CF1 feeder cells: Seed 170,000 cells per well on a gelatine-coated 6-well cell culture plate (adherent) 1 day before usage (*see* **Note 12**).

3 Methods

3.1 Isolation of PBMCs from Small Blood Volume

For the step-by-step procedure of isolating PBMCs from small volume blood, *see* Fig. 1.

1. Dilute whole blood with PBS/2 mM EDTA. Maximal dilution factor should be 1:3.
2. Prepare 15 ml tubes with a sufficient amount of Ficoll. The ratio of Ficoll:diluted blood should be 1:3 (*see* **Note 14**).
3. Very gently place the diluted blood from **step 1** onto Ficoll separation medium.
4. Centrifuge for 40 min (min) at 400 × *g*, room temperature with the lowest acceleration and no brake.
5. Carefully collect the ring of cells at the interface with a pipette and transfer to a 15 ml conical tube.
6. Fill up the tube with PBS/2 mM EDTA.

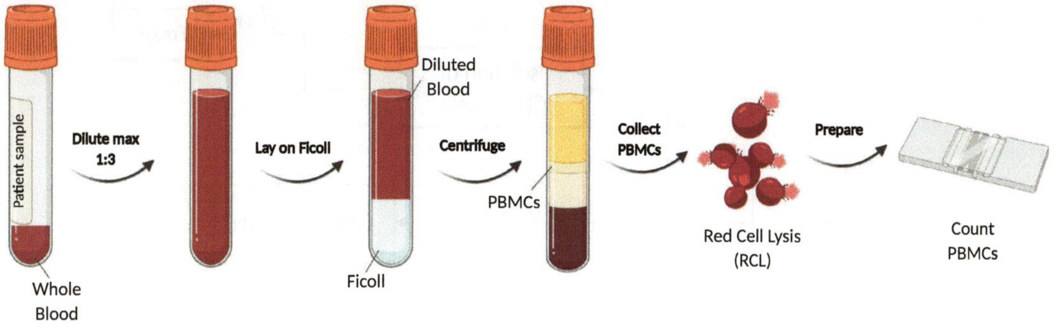

Fig. 1 Isolation of human PBMCs. Scheme for the isolation procedure. Whole blood is diluted with PBS/2 mM EDTA and layered on Ficoll separation solution. After centrifugation, PBMCs are collected, red cell lysis is performed, and cells are counted

7. Centrifuge for 10 min at 400 × *g*, at 4 °C with the lowest acceleration and no brake.
8. Aspirate supernatant.
9. Resuspend the cells with 1.5 ml of RCL buffer, and incubate at room temperature for 3 min (*see* **Note 15**).
10. Fill tube with PBS/2 mM EDTA/FCS 2% to stop lysis.
11. Spin for 10 min at 400 × *g*, at 4 °C.
12. Aspirate supernatant.
13. Resuspend all cells in 1 ml PBS/2 mM EDTA/FCS 2%.
14. Count the cells using a counting chamber (*see* **Note 16**).

3.2 Isolation of CD34$^+$ Cells from PBMCs

1. Whether or not PBMCs should be enriched for CD34$^+$ cells depends on the number of total PBMCs (Subheading 3.1, **step 14**; see also chart in Fig. 2). If the number of PBMCs exceeds 8 × 10^6 PBMCs in total, perform CD34$^+$ cell isolation (*see* **Note 17**).
2. For CD34$^+$ isolation, use a magnetic antibody labeling kit according to the manufacturing instructions.

3.3 Freezing of Donor PBMCs

1. Always freeze some donor PBMCs as a backup or for later experiments.
2. Freeze either aliquots of PBMCs or—after CD34$^+$ isolation—aliquots of the CD34$^-$ population.
3. For freezing resuspend 2 × 10^6 cells in 1 ml of freezing medium, pipette in a cryotube, and freeze the cells with the help of a suitable freezing container (e.g., with isopropanol) to achieve a cooling rate of 1 °C/min. Store the freezing container at −80 °C overnight and place the cryotube into liquid N_2 the next day.

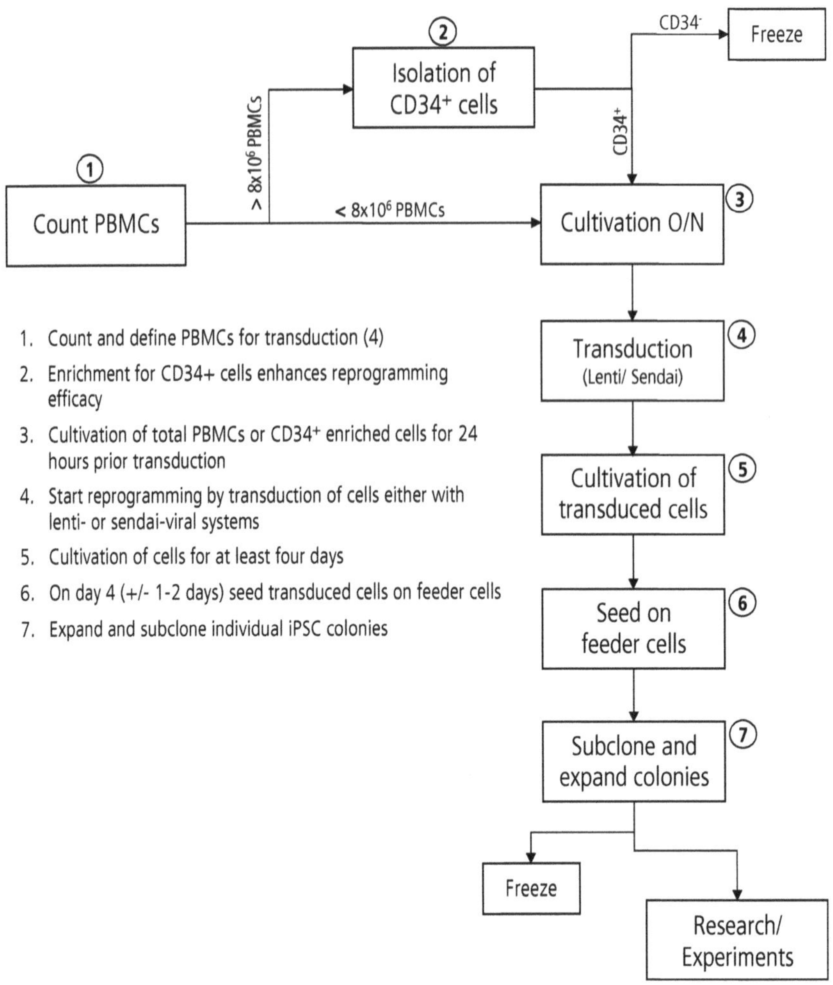

Fig. 2 Experimental workflow. Schematic representation of the reprogramming process. Whole PBMCs are first counted. Decision on whether CD34$^+$ cells are enriched is made on total PBMC number. Threshold for enrichment of CD34$^+$ cells is 8×10^6 PBMCs. Thereafter, either CD34$^+$ enriched cells or complete PBMCs are cultivated for 24 h before reprogramming. Subsequently, either CD34$^+$ enriched cells or complete PBMCs are reprogrammed by either lentiviral or Sendai-viral transduction. Four days post reprogramming, cells are seeded on CF1 mouse embryonic fibroblasts and cultivated until first human iPSC colonies appear. IPSC colonies are subcloned and expanded before they are frozen down or used for further experiments

3.4 Culturing of PBMCs/CD34$^+$ Cells

1. Transfer 2×10^5 PBMCs or all CD34$^+$ cells to one well of a 24-well cell culture plate in 1 ml of complete hematopoietic medium and place in an incubator at 37 °C, 5% CO_2 (*see* **Note 18**).
2. Incubate the cells overnight (Fig. 3a).

3.5 Transduction of PBMCs/CD34$^+$ Cells for Reprograming

The morphology of cells during reprogramming and appearance of single iPSC colonies is shown in Fig. 3.

1. Transfer the PBMCs/CD34$^+$ cells from 24-well cell culture plate to a 15 ml falcon tube.

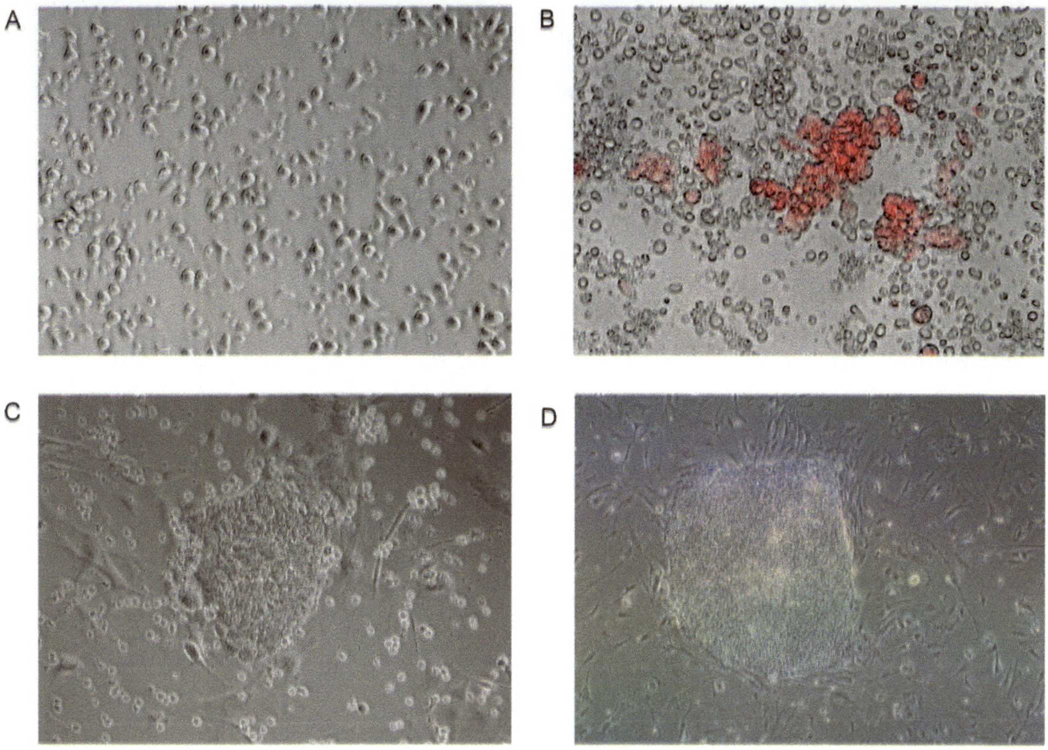

Fig. 3 Microscopic pictures of the reprogramming process. (**a**) PBMCs 24 h after isolation: PBMCs were isolated from whole peripheral blood and cultivated for 24 h in complete hematopoietic medium containing 100 ng/ml hSCF, 50 ng/ml hTPO and 100 ng/ml hFLT3L and 100 U/ml Pen/Strep. (magnification 20×) (**b**) PBMCs 96 h after transduction. PBMCs were transduced with a third-generation SIN lentiviral vector containing reprogramming factors hOkt4/5, hKlf4, hSox2, and hcMyc driven by a SFFV promotor and coupled to an IRESdTomato site. Picture shows brightfield image and transduction of PBMCs (red). (magnification 20×) (**c**) Human iPSC colony before subcloning. Representative picture of an iPSC colony 12 days after transduction. PBMCs were reprogrammed and seeded on CF1 mouse embryonic fibroblasts after 96 h. Cells were cultivated until first human iPSC colonies appear. (magnification 10×) (**d**) Subcloned iPSC colony: Representative picture of an iPSC colony after subcloning and expansion (magnification 4×)

2. Centrifuge for 10 min at 200 × *g* at room temperature.
3. Aspirate the supernatant.
4. Resuspend the cells in 1 ml of PBS/2 mM EDTA/FCS 2%.
5. Count cells (*see* **Note 15**).
6. Combine the components shown in Table 1 in a 1.5 ml tube for viral transduction (multiplicity of infection; MOI 20).
7. Place in an incubator at 37 °C, 5% CO_2, on a rotator at 6 rpm for 4 h (*see* **Note 19**).
8. Transfer the suspension into a well of a 24-well cell culture plate, and add 500 μl of complete hematopoietic medium.
9. Place the 24-well cell culture plate in an incubator at 37 °C, 5% CO_2 (*see* **Note 18**).
10. Incubate overnight.

Table 1
Composition of viral transduction

Description	Amount
Lentiviral particles	MOI of 20
Polybrene	4 μg
PBMCs/$CD34^+$	2×10^5
Complete hematopoietic medium	Up to 500 μl

3.6 Further Cultivation of Transduced PBMCs/$CD34^+$ Cells

1. Add 2.85 μl 0.7 M valproic acid (1:350) to the cells (final concentration 2 mM) 24 h after transduction.
2. Place back in an incubator at 37 °C, 5% CO_2 (*see* **Note 18**).
3. 48 h after transduction evaluate the cells using a fluorescent microscope to determine the transduction efficiency.
4. Collect the cells in a 15 ml tube, wash the well with 1× PBS and also add the washing flow to the 15 ml tube. Centrifuge cells for 5 min at 400 × *g* at room temperature. After aspirating the supernatant, resuspend cells in 2 ml of complete hematopoietic medium + 2 mM valproic acid.
5. Transfer the cells into one well of a 12-well cell culture plate for suspension cells.
6. Place the 12-well cell culture plate in an incubator at 37 °C, 5% CO_2 (*see* **Note 18**).

3.7 Transfer of Reprogrammed Cells for iPSC Colony Formation on Feeder Cells

1. Re-evaluate the transduction efficiency using a fluorescent microscope 96 h after transduction. Some cells might form clumps and some of the cells might be expanding (Fig. 3b).
2. Collect cells by pipetting the medium up and down a few times to wash the cells off the surface and transfer into a 15 ml Falcon tube.
3. Wash twice with 1 ml 1× PBS and add to the 15 ml Falcon tube.
4. Verify by microscope that all cells have detached from the well.
5. Centrifuge cells for 5 min at 200 × *g* at room temperature.
6. Prepare medium for iPSCs colony formation by mixing in a 1:1 ratio complete hematopoietic medium with complete hESC-medium supplemented with RI 1:1000.
7. Aspirate medium from plate with feeder cells.
8. Add 1 ml of 1× PBS to each well containing feeders and rock the plate to thoroughly wash the wells.
9. Aspirate 1× PBS from the plates containing feeder cells.
10. Aspirate supernatant carefully from the 15 ml falcon tube.

11. Resuspend the cells in 6 ml of 1:1 complete hematopoietic medium/complete hESC-medium supplemented with RI.
12. Transfer 2 ml of cell suspension per well into 3 wells of a 6-well cell culture plate containing feeder cells.
13. Place the plate in an incubator at 37 °C, 5% CO_2 (*see* **Note 20**).
14. Add 1 ml of complete hESC-medium after 2–3 days, afterwards change medium every second day (*see* **Note 21**).
15. Initial colonies are expected to appear approximately 12 days post transduction (Fig. 3c).
16. Let colonies of iPSCs grow until they are big enough to be picked from the plate.

3.8 Subcloning and Expansion of Individual iPSC Colonies

1. Aspirate medium from 12-well cell culture plate with feeder cells. Add 1 ml of 1× PBS to each well containing feeders and carefully rotate the plate to thoroughly wash the wells. Aspirate 1× PBS and add 1 ml of complete hESC-medium in each well.
2. For picking colonies, first scratch around the colony by using a 27 G needle to remove the feeder cells. Then scratch 2–3 vertical and horizontal lines across the colony to get smaller pieces. Afterwards use a 200 μl pipette to collect the scratched colony. It may be necessary to gently scrape the colony from the bottom with the pipette tip (*see* **Note 22**).
3. Transfer one colony per well of the 12-well cell culture plate with feeder cells previously prepared (*see* **Note 23**).
4. Place the plate in an incubator at 37 °C, 5% CO_2 for colony expansion.
5. Change medium every 2–3 days.
6. Passage the colonies into a 6-well cell culture plate after 5–7 days.
7. When passaging colonies, aspirate the medium from the wells and wash gently with 1 ml of 1× PBS/well. Add 500 μl collagenase IV to each well and incubate for 20 min at 37 °C, 5% CO_2 (*see* **Note 24**). Subsequently, add 1 ml of basic hESC-medium to each well and try to loosen the colonies from the well. Collect all fragments of the colonies in a 15 ml tube and wash the well again with 1 ml 1× PBS (*see* **Note 25**). After centrifugation for 1 min at 100 × *g* at room temperature, aspirate supernatant and loosen the pellet by adding 1 ml of complete hESC-medium (*see* **Note 26**). Distribute colony fragments to 1–6 wells of the 6-well cell culture plate containing feeder cells.
8. Again change medium every 2–3 days and passage subcloned colonies every 7–10 days.
9. Now cells are ready for characterization steps and freezing (*see* **Note 27**) (Fig. 3d).

4 Notes

1. It is very important to perform the PBMC Isolation at room temperature to avoid clumping of the blood in order to achieve high cell recovery. Prepare this buffer under sterile conditions. Buffer can be stored at 4 °C for up to 3 months.
2. Ficoll separating solution is commercially available. Store Ficoll in the dark.
3. Store RCL buffer at room temperature, prepare aliquots, and keep it sterile.
4. Buffer should be prepared under sterile conditions and can be stored at 4 °C for up to 3 months.
5. In our lab, we regularly use magnetic labeling of $CD34^+$ cells (positive selection) to enrich PBMCs for $CD34^+$ cells. Suitable kits are available from various companies. Other selection procedures may also be applicable (e.g., fluorescence-activated cell sorting).
6. You can use any basic hematopoietic stem cell medium of your choice. In our lab, we regularly use StemSpan™ SFEM from STEMCELL Technologies. Cytokines should always be used as freshly thawed aliquots, to avoid repeated freeze/thaw cycles.
7. We use a third-generation SIN lentivirus containing reprogramming factors hOkt4/5, hKlf4, hSox2, and hcMyc. Factors are driven by the SFFV promotor [14]. The virus also contains an IRES.dTomato site. Alternatively, you can also use a Sendai virus for reprogramming. Sendai reprogramming kits are commercially available. For using Sendai virus, please follow the manufacturer's manual.
8. Polybrene (Hexadimethrine bromide) can improve transduction efficiency. Store the stock solution at 4 °C.
9. Valproic acid increases reprogramming efficiency by inhibiting histone deacetylases. Stock solution should be aliquoted and stored at −20 °C.
10. Prepare hESC-medium in single-use plastic bottles and not in multi-use glass bottles, as ESCs/iPSCs are very sensitive to washing detergent residuals. Store medium at 4 °C and warm up aliquots with the amount of medium needed.
11. Rock inhibitor (RI) is an RHO/ROCK pathway inhibitor which increases survival of human embryonic stem cells when they are single cells by preventing dissociation-induced apoptosis. Store 10 μM stock at −20 °C. Always use freshly added RI in your medium.
12. CF1 mouse embryonic fibroblasts are commercially available. Otherwise fibroblasts can be prepared in-house (age of mouse embryos should be 13.5 days), expanded and irradiated at

30 Gy. After irradiation, freeze aliquots (e.g., 2×10^6, 4×10^6, and 6×10^6 irradiated cells), which are then ready to use after thawing. Feeder cells are growing much more homogenous when seeded on gelatine-coated plates and incubated at 37 °C and 8.5% CO_2.

13. Units/mg will vary between different contributors and/or LOTs. Always adapt the amount of 1× HBSS to the concentration of your collagenase IV powder. Stock should be aliquoted and stored at −20 °C. Diluted aliquots can also be stored at −20 °C until needed.
14. When adding Ficoll to the tube, be careful to avoid any droplets of Ficoll on the tube side. 50 ml tubes can also be used when receiving a bigger amount of blood.
15. It is important not to exceed the incubation time of RCL buffer, as PBMCs are also susceptible to cell lysis if they are exposed to the lysis buffer for extended time periods.
16. Perform live/dead staining using trypan blue for quality control by mixing the cells with trypan blue in a ratio of 1:10.
17. Isolation and use of $CD34^+$ cells will multiply the reprogramming efficiency as $CD34^+$ cells are much easier to reprogram.
18. As PBMCs/$CD34^+$ cells are very sensitive to environmental changes, they should be incubated in a separate wet chamber within the incubator.
19. Seal the lid of the 1.5 ml tube with PARAFILM® M.
20. After seeding transduced cells on feeder cells, it is no longer required to store the plate in an extra wet chamber at the incubator.
21. Feeder cells might detach if colonies grow very slowly. If so, seed additional 100,000 feeder cells per well on top after 1.5 weeks.
22. Select colonies under sterile conditions with the help of a stereo microscope. Try to choose colonies that are separated from other colonies, are mostly circular, and are without differentiating edges.
23. Select at least 10–12 subclones as not all of them will attach and grow. It is not necessary to keep all of them but it is preferable to have at least 4 subclones of every reprogramming at the end.
24. Incubation with collagenase IV can also be extended (up to 40 min), as it is preferable to incubate the colonies longer in collagenase IV rather than forcing detachment by pipetting, which will result in over-fragmentation of the colonies.
25. It is very important to pipette just a few times. This will minimize the colony fragmentation. If there are still iPSC colonies in the well, repeat wash with new 1× PBS until all colonies are collected.

26. Take care not to resuspend the pellet of fragmented iPSC colonies. Just add medium on top and tap the tube to loosen the pellet.
27. Before freezing iPSC lines, passage them at least five times. This will enhance the regrow efficacy after thawing.

Acknowledgments

The authors thank Dr. Miriam Hetzel (Hannover Medical School) for proofreading and critical comments on the manuscript.
This work was supported by the Federal Ministry of Education and Research in Germany (BMBF) [01EK1602A], the REBIRTH Cluster of Excellence (EXC62, Deutsche Forschungsgemeinschaft (DFG)), and the REBIRTH Research Center for Translational Regenerative Medicine [ZN3440, State of Lower Saxony, Ministry of Science and Culture (Nieders. Vorab)]. The project was further supported by the European Research Council (ERC) under the European Union's Horizon 2020 research and innovation program (Grant agreement No. 852178). Figure 1 was created with Biorender online software.

References

1. Takahashi K, Yamanaka S (2006) Induction of pluripotent stem cells from mouse embryonic and adult fibroblast cultures by defined factors. Cell 126(4):663–676. https://doi.org/10.1016/j.cell.2006.07.024
2. Yu J, Vodyanik MA, Smuga-Otto K, Antosiewicz-Bourget J, Frane JL, Tian S et al (2007) Induced pluripotent stem cell lines derived from human somatic cells. Science 318(5858):1917–1920. https://doi.org/10.1126/science.1151526
3. Al Abbar A, Ngai SC, Nograles N, Alhaji SY, Abdullah S (2020) Induced pluripotent stem cells: reprogramming platforms and applications in cell replacement therapy. Biores Open Access 9(1):121–136. https://doi.org/10.1089/biores.2019.0046
4. Malik N, Rao MS (2013) A review of the methods for human iPSC derivation. Methods Mol Biol 997:23–33. https://doi.org/10.1007/978-1-62703-348-0_3
5. Elitt MS, Barbar L, Tesar PJ (2018) Drug screening for human genetic diseases using iPSC models. Hum Mol Genet 27(R2): R89–R98. https://doi.org/10.1093/hmg/ddy186
6. Flinn AM, Gennery AR (2018) Adenosine deaminase deficiency: a review. Orphanet J Rare Dis 13(1):65. https://doi.org/10.1186/s13023-018-0807-5
7. Haake K, Neehus AL, Buchegger T, Kuhnel MP, Blank P, Philipp F et al (2020) Patient iPSC-derived macrophages to study inborn errors of the IFN-gamma responsive pathway. Cells 9(2). https://doi.org/10.3390/cells9020483
8. Haake K, Wustefeld T, Merkert S, Luttge D, Gohring G, Auber B et al (2020) Human STAT1 gain-of-function iPSC line from a patient suffering from chronic mucocutaneous candidiasis. Stem Cell Res 43:101713. https://doi.org/10.1016/j.scr.2020.101713
9. Neehus AL, Lam J, Haake K, Merkert S, Schmidt N, Mucci A et al (2018) Impaired IFNgamma-signaling and mycobacterial clearance in IFNgammaR1-deficient human iPSC-derived macrophages. Stem Cell Rep 10(1): 7–16. https://doi.org/10.1016/j.stemcr.2017.11.011
10. Lachmann N, Happle C, Ackermann M, Luttge D, Wetzke M, Merkert S et al (2014) Gene correction of human induced pluripotent stem cells repairs the cellular phenotype in pulmonary alveolar proteinosis. Am J Respir Crit Care Med 189(2):167–182. https://doi.org/10.1164/rccm.201306-1012OC

11. Merkert S, Schubert M, Haase A, Janssens HM, Scholte B, Lachmann N et al (2020) Generation of an induced pluripotent stem cell line (MHHi018-A) from a patient with Cystic Fibrosis carrying p.Asn1303Lys (N1303K) mutation. Stem Cell Res 44: 101744. https://doi.org/10.1016/j.scr.2020.101744
12. Pongpamorn P, Dahlmann J, Haase A, Ebeling CT, Merkert S, Gohring G et al (2020) Generation of three induced pluripotent stem cell lines (MHHi012-A, MHHi013-A, MHHi014-A) from a family with Loeys-Dietz syndrome carrying a heterozygous p.M253I (c.759G>A) mutation in the TGFBR1 gene. Stem Cell Res 43:101707. https://doi.org/10.1016/j.scr.2020.101707
13. Hoffmann D, Kuehle J, Lenz D, Philipp F, Zychlinski D, Lachmann N et al (2020) Lentiviral gene therapy and vitamin B3 treatment enable granulocytic differentiation of G6PC3-deficient induced pluripotent stem cells. Gene Ther 27(6):297–306. https://doi.org/10.1038/s41434-020-0127-y
14. Warlich E, Kuehle J, Cantz T, Brugman MH, Maetzig T, Galla M et al (2011) Lentiviral vector design and imaging approaches to visualize the early stages of cellular reprogramming. Mol Ther 19(4):782–789. https://doi.org/10.1038/mt.2010.314

Chapter 4

Distinguishing Between Endodermal and Pluripotent Stem Cell Lines During Somatic Cell Reprogramming

A. Moauro and A. Ralston

Abstract

Mouse somatic cell reprogramming using *Oct4, Sox2, Klf4 and c-Myc* (OSKM) induces formation of two stem cell types: induced pluripotent stem (iPS) cells and induced extraembryonic endoderm stem (iXEN) cells. Since both stem cells types routinely arise alongside one another during reprogramming, it is critical to distinguish between both cell types to ensure that the desired cell population is selected and analyzed. This chapter details, from start to finish, how to reprogram mouse embryonic fibroblasts (MEFs) using retrovirus and how to distinguish between iXEN and iPS cells at the colony and single-cell levels.

Key words iPS cells, iXEN cells, *Oct4*, *Sox2*, *Klf4*, *c-Myc*, OSKM, Reprogramming

1 Introduction

Somatic cell reprogramming using transcription factors *Oct4, Sox2, Klf4* and *c-Myc* (OSKM) has long been recognized to produce induced pluripotent stem (iPS) cells [1]. Like embryonic stem (ES) cells, iPS cells are pluripotent and are capable of forming all three germ layers as well as the germ line. Additionally, iPS cells provide therapeutic promise for regenerative medicine and enable novel research models related to personalized medicine.

In addition to iPS cells, a distinct stem cell type has more recently been discovered: induced extraembryonic endoderm stem (iXEN) cells, which routinely arise alongside iPS cells during OSKM reprogramming in mouse somatic cells [2, 3]. iXEN cells are similar to the extraembryonic endoderm (XEN) of the embryo. In mouse, the XEN lineage is important in forming extraembryonic structures such as the parietal and yolk sac endoderms, which take part in nutrient exchange, and the visceral endoderm, which plays an inductive role in establishment of anterior-posterior axis, blood, and germ cells [4]. Lastly, in mice XEN has been shown to contribute to the organs of the definitive endoderm [5] and iXEN cells have been induced to form hepatocyte-like cells in dogs [3],

Nagarajan Kannan and Philip Beer (eds.), *Stem Cell Assays: Methods and Protocols*, Methods in Molecular Biology, vol. 2429,
https://doi.org/10.1007/978-1-0716-1979-7_4, © Springer Science+Business Media, LLC, part of Springer Nature 2022

highlighting the similarities between extraembryonic and definitive endoderms. Like iPS cells, iXEN cells also possess therapeutic promise and provide a novel research alternative to the use of embryos for studying the roles of XEN in development.

Since both iPS and iXEN cells routinely arise in OSKM reprogramming, it is important to distinguish between the cell types. Here, we describe how to reprogram cells and identify putative iPS and iXEN cells as they arise. The first step in distinguishing between the two cell types is to sort out putative iPS and iXEN cells by isolating individual clonal populations or by flow cytometry-based cell sorting. Once the cells have been collected, additional analyses are recommended to confirm the cellular phenotypes. Confirmatory analyses may be completed by evaluating transcript and protein levels of markers that are specific to iPS and iXEN cell types. For example, Quantitative Reverse Transcriptase Polymerase Chain Reaction (qRT-PCR) and RNA-seq can be used for transcriptional analyses. Immunofluorescent imaging and flow cytometry analysis can enable protein expression analysis.

The protocols below provide details, from start to finish, on how to derive starting somatic cells: mouse embryonic fibroblasts (MEFs) and how to reprogram MEFs cells using retrovirus. We also provide recommendations for distinguishing between iXEN and iPS cells at both colony and single-cell levels.

2 Materials

2.1 Media Preparation

1. *293T Cell Medium*: DMEM, 15% Fetal bovine serum (FBS), 2 mM Glutamax, 1× Non-essential amino acids, 100 U/mL Penicillin/streptomycin, 0.1 mM Beta-mercaptoethanol.
2. *ESC Medium*: DMEM, 0.1 mM Beta-mercaptoethanol, 2 mM Glutamax, 1× Non-essential amino acids, 1 mM Sodium pyruvate, 100 U/mL Penicillin/streptomycin, 15% Knock out serum replacement, 10 ng/mL Leukemia Inhibitory Factor (LIF).
3. *MEF Medium*: DMEM, 2 mM Glutamax, 100 U/mL Penicillin/streptomycin, 10% Fetal bovine serum (FBS).
4. *Reprogramming Medium 1*: DMEM, 0.1 mM Beta-mercaptoethanol, 2 mM Glutamax, 1× Non-essential amino acids, 100 U/mL Penicillin/streptomycin, 15% Fetal bovine serum FBS, 10 ng/mL Leukemia Inhibitory Factor (LIF).
5. *Reprogramming Medium 2*: DMEM, 0.1 mM Beta-mercaptoethanol, 2 mM Glutamax, 1× Non-essential amino acids, 100 U/mL Penicillin/streptomycin, 15% Knock out serum replacement (KOSR), 10 ng/mL Leukemia Inhibitory Factor (LIF).

6. *XEN Medium*: DMEM, 0.1 mM Beta-mercaptoethanol, 2 mM Glutamax, 1× Non-essential amino acids, 1 mM Sodium pyruvate, 100 U/mL Penicillin/streptomycin, 15% FBS.

2.2 Preparing Mouse Embryonic Fibroblasts (MEFs)

1. 0.25%Trypsin-EDTA.
2. MEF medium (see above).
3. PBS.
4. 6-well flat bottom tissue culture treated polystyrene plates.
5. Sterile 4.5″ dissecting scissors.
6. Sterile fine point forceps.
7. Dissecting microscope.

2.3 Preparing Replication-Incompetent Retroviruses for Overexpression of OSKM for Reprogramming

1. HEK 293T cells (below passage 12).
2. 293T medium (*see* Subheading 2.1).
3. MEF medium (*see* Subheading 2.1) without antibiotics.
4. Lipofectamine 2000.
5. Opti-Mem.
6. 16 μg pMXs plasmid with gene of interest (the original plasmids used to establish OSKM reprogramming [1] can be found below).
   (a) pMXs-Oct3/4 (Addgene Plasmid #13366).
   (b) pMXs-Sox2 (Addgene Plasmid #13367).
   (c) pMXs-Klf4 (Addgene Plasmid #13370).
   (d) pMXs-c-Myc (Addgene Plasmid #13375).
7. 8 μg pCL-Eco packaging plasmid (Addgene Plasmid #12371).
8. 100 mm tissue culture treated polystyrene plate.
9. 0.45 μm filter.

2.4 Viral Titer of Retrovirus

1. MEFs.
2. MEF medium (*see* Subheading 2.1).
3. Replication-incompetent retrovirus (*see* Subheading 2.3).
4. 0.25%Trypsin-EDTA.
5. Polybrene.
6. TRIzol.
7. Chloroform.
8. Reverse transcription kit.
9. qPCR machine.
10. Primers specific to viral genome (sequences provided in Table 1).
11. qPCR fluorescent detector such as Sybr Green.

Table 1
qRT-PCR primers for detecting endogenous and viral transcripts

Gene target	Forward sequence (5′ to 3′)	Reverse sequence (5′ to 3′)
Oct4	GTTGGAGAAGGTGGAACCAA	CCAAGGTGATCCTCTTCTGC
Nanog	ATGCCTGCAGTTTTTCATCC	GAGGCAGGTCTTCAGAGGAA
Sox2	GCGGAGTGGAAACTTTTGTCC	CGGGAAGCGTGTACTTATCCTT
Gata6	ATGCTTGCGGGCTCTATATG	GGTTTTCGTTTCCTGGTTTG
Gata4	CTGGAAGACACCCCAATCTC	ACAGCGTGGTGGTGGTAGT
Sox7	GGCCAAGGATGAGAGGAAAC	TCTGCCTCATCCACATAGGG
Sox17	CTTTATGGTGTGGGCCAAAG	GCTTCTCTGCCAAGGTCAAC
ActinB	CTGAACCCTAAGGCCAACC	CCAGAGGCATACAGGGACAG
Viral Oct4	GAACCTGGCTAAGCTTCCAA	ACTTCCTTTCCACTCGTGCT
Viral Sox2	AACCAAGACGCTCATGAAGAA	GCTGTAGCTGCCGTTGCT
Viral Klf4	CTGAACAGCAGGGACTGTCA	GTGTGGGTGGCTGTTCTTTT
Viral c-Myc	GCCCAGTGAGGATATCTGGA	ATCGCAGATGAAGCTCTGGT

2.5 OSKM Viral Reprogramming

1. MEFs.
2. MEF medium (*see* Subheading 2.1).
3. Reprogramming medium 1 (*see* Subheading 2.1).
4. Reprogramming medium 2 (*see* Subheading 2.1).
5. Replication-incompetent retrovirus (*see* Subheading 2.3).
6. Polybrene.
7. Tissue culture treated polystyrene dishes.

2.6 Picking and Passaging iXEN and iPS Cell Colonies

1. Reprogrammed cells.
2. PBS without Ca^{2+} and Mg^{2+}.
3. 0.25%Trypsin-EDTA.
4. Reprogramming medium 1.
5. Stereomicroscope.
6. Fine point forceps.
7. 96-well tissue culture treated polystyrene.
8. Tissue culture treated polystyrene plates.

2.7 Confocal Imaging of Reprogramming Cells

1. Reprogrammed cells grown on confocal grade tissue culture treated plastic 4-wells, for example ibidi #80426.
2. 0.1% Gelatin.
3. 4% Formaldehyde in PBS without Ca^{2+} and Mg^{2+}.

Table 2
Antibodies for fluorescent imaging

Antigen	Antibody source
SOX17	R&D systems (AF1924)
SOX7	R&D systems (AF2766)
GATA6	R&D systems (AF1700)
GATA4	Santa Cruz biotech (sc-1237)
OCT4	Santa Cruz biotechnology (sc-5279)
SOX2	Neuromics (GT15098)
NANOG	Reprocell (RCAB0002P-F)
Anti-mouse IgG 488	Jackson Immuno research (715-545-140)
Anti-rabbit IgG 488	Invitrogen (A10040)
Anti-rabbit IgG 647	Jackson Immuno research (711-606-152)
Anti-goat 546	Invitrogen (A11055)

4. Pre-Block Solution (0.5% Triton X-100 in PBS without Ca^{2+} and Mg^{2+}).
5. Blocking Solution (0.1% Triton X-100 and 10% FBS in PBS, PBS without Ca^{2+} and Mg^{2+}).
6. Primary and secondary antibodies with conjugated fluorophores (*see* Table 2).
7. 0.001 mg/mL DAPI (final concentration).

2.8 Cell Sorting of Reprogramming Cells

1. MEF cells carrying a fluorescent reporter allele of interest (see below).
2. 0.2 μm filtered cell sorting buffer (PBS without Ca^{2+} and Mg^{2+}, 1 mM EDTA, 25 mM HEPES pH 7.0, 2% FBS).
3. Irradiated MEFs (E13.5 MEFs exposed to 6000 rads of γ-irradiation) [6].
4. 0.25%Trypsin-EDTA.
5. Reprogramming medium 1 or XEN medium (*see* Subheading 2.1).
6. 96-well tissue culture treated polystyrene.
7. 40 μm filter.
8. 0.001 mg/mL DAPI (final concentration).

3 Methods

3.1 Preparing Mouse Embryonic Fibroblasts (MEFs)

Reprogramming cells using OSKM can be completed using diverse somatic cell types; however, many researchers choose to use MEFs. MEFs are a great choice for primary cell line because they are easy to collect and expand. In addition, MEFs carrying specialized alleles, such as knock-ins, knock-outs, and fluorescent gene expression reporters can be easily derived by breading mice to produce the desired genotype. Our lab often uses fluorescent reporters such as GFP tagged to a gene interest. For example, the reporter allele *Nanog-GFP* [7] helps in visualizing putative iPS cells.

1. Sacrifice a pregnant female mouse 13 days after the morning plug was observed (E13.5).
2. Swab the abdomen of the mouse with 70% ethanol. Make a large incision in the shape of a U on the mouse's abdomen. Grab the body wall with fine forceps and make an incision with scissors. Remove both horns of the uterus by cutting at the oviduct and cervix, and then place the uterus in a dish with PBS.
3. Cut open the uterus longitudinally to expose all of the embryos. Remove all extraembryonic tissue (yolk sac and placenta) and transfer each embryo to a dish with fresh PBS. Embryos can be placed in individual wells of a 6-well dish or in a 10 cm plate with embryos arranged in a clock pattern around the dish.
4. Decapitate the embryo and remove all the internal organs of each embryo with forceps leaving behind only skin, muscle, and bone (*see* **Note 1**). Place the desired embryo tissue in a new well of a 6-well dish with 1 mL PBS (*see* **Note 2**).
5. Remove the PBS and add 0.5 mL Trypsin-EDTA per embryo. Mince the embryos with sterile scissors. Once minced as much as possible, incubate the embryo-trypsin mixture at 37 °C for 5 min.
6. Mince the mixture again and use a P1000 pipettor to break up the tissue further by gently pipetting up and down (*see* **Note 3**).
7. Once the tissue is disaggregated, add 2 mL MEF medium to each well to quench the Trypsin digestion.
8. Add the embryo mixture and 8 mL MEFs medium to a 15 mL conical tube. Centrifuge cells at 200 × *g* for 4 min. Carefully remove the medium and add 10 mL fresh MEF medium. Plate the cells onto a gelatin coated 10 cm dish and incubate at 37 °C in 5% CO_2.

9. Replace the medium every day until cells are confluent. Once confluent, freeze down the cells or continue on for reprogramming. These cells are passage #1.

3.2 Making Retrovirus for Reprogramming

In order to reprogram cells, many researchers use replication-incompetent retrovirus to deliver transcripts *Oct4, Sox2, Klf4,* and *c-Myc*. Making retrovirus is easy and produces a high yield. The protocol below details how to produce Moloney Murine Leukemia Virus (MMLV)-derived retrovirus (*see* **Note 4**).

1. Plate 293T cells in 293T medium onto a 10 cm plate. Change medium every other day and split cells once they reach ~80–90% confluence. Passage the cells twice before using for transfection, splitting at a ratio of 1:6 each passage. Make sure cells are no higher than passage 12 on the day of transfection.
2. On the day of plasmid transfection, cells should be 70–80% confluent. For transfection, set up two tubes per virus and wait for 5 min for the contents of each tube to mix:
   (a) Tube A: 1.5 mL Opti-MEM (*see* **Note 5**) with 50 μL Lipofectamine.
   (b) Tube B: 1.5 mL Opti-MEM with 16 μg pMXs gene plasmid and 8 μg pCL-Eco packaging plasmid (*see* **Note 6**).
3. Next, mix the contents of the tubes A and B, and then incubate at room temperature for 20 min. During this time, replace the medium on 293T cell plates with 12 mL fresh MEF medium that contain no antibiotics.
4. Add the 3 mL transfection solution (tubes A + B) in a dropwise manner to the 293T cells. Place cells in the incubator at 37 °C with 5% CO_2.
5. After 18–24 h, replace the medium with 10 mL fresh MEF medium.
6. After another 24–30 h, collect conditioned medium and filter through a 0.45 μm filter or spin medium at 200 × *g* for 4 min to remove any cells. Divide virus into 1 mL aliquots and immediately store at −80 °C until the virus is ready to be used for reprogramming or titers.

3.3 Viral Titer of Retrovirus

Once retrovirus has been prepared, it is important to titer the virus to determine the quantity of virus produced. This added step ensures each reprogramming experiment uses the same quantity of virus allowing for consistent results independent of virus preparations.

1. Plate MEFs onto gelatin coated wells at a density of 50–100 cells/mm^2. Prepare 4 wells of cells per virus to be titred.

2. Dilute virus of interest using DMEM to create 4 different virus concentrations (*see* **Note 7**) to be tested and add 4 mg/mL of Polybrene to each of the final virus solution. Add viral mixture to cells and incubate in the cell culture incubator at 37 °C with 5% CO_2.
3. After 48 h, collect the infected MEFs and harvest the RNA using TRIzol and chloroform (*see* **Note 8**).
4. Reverse transcribe the RNA to create cDNA and quantify viral transcript levels by qPCR (primer sequences are provided in Table 1).
5. Standard curves can be created using serial dilutions of the retroviral plasmids. The concentration of virus can then be calculated using the standard curve.

3.4 OSKM Viral Reprogramming

Once virus and MEFs are prepared, viral reprogramming can proceed (*see* **Note 9**). The protocol below details how to infect cells with retrovirus and how cells should be maintained through the 23-day protocol (Fig. 1a).

1. Day −1: Plate MEFs (passage 3 or lower) at a density of 50–100 cells/mm^2 onto gelatin coated dishes. The lower the density, the less likely the cells will be overgrown by the end of reprogramming.
2. Day 0: Mix together all OSKM viruses (1×10^8 transcripts/mL/cell) and 4 mg/mL of Polybrene. Using DMEM, bring the volume to the final well volume (e.g., 2 mL for one 6-well). Add viral mixture to cells and incubate for 24 h in the cell culture incubator at 37 °C with 5% CO_2.
3. Day 1: Replace virus medium with MEF medium and incubate for another 24 h in the incubator.
4. Day 2 and 4: Replace medium with Reprogramming medium 1.
5. Day 6: Replace medium with Reprogramming medium 2 and then every other day until the end of the experiment (~21 days) (*see* **Note 10**).

3.5 Picking and Passaging iXEN and iPS Cell Colonies

Once colonies have started to emerge during reprogramming, colonies can be picked and expanded. These colonies can then be analyzed for iXEN and iPS cells markers and appropriate cell morphology (*see* **Note 11**).

1. Once colonies have formed (*see* **Note 12**) and are ready for picking, wash well with PBS and replace with fresh PBS.
2. Using a microscope and fine point forceps, trace the colony of interest (*see* **Note 13**) to separate it from the underlying fibroblasts.

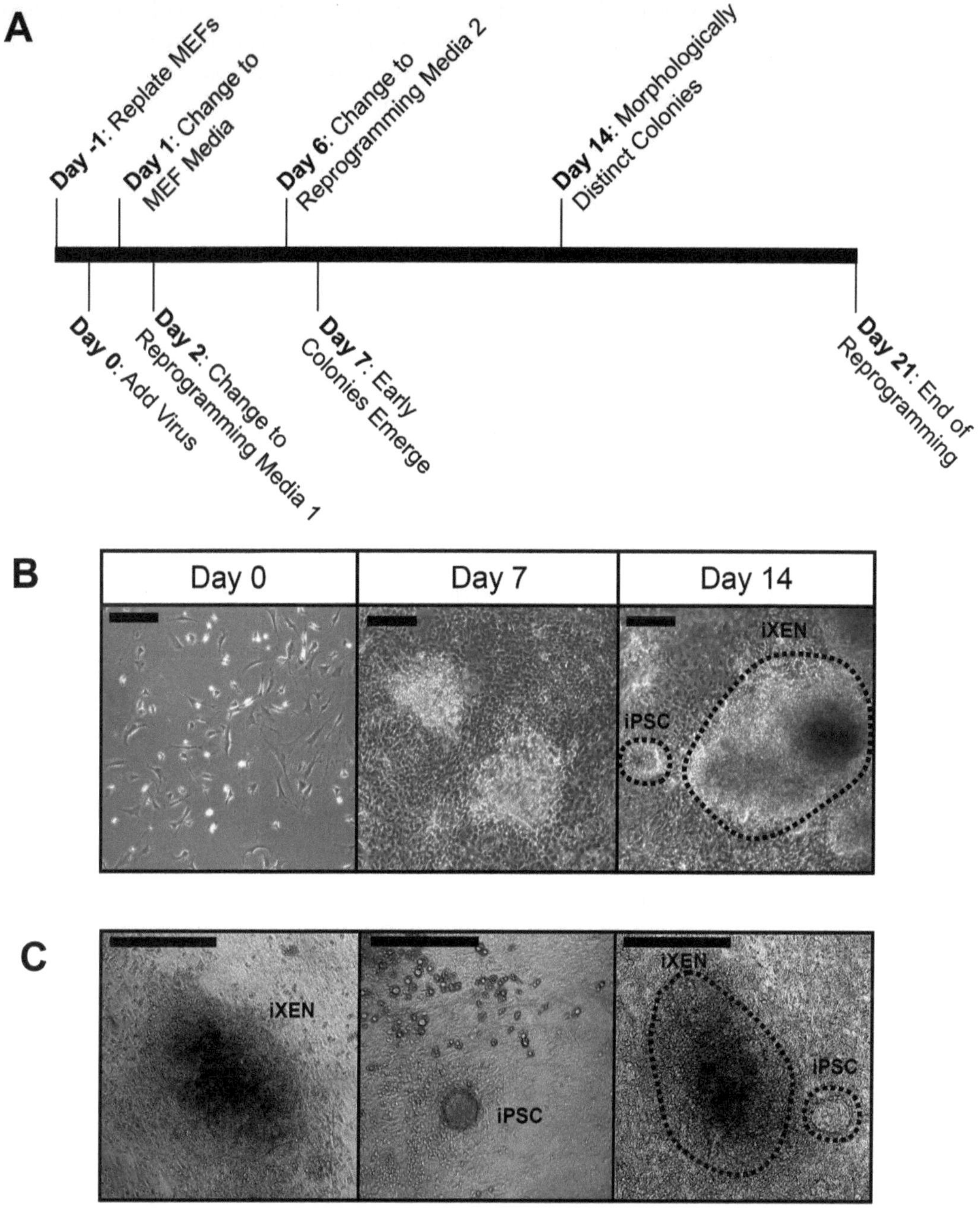

Fig. 1 iXEN and iPS cell formation in OSKM reprogramming. (**a**) Timeline of reprogramming. (**b**) Colonies of unknown identity first emerge around day 7 with distinct morphologies becoming clear by day 14. (**c**) Days 14 to 21 yield iXEN and iPS cell colonies. iXEN cell colonies are larger with ragged borders. iPS cell colonies are smaller with well-defined borders. Bars are equal to 200 μm.

3. Set a P20 pipettor to 5 μL, use it to lift the colony, and then transfer it onto a gelatin coated well of a 96-well dish containing 30 μL trypsin. Once colonies are picked, incubate the 96-well at 37 °C for 3 min.
4. Quench trypsin with 200 μL medium (medium should contain FBS) of choice. Gently pipet cell suspension up and down to break up the colony.
5. Replace medium every other day until cells are confluent. Once confluent, transfer cells to a 24-well. Continue to passage the cells for at least 11 passages [2] to ensure cells have settled into their final cell fate.
6. After 11 passages, cell lines should be analyzed using various methods (qPCR, morphology, and imaging) to determine whether they are iPS or iXEN cells.

3.6 Fluorescently Activated Cell Sorting (FACS) of Reprogramming Cells

Single-cell sorting allows for individual cell analysis which is useful for cell fate determination. Single-cell techniques can provide better resolution and detect subtle differences among cells that are lost in bulk analyses. The below protocol details how to sort single cells and expanding those single cells to form cell lines. Before starting, we recommend having a MEF line that expresses a fluorescent reporter that distinguishes between iPS and iXEN cells. For example, MEF lines carrying *Nanog-GFP* [7] can be used to select putative iPS cells.

1. Cells will be sorted into 96-well plates. The day before the sort, plate 8×10^6 cells/plate of irradiated fibroblasts onto 1–2 96-well plates (*see* **Note 43**).
2. On the day of the sort, replace the 96-well plate medium with 300 μL Reprogramming medium 1. Wrap the plates in parafilm and place on ice.
3. Harvest confluent cells with trypsin (incubation at 37 °C for 4 min) and wash twice with PBS. These cells should contain a fluorescent marker read out such as GFP (*see* **Note 15**) for sorting.
4. Wash collected cells with Cell Sorting Buffer and resuspend to $4–7 \times 10^6$ cells/mL (cell concentration may vary depending on the instrument). Filter samples using a 40 μm filter right before sorting. Keep cells on ice until you are ready to run the sort.
5. Right before running the sort, add 1 μg/mL of DAPI (sample can only sit in DAPI for 20 min). DAPI-positive cells indicate dead cells.
6. Proceed to sort living cells (*see* **Note 15**). For single-cell resolution, we recommend sorting 1 living cell into a single 96-well. Once a plate is filled, replace parafilm and place the plate back on ice.

7. Once all cells are sorted, remove parafilm and place plates in the incubator at 37 °C with 5% CO_2.
8. After 6 h, replace medium with 100 μL of desired medium.
9. Replace medium every 2 days and passage cells once colonies have formed.

3.7 Confocal Imaging of Colonies Undergoing Reprogramming

Fluorescent imaging using confocal microscopy (*see* **Notes 11 and 17**) allows for higher resolution imaging of multiple markers in combination with morphology while limiting background fluorescence and light scatter. Confocal microscopy provides spatial information on how different markers are expressed throughout a colony and how individual colonies express markers in comparison to one another. Confocal imaging allows for the analysis of multiple colonies on a whole plate which retains dynamic spatial reprogramming information. In comparison, single-cell sorting or colony picking only provides information about individual cells or individual colonies, respectively.

1. Complete reprogramming on gelatin coated confocal grade tissue culture treated 4-well dishes.
2. At the desired time point, carefully aspirate cell culture medium and fix cells for 10 min at room temperature with 200 μL 4% formaldehyde.
3. Carefully aspirate 4% formaldehyde and add 280 μL of pre-blocking solution and incubate for 30 min at room temperature.
4. Carefully aspirate pre-blocking solution. Add 280 μL of Blocking Solution and incubate for 1 h at room temperature.
5. Carefully aspirate blocking solution and add 280 μL of diluted primary antibodies to wells. Seal and incubate at 4 °C overnight.
6. Carefully aspirate primary antibodies solution. Add 280 μL of Blocking Solution and incubate for 30 min at room temperature.
7. Carefully aspirate the Blocking Solution and add diluted 280 μL of secondary antibodies for 1 h at room temperature in the dark.
8. Carefully aspirate secondary antibodies solution. Add 280 μL of Blocking Solution and incubate for 30 min at room temperature in the dark.
9. Aspirate the blocking solution and stain with 280 μL DAPI (*see* **Note 18**) for 5 min at room temperature in the dark.
10. Aspirate DAPI and add 280 μL PBS. Keep cells in the dark until they are ready for imaging (*see* **Note 19**).

4 Notes

1. Preserving the anatomy of the embryo is key to making sure all viscera are removed. The more the embryo morphology is disrupted, the more difficult it will be to interpret the anatomy.
2. The head can be saved and used for genotyping, for example when establishing MEFs which carry a specific reporter or allele of interest.
3. Disaggregate tissue as much as possible. This will help the cells proliferate.
4. MMLV-derived retrovirus will only infect mouse cells. In comparison lentivirus can infect human and mouse cells and extra precaution must be used when working with lentivirus. Both lentivirus and MMLV-derived retrovirus are replication-incompetent retroviruses. This means that both viruses lack the genes that allow infected cells to produce more virus.
5. DMEM can be substituted for Opti-MEM.
6. The same protocol can be used to make lentivirus. In order to make lentivirus use your preferred lentivirus envelope plasmid, packaging plasmid, and gene plasmid at a ratio of 2:3:4. For example, our lab uses 2.52 μg PMD2.g, 3.78 μg psPAX2, and 10 μg of gene plasmid when infecting a 10 cm plate.
7. When determining viral titers, we test three different concentrations of virus. This is to ensure that qRT-PCR measurements are detectable within the instrument's dynamic measurement range. For a 6-well transfection, we test the following viral volumes: 0 μL retrovirus, 50 μL retrovirus, 100 μL retrovirus, and 200 μL retrovirus. These volumes can be adjusted up or down depending on the viral yield.
8. An alternative method of titrating virus is to use a fluorescent protein-expressing virus instead of quantifying transcript levels qPCR. A fluorescent virus can be made in parallel to the virus of interest and infected at different concentrations onto MEFs. After 48 h, fluorescent MEFs are then counted under a fluorescent microscope to record the percentage of fluorescent cells. This percentage can then be used to back calculate the transduction units per mL (TU/mL). TU/mL = (Number of cells transduced x Percent fluorescent)/(Virus volume in mL).
9. Reprogramming can be completed with a doxycycline inducible system if MEFs carry at least one copy of *Rosa26-M2rtTA* and *mCol1a OSKM* [8]. Instead of adding virus at Day 0, 2 μg/mL doxycycline can be added to the medium and all subsequent medium changes.
10. Colonies will start to emerge at day 7. iPS cell colonies look like ES cell colonies, and are generally small, circumscribed,

with dome-shaped in morphology. iXEN cell colonies appear flatter with ragged borders and are ~3× larger than iPS cells (Fig. 1b, c).

11. Having the techniques to analyze emerging reprogramming cells is important for early and proper detection. When it comes to iPS and iXEN cells, morphology and marker expression are very important in distinguishing these two cell types.
12. Colonies can be observed as early as 7 days. Clear morphological differences can be observed between iXEN and iPS cells by 14 days.
13. It can be difficult to remove colonies as they are often well attached to the underlying fibroblasts. If the underlying fibroblast layer starts to peel off the dish when removing the colony, an alternative technique can be used. Rather than taking the whole colony, the surface of the colony can be scraped off and moved to the 96-well.
14. The number of plates used for sorting can be adjusted up or down depending on the cell sorting yield and the number of cells needed for downstream analysis. Note that one 6-well of reprogramming cells is enough to fill one to two 96-well plates.
15. Instead of using a mouse line that contains a fluorescent reporter, cell surface marker staining can be performed.
16. Make sure to have proper controls. These include a live sample with no fluorescence (negative control) to determine background fluorescence, a live sample expressing the fluorophore of interest (positive control) for gating, and a permeabilized and fixed sample with DAPI for live/dead exclusion.
17. Note that confocal imaging of colonies can be challenging for a number of reasons. As the colonies become large, it is harder to image them, especially iXEN. It is recommended to try imaging at different time points and different magnifications.
18. Other nuclear stains can be used, such as DRAQ5. It is important to look at the instrument's lasers and available fluorescent antibodies to determine the best nuclear stain and fluorophore panel for the experiment.
19. When imaging colonies, make sure to test antibody specificity using mouse blastocyst-derived ES and XEN cell lines can be used.

As discussed previously in the note section of "OSKM Viral Reprogramming" and demonstrated in Fig. 1b, iXEN and iPS cell colonies have very distinct morphologies from one another. As cells continue to be passaged to establish stable cell lines, iPS and iXEN cells begin to resemble their blastocyst-derived counterparts. iPS

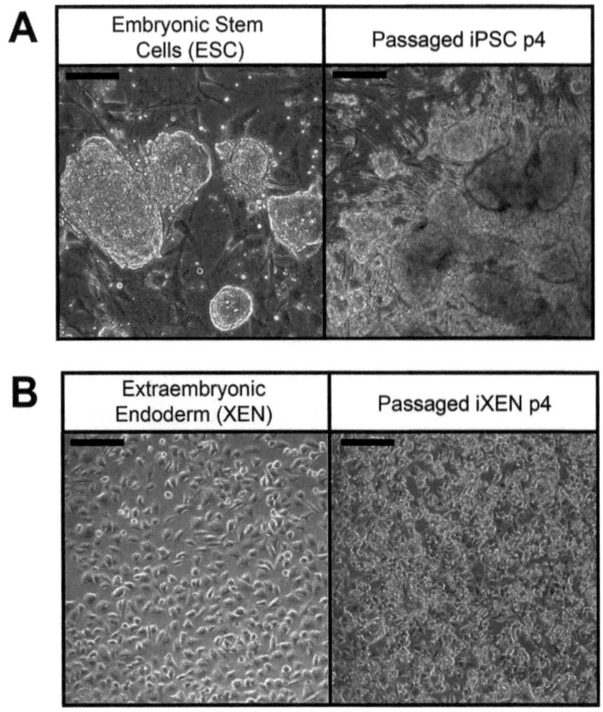

Fig. 2 Morphology of iXEN and iPS cell lines. As colonies are passaged, they acquire morphologies similar to their embryo-derived stem cell counterparts. (**a**) iPS cells appear epithelial and grow in dome-like colonies. (**b**) iXEN cells appear mesenchymal and individual cells are geometric in appearance. Bars are equal to 200 μm

cell lines continue to form tightly clustered colonies of rounded and small cells (Fig. 2a). By contrast, iXEN cell lines are mesenchymal, larger than ESCs and are often less rounded in appearance (Fig. 2b).

Morphology is a good indication of cellular identity, but follow-up with marker expression should always be performed. Evaluating marker expression can be performed at the protein level using fluorescent imaging with confocal imaging, flow cytometry, or western blotting. In addition, RNA levels can be quantified using qRT-PCR or RNA-seq. For iPS cells, common markers for pluripotency are NANOG, SOX2, and OCT4 [1]. For iXEN cells, reliable markers are GATA6, GATA4, SOX7, and SOX17 [2, 3, 9].

As a final note, before any analysis is performed, it is important to test all antibodies and primers using appropriate positive and negative controls, especially since there are many commercially available options. For positive controls, we recommend using blastocyst-derived ES and XEN cell lines [10]. Both lines can be derived from blastocysts. Tables 1 and 2 contain a list of qRT-PCR primers and antibodies that our lab has found to be reliable.

References

1. Takahashi K, Yamanaka S (2006) Induction of pluripotent stem cells from mouse embryonic and adult fibroblast cultures by defined factors. Cell 126(4):663–676
2. Parenti A, Halbisen MA, Wang K et al (2016) OSKM induce extraembryonic endoderm stem cells in parallel to induced pluripotent stem cells. Stem Cell Rep 6:447–455
3. Nishimura T, Unezaki N, Kanegi R et al (2017) Generation of canine induced extraembryonic endoderm-like cell line that forms both extraembryonic and embryonic endoderm derivatives. Stem Cells Dev 26:1111–1120
4. Ralston A (2018) XEN and the art of stem cell maintenance: molecular mechanisms maintaining cell fate and self-renewal in extraembryonic endoderm stem (XEN) cell lines. Adv Anat Embryol Cell Biol 229:69–78
5. Kwon GS, Viotti M, Hadjantonakis AK (2008) The endoderm of the mouse embryo arises by dynamic widespread intercalation of embryonic and extraembryonic lineages. Dev Cell 15: 509–520
6. Nagy A, Marina Gertsenstein KV, Behringer R (2006) Preparing feeder cell layers from STO or mouse embryo fibroblast (MEF) cells: treatment with γ-irradiation. Cold Spring Harb Protoc 2006(1):pdb.prot4400
7. Maherali N, Sridharan R, Xie W et al (2007) Directly reprogrammed fibroblasts show global epigenetic remodeling and widespread tissue contribution. Cell Stem Cell 1:55–70
8. Carey B, Markoulaki S, Beard C et al (2010) A single-gene transgenic mouse strain for reprogramming adult somatic cells. Nat Methods 3: 9–14
9. Zhao Y, Zhao T, Guan J et al (2015) A XEN-like state bridges somatic cells to pluripotency during chemical reprogramming. Cell 163(7):1678–1691
10. Kunath T, Arnaud D, Uy GD et al (2005) Imprinted X-inactivation in extra-embryonic endoderm cell lines from mouse blastocysts. Development 132:1649–1661

Chapter 5

Measuring Early Germ-Layer Specification Bias in Human Pluripotent Stem Cells

Alexander Keller, Nuša Krivec, Christina Markouli, and Claudia Spits

Abstract

Human pluripotent stem cells have a wide variety of potential applications, ranging from clinical translation to in vitro disease modeling. However, there is significant variation in the potential of individual cell lines to differentiate towards each of the three germ layers as a result of (epi)genetic background, culture conditions, and other factors. We describe here in detail a methodology to evaluate this bias using short directed differentiation towards neuroectoderm, mesendoderm, and definitive endoderm in combination with quantification by RT-qPCR and immunofluorescent stains.

Key words Human Pluripotent Stem Cells, Differentiation Bias, Quantification of differentiation, Mesendoderm, Definitive Endoderm, Neuroectoderm, Quantitative Real-Time PCR, Quantitative Immunofluorescent Stain

1 Introduction

Human pluripotent stem cells (hPSC), both embryonic (hESC) [1] and induced (iPSC) [2], offer endless therapeutic potential for regenerative medicine and serve as an invaluable research tool for in vitro disease and development modeling. However, for this potential to be fully realized, several important characteristics inherent to the cell type must be considered when included in any research question. Notably, while hPSC are defined by pluripotency, not all cell lines respond identically to differentiation cues. Genetic and epigenetic variation, as well as cell culture factors, can lead to wide ranging outcomes in differentiation capacity (reviewed in [3]). In order to evaluate potential biases in differentiation caused by these factors, a number of different approaches have been established in recent years, including the teratoma assay, TeratoScore [4] and ScoreCard assay [5, 6]. These assays rely on

Alexander Keller and Nuša Krivec contributed equally to this work.

Nagarajan Kannan and Philip Beer (eds.), *Stem Cell Assays: Methods and Protocols*, Methods in Molecular Biology, vol. 2429, https://doi.org/10.1007/978-1-0716-1979-7_5,

spontaneous differentiation in the form of teratomas or embryoid bodies. In the case of the teratoma-based assays, their main disadvantage is that the formation of teratomas is time-consuming, is difficult to standardize and requires the use of animals. Furthermore, in the case of the teratoma assay, it requires expertise in evaluating histological stains, which also lack a quantitative measurement. The ScoreCard assay largely reconciles these shortcomings by comparing gene-expression data of day-12 embryoid bodies to an established dataset. However, embryoid body standardization still poses a challenge, as different culture systems and embryoid body sizes can affect the balance in cell types, and the dataset used as reference is based on cells that were routinely cultured on mouse feeder layers, which may not be comparable to many current culture techniques. Here we describe a fast approach to evaluate differentiation outcome during early germ-layer specification that is based on directed differentiation. This method can be used and adapted to suit any number of research questions. For example, it can be used to evaluate the impact of a given gene on germ-layer specification by combining it with a genome-wide CRISPR-Cas9 knockout screens, or to assess the impact of specific (growth) factors and culture condition modifications on differentiation. Using this approach, we have previously identified a deficiency in ectoderm differentiation in hESC lines carrying the highly recurrent gain of 20q11.21 [7] and a missspecification process in a line endogenously activating WNT and BMP4 signaling [8].

The methods described here are now well established in the field of hPSC research, though often used in isolation. Used together, this approach provides a powerful tool for the evaluation hPSC differentiation outcome between different conditions. This method has three basic steps: directed differentiation, quantification of gene expression by qRT-PCR, and quantification of successfully differentiated cells by immunocytochemistry. The protocol for neuroectoderm is adapted from Chetty et al. [9] and Chambers et al. [10] and mesendoderm and definitive endoderm differentiation is carried out using a protocol based on Sui et al. [11].

2 Materials

2.1 hPSC Culture and Differentiation

1. Matrix: Laminin-521 (Biolamina) diluted in PBS containing Mg^{+}/Ca^{+}.
2. Culture medium: NutriStem® hESC XF (Biological Industries) medium supplemented with 100 U/mL penicillin/streptomycin.

3. Passaging: TrypLE™ Express (Thermo Fisher).
4. Equipment for cell counting: for example, TALI Image-Based Cytometer (Thermo Fisher Scientific).

2.2 hPSC Differentiation

1. The basic neuroectoderm differentiation medium can be stored at 4 °C for up to 1 month and is composed of KnockOut DMEM, 10% KnockOut Serum Replacement. This medium is supplemented prior to use with 500 ng/mL Recombinant Human Noggin Protein and 10 μM SB431542.
2. Mesendoderm differentiation medium contains Roswell Park Memorial Institute (RPMI) 1640 supplemented by GlutaMAX™ (Thermo Fisher). This medium is supplemented prior to use with 100 ng Recombinant Human/Mouse/Rat Activin A, 3 μM CHIR99021 and 0.5% B27 supplement.
3. Definitive endoderm differentiation medium contains Roswell Park Memorial Institute (RPMI) 1640 supplemented by GlutaMAX™. For the first 24 h of differentiation, it is supplemented with 100 ng Recombinant Human/Mouse/Rat Activin A, 3 μM CHIR99021 and 0.5% B27 supplement. For the following 48 h, it is supplemented only with 100 ng Recombinant Human/Mouse/Rat Activin A and 0.5% B27 supplement.

2.3 Gene-Expression Analysis by qRT-PCR

1. 2× qPCR Master Mix (Low ROX), taqman assays (Thermo Fisher) and primer-sets and probes (custom made). Table 1 lists taqman assays and primer sequences used for this protocol and specifies what lineage is marked by each. Custom primers are used at a working concentration of 50 μM and probes at 10 μM. Taqman assays are used according to manufacturer recommendations.

2.4 Immunocytochemistry

1. The primary antibodies and their working concentrations are listed in Table 2. The secondary antibodies are listed in Table 3.

3 Methods

3.1 hPSC Culture

1. Medium is changed daily with a suitable volume of NutriStem® medium pre-warmed to 37 °C. Cells are maintained at 37 °C at atmospheric O_2 and 5% CO_2.
2. hPSC (*see* **Note 1**) are cultured on dishes coated with 0.5 μg/cm^2 (routine culture) or 1 μg/cm^2 (during differentiation) laminin-521 [12] diluted in PBS with Mg^+/Ca (*see* **Note 2**). Cells are routinely passaged as single cells once cells have reached 70–90% confluency on freshly made laminin-coated dishes (*see* **Note 3**). Dishes should be warmed to 37 °C prior to use.

Table 1
Taqman assays and sequences

Gene	Taqman assay/sequence	Lineage
PAX6	Hs00240871_m1	Neuroectoderm
SOX1	Hs01057642_s1	Neuroectoderm
T	Hs00610080_m1	Mesendoderm
SOX17	Hs00751752_s1	Definitive Endoderm
FOXA2	Hs00232764_m1	Definitive Endoderm
POU5F1	Forward 5′-GGA-CAC-CTG-GCT-TCG-GAT-TT-3′ Reverse 5′-CAT-CAC-CTC-CAC-CAC-CTG-G-3′ Probe 6-FAM- GCC-TTC-TCG-CCC-CC-MGB	Undifferentiated hPSC
NANOG	Forward 5′-TGC-AAA-TGT-CTT-CTG-CTG-AGA-TG-3′ Reverse 5′-TCC-TGA-ATA-AGC-AGA-TCC-ATG-GA-3′ Probe 6-FAM- CAG-AGA-CTG-TCT-CTC-CTC-MGB	Undifferentiated hPSC
UBC	Forward 5′-CGC-AGC-CGG-GAT-TTG-3′ Reverse 5′-TCA-AGT-GAC-GAT-CAC-AGC-GA-3′ Probe 6-FAM- TCG-CAG-TTC-TTG-TTT-GTG-MGB	Endogenous control
GUSB	Hs99999908_m1	Endogenous control

Table 2
List of primary antibodies and their references

Antibody	Species	Provider	Reference	Concentration
PAX6	Mouse Monoclonal IgG	Abcam	Cat# ab78545	1:100
OCT3A	Mouse Monoclonal IgG	Santa Cruz	Cat# sc-5279	1:200
OCT3A	Rabbit Monoclonal IgG	Cell Signaling	Cat# C30A3	1:200
SOX17	Goat Polyclonal IgG	R&D Systems	Cat# AF1924	1:100
T	Goat Polyclonal IgG	R&D Systems	Cat# AF2085	1:200

Table 3
List of secondary antibodies and their references

Species	Fluorescent Dye	Provider	Reference
Goat anti-Mouse IgG (H+L)	Alexa Fluor 488	Thermo Fisher Scientific	Cat# A11001
Donkey anti-Goat IgG (H+L)	Alexa Fluor 488	Thermo Fisher Scientific	Cat# A-11055
Donkey anti-Rabbit IgG (H+L)	Alexa Fluor 546	Thermo Fisher Scientific	Cat# A10040
Donkey anti-Mouse IgG (H+L)	Alexa Fluor 594	Thermo Fisher Scientific	Cat# R37115

3. To passage, at approximately 70–90% confluency, aspirate medium, wash 1× with PBS and add 0.5 volumes of TrypLE dissociation reagent (i.e., 1 mL per well of a 6-well dish) and place at 37 °C for 5–7 min, or until visibly detached under a microscope. Add at minimum an equal volume of medium to deactivate the TrypLE and transfer to a 15 mL tube. Centrifuge 5 min at 180 × *g*, aspirate the supernatant, and resuspend the cells in 1 mL medium. Aspirate laminin from the coated plate and add an appropriate volume of medium, transfer cells to the dish at a ratio of 1:10–1:20 for routine passaging.

3.2 hPSC Differentiation

1. Accurate cell counting is vital to successful differentiation. If plating cells for differentiation, generate a cell suspension as described above and resuspend in 2–3 mL of pre-warmed medium. Count using the TALI Image-Based Cytometer (*see* **Note 4**). Optimal differentiation is achieved by using cells that are in the exponential growth phase between 1 and 2 days after passaging, typically at 70–90% confluency (*see* **Note 5**).
2. Calculate the necessary number of cells needed to reach the desired confluency per differentiation protocol (see below), and plate cells on a freshly coated 1 μg/cm^2 dish (*see* **Note 6**). Ensure even distribution of cells in the dish by mixing cells in pre-warmed NutriStem® medium prior to plating, adding cells directly to the dish will cause cells to disproportionally cluster in the center, giving an uneven plating density.
3. Neuroectoderm differentiation (Fig. 1a). Cells should be 80–90% confluent (Fig. 1b) on the starting day, which equates to plating 60,000–80,000 cells per cm^2 (*see* **Note 7**). When at the appropriate density, cells are washed with PBS and transitioned to neuroectoderm medium supplemented with 500 ng/mL Noggin and 10 μM SB431542. Medium is changed daily for 4 days, starting at D0. Factors should be added fresh prior to feeding (*see* **Note 8**).
4. Mesendoderm differentiation (Fig. 2a). Cells should be 40–50% confluent (Fig. 2b) on the starting day, which equates to plating 45,000–65,000 cells per cm^2 (*see* **Note 7**). When at the appropriate density, cells are washed with PBS and transitioned to mesendoderm medium supplemented with 100 ng/mL Activin A, 3 μM CHIR99021, and 0.5% B27 supplement. Cells are collected after 24 h. Factors should be added fresh prior to feeding (*see* **Note 8**).
5. Definitive Endoderm (Fig. 3a). Cells should be 40–50% confluent (Fig. 2b) on the starting day, which equates to plating 45,000–65,000 cells per cm^2 (*see* **Note 7**). When at the appropriate density, cells are washed with PBS and transitioned to definitive endoderm medium supplemented with 100 ng/mL

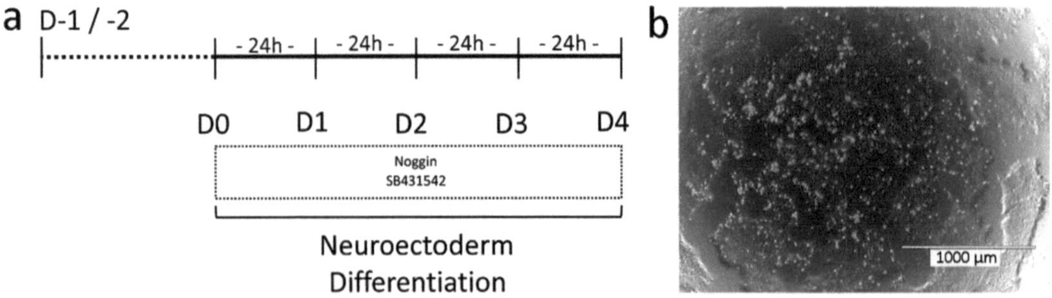

Fig. 1 (**a**) Schematic representation of the neuroectoderm differentiation timeline. Cells are seeded at day −2 or −1 and differentiation is initiated at day 0. (**b**) Representative phase contrast image of starting cell density at D0

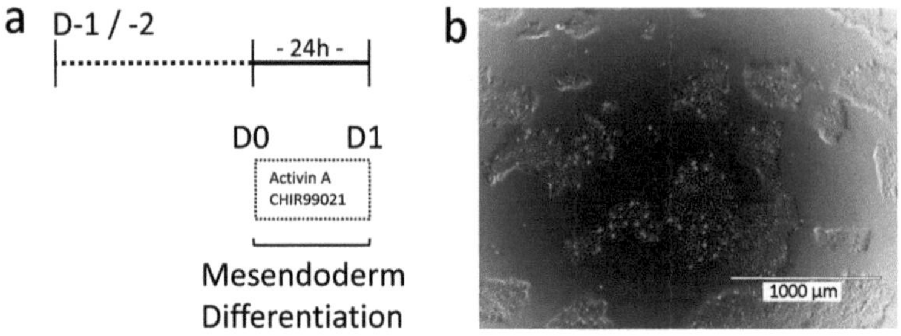

Fig. 2 (**a**) Schematic representation of the mesendoderm differentiation timeline. Cells are seeded at day −2 or −1 and differentiation is initiated at day 0. Activin A and CHIR99021 are administered for 24 h. (**b**) Representative phase contrast image of starting cell density at D0

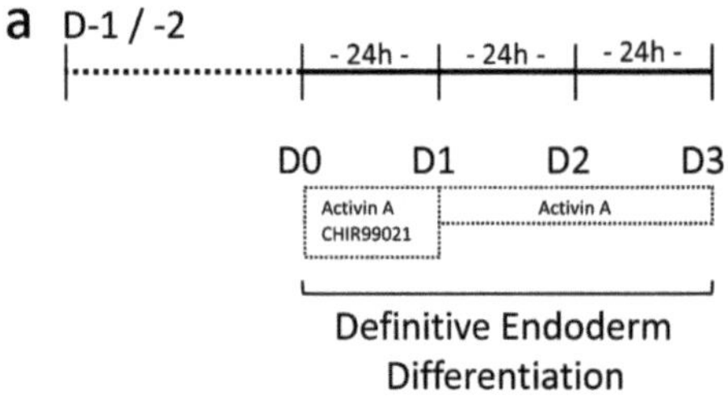

Fig. 3 (**a**) Schematic representation of the definitive endoderm differentiation timeline. Cells are seeded at day −2 or −1 and differentiation is initiated at day 0. Activin A and CHIR99021 are administered for the first 24 h, after which only Activin A is given to the cells for another 48 h

Activin A, 3 μM CHIR99021, and 0.5% B27 supplement for 24 h, and only Activin A and 0.5% B27 supplement for the following 48 h. Factors should be added fresh prior to feeding (*see* **Note 8**).

3.3 Gene-Expression Analysis by Quantitative Real-Time PCR

1. After differentiation is complete, cells are harvested into a suspension as described above and spun down into a pellet that is snap-frozen by immersion in a LN_2 before storage at −80 °C.
2. RNA extraction is carried out using the RNA Mini Kit (for example Qiagen) using the recommended protocol of the manufacturer. cDNA conversion is performed using the first-strand cDNA synthesis kit (for example Cytiva Life Sciences) using the procedure described in the manufacturer's protocol (*see* **Note 9**). RNA input concentration should be normalized across samples prior to cDNA conversion.
3. Quantitative Real-Time Polymerase Chain Reaction (qRT-PCR) is carried out on a ViiA 7 thermocycler (Thermo Fisher Scientific), using the standard protocol provided by the manufacturer. Details on the probes, assays, and primers are listed in the tables bellow. A total cDNA input of 20 ng is used per reaction in a reaction volume of 20 μL (*see* **Note 10**). All reactions are performed in triplicate with UBC and GUSB used in a geometric mean as housekeeping genes and 2 no template controls (*see* **Note 11**).
4. Gene expression is initially interpreted as a relative quantification in QuantStudio™ Real-Time PCR Software before being offloaded as an excel or txt file for additional downstream processing (*see* **Notes 6** and **11**).

3.4 Immunocytochemistry and Quantification

1. On the final day of differentiation, the cells are washed with PBS and fixed with 3.7% Paraformaldehyde for 10 min at room temperature. After three additional washes, they are stored in PBS in order to protect them from drying out. The fixed dishes are stored at 4 °C.
2. Fixed cells are permeabilized with 0.1% Triton for 10 min at room temperature and then blocked with 10% Fetal Bovine Serum or in serum of the species of the chosen secondary antibody (i.e., donkey serum) for 1–2 h. The primary antibodies are diluted in 10% FBS or appropriate serum to the desired concentration and incubated on the cells overnight at 4 °C. On the second day the cells are washed three times with PBS on a plate shaker at room temperature. Secondary antibodies are diluted 1:200 in 10% FBS or appropriate serum and incubated for 1–2 h at room temperature. DNA is stained with Hoechst 33342 simultaneously with the secondary antibody.
3. Imaging is performed on a confocal microscope (*see* **Note 12**) and quantification of the images can be done using for instance Zen 2 (blue edition) imaging software or ImageJ (*see* **Note 13**).
4. To quantify the stainings using Zen Blue, import the image in CZI format. Adjust minimum and maximum values in

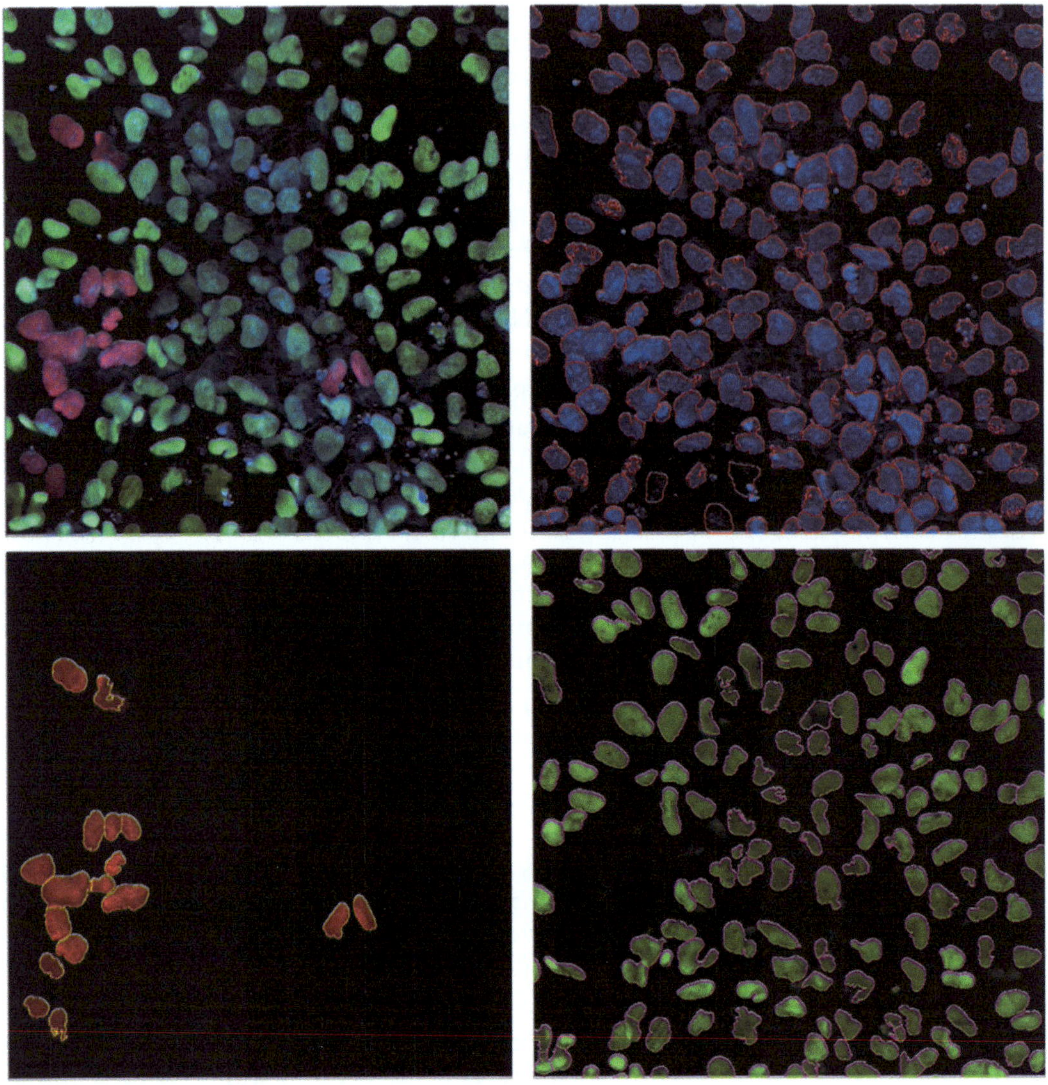

Fig. 4 Example of segmentation with Zen Blue. Immunofluorescent stain of cells differentiated to definitive endoderm. Nuclear staining with Hoechst (blue), POU5F1 (red), and SOX17 (green). Segmentation is represented with the outline of the masks

histogram for each fluorescence channel in a way that all positive cells have a high intensity that is visible and clearly different from the negative cells that might have lower intensity. The setup of the image analysis includes pre-processing and segmentation. Step 1/7—Classes. For each fluorescence channel used in the image acquisition, create a separate class. In the example in Fig. 4 we created Class 1 for nuclear stain with Hoechst (450 nm, blue channel), Class 2 for SOX17 (488 nm, green channel), and Class 3 for OCT4 (594 nm, red channel). Assign appropriate fluorescent channel to each class and choose the color of the mask. Step 2/7—Frame. Choose the region of interest if only a part of the image will be quantified. In the example, we quantify the nuclei of the whole image. Step 3/7—Automatic segmentation. For each class individually choose the necessary filters such as smoothing (lowpass, Gauss or median) and sharpen (delineate or unsharp masking). With minimum area slider, adjust the minimum area of pixels of the object you would like to include in the segmentation. In the histogram, define lower and upper threshold values of the nuclei intensity. Begin with automatic function and adjust it manually for the details. Another possible method for setting the limits is using values in fluorescent channel histogram that you adjusted at the beginning. Copy values from the histogram to the threshold minimum and maximum. Activate "Fill holes" function to fill the holes in selected objects, if necessary. For additional adjustment of the mask, choose between binary options (open, close, dilate, and erode). Mask of nuclei can be further separated with watershed or morphology function. This step is highly dependent on factors such as cell density, whether the immunostaining is nuclear or cytoplasmic, clear distinction between negative and positive cells, quality of immunostaining and acquisition. For the reproducible data of the same cell line, use same laser power and gain during acquisition and same histogram limits of each fluorescent channel in all the experiments. Other cell lines and differentiations require individual settings. Step 4/7—Condition. Make segmentation more specific by adding more conditions for nuclei you want to quantify, e.g., roundness, intensity, maximum size, position. In this step, you can eliminate artifacts of immunostaining that should not be quantified. Step 5/7—Interactive segmentation. Manually add, subtract, or modify any nuclei or background that was not selected correctly with automatic segmentation. Step 6/7—Features. Choose which data you would like to be displayed in the table after analysis. Step 7/7—Measure. Last step is a preview of results that will be displayed after analysis. An example of expected outcomes can be found in **Note 14**.

4 Notes

1. When evaluating differentiation bias, cell line selection is crucial. It is recommended to use a minimum of three lines to avoid drawing conclusions based on the unique characteristics of any single-cell line. Further considerations should also be made based on experimental setup, for example, when evaluating a mutation, isogenic pairs should be used whenever possible. Proper evaluation of the genetic content of each hPSC line

used is also crucial. While metaphase-spread banding techniques have long been a gold standard for genetic screening in hPSC, these should be avoided in favor of methods such as array-based comparative genomic hybridization or shallow (low-pass) whole genome sequencing, as a significant number of hPSC-associated copy number variations (CNVs) are below the detection limit of a regular metaphase-based method [13]. Large stocks of frozen cells should be generated after determining the genetic content of each line. New cells should be thawed from this bulk after no more than 10 passages to avoid the takeover of highly recurrent CNVs, which are very likely to skew results [3].

2. Typical coating volumes are 0.5 volumes of the dish volume (i.e., 1 mL in a single well of a 6-well). However, smaller wells should use larger volumes to ensure better quality coatings (i.e., 300 μL/well total coating volume instead of 250 μL/well in a 24-well). Coated dishes can be prepared either the day (s) prior to plating and stored at 4 °C or alternatively at least 2 h before plating and kept at 37 °C. Laminin-coated dishes may be used for up to 14 days after coating.
3. Alternative hPSC culture systems can be used but have not been tested by us for these differentiation protocols.
4. Cells that are too highly concentrated can lead to inaccurate cell counts; if a large cell pellet is generated, resuspend in a larger volume of medium to avoid incorrect seeding densities the following day. Furthermore, cells will settle to the bottom of tubes, as such thorough, but gentle, as to not kill the cells, resuspension is necessary prior to taking cells for counting or for plating. With the TALI Image-Based Cytometer, cells should be gated to a diameter of 6–20 μm, everything below this range is considered as debris and anything beyond are cell clusters. Several viable alternative counting methods exist for this step.
5. Directed differentiations rely heavily on a number of factors, such as initial cell density at the onset of differentiation and factor concentration and timing, each of which has a significant impact on successful differentiation. Differentiations should be performed in triplicate across multiple passages of the same hPSC line to mitigate for experiment to experiment variability.
6. Sufficient cells should be plated to ensure enough material for downstream analysis. A minimum of 2 wells of a 24-well plate should be used, one that will be collected for RT-qPCR and the other which will be fixed for immunostaining. If larger amounts of RNA are needed, multiple wells can be pooled together on the day of collection. Establishment of

a

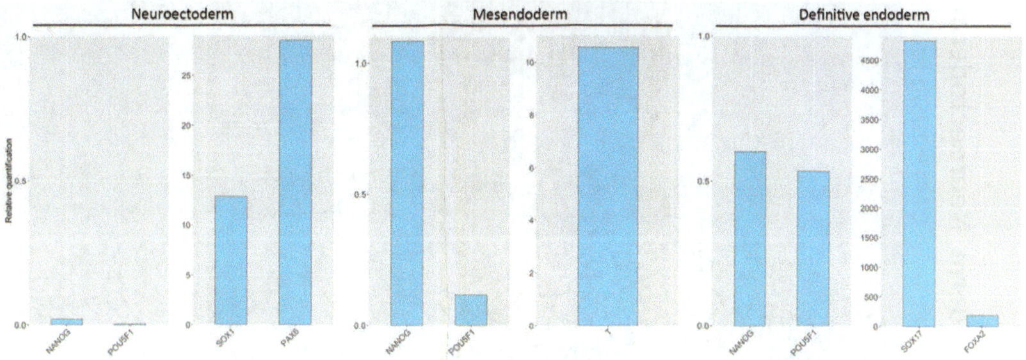

b

	hESC	NE
NANOG	24.7	30.8
POU5F1	21.5	30.1
SOX1	32.0	28.8
PAX6	26.7	22.4
UBC	20.7	22.0
GUSB	23.8	23.5

	hESC	ME
NANOG	25.5	24.8
POU5F1	22.9	25.4
T	33.2	29.2
UBC	23.0	21.9
GUSB	25.0	24.8

	hESC	DE
NANOG	23.5	25.1
POU5F1	21.0	22.7
SOX17	35.4	23.9
FOXA2	37.5	30.8
UBC	19.8	21.6
GUSB	23.8	23.7

Fig. 5 (**a**) Representative relative quantifications of differentiations towards Neuroectoderm (NE), Mesendoderm (ME), and Definitive Endoderm (DE), set against the undifferentiated state (**b**) Corresponding C_t values

differentiation outcome should incorporate quantitative measures of both mRNA and protein for relevant markers. qRT-PCR, while fast and reliable, will drown out the variability seen at the single-cell level (*see* **Notes 11** and **14** and Figs. 5 and 6 for examples). It is possible for two samples to have comparable mRNA expression at the bulk level but differ significantly in the total number of cells that account for that expression. Combining expression data with quantitative evaluation of immunofluorescent stains paints a more robust picture of differentiation efficiency.

7. Starting differentiation 1 or 2 days after passaging, and at the correct cell density is key for an efficient differentiation. To achieve this, on the day of passaging, 45,000–100,000 cells per cm^2 of coated area should be seeded. Achieving the optimal density at maximum 2 days after passaging may require fine-tuning based on the growth rate of each cell line. For fast growing cell lines, choose the lower seeding range, for slow growing lines the higher range. The days after the seeding, monitor the cells in order to detect the optimal moment to initiate the differentiation.
8. Refreshing the differentiation medium consistently every 24 h (not several hours earlier or later) is crucial for a reproducible induction of differentiation.
9. Alternative kits can be used but have not been tested by us.
10. This protocol makes use of Taqman probes and the ViiA 7 platform. The probes and primer-sets provided in this protocol have been tested for optimal efficiency and correct protein

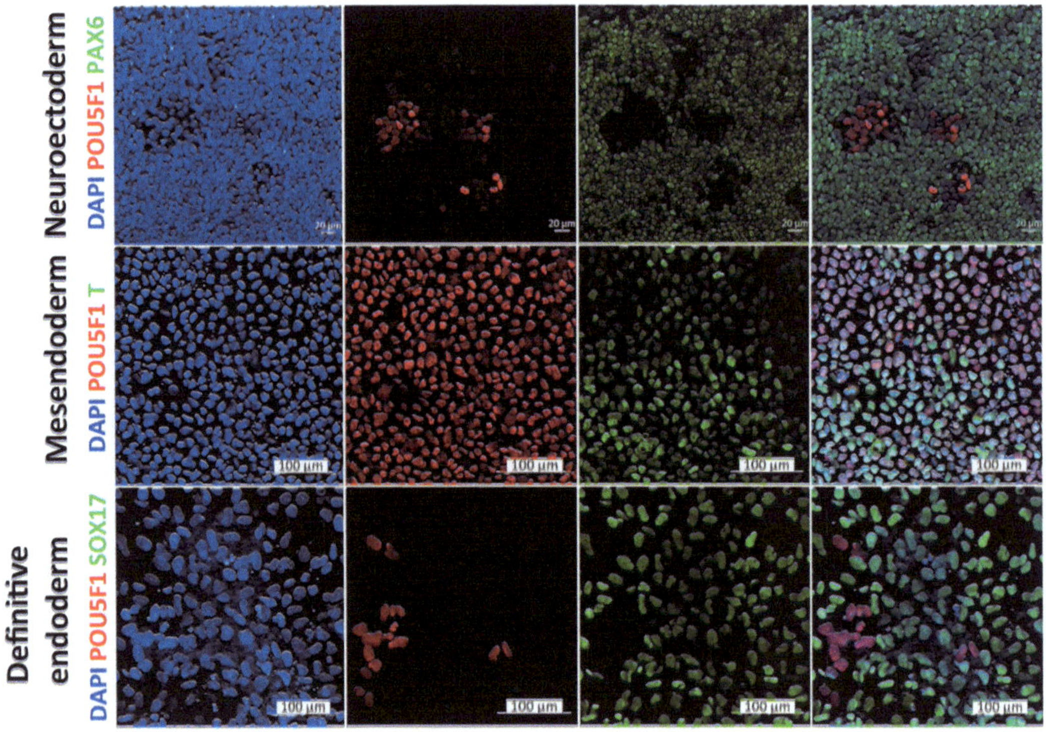

Fig. 6 Representative immunofluorescent stains towards Neuroectoderm (PAX6), Mesendoderm (T) and Definitive Endoderm (SOX17), and POU5F1

isoforms when necessary. Alternatively, other methods can be used, such as SYBRgreen-based qPCR. In all cases, it is key to test the efficiency of every primer-set to ensure that the quantification is fully reliable.

11. Real-time quantitative PCR results can be represented in several ways, including relative quantification (fold change), Log_2 fold change, ΔC_t and $\Delta\Delta C_t$, and should be chosen on a case by case basis. Typical C_t values for housekeeping genes range from 19 to 26, for pluripotency genes in the undifferentiated state from 22 to 25 and with a wider 24–30+ range in the differentiated state, depending on the lineage. Differentiation-related genes are typically seen between 22 and 28 upon differentiation. Proper induction of *SOX17* for example is seen at approximately C_t 24. Ct values are directly correlated to cDNA input, and so these values serve only as guidelines. It should be noted that when selecting samples for relative quantification, especially for differentiation-associated genes in the undifferentiated state, C_t values may beyond the reliable range of detection (C_t 36–40), leading to inaccurate determination of expression increase when used as a reference for fold change. In such cases, it is not advised to represent changes in gene

expression as a fold change, but instead as a ΔC_t or $\Delta\Delta C_t$. For greater detail than can be addressed here, refer to the MIQE guidelines for Quantitative Real-Time PCR Experiments [14]. Figure 5 shows representative relative quantifications and the corresponding C_t values for a single differentiation to each lineage.

12. Obtaining the optimal images is key for reproducible quantification. Use culture plates with glass or plastic bottom that are compatible with the microscope. There is a variability within one plate of the same experiment and even higher variability between different plates and different experiments. Reducing technical variables will provide quantification that is more accurate. Using wide field acquisition instead of confocal is advised. Less detail in the image will result in more homogeneous intensity of the nuclei, which will make thresholding and segmentation in later steps more precise. Some important features of acquisition for reproducible counting are equally lit field of the image, low signal-to-noise ratio, and similar laser power and gain. The latest two need to be adjusted specifically for each fluorescence channel and experiment. Zen Blue uses CZI file in which histogram can be adjusted before counting. ImageJ/Fiji uses TIFF or PNG files, because they preserve any calibration made during acquisition. For the same reason, JPEG files are not recommended.

13. The quantification of the staining of differentiated cells using ImageJ has 3 steps: pre-processing, segmentation, and analysis. (1) Pre-processing of the image: The goal of pre-processing is to make clear distinction between background and the nuclei you are quantifying and to level the intensities within nuclei. This is done by using filters such as Gaussian blur, Sharpen, and Linear Kuwahara. It is important not to manipulate brightness and contrast settings prior to the quantification. This can create false negative or false positive signal, which can interfere with the final result. (2) Segmentation and creating binary mask: Segmentation of particles is done by creating binary masks that can label and separate individual nuclei. This step is the most challenging, because there is no universal method that could be used in all the images. There are many different segmentation methods available in the ImageJ/Fiji, e.g., region-based methods, clustering, shape-based method, morphological methods, thresholding, machine learning based methods. In the example represented in Fig. 7, we use thresholding and maxima to segment by intensity and watershed for morphological separation. Mask 1 is generated with "Find Maxima" plugin. It marks maximum intensities on the image based on the set threshold. As the output we choose "Segmented Particles." Next, use manual or automated threshold to create a binary mask of the

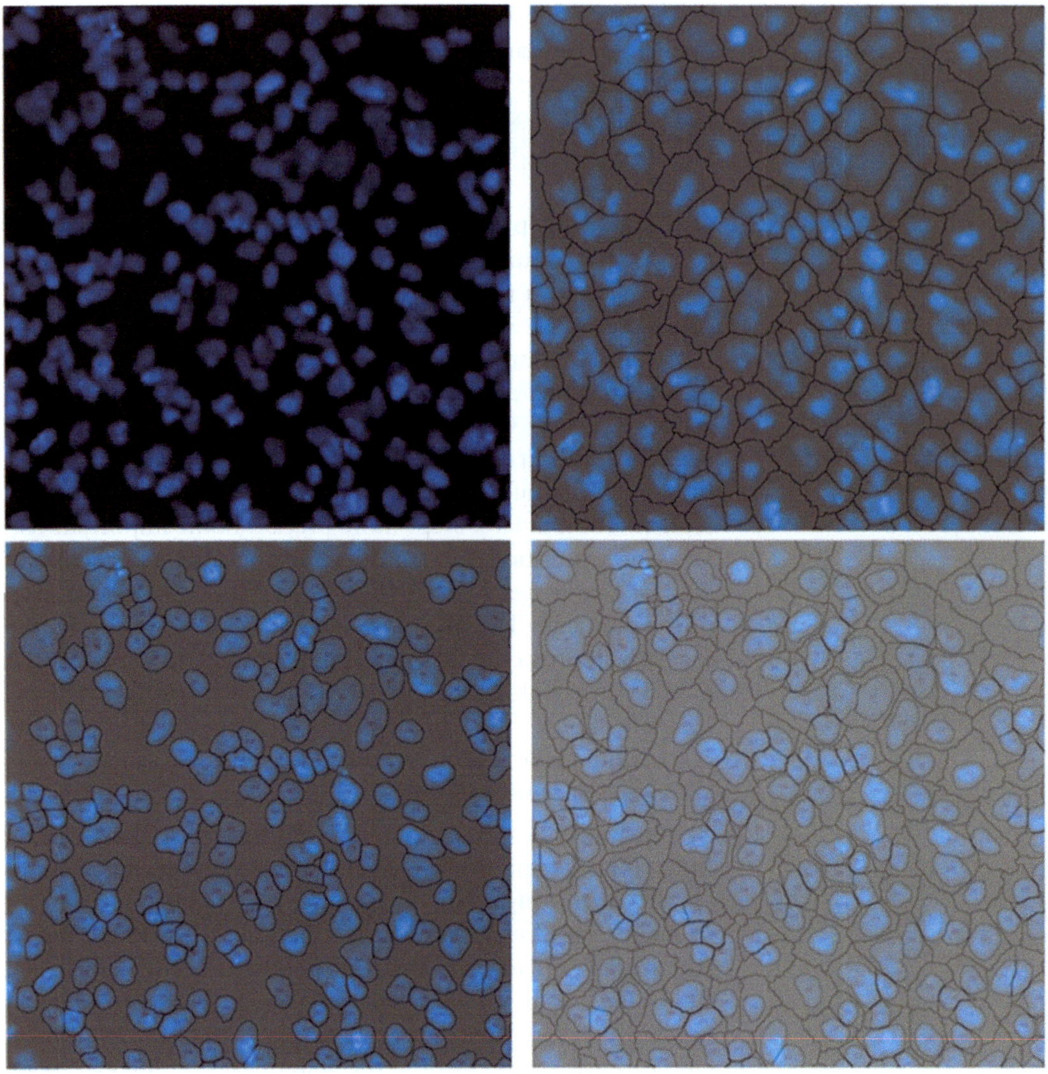

Fig. 7 Example of segmentation with ImageJ. Nuclear immunofluorescent stain with an overlay of Mask 1, Mask 2, and the combination of both

nuclei. Manual threshold has a low reproducibility and a high user bias. Using fixed value of the threshold will not result in reproducible data because of the variability between images. Although automated threshold is not strictly objective and bias-free, it gives a good starting point that can be adjusted to fit the individual image. In example, we use "Huang dark" autothreshold to create Mask 2. We combine both, Mask 1 and Mask 2, by creating Mask 3 with Boolean operations in the image calculator. Additional operations of binary mask such as "Fill holes, Erode, Dilate, Skeletonize, Erode" can be used to improve the separation of the nuclei. In the example, we move directly to morphological segmentation with watershed, which separates any clumps that were created during making a binary mask. We set the measurements to redirect Mask 3 to the original image. (3) Analysis: The last step is analyzing the particles. You can limit the size of nuclei you want to include in the quantification and choose the data you would like in the summary of the results. In the example in Fig. 7, we excluded any objects smaller than 500 pixels.

14. Figure 6 shows representative, single field of view, immunofluorescent stains of expected outcomes of neuroectoderm, mesendoderm, and definitive endoderm. It is not unusual to have POU5F1+ cells as differentiation progresses, most notably at the mesendoderm stage. An example of the importance of cross referencing both expression data and stains can be seen in the definitive endoderm POU5F1 panel, where relatively few cells are POU5F1+ at the protein level, but high levels can still be seen at the mRNA level (Fig. 5).

References

1. Thomson JA, Itskovitz-Eldor J, Shapiro SS, Waknitz MA, Swiergiel JJ, Marshall VS, Jones JM (1998) Embryonic stem cell lines derived from human blastocysts. Science 282: 1145–1147
2. Takahashi K, Tanabe K, Ohnuki M, Narita M, Ichisaka T, Tomoda K, Yamanaka S (2007) Induction of pluripotent stem cells from adult human fibroblasts by defined factors. Cell 131: 861–872
3. Keller A, Dziedzicka D, Zambelli F, Markouli C, Sermon K, Spits C, Geens M (2018) Genetic and epigenetic factors which modulate differentiation propensity in human pluripotent stem cells. Hum Reprod 24: 162–175
4. Avior Y, Biancotti JC, Benvenisty N (2015) TeratoScore: assessing the differentiation potential of human pluripotent stem cells by quantitative expression analysis of Teratomas. Stem cell Rep 4:967–974
5. Bock C, Kiskinis E, Verstappen G, Gu H, Boulting G, Smith ZD, Ziller M, Croft GF, Amoroso MW, Oakley (2011) Reference maps of human ES and iPS cell variation enable high-throughput characterization of pluripotent cell lines. Cell 144:439–452
6. Tsankov AM, Akopian V, Pop R, Chetty S, Gifford CA, Daheron L, Tsankova NM, Meissner A (2015) A qPCR ScoreCard quantifies the differentiation potential of human pluripotent stem cells. Nat Biotechnol 33:1182–1192
7. Markouli C, Couvreu De Deckersberg E, Regin M, Nguyen HT, Zambelli F, Keller A, Dziedzicka D, De Kock J, Tilleman L, Van Nieuwerburgh F, Franceschini L, Sermon K, Geens M, Spits C (2019) Gain of 20q11.21 in human pluripotent stem cells impairs TGF-β-dependent Neuroectodermal commitment. Stem Cell Rep 13:163–176
8. Markouli C, De Deckersberg EC, Dziedzicka D, Regin M, Franck S, Keller A, Gheldof A, Geens M, Sermon K, Spits C (2021) Sustained intrinsic WNT and BMP4 activation impairs hESC differentiation to definitive endoderm and drives the cells towards extra-embryonic mesoderm. Sci Rep 11(1):8242. https://doi.org/10.1038/s41598-021-87547-7. PMID: 33859268
9. Chetty S, Pagliuca FW, Honore C, Kweudjeu A, Rezania A, Melton DA (2013) A simple tool to improve pluripotent stem cell differentiation. Nat Methods 10:553–556
10. Chambers SM, Fasano CA, Papapetrou EP, Tomishima M, Sadelain M, Studer L (2009) Highly efficient neural conversion of human ES and iPS cells by dual inhibition of SMAD signaling. Nat Biotechnol 27:275–280
11. Sui L, Mfopou JK, Geens M, Sermon K, Bouwens L (2012) FGF signaling via MAPK is required early and improves Activin A-induced definitive endoderm formation from human embryonic stem cells. Biochem Biophys Res Commun 426:380–385

12. Rodin S, Antonsson L, Hovatta O, Tryggvason K (2014) Monolayer culturing and cloning of human pluripotent stem cells on laminin-521-based matrices under xeno-free and chemically defined conditions. Nat Protoc 9:2354–2368
13. Baker D, Hirst AJ, Gokhale PJ, Juarez MA, Williams S, Wheeler M, Bean K, Allison TF, Moore HD, Andrews PW, Barbaric I (2016) Detecting genetic mosaicism in cultures of human pluripotent stem cells. Stem Cell Rep 7:998–1012
14. Bustin SA, Benes V, Garson JA, Hellemans J, Huggett J, Kubista M, Mueller R, Nolan T, Pfaffl MW, Shipley GL, Vandesompele J, Wittwer CT (2009) The MIQE guidelines: minimum information for publication of quantitative real-time PCR experiments. Clin Chem 55:611–622

Chapter 6

Detection of Soluble and Insoluble Protein Species in Patient-Derived iPSCs

Stephanie Santarriaga, Ian Luecke, and Allison D. Ebert

Abstract

Protein aggregation is one of the hallmarks of many neurodegenerative diseases. While protein aggregation is a heavily studied aspect of neurodegenerative disease, methods of detection vary from one model system to another. Induced pluripotent stem cells (iPSCs) present an opportunity to model disease using patient-specific cells. However, iPSC-derived neurons are fetal-like in maturity, making it a challenge to detect key features such as protein aggregation that are often exacerbated with age. Nevertheless, we have previously found abnormal soluble and insoluble protein burden in motor neurons generated from amyotrophic lateral sclerosis (ALS) iPSCs, though protein aggregation has not been readily detected in iPSC-derived neurons from other neurodegenerative diseases. Therefore, here we present an ultracentrifugation method that detects insoluble protein species in various models of neurodegenerative disease, including Huntington's disease, Alzheimer's disease, and ALS. This method is able to detect soluble, insoluble, and SDS-resistant species in iPSC-derived neurons and is designed to be flexible for optimal detection of various aggregation-prone proteins.

Key words Induced Pluripotent Stem Cells, Neurons, Protein Aggregation, Neurodegeneration, Amyotrophic Lateral Sclerosis, Huntington's Disease, Alzheimer's Disease

1 Introduction

Protein aggregation is one of the hallmarks of many neurodegenerative diseases, including Huntington's disease (HD), amyotrophic lateral sclerosis (ALS), Parkinson's disease (PD), and Alzheimer's disease (AD) [1–3]. While animal models of neurodegenerative diseases exist, there are often unique and fundamental differences in the organization and function of neurons in the human brain, and the findings from animal models do not always translate well from mice to humans [4–6]. With the advent of human induced pluripotent stem cells (iPSCs), we now have the opportunity to generate neurons from individual patients to develop better models of disease. However, one challenge of using iPSCs to model neurodegenerative disease remains. The core reprogramming process

Nagarajan Kannan and Philip Beer (eds.), *Stem Cell Assays: Methods and Protocols*, Methods in Molecular Biology, vol. 2429, https://doi.org/10.1007/978-1-0716-1979-7_6,

for generating iPSCs erases many of the features associated with aging, making it difficult to detect features that are often exacerbated with age [7–10]. As such, protein aggregation may not be readily detected in iPSC models. While fluorescence microscopy currently serves as the main method of detecting protein aggregates in iPSCs, few studies have been able to detect aggregates without additional chemical or cellular manipulation [11–15].

Ultracentrifugation serves to isolate smaller insoluble species by generating a soluble and insoluble fraction. This technique may be particularly useful in model systems where protein aggregates cannot be detected by fluorescence microscopy. While ultracentrifugation has been commonly used to detect protein aggregation in other model systems, protocols vary widely, and its use has been much less frequent in iPSC models [16–20]. In addition, this technique is highly dependent on the model system and the target protein being studied, with aggregates typically being detected in overexpression and transgenic mouse models. Here, we present an ultracentrifugation protocol that detects soluble and insoluble protein species as well as SDS-resistant protein species in various iPSC-derived models of neurodegenerative disease (Fig. 1). Using this technique, we have detected insoluble protein species in iPSC models of ALS [21]. Here, we report the detection of insoluble huntingtin species in HD iPSC-derived neurons, which have previously remained undetected without manipulation (Fig. 2a) [11, 12]. Additionally, we find high levels of soluble amyloid beta in AD iPSC-derived cortical organoids, though some insoluble amyloid beta was also detected in AD iPSC colonies (Fig. 2b). Through western blot analysis, this technique is then amenable to quantification and can be used to yield semi-quantitative results. In addition to the core technique presented, we present a wide variety of modifications that may be utilized for experiment optimization.

2 Materials

2.1 Cell Lysis

1. 1× PBS: 137 mM NaCl, 2.7 mM KCl, 8 mM Na_2HPO_4, 2 mM KH_2PO_4. Store at room temperature.
2. Triton X-100 Buffer: 1% Triton X-100, 150 mM NaCl, 50 mM Tris–Cl pH 8.0, 1 mM EDTA. All components are made up to desired final volume. Store at 4 °C (*see* **Note 1**).
3. Final Triton X-100 Buffer: Before use, add 1:100 Protease Inhibitor Cocktail (100×) to Triton X-100 Buffer described above. Keep cold.
4. Sonicator, for example Fisherbrand 120 Sonic Dismembrator.

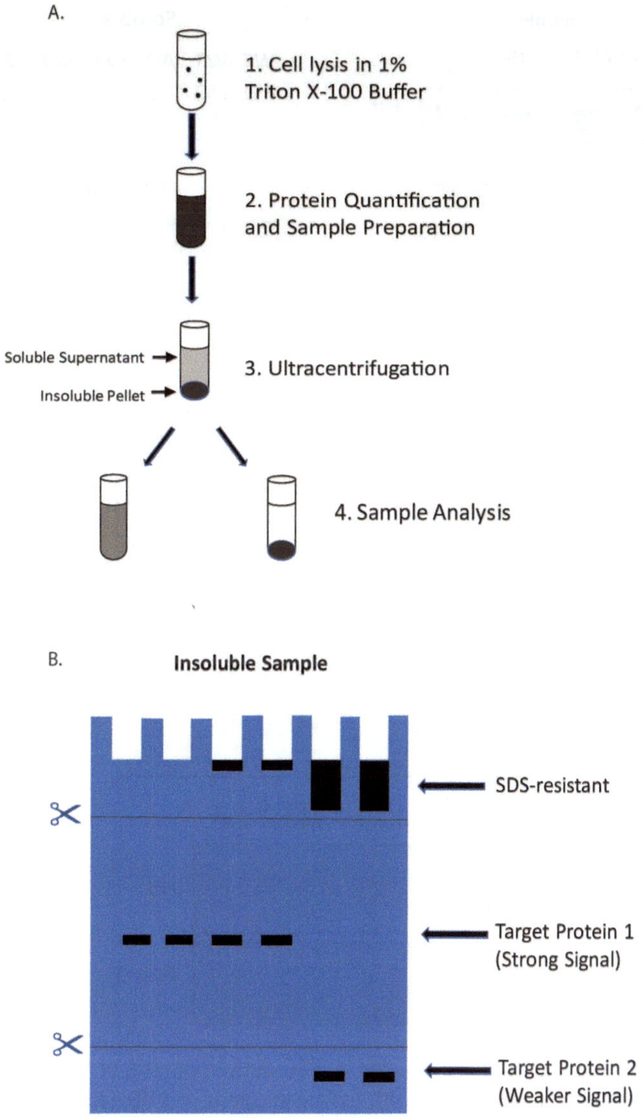

Fig. 1 Detection of soluble and insoluble species using ultracentrifugation. (**a**) Schematic of fractionation protocol for generating soluble and insoluble fractions. Cells are directly lysed in 1% Final Triton X-100 Buffer and sonicated to ensure cell lysis. BCA assay is used to quantify protein and ensure that all samples contain the same starting amount of protein, volume, and concentration. Samples are then spun down using ultracentrifugation in order to obtain a soluble, supernatant and insoluble, pellet fraction. Fractions are then imaged by western blot and analyzed for soluble, insoluble, and SDS-resistant species. (**b**) Visualization of western blot of insoluble sample with expected results. Soluble and insoluble samples are run on their own respective gels. For insoluble species, signal can be expected where target protein would normally appear. If SDS-resistant species are present, signal may also appear on the stacking gel. If SDS-resistant species are expected, cutting the membrane is recommended. If target protein and antibody typically yield a strong signal, no additional cutting is necessary. However, if target protein and antibody typically yield a weak signal, cut membrane to confine antibody to region of interest and reduce non-specific background

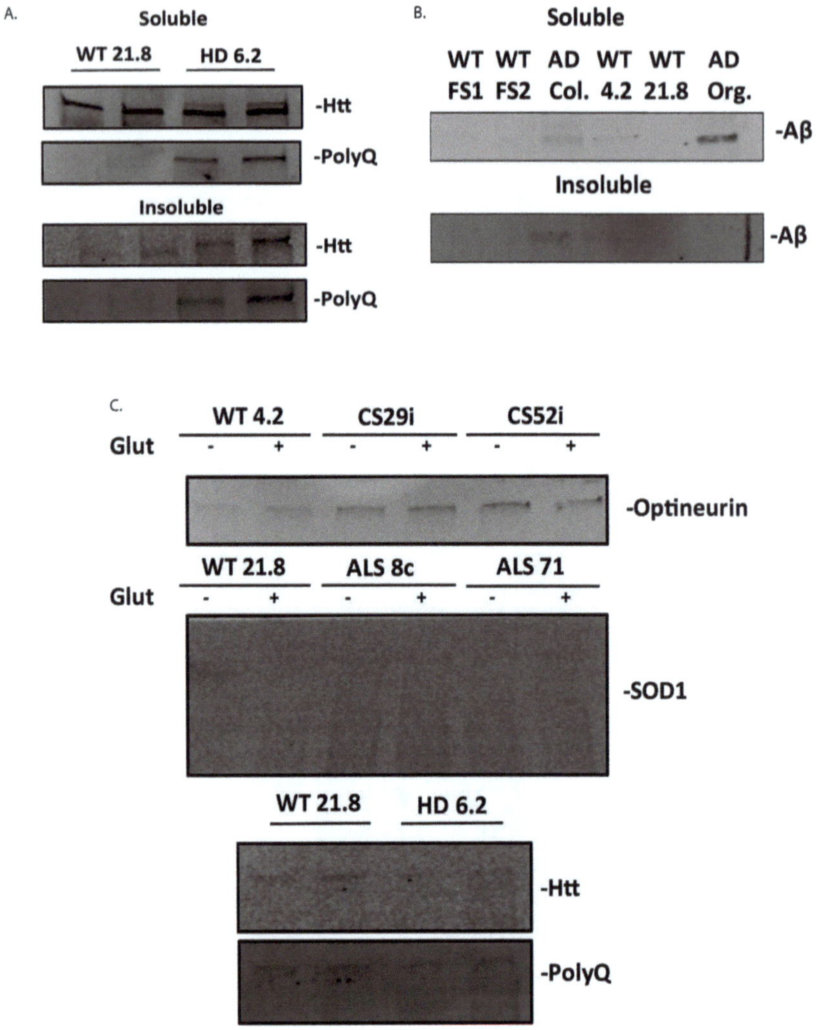

Fig. 2 Detection of soluble and insoluble species in iPSC-derived models of neurodegenerative disease. (**a**) Detection of soluble and insoluble huntingtin (Htt) protein in control and HD neurons. Control (WT 21.8) and HD (HD 6.2) neurons were harvested and fractionated using ultracentrifugation. Soluble and insoluble fractions were run on their respective gels and imaged by western blot using antibodies for Htt and polyglutamine-expanded proteins (polyQ). Note the insoluble Htt protein is only detected in HD lines. (**b**) Detection of soluble and insoluble Aβ in control iPSC-derived neural progenitors (WT FS1 and WT FS2), neurons from controls (WT 4.2 and WT 21.8), iPSC colonies from an AD patient, and cortical organoids from an AD patient. Control and AD samples were harvested and fractionated using ultracentrifugation. Soluble and insoluble fractions were run on their respective gels and imaged by western blot using antibodies for Aβ. High levels of Aβ were detected in the soluble fraction from AD organoids. Low levels of Aβ were detected in both the soluble and insoluble fractions from AD iPSC colonies. (**c**) Appearance of SDS-resistant aggregates varies per protein of interest. Neurons from ALS models (CS29i and CS52i; ALS 8c and ALS 71) and HD lines were subjected to fractionation and compared to control lines (WT 4.2 and WT 21.8). Insoluble fractions were imaged by western blot using antibodies against Htt and polyQ for HD lines and optineurin and SOD1 for ALS lines. Optineurin antibody depicts clear SDS-resistant aggregates in the stacking gel, whereas SOD1 antibody depicts SDS-resistant aggregates that are present as a smear. The effect of exogenous cellular stressors (e.g., glutamate) on protein solubility can also be assessed by this method. HD lines depict SDS-resistant species that cannot be differentiated from control lines

2.2 Protein Quantification and Centrifugation

1. Detergent Compatible Protein Quantification Assay (Bicinchoninic acid, BCA).
2. Beckman Coulter Tubes (RCF Limit>100,000 × *g*).
3. SDS Lysis Buffer: Triton X-100 Buffer, 1:100 Protease Inhibitor Cocktail (100×), 2% SDS.
4. Beckman TL-100 Ultracentrifuge.

2.3 SDS-PAGE

1. SDS-Running Buffer: 25 mM Trizma base, 192 mM Glycine, and 650 ml distilled water, then pH to 8.8. Add 100 ml of 10% SDS for a final concentration of 0.1% and bring up to 1 l with water. Store at room temperature.
2. Mini PROTEAN TGXT precast gels.
3. 4× Laemmli Buffer: 200 mM Tris–Cl (pH 6.8), 8% SDS (sodium dodecyl sulfate), 0.02% Bromophenol blue, 40% glycerol, 10% β-mercaptoethanol (added fresh).
4. Precision Plus Protein Standards-Dual Color.
5. SDS-PAGE Equipment: Gel Apparatus.

2.4 Transfer

1. Transfer Buffer: For 1 liter, combine 25 mM Trizma base, 192 mM glycine, 500 ml of distilled water, and 20% (v/v) methanol, then pH to 8.3–8.4 and bring up to 1 l with water. Store at 4 °C.
2. Methanol.
3. Polyvinylidene difluoride (PVDF) (optimized for fluorescence or film, depending on protocol).
4. Western Blot Equipment: Cassettes, filter pads, sponges, ice block.

2.5 Fluorescent Western Blot (See Note 2)

1. Total Protein Stain.
2. Blocking Buffer.
3. Primary Antibody Diluent: Primary, Blocking Buffer diluted 1:1 with TBS, 0.2% Tween-20.
4. Secondary Antibody Diluent: Secondary, Blocking Buffer diluted 1:1 with TBS, 0.2% Tween-20, 0.02% SDS.
5. TBS: 20 mM Trizma Base, 150 mM NaCl, and 800 ml distilled water, then pH to 7.5. Add water to 1 l. Store at room temperature.
6. TBS-T: Add 500 μl Tween-20 to 1 l of TBS. Store at room temperature.

3 Methods

3.1 Triton X-100 Cell Lysis

1. For adherent cells, prepare for harvesting by aspirating off media and gently washing with 1× PBS. Repeat for a total of two washes (*see* **Note 3**).
2. Lyse cells directly by adding 50–150 μl per well (recommendations here are for a 6-well plate) of Final Triton X-100 Buffer straight to the well (*see* **Notes 4** and **5**). Scrape cells using a cell scraper or the base of a 200 μl pipette tip. Transfer lysed cells to an Eppendorf tube.
3. Ensure lysis of cells and shearing of DNA by sonication (*see* **Note 6**). For samples of ~100 μl, sonicate twice for 3 s at 30% output (*see* **Notes 7** and **8**). Ensure that samples are kept on ice when not actively using sonicator.
4. To quickly verify cell lysis, spin lysate in a mini-centrifuge for ~1 s. The appearance of a pellet after a brief spin at low speed typically represents unlysed cells. If pellet appears, either add additional lysis buffer or repeat sonication step until pellet is no longer apparent after low-speed spin (*see* **Note 9**).

3.2 Total Protein Quantification and Sample Preparation

1. Following the protocol for BCA assay, quantify protein concentration in order to ensure equal starting quantities of protein samples (*see* **Notes 10** and **11**). This step is important as the starting protein content in samples may differ due to differences in seeding and technical variation in the harvesting process.
2. Select a starting protein quantity between 50 and 200 μg of protein based on lowest BCA result. Calculate necessary volume of lysate to obtain the select quantity across samples. Dilute all samples to a concentration of 0.5–1 μg/μl with additional Final Triton X-100 Buffer (*see* **Notes 12** and **13**). All samples should ultimately have the same quantity of protein, concentration, and volume. Samples must be prepared in or transferred to microcentrifuge tubes that can withstand ultracentrifugation.

3.3 Ultra-centrifugation

1. Centrifuge samples at 45,000 rpm (or >100,000 × *g*) for 30 min using a Beckman TL-100 Ultracentrifuge and a TLA-45 rotor at 4 °C (*see* **Note 14**).
2. Remove supernatant and save for western blot. Pellet will be very small but should be visible to the eye. You must be very careful to not disturb pellet. Pellet disturbance could lead to inappropriate removal of insoluble proteins.
3. Wash pellets by adding 100 μl of Final Triton X-100 Buffer, pipette up and down, and spin at 45,000 rpm for 5 min using a

Beckman TL-100 Ultracentrifuge and a TLA-45 rotor at 4 °C. Remove buffer and repeat for a total of three washes with Final Triton X-100 Buffer (*see* **Note 15**).

4. Resuspend pellets in 120 μl of SDS Lysis Buffer and sonicate briefly ~1 s (*see* **Note 16**).
5. For this protocol, BCA assay was used to quantify protein concentration of pellet fraction. This step ensures sufficient protein was isolated in pellet for detection by western blot.
6. At this point, you may proceed directly to SDS-PAGE or freeze samples at −20 °C for short-term storage or −80 °C for long-term storage.

3.4 SDS-PAGE and Transfer

1. For loading, select a volume between 10 and 30 μl (~10–20 μg depending on initial concentration) for the soluble, supernatant sample and between 20 and 50 μl (minimum of 5 μg) for the insoluble, pellet sample (*see* **Note 17**).
2. To prepare samples for SDS-PAGE, calculate amount of sample that will be mixed with 4× Laemmli Buffer (diluted to 1×) and boil for 5 min (*see* **Note 18**).
3. Load 10 μl of ladder and samples onto a Bio Rad, Mini PROTEAN TGXTM precast gel (*see* **Note 19**). Run samples until the loading dye is no longer visible. Settings utilized here were constant 105 V for about 90 min at room temperature. Voltage settings and time can vary.
4. Once gel has finished, transfer onto a PVDF membrane. For this, remove wells from gel but ensure that the top of the gel remains intact.
5. Transfer was performed with cold transfer buffer and ice block within the apparatus at constant 105 V for 1 h (*see* **Note 20**).
6. Once transfer is complete, dry membrane for 1 h or overnight.

3.5 Western Blot

1. Block membrane for at least 1 h at room temperature or overnight at 4 °C.
2. Following blocking, cut the top of the membrane to separate the stacking gel (typically above the first ladder marking). This portion of the membrane will be used to identify SDS-resistant species (Figs. 1b and 2c) (*see* **Note 21**). If necessary, cut membrane for other targets (Fig. 1b) (*see* **Note 22**).
3. Incubate membrane in primary antibody overnight at 4 °C. For antibodies yielding a strong signal, a 1–2 h incubation at room temperature may be sufficient (*see* **Notes 23** and **24**).
4. Wash membrane with TBS-T for 5 min at room temperature. Repeat for a total of three washes.

5. Incubate membrane in secondary antibody for 30–60 min at room temperature (*see* **Note 24**). If using fluorescent secondaries, use dark container or cover with foil.
6. Wash membrane with TBS-T for 5 min at room temperature. Repeat for a total of three washes.
7. Wash membrane once with TBS to remove Tween. Keep in TBS until imaging.
8. Image membrane. Here, membrane was imaged using a Licor Odyssey. Imaging settings will vary based on antibody quality and protein expression.

3.6 Quantification Using Image Studio Lite (See Note 25)

1. To quantify bands, utilize the analysis section in Image Studio Lite. To begin, auto-detect bands or manually enclose bands by drawing a rectangle (*see* **Note 26**).
2. Subtract background. Corrected values will appear in a table in the program.
3. Quantify signal of total protein staining (or loading control) and target protein bands.
4. Transfer values to Excel or a similar program. Normalize target protein band to total protein staining or loading control values.
5. For improved visualization, calculate fold change of normalized samples.

4 Notes

1. NP-40 detergent can be substituted for Triton X-100. Protocol can also be modified to detect SDS-resistant species by adding 2% SDS or utilizing RIPA Buffer.
2. While fluorescence was the primary method of detection, certain proteins will be better detected by chemiluminescent westerns visualized on film. Particularly, visualization by film may be useful if the target species are the SDS-resistant aggregates, which are often difficult to detect. Signal varies on the antigen being detected and antibody quality.
3. A third wash may be helpful to further remove dead cells.
4. Alternatively, cells may be lysed after pelleting. In general, cells may be lifted off by pipetting up and down gently and dispensed into a 15 ml conical tube. For very adherent cells, an enzyme may be necessary. Pellet cells by spinning at 1000 rpm for 5 min. Remove media and wash gently with the addition of 1× PBS and spin down again to pellet. Repeat for a total of two washes. Lyse pellet with Final Triton X-100 Buffer using recommended amounts in **step 2**. Resume protocol at **step 3**. This harvesting method may also be used for non-adherent cells.

5. Lysis volume will require optimization based on cell density and lysis method. Here, ~300,000 cells/well were seeded and grown for 2–4 weeks to generate mature neurons. Sufficient lysis buffer must be added to ensure efficient lysis. Alternatively, excess lysis buffer will result in protein concentrations too dilute for SDS-PAGE loading.
6. For samples less than 100 μl, lyse by freeze-thawing with dry ice or liquid nitrogen (at least three times) and heavy vortexing in between freeze–thaw cycles.
7. Bring probe to bottom of tube otherwise the likelihood of sample foaming and sample loss increases. However, do not press against tube too hard or tube will melt.
8. Sonication step was written for a probe sonicator. Other sonication methods may be used but may require optimization. Sonication must be minimal and solely for the purpose of ensuring cell lysis. Over-sonication may lead to false positives.
9. This step may be added if interested solely in smaller insoluble species: Spin lysate at 14,000 rpm for 1 min to remove large insoluble material. After spinning, continue with supernatant and discard pellet (large insoluble material).
10. While it is best to continue onto the ultracentrifugation step, samples may be frozen either before or after BCA assay. Some proteins are more sensitive to freezing, which may contribute to aggregation artifacts. If samples are frozen, very brief sonication may be useful (~1 s) prior to ultracentrifugation.
11. Protein quantification reagents must be compatible with detergent. Common reagents (such as Bradford) will not accurately quantify protein with high levels of detergent.
12. In regard to starting protein content, a higher protein content increases the likelihood of detecting protein aggregates. However, the recommended concentrations must be maintained in order to avoid supersaturation, which influences aggregation.
13. If possible, a higher concentration of 1 μg/μl is preferred. This simplifies the removal of supernatant steps and reduces the loading volume necessary for SDS-PAGE. However, if the starting protein content is low, then a concentration of 0.5 μg/μl may be used.
14. Speed and time can be optimized to maximize detection.
15. Insufficient washing could result in contamination of soluble protein in the pellet sample. If soluble proteins consistently appear in insoluble fraction, then increase number of wash steps.
16. BCA protein assay is compatible with 2% SDS and may be used to quantify protein content of pellet fraction. Typical yield for a pellet fraction with a starting quantity of 50 μg ranged from 0.25 to 0.75 μg/μl.

17. We typically do not load equal volumes of supernatant and pellet as volumes loaded depend highly on protein expression, quantity of starting material, and antibody quality. If desired, equal volumes of both supernatant and pellet may be loaded for a more direct comparison between the two fractions. However, this would require additional optimization as signal varies greatly between soluble and insoluble fractions.
18. For some proteins, boiling may result in aggregation. In these scenarios, sample may be heated at a lower temperature for an extended period of time. This information, if not general knowledge, may be provided by the company where the antibody was purchased.
19. When determining loading order, detection may improve if all soluble and all insoluble samples are loaded separately in different gels rather than on the same gel.
20. Voltage settings and time can vary. For larger protein and SDS-resistant species, transfer may need to be performed for an extended period of time or overnight.
21. Detection of SDS-resistant aggregates varies per protein of interest. It is recommended that western blot protocol be optimized for proteins of interest. Modifications to lysis buffer and ultracentrifugation conditions may also improve detection. Note that depending on levels of protein aggregation, the appearance of SDS-resistant aggregates may differ. Large species unable to enter SDS-PAGE gel will appear as clear bands where loading wells would have been present, whereas large species that began to enter gel may appear as smears on the stacking gel (Figs. 1b and 2c). For the detection of SDS-resistant aggregates, it is highly recommended to compare to a control sample as disease lines may not always show a greater number of SDS-resistant species than the control (Fig. 2c).
22. If an antibody is weaker, the membrane can be cut so that only the area where the target protein is expected is exposed to primary antibody (Fig. 1b).
23. To confirm efficient lysis and clean separation of insoluble species, probe for a non-aggregating protein as a control. Depending on context, this control may be a loading control (e.g., actin, GAPDH, tubulin) or known soluble proteins (Fig. 3).
24. Antibody concentrations are based on manufacturer's instructions. Primary antibody concentrations typically range from 1: 50 to 1:10,000. Secondary antibody concentrations can range from 1:1000 to 1:40,000.
25. A similar process may be followed using Image J.

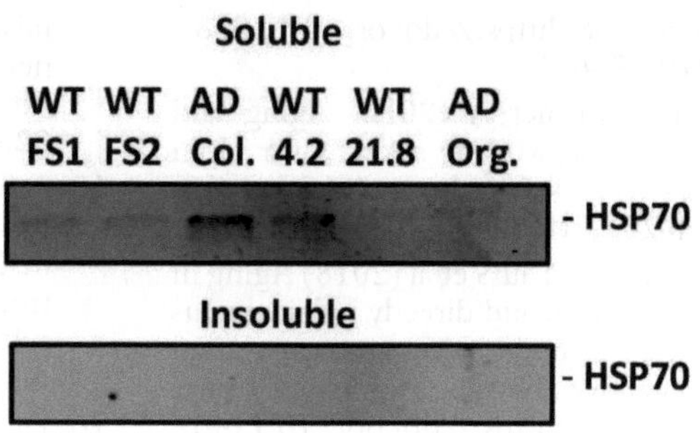

Fig. 3 Detection of control protein to ensure efficient separation between soluble and insoluble fractions. Control iPSC-derived neural progenitors (WT FS1 and WT FS2), neurons from control iPSCs (WT 4.2 and WT 21.8), iPSC colonies from an AD patient, and cortical organoids from an AD patient were harvested and fractionated using ultracentrifugation. Soluble and insoluble fractions were run on their respective gels and imaged by western blot using antibodies for HSP70. Here, HSP70 serves as a control protein that should only be present in the soluble fraction

26. Size of enclosure should not matter as long as background is correctly subtracted and enclosures do not overlap in regions of high signal.

Acknowledgments

This work was funded by the Pick Charitable Innovation Fund for ALS research (A.D.E.). The authors thank E. Seminary and G. Lin for early technical assistance.
Author Contributions: S.S. developed and wrote the protocol and contributed to data collection. I.L. contributed to data collection and protocol writing. A.D.E. provided resources and edited the protocol.

References

1. Ross CA, Poirier MA (2004) Protein aggregation and neurodegenerative disease. Nat Med 10(Suppl):S10–S17. https://doi.org/10.1038/nm1066
2. Kumar V, Sami N, Kashav T et al (2016) Protein aggregation and neurodegenerative diseases: from theory to therapy. Eur J Med Chem 124:1105–1120. https://doi.org/10.1016/j.ejmech.2016.07.054
3. Hipp MS, Park SH, Hartl FU (2014) Proteostasis impairment in protein-misfolding and -aggregation diseases. Trends Cell Biol 24(9): 506–514. https://doi.org/10.1016/j.tcb.2014.05.003
4. Dawson TM, Golde TE, Lagier-Tourenne C (2018) Animal models of neurodegenerative diseases. Nat Neurosci 21(10):1370–1379. https://doi.org/10.1038/s41593-018-0236-8
5. van der Worp HB, Howells DW, Sena ES et al (2010) Can animal models of disease reliably inform human studies? PLoS Med 7(3): e1000245. https://doi.org/10.1371/journal.pmed.1000245
6. Hodge RD, Bakken TE, Miller JA et al (2019) Conserved cell types with divergent features in human versus mouse cortex. Nature

573(7772):61–68. https://doi.org/10.1038/s41586-019-1506-7

7. Mahmoudi S, Brunet A (2012) Aging and reprogramming: a two-way street. Curr Opin Cell Biol 24(6):744–756. https://doi.org/10.1016/j.ceb.2012.10.004
8. Mertens J, Reid D, Lau S et al (2018) Aging in a dish: iPSC-derived and directly induced neurons for studying brain aging and age-related neurodegenerative diseases. Annu Rev Genet 52:271–293. https://doi.org/10.1146/annurev-genet-120417-031534
9. Miller JD, Ganat YM, Kishinevsky S et al (2013) Human iPSC-based modeling of late-onset disease via progerin-induced aging. Cell Stem Cell 13(6):691–705. https://doi.org/10.1016/j.stem.2013.11.006
10. Studer L, Vera E, Cornacchia D (2015) Programming and reprogramming cellular age in the era of induced pluripotency. Cell Stem Cell 16(6):591–600. https://doi.org/10.1016/j.stem.2015.05.004
11. The HD iPSC Consortium (2012) Induced pluripotent stem cells from patients with Huntington's disease show CAG-repeat-expansion-associated phenotypes. Cell Stem Cell 11(2):264–278. https://doi.org/10.1016/j.stem.2012.04.027
12. Jeon I, Lee N, Li J-Y et al (2012) Neuronal properties, in vivo effects, and pathology of a Huntington's disease patient-derived induced pluripotent stem cells. Stem Cells 30(9):2054–2062. https://doi.org/10.1002/stem.1135
13. Koyuncu S, Saez I, Lee HJ et al (2018) The ubiquitin ligase UBR5 suppresses proteostasis collapse in pluripotent stem cells from Huntington's disease patients. Nat Commun 9(1):2886. https://doi.org/10.1038/s41467-018-05320-3
14. Chen H, Qian K, Du Z et al (2014) Modeling ALS with iPSCs reveals that mutant SOD1 misregulates neurofilament balance in motor neurons. Cell Stem Cell 14(6):796–809. https://doi.org/10.1016/j.stem.2014.02.004
15. Sun X, Song J, Huang H et al (2018) Modeling hallmark pathology using motor neurons derived from the family and sporadic amyotrophic lateral sclerosis patient-specific iPS cells. Stem Cell Res Ther 9(1):315. https://doi.org/10.1186/s13287-018-1048-1
16. Dhillon J-KS, Riffe C, Moore BD et al (2017) A novel panel of α-synuclein antibodies reveal distinctive staining profiles in synucleinopathies. PLoS One 12(9):e0184731. https://doi.org/10.1371/journal.pone.0184731
17. Bandopadhyay R (2016) Sequential extraction of soluble and insoluble alpha-synuclein from Parkinsonian brains. J Vis Exp 107:e53415. https://doi.org/10.3791/53415
18. Basso M, Samengo G, Nardo G et al (2009) Characterization of detergent-insoluble proteins in ALS indicates a causal link between Nitrative stress and aggregation in pathogenesis. PLoS One 4(12):e8130. https://doi.org/10.1371/journal.pone.0008130
19. Karch CM, Borchelt DR (2008) A limited role for disulfide cross-linking in the aggregation of mutant SOD1 linked to familial amyotrophic lateral sclerosis. J Biol Chem 283(20):13528–13537. https://doi.org/10.1074/jbc.M800564200
20. Kazantsev A, Preisinger E, Dranovsky A et al (1999) Insoluble detergent-resistant aggregates form between pathological and nonpathological lengths of polyglutamine in mammalian cells. Proc Natl Acad Sci U S A 96(20):11404. https://doi.org/10.1073/pnas.96.20.11404
21. Seminary ER, Santarriaga S, Wheeler L et al (2020) Motor neuron generation from iPSCs from identical twins discordant for amyotrophic lateral sclerosis. Cells 9(3):571

Chapter 7

Extracellular Flux Analysis of Mitochondrial Function in Pluripotent Stem Cells

Enkhtuul Tsogtbaatar, Katherine Minter-Dykhouse, Alicia Saarinen, and Clifford D. L. Folmes

Abstract

Mitochondrial function and energy metabolism are increasingly recognized not only as regulators of pluripotent stem cell function and fate, but also as critical targets in disease pathogenesis and aging. Therefore across the downstream applications of pluripotent stem cells, including development and disease modeling, drug screening, and cell-based therapies, it is crucial to be able to measure mitochondrial function and metabolism in a high-throughput, real-time and label-free manner. Here we describe the application of Seahorse extracellular flux analysis to measure mitochondrial function in pluripotent stem cells and their derivatives. Specifically, we highlight two assays, the Mitochondrial Stress Test, which quantifies overall mitochondrial function including basal, maximal and ATP-couple oxygen consumption rates, and the Electron Transport Chain Complex Specific assay, that quantifies function of individual complexes within the electron transport chain.

Key words Embryonic stem cells, Induced pluripotent stem cells, Differentiation, Oxidative metabolism, Oxidative phosphorylation, Mitochondrial respiration

1 Introduction

Pluripotent stem cells (PSCs) can both self-renew indefinitely and differentiate into all cell types found within the three germ layers of the developing embryo. These characteristics make PSCs a unique resource for a variety of downstream applications including modeling development and disease in an in vitro patient specific manner, and generation of specific cell types for drug discovery and regenerative applications. Mitochondrial function and energy metabolism are increasingly recognized as not only drivers of stem cell function and fate during normal development [1, 2], but also represent critical targets that contribute to aging and disease pathogenesis [3–5]. Therefore the ability to quantitatively assess mitochondrial function in PSCs and their derivatives is vital for their downstream applications. Mitochondrial structure and function

Nagarajan Kannan and Philip Beer (eds.), *Stem Cell Assays: Methods and Protocols*, Methods in Molecular Biology, vol. 2429, https://doi.org/10.1007/978-1-0716-1979-7_7,

can be assessed using a variety of techniques, including: (a) imaging (ultrastructure and mitochondrial membrane potential), (b) biochemical detection of specific enzyme activities (e.g., cytochrome c oxidase or succinate dehydrogenase activity) and metabolites (adenine nucleotides, tricarboxylic acid cycle intermediates, and reactive oxygen species) and (c) protein-based methods including Western blotting, mitochondrial-specific proteomics, and Blue Native Polyacrylamide Gel Electrophoresis [6, 7]. However, these assays have inherent limitations such as having to isolate mitochondria from their intracellular environment, restricting investigation to a single pathway/process per time point, and the inability to assess the real-time impact of small molecules and the microenvironment on mitochondrial function. Clark electrode-based high resolution respirometry of isolated mitochondria or permeabilized cells represents the gold standard for assessing mitochondrial respiration rates [8–10]. However, this method requires a relatively large number of cells in suspension to ensure sufficient signal to noise and is low throughput. By comparison, Seahorse extracellular flux analysis enables high-throughput, real-time, and label-free kinetic measurements of oxygen consumption and changes in pH of the extracellular medium in a multiwell format (8, 24 or 96 wells at a time) [11–13]. This is achieved by the creation of a microwell to increase the cell number to medium volume ratio, enabling the measurement of oxygen consumption (OCR) and extracellular acidification rate (ECAR, a surrogate measure of glycolysis) in a small number of cells over a short period of time (2–5 min).

In this chapter, we describe how Seahorse XF technology can be applied to assess mitochondrial function in PSCs. Specifically we will highlight two assays, the Mitochondrial Stress Test, which provides an overview of mitochondrial function, and secondly the Electron Transport Chain (ETC) Complex Specific Assay, which provides a more detailed view of the contributions of individual ETC complexes to overall mitochondrial function. The Mitochondrial Stress Test is performed using intact cells and uses sequential injections of oligomycin, 2-[2-[4-(trifluoromethoxy)phenyl]hydrazinylidene]-propanedinitrile (FCCP), and rotenone/antimycin A to quantify indices of mitochondrial oxidative phosphorylation, including spare respiratory capacity, basal, ATP-coupled, and maximal respiration rates. The ETC Complex Specific Assay permeabilizes cells in order to provide specific combinations of respiratory substrates/inhibitors to dissect the complex specific contributions to the ETC and oxidative phosphorylation. In addition, we discuss options for data normalization and highlight the application of a Cytation 1 Cell Imaging Multi-Mode Reader for nuclear counting to normalize the Seahorse XF data. In combination these assays provide a comprehensive overview of mitochondria function that can be applied to PSCs and their differentiated counterparts to address questions across development, disease, and regeneration.

2 Materials

2.1 Equipment

1. XF HS Mini, XFe24 or XFe96 Extracellular Flux Analyzer (Seahorse Agilent, MA, USA).
2. Cytation 1 or 5 Cell Imaging Multi-Mode Reader.
3. Non-CO_2 incubator.

2.2 Cell Seeding

1. mTeSR1 Cell Culture medium (STEMCELL Technologies): 400 mL base medium, 100 mL 5× mTeSR supplement, and 1% penicillin–streptomycin.
2. Accumax dissociation reagent (Innovative Cell Technologies, Inc).
3. Geltrex LDEV-free reduced growth factor basement membrane (Gibco).
4. 10 mM Rho Kinase (ROCK) inhibitor Y-27632: Dissolve 200 mg Y-27632 in 62.35 mL Milli-Q water. Filter sterilize, aliquot and store at −20 °C.
5. 4400 Units/mL DNase I recombinant in Milli-Q water. Aliquot and store at −80 °C.

2.3 Mitochondrial Stress Test

1. Mitochondrial Stress Test Assay Medium, pH 7.4: XF Base Medium (with or without phenol red), 1 mM pyruvate, 2 mM glutamine, 17.5 mM glucose, 5 mM HEPES. Heat medium on a stirring hotplate until the medium temperature reaches 37 °C and adjust the pH to 7.4 using 5 M NaOH (*see* **Note 1**). Bring volume up to 500 mL, filter sterilize and store it at 4 °C until the day of the assay.
2. Seahorse FluxPak with sensor cartridges, cell culture microplates and XF Calibrant.
3. 12 mM oligomycin concentrated stock: Dissolve 5 mg of oligomycin (for example Sigma O4876) in 0.5 mL of cell culture grade DMSO, aliquot and store at −80 °C. Thaw once and then discard.
4. 10× oligomycin for injection (final working concentration determined per cell line): Dilute 12 mM oligomycin concentrated stock to 10× final working concentration in Mitochondrial Stress Test Assay Medium. Ensure pH is 7.4 with pH test paper, then aliquot and store at −80 °C. Thaw once and then discard.
5. 10 mM FCCP [carbonyl cyanide-4-(trifluoromethoxy)-phenylhydrazone] concentrated stock: Dissolve 10 mg FCCP in cell culture grade DMSO, aliquot and store at −80 °C. Thaw once and then discard.

6. 10× FCCP for injection (final working concentration determined per cell line): Dilute 10 mM FCCP concentrated stock to 10× final working concentration in Mitochondrial Stress Test Assay Medium. Ensure pH is 7.4 with pH test paper, then aliquot and store at −80 °C. Thaw once and then discard.
7. 10 mM rotenone concentrated stock: Dissolve 19.7 mg of rotenone in 5 mL of cell culture grade DMSO, aliquot and store at −80 °C. Thaw once and then discard.
8. 20 mM antimycin A concentrated stock: Dissolve 25 mg antimycin A in 2.28 mL of cell culture grade DMSO, aliquot and store at −80 °C. Thaw once and then discard.
9. Hoechst 33342: 10 mg/mL solution. Store at 4 °C.
10. Antimycin A/Rotenone/Hoechst 33342 injection: 5 μM rotenone, 11 μM antimycin A, 16.4 μM Hoechst 33342. Dilute 6 μL of 10 mM rotenone, 6.6 μL of 20 mM antimycin A, and 120 μL of 16.4 mM Hoechst 33342 in Mitochondrial Stress Test Assay Medium up to 12 mLs. Ensure pH is 7.4 with pH test paper, then aliquot and store at −80 °C. Thaw once and then discard (*see* **Notes 2-4**).
11. 3% Paraformaldehyde: 3% (w/v) PFA, 58 mM sucrose dissolved in PBS. Aliquot and store at −20 °C.

2.4 ETC Complex Specific Assay

1. 3× *M*itochondrial *A*ssay *S*olution (MAS, pH 7.4 at 37 °C): 660 mM mannitol, 210 mM sucrose, 30 mM KH_2PO_4, 15 mM $MgCl_2$, 6 mM HEPES, 3 mM EGTA, and 0.6% (w/v) fatty-acid free BSA (optional). Adjust pH to 7.4 at 37 °C using KOH. Filter sterilize the medium and store it at 4 °C until the day of the assay.
2. ETC Complex Specific Assay Medium (pH 7.4 at 37 °C): 100 mL of 1X MAS-based assay medium containing 1 mM Malate, 10 mM Pyruvate, 4 mM ADP and 4 μM FCCP. Adjust pH to 7.4 at 37 °C using KOH (not NaOH, *see* **Note 5**).
3. Ascorbate/TMPD/Hoechst 33342; 100 mM ascorbate, 1 mM TMPD and 16.4 μM Hoechst 33342 dissolved in 1× MAS. Adjust pH to 7.4 using KOH (not NaOH). Alternatively, ascorbate and TMPD solutions can be prepared separately, but ensure that each solution has pH of 7.4.
4. Rotenone (refer to Subheading 2.3, **item** 7).
5. Antimycin A (refer to Subheading 2.3, **item 8**).
6. XF Plasma Membrane Permeabilization (PMP) agent (Agilent): 10 μM. Keep on ice.

3 Methods

The cell density and medium/inhibitor volumes are optimized for using an XFe96 instrument; however, they can be adapted for an XFe24 instrument. Please *see* Fig. 1 for flowchart of complete assay.

3.1 Day 1: Cell Seeding and Sensor Cartridge Hydration

1. All procedures are performed using sterile techniques in a biosafety cabinet.
2. Ensure good quality hPSC cultures with less than 10% differentiation from either 2-dimensional (2-D) colonies (minimum of 2 × 60 mm plates at 70–80% confluency) or 3-dimensional (3-D) spheroids (250–300 μm in diameter) to seed a complete XFe96 plate. We recommend plating at least 8 wells per cell line or condition.
3. Coat the XFe96 cell culture microplate (Agilent) with a basement membrane using the thin gel method by adding 100 μL Geltrex (Gibco; 0.1 mg/mL) in DMEM/F12 (Gibco) to each well (*see* **Note 6**). Incubate the microplate for 1 h at 37 °C, followed by a minimum of 10 min at room temperature prior to use.
4. Generate a single cell suspension of hPSCs. 3-D hPSC spheroids are collected via centrifugation at 200 × *g* for 3 min at room temperature. Medium is aspirated from either the cell pellet or 2-D hPSC colonies and replaced with 10 mL Accumax containing 220 U of DNase I and 10 μM Y-27632. Incubate at 37 °C for 8 min, then dissociate cells by gently pipetting up and down with a serological pipette. Confirm predominantly single cells under a microscope, then dilute Accumax with an equal volume of 1× DPBS and collect cells by centrifugation at 200 × *g* for 3 min at room temperature. Aspirate supernatant, resuspend cells in mTeSR1 medium including 10 μM Y-27632 inhibitor and then filter cells through a 0.1 μm sterile strainer. Count cells with a hemocytometer or automated cell counter (e.g., Countess II™ (Invitrogen)) and calculate volume of cell suspension required to make 10 mL of 600,000 cells/mL cell suspension in mTeSR1 (to seed a complete XFe96 plate, *see* **Note 7**). Transfer cell suspension to reagent reservoir.
5. Seed XFe96 cell culture plate. Remove remaining Geltrex from the plate (*see* **Note 8**) and immediately use a multichannel pipette to transfer 100 μL of the cell suspension to each well by placing the pipette tip along the side wall to ensure even distribution of cells. Do not seed cells into the four corner wells as these are utilized for background correction. Let cells incubate for 10 min at room temperature and then transfer to a cell culture incubator.

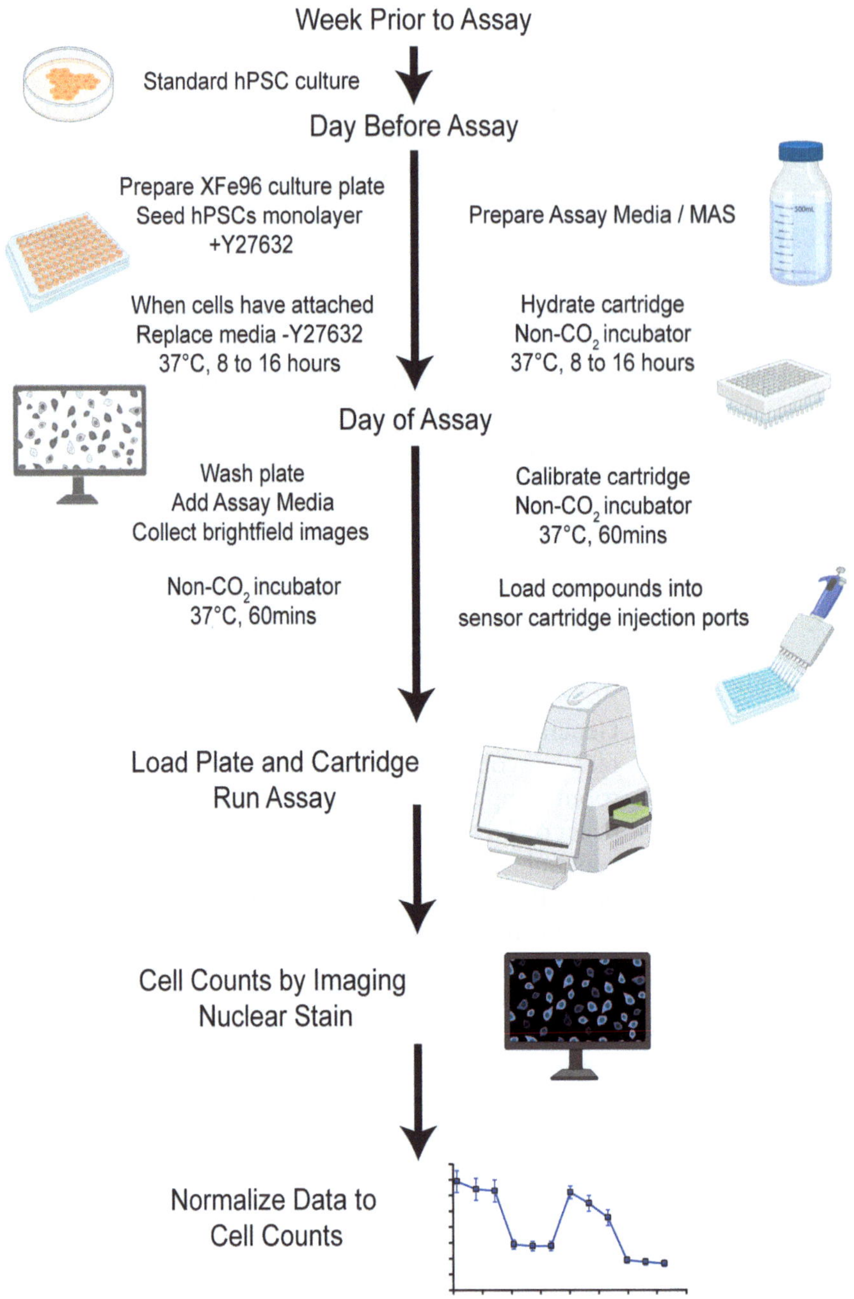

Fig. 1 Flowchart for performing Seahorse extracellular flux analysis using either the Mitochondrial Stress Test or the Electron Transport Chain Complex Specific Assay

6. When cells have adhered to the plate (typically 6 h post-split), medium is removed and replace with 100 μL of mTeSR1 without Y-27632, as it has been previously shown to influence energy metabolism [14].

7. Hydrate sensor cartridge by adding 200 μL of Milli-Q water into each well of the utility plate (*see* **Note 9**) and lowering the sensor cartridge onto the utility plate to submerge the sensors. Incubate the sensor cartridge overnight in a humidified non-CO_2 incubator at 37 °C.

3.2 Day 2: Extracellular Flux Assays

For clarity we have separated protocols for the Mitochondrial Stress Assay (Subheading 3.2.1) and the Electron Transport Chain Complex Specific Assay (Subheading 3.2.2).

3.2.1 Mitochondrial Stress Assay

1. Ensure XFe96/Cytation 1 is powered on and temperature has reached 37 °C (*see* **Note 10**).
2. Warm Mitochondrial Stress Test Medium to 37 °C in water/bead bath and thaw aliquots of oligomycin, FCCP, and rotenone/antimycin A on ice. Transfer thawed aliquots to water/bead bath just prior to loading the sensor cartridge to minimize time at 37 °C.
3. Prepare sensor cartridge by replacing water in utility plate with 200 μL of pre-warmed XF Calibrant and incubate the sensor cartridge in a humidified non-CO_2 incubator at 37 °C for at least 60 min. Following incubation, use a multichannel pipette to load each compound into the respective injection ports in the sensor cartridge (*see* Table 1, **Notes 11** and **12** for specific injection volumes, ports, and recommended concentrations). Compounds or an equivalent volume of assay medium must be added to all injection ports (even if they do not contain cells) otherwise incomplete injections will occur. Return sensor cartridge to non-CO_2 incubator at 37 °C until starting the assay.
4. Prepare cell plate by discarding the growth medium, washing with Mitochondria Stress Test Medium (*see* **Note 13**) and adding 150 μL of pre-warmed MitoStress Test Medium. Immediately proceed to generating brightfield images of the wells (*see* **Note 14**) using the Cell Imaging software on the XFe96. To initiate scan, first scan the barcode on a side of the cell culture microplate and load the plate on the Cytation 1 tray, ensuring correct orientation of the cell plate in the equipment tray. Follow onscreen prompts to assign a file name, select "Brightfield" and then click "Scan all wells." Upon completion of scanning, remove the plate from the instrument and return plate to a 37 °C non-CO_2 incubator for 1 h (*see* **Note 13**). Manually examine the image of each well to ensure proper image acquisition and cell seeding.
5. Set up assay in the Seahorse Wave software by selecting the preexisting "XF Cell Mito Stress Test" template, and assign the following parameters: (a) groups according to cell type/treatment in the "Plate Map" view on the Assay Navigation Panel.

Table 1
Injection port setup and recommended doses for Mitochondrial Stress Test

Port	Compound	Biologic target	Injected concentration (μM) [10×]	Injection volume (μL)	Final well concentration (μM) [1×]
A	Oligomycin	Complex V (ATP synthase)	5–15	20	0.5–1.5
B	FCCP	Mitochondrial membrane uncoupler	1.25–20	22	0.125–2
C	Rotenone/AA	Complex I/III	5/10	25	0.5/1

Ensure that four corner wells that do not contain cells are included as Background wells; (b) compounds to their respective injection ports and ensure correct injection order on the "Protocol" view on the Assay Navigation Panel; (c) assay parameters under "Edit Measurement Details" located below each port map. Our protocol is optimized to utilize assay times of 3 min mix, 0 min wait and 3 min measure with 3 replicate measurements following each compound injection, but may differ depending on cell seeding density and oxidative capacity of cell line. Select "Run Assay" on the Navigation Panel and enter all the essential information in the "Note" box.

6. Start assay when the cell plate and sensor cartridge have incubated for a minimum of 60 min in the non-CO_2 incubator by selecting "Start Run" and saving the assay to the designated folder. Remove the lid and place the loaded sensor cartridge onto heatblock on the instrument tray, verify correct plate orientation and stability (*see* **Note 15**), and press "I'm Ready" to load sensor cartridge into instrument. The first ~20 min of the assay is used to calibrate the cartridge sensors. Following successful calibration, the XFe96 will prompt to load the cell plate. Just prior to loading the cell plate, the Mitochondrial Stress Test medium is replaced with 180 μL of fresh pre-warmed assay medium (*see* **Note 13**). Upon changing the medium in the microplate, press "I'm ready" shown on the screen to eject the sensor cartridge utility plate (while keeping the sensor cartridge in the instrument), which is replaced with the cell plate to begin the assay. Completion of the assay is indicated by a message on the Wave software and by the indicator light on the top of the instrument changing color from blue to yellow, whereupon the cell culture plate and sensor cartridge can be removed from the instrument.

7. Normalization can be performed immediately or cells can be fixed in 3% paraformaldehyde to normalize at a later time. While normalization of extracellular flux can be performed in a number of ways (*see* **Note 16**), we utilize cell counts based upon staining of nuclei with Hoechst 33342. We add Hoechst 33342 to the final injection (as it does not impact metabolic function); however, cells can also be stained following the assay. Fluorescent images are collected using a Cytation 1 using the Seahorse/BioTek Cell Imaging software and are linked to the pre-assay brightfield images. When the software is opened, either select the file containing the pre-assay brightfield images OR select "Plate Menu" and "Fluorescent Image." Load the plate on the Cytation 1 tray, ensuring correct orientation in the equipment tray and select "Scan all wells". Upon completion of image acquisition, ensure images were obtained for each well and define whether cells are properly segmented (flag wells where there is poor segmentation or poor cell morphology by clicking on the upper right corner of the individual image window, *see* **Note 5**). If the overall quality of the fluorescent image is satisfactory, the cell plate can be fixed with 3% PFA and stored at 4 °C in case alternative modes of normalization are required. If poor staining has occurred, the cells can be restained to improve the image quality. Cell counts are applied to the assay within the Wave software by selecting "Normalize" and then "import." Once these values have been applied, the OCR values should display as per 1000 cells.
8. Data can be analyzed within the Wave software and exported using Report Generators or manually into Excel to generate figures and perform statistical analysis. Alternatively, data can be uploaded to the cloud-based Agilent Seahorse Analytics (https://seahorseanalytics.agilent.com/), where analysis can be performed and exported directly into Excel or Prism. Individual wells are examined to ensure linearity of changes in pH and O_2 (*see* **Note** 7), that they respond to each injected compound and have normalization data. If these conditions are not met, then individual wells can be obmitted from the analysis by selecting the well within the Wave software. This assay provides quantitative measures of fundamental parameters of mitochondrial function (*see* Fig. 2, **Note 17**):
   (a) *Basal respiration*—steady state OCR of intact cells, calculated as the baseline OCR—non-mitochondrial respiration.
   (b) *ATP-coupled respiration*—OCR that is coupled with ATP synthesis, calculated by the difference between basal respiration and the oligomycin sensitive respiration.

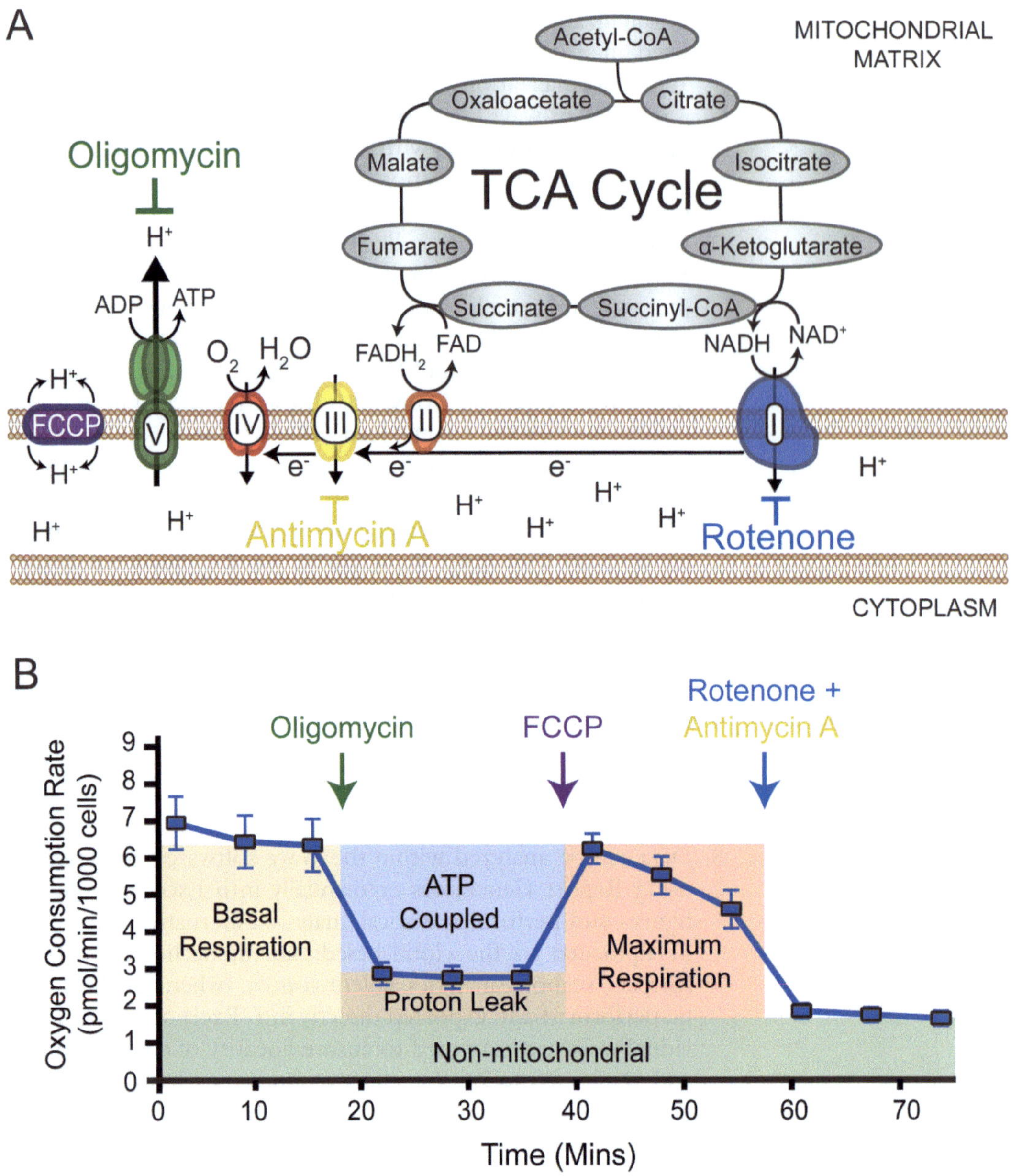

Fig. 2 Mitochondrial Stress Test. (**a**) Scheme of electron transport chain denoting biological targets of injection compounds. (**b**) Representative Mitochondrial Stress Test OCR trace from hPSC highlighting basal, maximal, and ATP-coupled OCRs are calculated. Note, hPSCs typically display very low levels of reserve capacity [17, 18]

(c) *Proton leak*—basal respiration not coupled to ATP production, calculated by the difference between the basal respiration and ATP-coupled respiration.

(d) *Maximal respiration*—maximal uncoupled OCR that indicates the maximum rate of respiration that cells can achieve in the case of increased energy demand.

(e) *Spare respiratory capacity*—capacity to increase OCR in response to uncoupling of the ETC from ATP synthesis (indicates the capacity of the cell to respond to an energetic demand, and how close the respiration of cells is to its theoretical maximum), calculated as the difference between the maximal respiration and basal respiration.

(f) *Non-mitochondrial respiration*—oxygen consumption associated with non-mitochondrial functions following injection of rotenone and antimycin A.

3.2.2 ETC Component Specific Assay

1. Ensure XFe96/Cytation 1 is powered on and temperature has reached 37 °C (*see* **Note 10**).
2. Prepare ETC Component Specific Assay Medium fresh (*see* Subheading 2.4, **item 2**) and warm to 37 °C in water/bead bath. Thaw stock solutions of rotenone and antimycin A and make 10× injection stocks in ETC Component Specific Assay Medium. Prepare fresh stocks of 90 mM succinate and 100 mM ascorbate/1 mM TMPD/16.4 μM Hoechst 33342 in ETC Component Specific Assay Medium. Transfer all compounds to water/bead bath just prior to loading the sensor cartridge to minimize time at 37 °C.
3. Prepare sensor cartridge as described in Subheading 3.2.1, **step 3**; however, load injection ports according to Tables 2 and 3.
4. Collect brightfield images as described in Subheading 3.2.1, **step 4**.
5. Set up assay in the Seahorse Wave software using a blank template (Wave does not come with a preloaded template for this assay, but a template can be created the first time the assay is run) and assign the following parameters: (a) groups according to cell type/treatment in the "Plate Map" view on the Assay Navigation Panel. Ensure that the four corner wells that do not contain cells are included as Background wells; (b) compounds to their respective injection ports (*see* Table 3) and ensure correct injection order on the "Protocol" view on the Assay Navigation Panel; (c) assay parameters under "Edit Measurement Details" located below each port map. Our protocol is optimized to utilize assay times of 0.5 min mix, 0.5 min wait and 2 min measure with 3 replicate measurements following each compound injection (*see* **Note 18**). Select "Run Assay" on the Navigation Panel and enter all the essential information in the "Note" box.
6. Start assay by selecting "Start Run" and saving the assay to the designated folder. Remove the lid and place the loaded sensor cartridge onto heatblock on the instrument tray, verify correct plate orientation and stability (*see* **Note 15**) and press "I'm Ready" to load sensor cartridge into instrument. The first

Table 2
Inhibitors, targets, and substrates required for ETC complex activity assay

Substrates	Final concentrations (mM)	Inhibitors	Final concentrations (μM)	Biologic targets
Pyruvate/ Malate	10/1	Rotenone	2	Complex I
Succinate	10	Antimycin A	2	Complex II/complex III
TMPD/ Ascorbate	0.1/10	Azide	20000	Complex IV

Table 3
Injection port setup for ETC complex activity assay

Port	Compounds	Injected concentration (μM)	Injection volume (μL)	Final well concentration (μM) [1×]
A	Rotenone	18	20	2
B	Succinate	90,000	22	10,000
C	Antimycin A	18	25	2
D	Ascorbate/ TMPD	100,000/1000	25	10,000/100

~20 min of the assay is used to calibrate the cartridge sensors. Following successful calibration, the XFe96 will prompt to load the cell plate. Just prior to loading the cell plate, the medium is removed, the plate is washed twice with ETC Complex Specific Assay Medium as quickly as possible (*see* **Note 19**) and 160 μL of ETC Complex Specific Assay Medium supplemented with 1 nM PMP (keep on ice until used) is added into each well (*see* **Note 20**). Upon changing the medium in the microplate, press "I'm ready" shown on the screen to eject the sensor cartridge utility plate (while keeping the sensor cartridge in the instrument), which is replaced with the cell plate to begin the assay. Completion of the assay is indicated by a message on the Wave software and by the indicator light on the top of the instrument changing color from blue to yellow, whereupon the cell culture plate and sensor cartridge can be removed from the instrument.

7. Normalization is performed as described in Subheading 3.2.1, **step** 7.
8. Initial data can be analyzed within the Wave software or the cloud-based Agilent Seahorse Analytics (https://seahorseanalytics.agilent.com/), to ensure linearity of changes

in O_2 (*see* **Note 7**), that they respond to each injected compound and have normalization data. If these conditions are not met, then individual wells can be obmitted from the analysis by selecting the well within the Wave software. Report generators are not available for this assay, therefore data must be exported into Excel and calculations performed manually. This assay provides a quantitative measure of OCR that is dependent on specific subunits of the electron transport chain (*see* Fig. 3):

(a) *Basal respiration*—steady state OCR of permeabilized cells upon receiving a complex I-linked substrate such as pyruvate.

(b) *Complex I activity*—Rotenone injection inhibits the activity of the complex I and halts NADH-linked respiration. Thus, a decline in the OCR value upon rotenone injection indicates the activity of complex I.

(c) *Complex II/III activity*—Succinate drives respiration through the complex II (succinate dehydrogenase) thus by-passing the complex I inhibition. Antimycin A inhibits complex III and abolishes this rate. Thus, a change in the OCR value between succinate and antimycin A injection will elucidate respiration through complexes II and III.

(d) *Complex IV*—The injection of ascorbate with TMPD bypasses the block at complex III and delivers electrons directly to cytochrome C oxidase. Thus, OCR value resulting from this injection reveals complex IV activity.

4 Notes

1. The assay medium must have a pH of 7.4 at 37 °C as this impacts the algorithm to calculate OCR. We recommend confirming the pH of 7.4 prior to each assay and routinely add an additional medium change just prior to running the assay (*see* **Note 11**).
2. Simultaneous injection of rotenone and antimycin A is utilized to inhibit the ETC by acting on complex I and III, respectively.
3. Hoechst 33342 is light sensitive so it should be protected from light and loaded as quickly as possible.
4. The optimal staining concentration of Hoechst 33342 is dependent on cell type and seeding density. Thus, it is highly recommended to optimize the concentration of Hoechst 33342 stain, which will in turn impact data normalization. Weak staining intensity can be rectified by fixing the cells with 3% PFA followed by re-staining with Hoechst. For the detailed instruction on staining optimization, please refer to "*Hoechst*

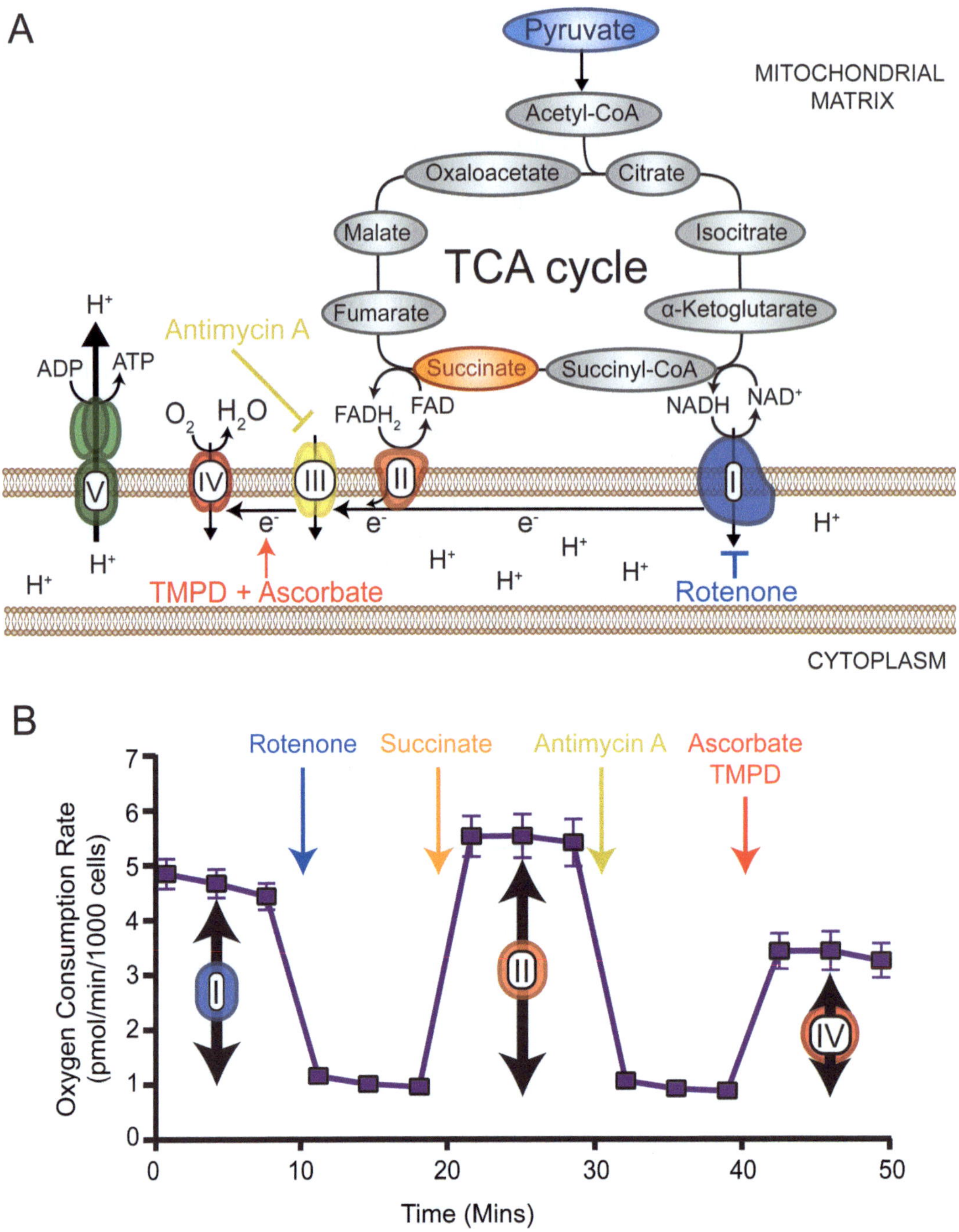

Fig. 3 Electron Transport Chain Complex Specific Assay. (**a**) Scheme of the electron transport chain indicating how substrates feed specific complexes and the biologic targets of compounds. (**b**) Representative Electron Transport Chain Complex Specific Assay trace from human pluripotent stem cells highlighting the contribution of complex I, II, and IV to overall mitochondrial oxygen consumption

33342 Staining Optimization Guidelines" (www.agilent.com/cs/library/troubleshootingguide/public/troubleshootingguide-hoechst-guidelines-XFe96-cell-analysis-5994-0628en-agilent.pdf.pdf).

5. The pH of assay medium is adjusted with KOH, not NaOH, as Na+ leads to mitochondrial swelling and loss of mitochondrial membrane integrity [15, 16].
6. Geltrex is temperature sensitive and solidifies as the temperature rises, thus it is critical to maintain Geltrex at 4 °C on ice or in a cold block and to dilute it with cold medium.
7. We routinely utilize 60,000 cells per well for extracellular flux analysis across many human pluripotent stem cell lines (baseline OCR readings between 100 and 400 pmol/min). Under some cases, new cell lines or medium conditions produce OCR slopes that are not linear (which can be observed within the Wave software by selecting "Well" and "Level" to examine the slope of oxygen utilization and to ensure that oxygen is not completely utilized within the microwell during the measurement period and returns to baseline during the mix period). Under these circumstances it is recommended to empirically define the seeding density (30,000–90,000 cells per well) of the particular cell line required to produce basal OCR levels in the linear detection range.
8. Multiple methods can be used to remove cell culture medium. We find that removal of medium by inverting the plate onto a sterile WYPALL quarter-fold wiper is time efficient, and works well if cells are well adhered to the plate. For sensitive cells or those not well adhered, medium can be carefully removed with a either multichannel pipette or vacuum aspirator. In all cases, the wells should be washed with the assay medium to ensure the removal of all culture medium before the addition of the final assay medium.
9. It is recommended to use sterile water for the initial overnight hydration of the sensor cartridge in place of XF calibrant due to the evaporation that can occur overnight (this can be mitigated by humidifying the non-CO_2 incubation). It is recommended to utilize the hydrated sensor cartridge within 48 h of hydration but we have successfully utilized sensor cartridges up to a week following hydration if the sensors remain submerged in water or XF calibrant.
10. Metabolic rates are highly dependent on temperature, therefore it is critical that the XF instrument is stable at 37 °C. To ensure temperature stability, we typically keep our instrument at 37 °C unless we are not running assays for a number of days.
11. Optimal concentrations of oligomycin and FCCP have to be empirically defined as it is dependent on cell line and seeding

Table 4
Example oligomycin FCCP matrix

A

Oligomycin concentrations for port A		FCCP concentrations for port B: 0 μM		0.125 μM		0.25 μM		0.50 μM		1.0 μM		2.0 μM	
		1	2	3	4	5	6	7	8	9	10	11	12
0 μM	A												
	B												
0.5 μM	C												
	D												
1.0 μM	E												
	F												
2.0 μM	G												
	H												

density. In addition, the oligomycin concentration used also impacts the FCCP concentration required for maximal OCR. To address this we typically run a matrix of oligomycin vs FCCP concentrations as shown in Table 4. Once optimized concentrations are defined, we make 10× stocks, which can be aliquoted and stored at −80 °C.

12. While the sensor cartridge is supplied with a loading guide, we find loading more consistent without them. To ensure that no bubbles form and that inhibitors are not pushed through the injection port, we slowly dispense the inhibitors along the side of the injection port.
13. Cell culture medium contains many components that can interfere with extracellular flux analysis (such as FBS and buffers), therefore it is essential to completely remove the cell culture medium. To ensure this, we wash the cell culture microplates with XF assay medium and perform two medium changes, one prior to the hour incubation in the non-CO_2 incubator and another just prior to running this assay. Incubation in a non-CO_2 incubator allows CO_2 diffusion from cells and medium, thus eliminating confounding CO_2 effects on medium pH. As CO_2 diffused from the cells can acidify the surrounding medium, we perform an additional medium change just prior to the assay.
14. Brightfield images obtained prior to the assay can be utilized to detect wells which have low or non-homogenous cell seeding and wells where cells become detached during the course of the assay. These wells can then be excluded from subsequent OCR calculations.

15. To ensure proper loading of the sensor cartridge, the blue triangle on the plate must be on the bottom left corner of the plate and the plate must be secured on the heat sink.
16. Post-assay normalization of XF data is critical in order to compare metabolic rates across cell lines, medium conditions, and treatment groups. A number of approaches have been utilized for normalization, including quantifying levels of genomic DNA and total cellular protein; however, these approaches are time consuming, can introduce error by transferring samples to other plates and can be incompatible with the basement membranes utilized for coating the wells. We have adopted a method where Hoechst 33342 is added directly to the final injection of the assay, so that immediately following the completion of the assay the cell plate can be transferred to and imaged with a Cytation 1 to generate a total cell count per well.
17. When running these assays we ensure that the instrument is collecting data on both OCR and extracellular acidification rates (ECAR), which is a surrogate marker of glycolytic flux. This proves a relative rate of glycolytic/oxidative function of cells.
18. The equilibration step is eliminated from the protocol to reduce assay duration to as short as possible to avoid cells lifting from the plate. If cells are lifting from the plate, replicate measurements can be reduced from 3 to 1 to minimize the time that cells are in contact with PMP.
19. Perform the wash step as quickly (but gently) as possible to minimize the cell-exposure time to the 1× MAS buffer that does not contain the complete components required for maintaining cells in an optimal metabolic state.
20. PMP is added immediately before transferring the plate into the XFe96 as it permeabilizes the outer cell membrane and prolonged exposure leads to cell death.

References

1. Tsogtbaatar E, Landin C, Minter-Dykhouse K, Folmes CDL (2020) Energy metabolism regulates stem cell pluripotency. Front Cell Dev Biol 8:87. https://doi.org/10.3389/fcell.2020.00087
2. Khacho M, Harris R, Slack RS (2019) Mitochondria as central regulators of neural stem cell fate and cognitive function. Nat Rev Neurosci 20(1):34–48. https://doi.org/10.1038/s41583-018-0091-3
3. Zhang H, Menzies KJ, Auwerx J (2018) The role of mitochondria in stem cell fate and aging. Development 145(8). https://doi.org/10.1242/dev.143420
4. Ahlqvist KJ, Suomalainen A, Hamalainen RH (2015) Stem cells, mitochondria and aging. Biochim Biophys Acta 1847(11):1380–1386. https://doi.org/10.1016/j.bbabio.2015.05.014
5. Murphy MP, Hartley RC (2018) Mitochondria as a therapeutic target for common pathologies. Nat Rev Drug Discov 17(12):865–886. https://doi.org/10.1038/nrd.2018.174

6. Perry CG, Kane DA, Lanza IR, Neufer PD (2013) Methods for assessing mitochondrial function in diabetes. Diabetes 62(4): 1041–1053. https://doi.org/10.2337/db12-1219
7. Marín-García J (2013) Methods to study mitochondrial structure and function. In: Mitochondria and their role in cardiovascular disease. Springer US, Boston, MA, pp 13–27. https://doi.org/10.1007/978-1-4614-4599-9_2
8. Li Z, Graham BH (2012) Measurement of mitochondrial oxygen consumption using a Clark electrode. Methods Mol Biol 837: 63–72. https://doi.org/10.1007/978-1-61779-504-6_5
9. Silva AM, Oliveira PJ (2012) Evaluation of respiration with Clark type electrode in isolated mitochondria and permeabilized animal cells. Methods Mol Biol 810:7–24. https://doi.org/10.1007/978-1-61779-382-0_2
10. Divakaruni AS, Rogers GW, Murphy AN (2014) Measuring mitochondrial function in Permeabilized cells using the seahorse XF analyzer or a Clark-type oxygen electrode. Curr Protoc Toxicol 60:25 22 21-16. https://doi.org/10.1002/0471140856.tx2502s60
11. Wu M, Neilson A, Swift AL, Moran R, Tamagnine J, Parslow D, Armistead S, Lemire K, Orrell J, Teich J, Chomicz S, Ferrick DA (2007) Multiparameter metabolic analysis reveals a close link between attenuated mitochondrial bioenergetic function and enhanced glycolysis dependency in human tumor cells. Am J Physiol Cell Physiol 292(1): C125–C136. https://doi.org/10.1152/ajpcell.00247.2006
12. Divakaruni AS, Paradyse A, Ferrick DA, Murphy AN, Jastroch M (2014) Analysis and interpretation of microplate-based oxygen consumption and pH data. Methods Enzymol 547:309–354. https://doi.org/10.1016/B978-0-12-801415-8.00016-3
13. Gerencser AA, Neilson A, Choi SW, Edman U, Yadava N, Oh RJ, Ferrick DA, Nicholls DG, Brand MD (2009) Quantitative microplate-based respirometry with correction for oxygen diffusion. Anal Chem 81(16):6868–6878. https://doi.org/10.1021/ac900881z
14. Vernardis SI, Terzoudis K, Panoskaltsis N, Mantalaris A (2017) Human embryonic and induced pluripotent stem cells maintain phenotype but alter their metabolism after exposure to ROCK inhibitor. Sci Rep 7:42138. https://doi.org/10.1038/srep42138
15. Salabei JK, Gibb AA, Hill BG (2014) Comprehensive measurement of respiratory activity in permeabilized cells using extracellular flux analysis. Nat Protoc 9(2):421–438. https://doi.org/10.1038/nprot.2014.018
16. Javadov S, Chapa-Dubocq X, Makarov V (2018) Different approaches to modeling analysis of mitochondrial swelling. Mitochondrion 38:58–70. https://doi.org/10.1016/j.mito.2017.08.004
17. Folmes CD, Nelson TJ, Martinez-Fernandez A, Arrell DK, Lindor JZ, Dzeja PP, Ikeda Y, Perez-Terzic C, Terzic A (2011) Somatic oxidative bioenergetics transitions into pluripotency-dependent glycolysis to facilitate nuclear reprogramming. Cell Metab 14(2): 264–271. https://doi.org/10.1016/j.cmet.2011.06.011
18. Folmes CD, Arrell DK, Zlatkovic-Lindor J, Martinez-Fernandez A, Perez-Terzic C, Nelson TJ, Terzic A (2013) Metabolome and metaboproteome remodeling in nuclear reprogramming. Cell Cycle 12(15):2355–2365. https://doi.org/10.4161/cc.25509

Chapter 8

Assessment of Endothelial-to-Hematopoietic Transition of Individual Hemogenic Endothelium and Bulk Populations in Defined Conditions

Gene I. Uenishi, Ho Sun Jung, and Igor I. Slukvin

Abstract

Endothelial-to-hematopoietic transition (EHT) is a unique morphogenic event in which flat, adherent hemogenic endothelial (HE) cells acquire round, non-adherent blood cell morphology. Investigating the mechanisms of EHT is critical for understanding the development of hematopoietic stem cells (HSCs) and the entirety of the adult immune system, and advancing technologies for manufacturing blood cells from human pluripotent stem cells (hPSCs). Here we describe a protocol to (a) generate and isolate subsets of HE from hPSCs, (b) assess EHT and hematopoietic potential of HE subsets in bulk cultures and at the single-cell level, and (c) evaluate the role of NOTCH signaling during HE specification and EHT. The generation of HE from hPSCs and EHT bulk cultures are performed in xenogen- and feeder-free system, providing the unique advantage of being able to investigate the role of individual signaling factors during EHT and the definitive lympho-myeloid cell specification from hPSCs.

Key words Endothelial-to-hematopoietic transition, Hemogenic endothelium, Hematopoiesis, Human pluripotent stem cells, Notch signaling

1 Introduction

During development, hematopoietic cells arise from endothelial cells lining embryonic vasculature via endothelial-to-hematopoietic transition (EHT). Initially, the concept of EHT was established based on the observation of hematopoietic stem cell (HSC) formation from specialized subsets of endothelium (hemogenic endothelium; HE) in the ventral floor of dorsal aorta in the embryo. Later studies demonstrated that earlier, more primitive waves of hematopoiesis that emerge in the yolk sac and extra-embryonic vessels are also generated through EHT (reviewed in [1, 2]). Human pluripotent stem cells (hPSCs) opened a unique opportunity to investigate the earliest stages of hematopoietic development, including the process of EHT, that was previously inaccessible. In the last two decades, significant progress has been made in defining cellular

Nagarajan Kannan and Philip Beer (eds.), *Stem Cell Assays: Methods and Protocols*, Methods in Molecular Biology, vol. 2429,
https://doi.org/10.1007/978-1-0716-1979-7_8,

pathways of hematopoietic development from hPSCs. The discovery of CD43 as a marker for early blood cells [3] and CD73 as a marker for non-HE [4] allowed for the identification and separation of HE from non-HE, and HE from blood cells based on $CD144^{+}CD34^{+}CD43^{-}CD73^{-}$ phenotype [4, 5]. Recently, we distinguished two discrete subsets within the HE population: $DLL4^{+}CXCR4^{+/-}NOTCH^{hi}$ and $DLL4^{-}CXCR4^{-}NOTCH^{lo}$. Because $DLL4^{+}CXCR4^{+/-}NOTCH^{hi}$ HE have molecular signatures indicative of arterial endothelium, they were defined as arterial-type HE (AHE). AHE gave rise to hematopoietic progenitors with significantly increased T-lymphoid potential, as well as erythrocytes expressing increased levels of adult globins. AHE required stroma and was dependent on NOTCH signaling to undergo EHT and give rise to hematopoietic progenitors. In contrast, the population of $DLL4^{-}CXCR4^{-}NOTCH^{lo}$ HE (non-AHE) did not have molecular signatures of arterial endothelium, and proceed through EHT independent of NOTCH activation. Hematopoietic progeny of non-AHE closely resemble the primitive wave of hematopoiesis and have no lymphoid potential [6]. AHE specification itself was NOTCH signaling-dependent from an earlier, more immature or primordial HE (PHE) population which emerge about 24 h before AHE in hPSC culture. These findings provide the capability to distinguish specific stages and different subsets of HE and their corresponding (1) primitive, myeloid-restricted and (2) definitive, lympho-myeloid waves of hematopoietic progeny in hPSC cultures.

Here, we describe the platforms that were developed to analyze EHT and interrogate the role of NOTCH signaling during HE specification and EHT. This protocol will cover (1) culture and maintenance of hPSCs on vitronectin in E8 media, which have shown improved differentiation in our system, (2) xenogen- and feeder-free differentiation of hPSCs into PHE (Days −1 to 4), (3) magnetic purification of PHE on Day 4 and continued xenogen-/feeder-free culture through EHT and differentiation into hematopoietic progenitors, (4) single-cell EHT assay for PHE to assess endothelial versus hematopoietic potential, and (5) FACSorting of $CD144^{+}CD43^{-}CD73^{-}DLL4^{+}CXCR4^{+/-}NOTCH^{hi}$ AHE and $CD144^{+}CD43^{-}CD73^{-}DLL4^{-}CXCR4^{-}NOTCH^{lo}$ non-AHE for bulk culture on OP9 or OP9-DLL4 feeders.

While other hPSC to HE differentiation and EHT evaluation protocols exist, the major advantages of the protocol described here are that (1) it is serum- and albumin-free, which abrogates the question of unknown and variable components of serum and purified albumin, and (2) it is 2-dimensional, which minimizes the inherent variability of embryoid body (EB) size, composition, and asynchronous differentiation [7, 8]. Compared to 3-dimensional EB-based protocols, our protocol generates HE faster and allows for efficient dissociation and preparation into single-cell suspension for cell sorting. Subsequent cultures of isolated HE in bulk or single-cell cultures allow for directly observing EHT and assessing the effect of individual signaling factors on EHT. Although this protocol as presented can produce HE with T-lymphoid potential at high frequency, it still lacks the ability to generate HE with long-term HSC engraftment potential but provides the platform to identify the required factors for it [9, 10].

Overall, this 2D xenogen- and feeder-free protocol has the unique advantage of being able to investigate the specific stage- and dose-dependent role of signaling factors such as NOTCH, Hedgehog, Hippo, hormone, and inflammatory signaling pathways during HE specification, EHT, definitive lympho-myeloid hematopoiesis, and HSC development.

2 Materials

2.1 Supplies/Reagents for hPSC Maintenance

1. Vitronectin-coated plates: Coat 6-well plates with 0.5 μg/cm^2 of vitronectin diluted in phosphate-buffered saline (PBS). Store at 4 °C for up to 1 week.
2. Complete Essential 8 Medium. Store at 4 °C for up to 1 month.
3. PBS-EDTA: Prepare 0.5 mM EDTA in PBS without Ca or Mg. Store at room temperature.
4. TrypLE Express.
5. Y-27632. Prepare a 10 mM stock concentration.

2.2 Supplies/Reagents for hPSC Differentiation into PHE

1. Collagen IV-coated plates: Coat 6-well plates with 0.5 μg/cm^2 of Collagen IV diluted in PBS. Store at 4 °C for up to 1 week.
2. Tenascin C-coated plates: Coat 6-well plates with 0.5 μg/cm^2 of Tenascin C diluted in PBS. Store at 4 °C for up to 1 week (*see* **Note 1**).
3. FGF2: Prepare a 100 μg/mL stock solution in 5 mM Tris. Aliquot and store at −80 °C for up to 1 year.
4. BMP4: Prepare a 100 μg/mL stock solution in 5 mM HCl. Aliquot and store at −80 °C for up to 3 months (*see* **Note 2**).
5. Activin A: Prepare a 10 μg/mL stock solution in ddH_2O. Aliquot and store at −80 °C for up to 1 year.
6. VEGF: Prepare a 100 μg/mL stock solution in ddH_2O. Aliquot and store at −80 °C for up to 1 year.
7. SB-431542: Prepare a 10 mM stock solution in 100% ethanol. Aliquot and store at −80 °C for up to 1 year.
8. Lithium Chloride: Prepare 2 M stock solution in ddH_2O. Store at 4 °C for up to 1 year.

9. Sodium Selenite: Prepare a 0.7 mg/mL stock solution in ddH_2O. Store at 4 °C for up to 1 year.
10. Holo-Transferrin: Prepare a 10.6 mg/mL stock solution in ddH_2O. Aliquot and store at −80 °C for up to 1 year.
11. Insulin.
12. Polyvinyl Alcohol: Prepare a 20 g/L stock solution in ddH_2O. Store at 4 °C for up to 1 year (*see* **Note 3**).
13. IF6S Differentiation Medium (10×): Weigh 21 g of sodium bicarbonate and 640 mg of L-ascorbic acid 2-phosphate Mg^{2+} salt, and combine with 5 packets of 1 L powdered IMDM and 5 packets of 1 L powdered F-12, and bring up to 500 mL ddH_2O. Add 400 μL of monothioglycerol, 120 μL of the sodium selenite solution, 100 mL of Glutamax, 100 mL of Non-Essential Amino Acids, and 20 mL of Chemically Defined Lipid Concentration. Bring up to 1 L with ddH_2O, mix, aliquot in 50 mL conical tubes, and store at −80 °C for up to 6 months.
14. IF9S Differentiation Medium (1×): Thaw 10× IF6S aliquot and dilute with 200 mL ddH_2O. Add 500 μL stock solution of Holo-Transferrin and 10 mg of insulin and sterile filter through a 0.22 μm filter. Add 250 mL of stock solution of polyvinyl alcohol. Store at 4 °C.
15. Primitive Mesoderm Induction Media: 1× IF9S Differentiation Media with 50 ng/mL of BMP4, 50 ng/mL of FGF2, 15 ng/mL of Activin A, 2 mM of LiCl, and 1 μM of Y-27632. Use immediately (*see* **Note 4**).
16. Hematovascular Mesoderm and Primordial Hemogenic Endothelium Induction Media: 1× IF9S Differentiation Media with 50 ng/mL of FGF2, 50 ng/mL of VEGF, 10 μM of SB-431542. Use immediately.

2.3 Supplies/Reagents to Continue hPSC Differentiation into HSPC

1. SCF: Prepare a 100 μg/mL stock solution in ddH_2O. Aliquot and store at −80 °C for up to 1 year.
2. TPO: Prepare a 100 μg/mL stock solution in ddH_2O. Aliquot and store at −80 °C for up to 1 year.
3. IL-3: Prepare a 10 μg/mL stock solution in ddH_2O. Aliquot and store at −80 °C for up to 1 year.
4. IL-6: Prepare a 100 μg/mL stock solution in ddH_2O. Aliquot and store at −80 °C for up to 1 year.
5. FLT3L: Prepare a 100 μg/mL stock solution in ddH_2O. Aliquot and store at −80 °C for up to 1 year.
6. HE and Hematopoiesis Media: 1× IF9S Differentiation Media with 50 ng/mL of FGF2, 50 ng/mL of VEGF, 50 ng/mL of TPO, 20 ng/mL of SCF, 10 ng/mL of IL-3, and 50 ng/mL of IL-6. Use immediately.

2.4 Supplies/Reagents for Bulk Population EHT Cultures

1. MACS Buffer.
2. MACS LS Columns.
3. 70 μm cell strainer.
4. LS magnet and stand.
5. Anti-CD31 magnetic beads.
6. EGF: Prepare a 100 μg/mL stock solution in ddH_2O. Aliquot and store at −80 °C for up to 1 year.
7. IGF-I: Prepare a 100 μg/mL stock solution in ddH_2O. Aliquot and store at −80 °C for up to 1 year.
8. IGF-II: Prepare a 100 μg/mL stock solution in ddH_2O. Aliquot and store at −80 °C for up to 1 year.
9. Delta Like Ligand 1 conjugated to an—Fc fragment (DLL1-Fc): Prepare a 1 mg/mL stock solution in ddH_2O. Aliquot and store at −80 °C for up to 1 year.
10. DAPT ((2S)-*N*-[(3,5-Difluorophenyl)acetyl]-L-alanyl-2-phenyl]glycine 1,1-dimethylethyl ester), γ-secretase inhibitor: Prepare a 10 mM stock solution in DMSO. Aliquot and store at −80 °C for up to 1 year.
11. DLL1-Fc-coated plates: Coat 6-well plates with 0.5 μg/cm^2 of Collagen IV with 0.5 μg/cm^2 of DLL1-Fc diluted in PBS. Store at 4 °C for up to 1 week.
12. EHT and Hematopoiesis Plating Media: 1× IF9S Differentiation Media with 50 ng/mL of FGF2, 50 ng/mL of VEGF, 50 ng/mL of TPO, 20 ng/mL of SCF, 10 ng/mL of IL-3, 50 ng/mL of IL-6, 50 ng/mL of EGF, 50 ng/mL of IGF-I, 50 ng/mL of IGF-II, and 5 μM of Y-27632. Use immediately.
13. EHT and Hematopoiesis Support Media: 1× IF9S Differentiation Media with 50 ng/mL of FGF2, 50 ng/mL of VEGF, 50 ng/mL of TPO, 20 ng/mL of SCF, 10 ng/mL of IL-3, 50 ng/mL of IL-6, 50 ng/mL of EGF, 50 ng/mL of IGF-I, and 50 ng/mL of IGF-II. Use immediately.

2.5 Supplies/Reagents for Single-Cell EHT Cultures

1. Mitomycin C: Prepare a 10 mg/mL stock solution in ddH_2O. Aliquot and store at −80 °C for up to 1 year.
2. Doxycycline: Prepare a 1 mg/mL stock solution in ddH_2O. Aliquot and store at −80 °C for up to 1 year.
3. Gelatin-coated plates: Coat 96-well plates with gelatin by adding 100 μL of 0.1% gelatin solution. Store at 4 °C for up to 1 year.
4. OP9 Maintenance Media: 80% α-MEM and 20% FBS. Store at 4 °C for up to 1 month.
5. OP9-iDLL4 seeded 96-well plates for single-cell deposition assay: 24 h before use, treat confluent OP9-iDLL4 with

Mitomycin C for 2 h at 10 μg/mL. Wash cells three times with OP9 maintenance media before washing once with EDTA-PBS. Singularize OP9-iDLL4 with TrypLE Select, collect cells into a 15 mL or 50 mL conical tube, and add OP9 maintenance media up to maximum volume of vessel. Centrifuge at 350 × *g* for 5 min, aspirate supernatant, resuspend in 5 mL OP9 maintenance media, count cells on a hemocytometer, and plate cells at a density of 12,500 cells/cm^2, or 4000 cells/well of a GTN-coated 96-well plate. For NOTCH activation conditions, add 1 μg/mL final concentration of Dox during plating. Place plates into normoxic incubator and allow cells to reattach overnight.

6. OP9 Differentiation Media: 90% α-MEM and 10% FBS with 100 μM of monothioglycerol, 50 ng/mL of TPO, 50 ng/mL of SCF, 10 ng/mL of IL-3, 50 ng/mL of IL-6, and 10 ng/mL of FLT3L. Use immediately.

2.6 Supplies/Reagents for AHE-Specific Cultures

1. Gelatin-coated plates: Coat 6-well plates with gelatin by adding 1 mL of 0.1% gelatin solution. Store at 4 °C for up to 1 month.
2. OP9/OP9-DLL4 seeded 6-well or 10 cm plates for D5 AHE EHT assay: 24 h before use. Maintain OP9-iDLL4 as previously described. Twenty four hours before use, treat confluent OP9/OP9-DLL4 with Mitomycin C for 2 h at 10 μg/mL. Wash cells three times with OP9 maintenance media before washing once with EDTA-PBS. Singularize OP9/OP9-DLL4 with TrypLE Select, collect cells into a 15 mL or 50 mL conical tube, and add OP9 maintenance media up to maximum volume of vessel. Centrifuge at 350 × *g* for 5 min, aspirate supernatant, resuspend in 5 mL OP9 maintenance media, count cells on a hemocytometer, and plate cells at a density of 12,500 cells/cm^2, or 125,000 cells/well of a 6-well GTN-coated plate, 1,000,000 cells/dish of a 10 cm GTN-coated plate. Place plates into normoxic incubator and allow cells to reattach overnight.

2.7 Supplies/Reagents for Measuring EHT and Hematopoiesis

1. Permeabilization Buffer for Immunofluorescence: PBS without Ca and Mg with 0.2% Triton X-100. Store at 25 °C up to 1 year.
2. Wash Buffer for Immunofluorescence: PBS without Ca and Mg with 0.1% Tween-20. Store at 25 °C up to 1 year.
3. Fixation Buffer for Immunofluorescence: PBS without Ca and Mg with 4% paraformaldehyde. Use immediately.
4. Blocking Buffer for Immunofluorescence: PBS without Ca and Mg with 0.1% Tween-20 and 5% normal goat serum. Use immediately.

2.8 General Supplies/Equipment

1. Incubator capable of sustaining 5% O_2 and 5% CO_2.
2. Centrifuge.
3. MACSmix tube rotator.
4. Flow Cytometer.
5. Fluorescence Microscope.
6. Fluorescence Activated Cell Sorter.
7. *See* Table 1 for a list of antibodies and reagents used for flow cytometry, magnetic enrichment, and immunofluorescence.

3 Methods

3.1 Feeder-Free Maintenance of hPSCs on Vitronectin in E8 Media

1. Thaw cryovial of hPSC in 37 °C water bath for 3 min.
2. Transfer to 15 mL conical tube, then add 10× volume of thawed cells, dropwise.
3. Spin cells for 5 min at 350 × *g*. During the spin, prepare 12 mL of warm E8 and add 12 μL of stock Rock inhibitor to make a 1× E8 with 10 μM Rock inhibitor mixture.
4. Aspirate supernatant without disturbing the cells, then resuspend hPSC colonies in prepared 12 mL of E8 with Rock inhibitor.
5. Add 2 mL of E8 containing hPSC colonies into each well of a VTN-coated 6-well plate.
6. Allow cells to attach overnight (*see* **Note 5**).
7. Replace spent E8 media with fresh, pre-warm E8 media daily.
8. Grow cells to 80% confluency, then passage cells using EDTA-PBS (*see* **Note 6**).
9. Aspirate spent E8 and wash once with room temperature EDTA-PBS by adding 1 mL to each well and aspirate immediately.
10. Add fresh room temperature EDTA-PBS and incubate at room temperature for 3–5 min. During this time, prepare another VTN-coated plate by aspirating VTN-coating solution and adding 1 mL fresh, pre-warmed E8.
11. Carefully aspirate EDTA-PBS (*see* **Note** 7).
12. Suspend cells by quickly and firmly pipetting 6 mL of fresh pre-warmed E8 directly at the bottom of the well (*see* **Note 8**).
13. Add 1 mL resuspended hPSC colonies into each well of the prepared VTN-coated 6-well plate, place back into normoxic incubator, and shake plate to distribute cells evenly.
14. Repeat every 3–5 days when cells are at 80% confluency again.

Table 1
List of antibodies used for flow cytometry, magnetic enrichment, and immunofluorescence

	Antigen	Conjugate	Clone	Vendor	Cat. No.
PHE Magnetic Enrichment	CD31	MicroBead	N/A	Miltenyi Biotec	130-091-935
PHE Antibody Cocktail	CD309/KDR	PE	89106	BD Biosciences	560494
	CD144	FITC	55-7H1	BD Biosciences	560411
	NOTCH1	APC	527425	R&D Systems	FAB5317A
	DLL4	PE-Vio770	MHD4-46	Miltenyi Biotec	130-101-563
	Live/Dead	Ghost Red 780		Tonbo Biosciences	13-0865
AHE Antibody Cocktail	CD144	FITC	55-7H1	BD Biosciences	560411
	CD73	BV421	AD2	BD Biosciences	562430
	CD43	BV510	1G10	BD Biosciences	563377
	DLL4	PE	MHD4-46	Miltenyi Biotec	130-096-567
	NOTCH1	APC	527425	R&D Systems	FAB5317A
	CXCR4	PerCP-Cy5.5	12G5	BD Biosciences	560670
	Live/Dead	Ghost Red 780		Tonbo Biosciences	13-0865
EHT Single Cell Assay	CD144	Polyclonal	Polyclonal	Invitrogen	36–1900
(Primary)	CD43	Purified	1G10	BD Biosciences	551457
	Anti-Mouse	AlexaFluor488	Polyclonal	Life Technologies	A11001
	Anti-Rabbit	AlexaFluor594	Polyclonal	Life Technologies	A11012
EHT Single Cell Assay	CD144	FITC	55-7H1	BD Biosciences	560411
(Alternative)	CD43	BV510	1G10	BD Biosciences	563377
	Live/Dead	Ghost Red 780		Tonbo Biosciences	13-0865
HP Antibody Cocktail	CD34	FITC	581	BD Biosciences	555821
	CD43	BV510	1G10	BD Biosciences	563377
	CD45	APC	HI30	BD Biosciences	555485
	CD235a	PE	GA-R2	BD Biosciences	555570
	CD41a	PE	HIP8	BD Biosciences	555467
	CD144	PE-Vio770	REA199	Miltenyi Biotec	130-100-720
	Live/Dead	Ghost Red 780		Tonbo Biosciences	13-0865

3.2 Preparation of hPSC Differentiation on TenC- or ColIV-Coated Plates (Day −1)

1. When hPSCs are at 50–80% confluency, prepare cells for singularization. Aspirate spent media and wash once with room temperature EDTA-PBS.
2. Add 1 mL room temperature TrypLE Select and incubate for 5 min.
3. Transfer cells in 1 mL TrypLE Select into a 15 mL conical tube with 9 mL E8 and centrifuge 350 × *g* for 5 min.
4. During the spin, mix in a 15 mL conical tube 12 μL stock Rock Inhibitor with 6 mL fresh pre-warmed E8 media to make a 2× E8+Rock inhibitor mixture. Aspirate TenC-coating solution from each well of a TenC-coated 6-well or 10 cm dish and add 1 mL of 2× E8+Rock inhibitor each well, or all 6 mL into the 10 cm dish.
5. Once the singularized cells are done spinning, aspirate the supernatant and resuspend hPSC pellet in 1 mL fresh E8 media.
6. Use a hemocytometer to determine the concentration of cells.
7. Transfer optimized number of cells to new 15 mL conical tube and add E8 up to 6 mL (*see* **Note 9**).
8. Add 1 mL of E8 containing hPSC into each well of ColIV or TenC-coated plates or all 6 mL of cells into 10 cm dish and place back in normoxic incubator overnight to allow cells to reattach (*see* **Note 5**).

3.3 Differentiation into Primordial Hemogenic Endothelium (PHE)

To induce the hematoendothelial program, undifferentiated hPSCs are cultured on collagen IV-coated plates in serum- and xenogeny-free conditions in the presence of BMP4, Activin A, and FGF2 for 2 days, followed by FGF2 and VEGF for another 2 days (Fig. 1a). The first $KDR^{hi}CD144^{+}CD31^{+}CD34^{+}$ PHE in this culture system appear on day 4 of differentiation. These PHE still express high levels of mesodermal gene HAND1 but lack arterial (DLL4, CXCR4), venous (NR2F2), hematopoietic ($CD43^{-}CD235a/41a^{-}$), and CD73 cell markers.

1. Differentiation Day 0: Initiate Primitive Mesoderm specification. Prepare fresh, pre-warmed Primitive Mesoderm Induction media. Retrieve cells plated 24 h prior on TenC- or ColIV-coated plate from normoxic incubator. Gently aspirate spent E8 media and replace with 2 mL prepared Primitive Mesoderm Induction media to each well (*see* **Note 10**). Place cells in hypoxic (5% CO_2, 5% O_2) incubator for 48 h (*see* **Note 11**).
2. Differentiation Day 2: Maturation into Hematovascular Mesoderm Precursors and PHE. Prepare fresh, pre-warmed Hematovascular Mesoderm Induction media. Retrieve cells from hypoxic incubator and gently aspirate spent Primitive Mesoderm Induction media and replace with 2 mL prepared Hematovascular Mesoderm Induction media. Place cells back in hypoxic incubator.

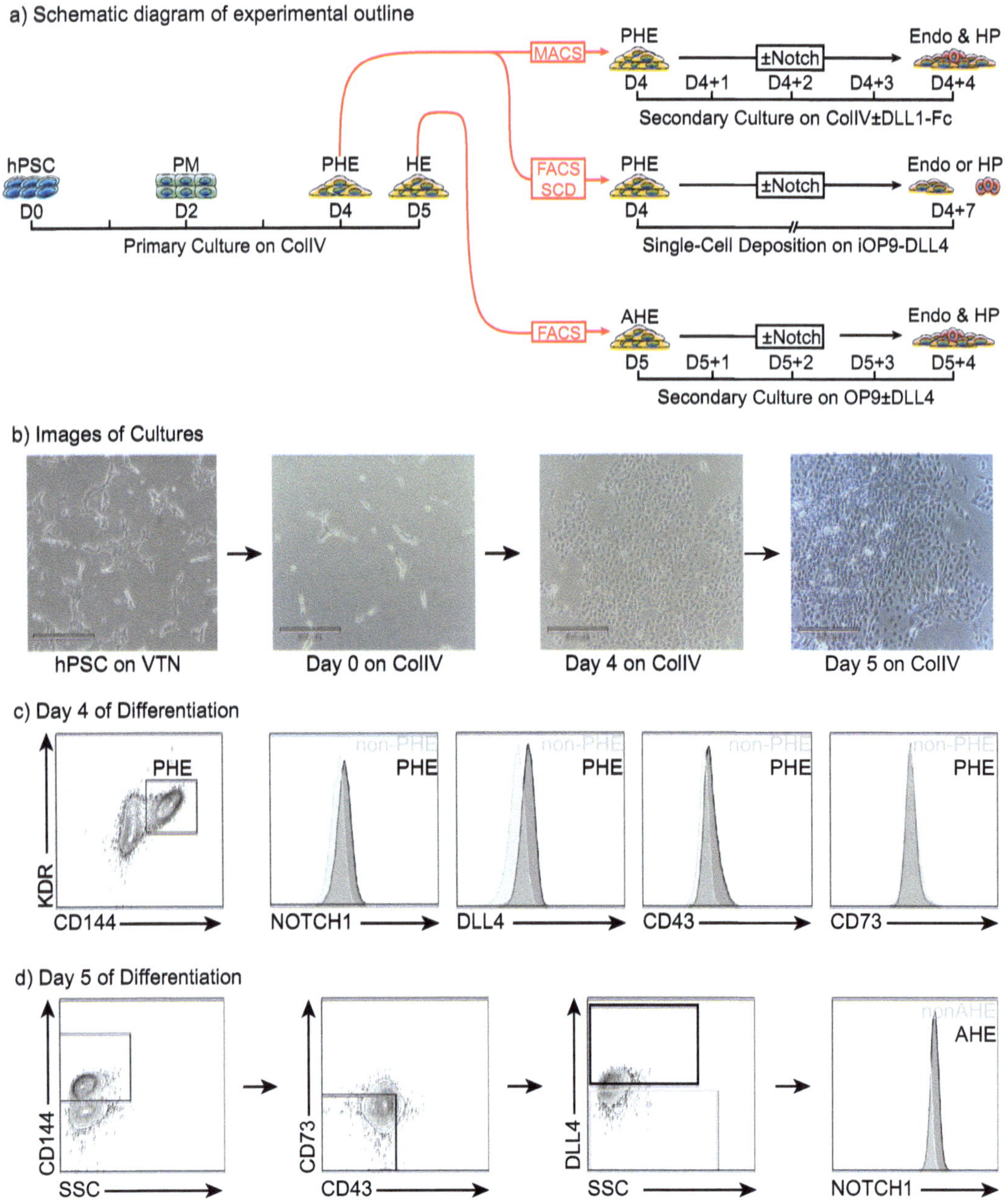

Fig. 1 hPSC differentiation into bulk primitive and definitive hematopoietic progenitors. (**a**) Schematic diagram of experimental outline for D4 PHE secondary culture, D4 PHE single-cell deposition EHT assay, and D5 AHE secondary coculture on OP9. *hPSC* human pluripotent stem cells, *PM* primitive mesoderm, *PHE* primordial hemogenic endothelium, *HE* hemogenic endothelium, *MACS* Magnetically activated cell sorting, *FACS SCD* fluorescence activated cell sorting single-cell deposition, *Endo* endothelium, *HP* hematopoietic progenitors. (**b**) Images of cultures from maintenance, D0, D4, and D5 of differentiation. Scale bar = 300 μm. (**c**) Flow analysis of D4 of differentiation. (**d**) Flow analysis of D5 of differentiation. All flow cytometry plots are gated on live cells unless otherwise noted

3.4 Continue Differentiation in Bulk Culture Without Purification of HE Subsets

$CD144^{+}CD43^{-}CD73^{-}$ PHE specifies into two major HE populations by day 5 of differentiation: $DLL4^{+}CXCR4^{+/-}NOTCH^{hi}$ AHE and $DLL4^{-}CXCR4^{-}NOTCH^{lo}$ non-AHE. In addition to these HE subsets, $CD144^{+}CD31^{+}$ population on day 5 of differentiation include $CD73^{+}CD43^{-}$ non-HE vascular endothelium and $CD43^{+}CD235a^{+}CD73^{-}$ hematopoietic progenitors with FGF2-dependent hematopoietic colony-forming potential [4, 11] (Fig. 1b–d). Continued culture with hematopoietic cytokines leads to the formation of blood cells from bulk HE and can be observed in culture from day 6 (day 4+2 or day 5+1). By day 8, numerous floating, round HPs are clearly visible in the culture. These HPs are a mixture of $CD34^{+}CD43^{+}CD45^{+/-}$ multipotent progenitor cells with high CFC-GEMM and T-lymphoid potential, $CD43^{+}CD235a^{+}41a^{+}CD45^{-}$ cells enriched in erythromegakaryocytic progenitor cells, and $CD43^{+}CD235a^{+/-}41a^{lo}CD45^{+}$ myeloid progenitors enriched in G- and M-CFCs [11, 12].

1. Differentiation Day 4: Induction of HE specification and Endothelial-to-Hematopoietic Transition. Prepare fresh, pre-warmed Hemogenic Endothelium and Hematopoiesis Media. Retrieve cells from hypoxic incubator and gently aspirate spent Hematovascular Mesoderm Induction media and replace with 2 mL prepared HE+H media. Place cells back in hypoxic incubator (*see* **Note 12**).
3. Differentiation Day 6: Continue Hematopoietic Progenitor Maturation. Prepare more HE+H Media. Retrieve cells from incubator and add 2 mL prepared HE+H without aspirating spent media. Place cells back in incubator for another 2 days.
4. Differentiation Day 8 or 9: Prepare cells for the assessment of hematopoietic specification for flow cytometry, CFU assay, and preparation for T-lymphocyte assay. Retrieve cells from incubator on day 8 or 9 and collect spent HE+H media with floating hematopoietic cells into 50 mL conical tubes, then wash with 6 mL EDTA-PBS twice. Add 1 mL per well of a 6-well plate, or 6 mL room temperature TrypLE Select and incubate for 5 min. Aspirate supernatant from 50 mL conical tubes and transfer cells in TrypLE Select into corresponding tubes. Centrifuge at 350 × *g* for 5 min. After centrifugation, aspirate the supernatant and resuspend cell pellet in 1/100 original culture volume of IMDM. Determine cell density and total cell number.
5. Collect 10,000 cells per CFU assay condition and follow manufacturer's recommended protocol for plating (*see* **Note 13**). Transfer cells to FACS tube and add up to 5 mL FACS Buffer. Centrifuge at 350 × *g* for 5 min. After centrifugation, aspirate the supernatant and resuspend cell pellet in HP staining antibody cocktail (Table 1) at a 1:100 dilution and incubate at 4 °C

for 30 min. Add 5 mL FACS buffer and centrifuge at 350 × *g* for 5 min. Decant the supernatant and resuspend the cell pellet in 100 μL FACS buffer for flow cytometry analysis.

3.5 Isolate PHE on D4 and Culture on Collagen IV+IgG-Fc or Collagen IV +DLL1-Fc Coated Plates to Enhance AHE Specification and Bulk EHT Assay

To assess HE specification and EHT, day 4 PHE are isolated by magnetic enrichment and cultured in defined conditions on collagen IV-coated plates in the presence of EHT and hematopoiesis-specific cytokine cocktail (Fig. 2a i). Specification of AHE can be evaluated on day 4+1 by flow cytometry using CD144, DLL4, and CD73 antibodies. EHT can be observed and recorded in a dish from day 4+1 through day 4+5 using time-lapse recording, and by evaluating CD144 and CD43 expression kinetics by flow cytometry [6] or using dual VEC-tdTomato/CD43-eGFP reporter hESC line [13]. The assessment of the role of NOTCH signaling on HE specification and EHT requires purifying the day 4 PHE and continue culturing on plates coated with immobilized DLL1 or DLL4 NOTCH ligands. The NOTCH ligands must be immobilized on the tissue culture plates as activation requires a physical interaction and mechanical stress. Adding soluble NOTCH ligands has been shown to inhibit NOTCH activation as the ligands would bind to the receptor but would not induce a mechanical stress strong enough for activation (Fig. 2a iii). In our prior studies, we used this assay to demonstrate the critical role of NOTCH signaling in specification of AHE and regulation of EHT.

1. Differentiation Day 4: Magnetic isolation of D4 Primordial Hemogenic Endothelium. Retrieve cells from hypoxic incubator and gently aspirate spent Hematovascular Mesoderm Induction media and wash with EDTA-PBS. Add 1 mL per well of a 6-well plate, or 6 mL for 10 cm dish, room temperature TrypLE Select and incubate for 5 min. Transfer cells in TrypLE Select into a 15 mL conical tube and add cold MACS Buffer up to 15 mL and centrifuge 350 × *g* for 5 min. After centrifugation, aspirate the supernatant and resuspend cell pellet in 200 μL cold MACS Buffer and keep on ice.
2. Use a hemocytometer to determine the concentration of cells. Resuspend cells up to 1.5×10^8 cells/mL, then add 300 μL/mL of CD31 MicroBeads. Place tube on the MACS mixer, set on the highest rotation speed, and incubate at 4 °C for 20 min.
3. During the incubation time, set up the MACS LS column according to manufacturer's instructions. Place a 70 μm pre-separation filter on top of the LS column and rinse the filter and column with 2 mL cold MACS buffer.
4. After the 20-min incubation, add MACS buffer up to 10× original volume of cells and centrifuge at 350 × *g* for 5 min. Resuspend the cell pellet in 1 mL MACS Buffer and transfer to LS column, passing cells through the pre-separation filter first.

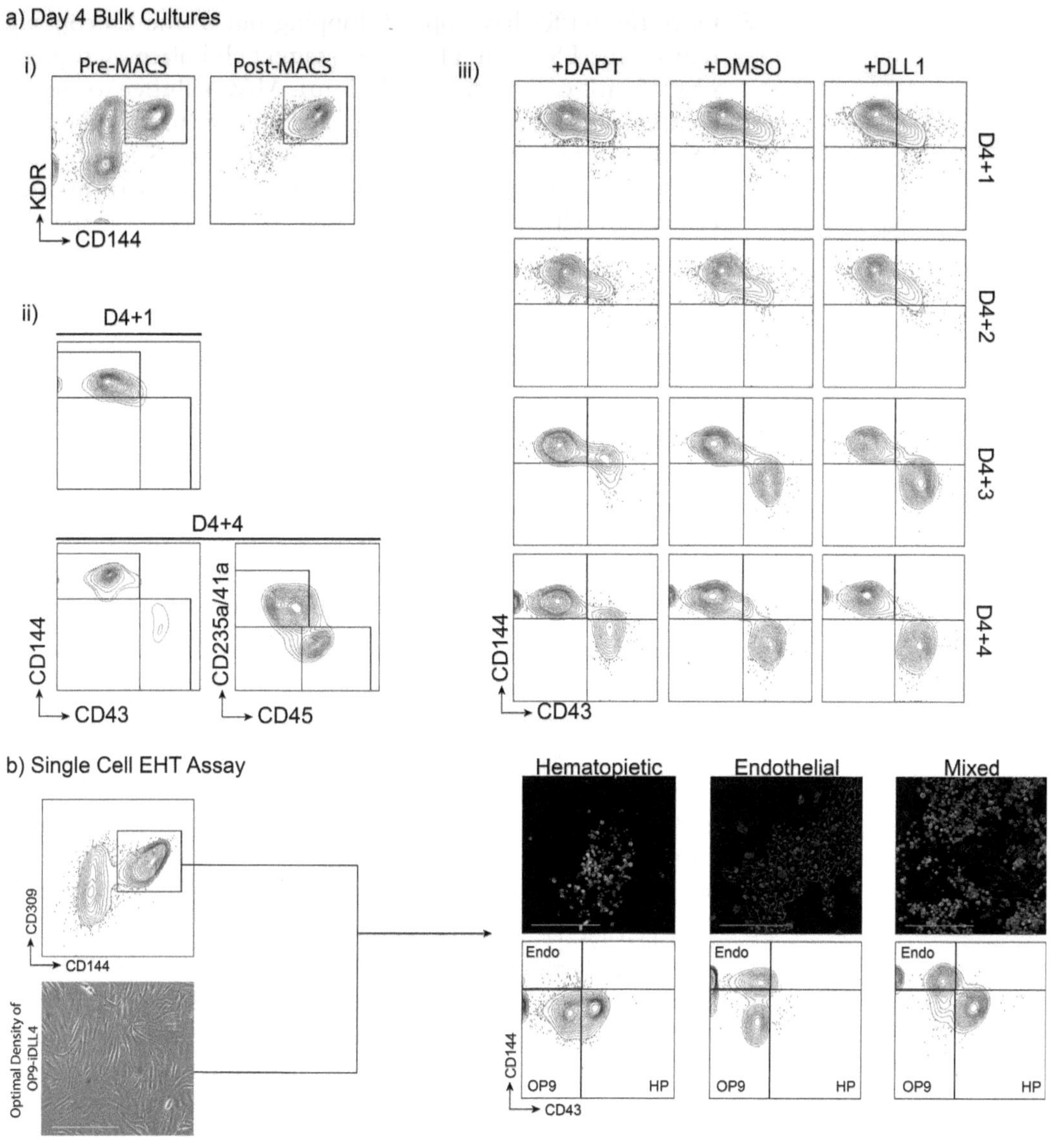

Fig. 2 D4 PHE purification and assessment of EHT potential in serum- and feeder-free conditions. (**a**) Analysis of D4 differentiation cultures (i) pre- and post-MACS, and (ii) 1 and 4 days post-secondary culture. (iii) images and flow analysis of secondary cultures from D4+1 to D4+4 in NOTCH modulating conditions. (**b**) Single-cell EHT assay on OP9 with inducible DLL4 expression. Fluorescent images and flow cytometry analysis of endothelial, hematopoietic, and mixed colonies. Scale bar = 300 μm. All flow cytometry plots are gated on live cells unless otherwise noted. Fluorescence image channels: DAPI in blue, CD43 in green, and CD144 in red

Once the buffer has stopped dripping out of the LS column, add 2 mL MACS Buffer to rinse the LS column. Repeat the wash step when the buffer has stopped dripping out of the LS column again.

5. Once the buffer has stopped dripping out of the LS column, remove the LS column from the magnet and place on top of a new 15-mL conical tube. Add 5 mL MACS Buffer to the LS column, and quickly add the supplied plunger to the top of the LS column and gently push the cells out of the LS column into the 15 mL conical tube. Add MACS buffer up to 15 mL and centrifuge cells at 350 × *g* for 5 min.
6. During the centrifugation, prepare EHT+H plating media, aspirate the DLL1-Fc- or IgG-Fc-coating solution out of the DLL1-Fc- or IgG-Fc-coated plate, and add 1.25 mL/well of a 6-well plate or 0.5 mL/well of a 12-well plate. After centrifugation, aspirate the supernatant and resuspend cell pellet in 1 mL prepared EHT+H Plating media and keep on ice.
7. Use a hemocytometer to determine the concentration of cells. Resuspend cells to 160,000 to 320,000 cells/mL of EHT+H plating media, and plate either 1.25 mL/well of a 6-well, or 0.5 mL/well in a 12-well plate. Place in normoxic incubator overnight to allow cells to reattach (*see* **Note 5**).
8. Differentiation Day 4+1: Prepare EHT+H Support media and retrieve cells from the normoxic incubator. Aspirate spent EHT +H plating media, and replace with EHT+H Support media, and place back in normoxic incubator for another 48 h.
9. Differentiation Day 4+2: Prepare fresh EHT+H Support media and effectively double the total media in culture (*see* **Notes 14** and **15**).
10. Preparation of cells for the assessment of AHE specification by flow cytometry. Retrieve cells from incubator on days 4+1 and 4+2, and gently collect spent EHT+H media into FACS tubes, then wash with EDTA-PBS. Centrifuge FACS tubes at 350 × *g* for 5 min. During the spin, add 1 mL per well of a 6-well plate, room temperature TrypLE Select and incubate for 5 min. Dump supernatant from FACS tubes and transfer cells in TrypLE Select into corresponding FACS tubes. Centrifuge at 350 × *g* for 5 min. After centrifugation, decant the supernatant and resuspend cell pellet by grating them on a tube rack. Add AHE staining antibodies (Table 1) at a 1:100 dilution and incubate at 4 °C for 30 min. Add 5 mL FACS buffer and centrifuge at 350 × *g* for 5 min. Decant the supernatant and resuspend the cell pellet in 100 μL FACS buffer for flow cytometry analysis.

3.6 Single-Cell Deposition Assay of D4 PHE on OP9-iDLL4

In order to study EHT at the single-cell level, stromal coculture system has to be used. While the completely chemically defined, xenogeny- and feeder-free culture system worked with bulk cultures, a single-cell deposition assay was not feasible with the feeder-free culture system since HE requires a minimal plating density for survival, presumably for cell–cell interaction [6, 9]. Wild-type OP9 or dox-inducible OP9-iDLL4 cell line can be used as feeders to enable EHT and assess the role of NOTCH signaling on EHT. OP9 feeders have to be pretreated with Mitomycin C to inhibit overgrowth and control for feeder density between all conditions (NOTCH activation with dox via DLL4 expression, controls with no dox, and NOTCH inhibition with the addition of DAPT, a small molecule gamma-secretase inhibitor). The 96-well plates containing the feeder cells were prepared 24 h prior to FACSorting single D4 PHE cells into each well. The cultures were grown out for 7 days, then fixed and stained for CD43 and CD144 via immunofluorescence. The frequency of hematopoietic versus endothelial colonies were manually counted with a fluorescence microscope to determine the frequency of EHT. Alternatively, cultures can be dissociated into single-cell suspension and prepared for flow cytometry if access to a sensitive and high throughput flow cytometer is available. In this case, human CD43 and CD144 can be used against mouse CD29 to gate out the OP9 (Fig. 2b).

1. Differentiation Day 3: Prepare OP9-iDLL4 seeded 96-well plates for single-cell deposition assay. Prepare fresh OP9 Differentiation media Retrieve OP9-iDLL4 seeded 96-well plate from incubator, gently aspirate spent OP9 maintenance media, and replace with 200 μL of OP9 Differentiation media (*see* **Note 16**).
2. Differentiation Day 4: Retrieve cells from hypoxic incubator and gently aspirate spent Hematovascular Mesoderm Induction media and wash with EDTA-PBS. Add 1 mL per well of a 6-well plate, or 6 mL for 10 cm dish, room temperature TrypLE Select and incubate for 5 min. Transfer cells in TrypLE Select into a 15 mL conical tube and add cold MACS Buffer up to 15 mL and centrifuge 350 × *g* for 5 min. After centrifugation, aspirate the supernatant and resuspend cell pellet in 500 μL cold MACS Buffer and keep on ice. Add PHE staining antibodies (Table 1) at a 1:100 dilution and incubate at 4 °C for 30 min. Add MACS buffer up to 10× original volume of cells and centrifuge at 350 × *g* for 5 min. Aspirate the supernatant and resuspend the cell pellet in 1 mL MACS Buffer and strain cells through a 70 μm pre-separation filter.
3. FACSort single KDR^{hi}CD144+CD73-CD43-D4 IHE according to manufacturer's recommendation into each well of the prepared mitomycin treated OP9-iDLL4 seeded 96-well plates. Centrifuge plates at 250 × *g* for 3 min and place into normoxic incubator (*see* **Note 17**).
4. Differentiation Day 4+1: Continued Culture of single-cell deposition assay. Twenty four hours after IHE were plated (D4+1), prepare fresh OP9 Differentiation media without

DAPT or DMSO. Gently aspirate spent media and replace with prepared OP9 Differentiation media. Place back into normoxic incubator.

5. Differentiation Day 4+4: Prepare fresh OP9 Differentiation media and gently add another 100 μL of prepared fresh OP9 Differentiation media, careful not to overflow the wells.
6. Differentiation Day 4+7: The hematoendothelial colonies should have grown large enough to be visually detectable.
7. Immunofluorescence staining for endothelial vs hematopoietic colony detection. Centrifuge culture plates at 350 × *g* for 5 min. Dump the supernatant by violently flicking the plates upside-down into a sink and dry off droplets on the edge by touching a paper towel, still upside-down. Add 200 μL Fixation buffer to each well and incubate for 30 min at room temperature.
8. Dump the spent Fixation buffer and gently add 200 μL Permeabilization buffer and incubate for 10 min at room temperature.
9. Dump the spent permeabilization buffer and wash with 200 μL of PBST twice. Add 200 μL of Blocking buffer and incubate for 1 h at room temperature.
10. Dump the spent permeabilization buffer and wash with 200 μL of PBST twice. Make a primary antibody staining solution for CD144 and CD43 and add 100 μL to each well and incubate overnight at 4 °C.
11. The next morning, wash out the primary antibodies by dumping the spent staining solution and washing with 200 μL of PBST twice. Make a secondary antibody staining solution and add 100 μL to each well and incubate at room temperature for 2 h.
12. Wash out the secondary antibodies by dumping the spent staining solution and washing with 200 μL of PBST twice. Add 100 μL of PBST with DAPI, and the plates are ready to be analyzed under a fluorescent microscope for hematoendothelial colony detection. Sample plates can be parafilmed and stored in 4 °C up to 1 week.

3.7 Isolate AHE on D5 and Culture on OP9/OP9-DLL4 for the Assessment of Hematopoietic Potential

In contrast to day 4 PHE or day 5 non-AHE, day 5 AHE requires stromal feeders and NOTCH ligands for the efficient EHT, even for bulk cultures. We reported that in the presence of high NOTCH activation using OP9-DLL4, D5 AHE underwent EHT to produce HPs with definitive-type hematopoietic phenotypic and functional characteristics within 5 days [6] (Fig. 3).

1. Differentiation Day 4: Prepare OP9/OP9-DLL4 seeded 6-well or 10 cm plates for D5 AHE. Prepare fresh OP9 Differentiation media. Retrieve OP9/OP9-DLL4 seeded plates from

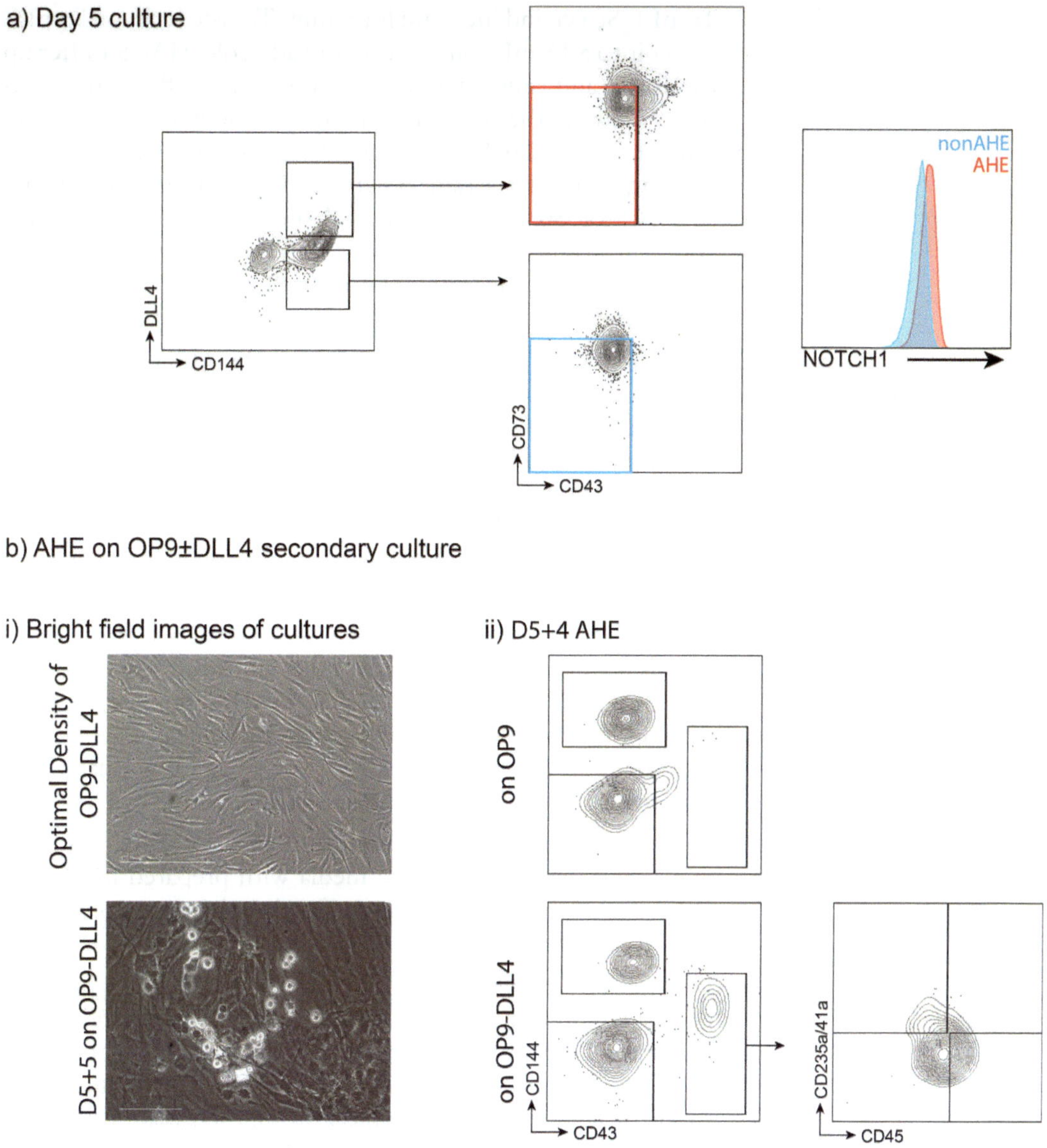

Fig. 3 D5 AHE purification and assessment of EHT potential. (**a**) Gating strategy to sort AHE, showing confirmation of enrichment of NOTCH+ population. (**b**) (i) Images of optimal density of OP9-DLL4 (Scale bar = 300 μm) and D5+5 coculture of AHE on OP9-DLL4 (Scale bar = 100 μm). (ii) Flow cytometry analysis of D5+4 secondary coculture on OP9 and OP9-DLL4. All flow cytometry plots are gated on live cells unless otherwise noted

incubator, gently aspirate spent OP9 maintenance media and replace with half the amount of OP9 Differentiation media (*see* **Note 18**).

2. Differentiation Day 5: Prepare cells for FACSorting. Retrieve cells from hypoxic incubator and gently aspirate spent HE+H media and wash with EDTA-PBS. Add 1 mL per well of a 6-well plate, or 6 mL for 10 cm dish, room temperature

TrypLE Select and incubate for 5 min. Transfer cells in TrypLE Select into a 15 mL conical tube and add cold MACS Buffer up to 15 mL and centrifuge 350 × *g* for 5 min. After centrifugation, aspirate the supernatant and resuspend cell pellet in 500 μL cold MACS Buffer and keep on ice. Add D5 AHE staining antibodies (Table 1) at a 1:100 dilution and incubate at 4 °C for 30 min. Add MACS buffer up to 10× original volume of cells and centrifuge at 350 × *g* for 5 min. Aspirate the supernatant and resuspend the cell pellet in 1 mL MACS Buffer and strain cells through a 70 μm pre-separation filter.

3. FACSort CD31+CD73-CD43-DLL+ D5 AHE according to manufacturer's recommendation into prepared 5 mL FACS tubes filled with 1 mL OP9 Differentiation media (*see* **Note 19**). Spin cells at 350 × *g* for 5 min and resuspend in OP9 Differentiation media to 20,000 cells/mL based on the cell count from the cell sorter. Add half of the total recommended volume to each OP9/OP9-DLL4 seeded plate prepared with OP9 Differentiation media, and place into normoxic incubator (*see* **Note 20**).
4. Differentiation Day 5+1: Continue culturing D5 AHE on OP9/OP9-DLL4. Prepare fresh OP9 Differentiation media with appropriate small molecule and/or vehicle. Gently aspirate spent media and replace with prepared OP9 Differentiation media. Place back into normoxic incubator.
5. Differentiation Day 5+3: Prepare fresh OP9 Differentiation media and gently double the media with prepared fresh OP9 Differentiation media. Round, floating hematopoietic cells should now be visible.
6. Preparation of cells for the assessment of EHT and hematopoietic specification: Flow Cytometry, CFU Assay, and preparation for T-lymphocyte assay. Retrieve cells from incubator on day 5+4, and collect spent EHT+H media with floating hematopoietic cells into 15 mL conical tubes, then wash with 6 mL EDTA-PBS twice. Centrifuge FACS tubes at 350 × *g* for 5 min. During the spin, add 1 mL per well of a 6-well plate, or 6 mL room temperature TrypLE Select and incubate for 5 min. Aspirate supernatant from 15 mL conical tubes and transfer cells in TrypLE Select into corresponding tubes. Centrifuge at 350 × *g* for 5 min. After centrifugation, aspirate the supernatant and resuspend cell pellet in 1/100 original culture volume of IMDM media. Determine cell density and total cell number.
7. Collect 10,000 cells per CFU assay condition and follow manufacturer's recommended protocol for plating (*see* **Note 21**). Transfer cells to FACS tube and add up to 5 mL FACS Buffer. Centrifuge at 350 × *g* for 5 min. After centrifugation, aspirate

the supernatant and resuspend cell pellet in AHE staining antibody cocktail (Table 1) at a 1:100 dilution and incubate at 4 °C for 30 min. Add 5 mL FACS buffer and centrifuge at 350 × *g* for 5 min. Decant the supernatant and resuspend the cell pellet in 100 μL FACS buffer for flow cytometry analysis.

4 Notes

1. Full-length Tenascin C is necessary for differentiation. If necessary, sterile filter with a 0.22 μm PVDF membrane after diluting the Tenascin C to working concentration.
2. BMP-4 is unstable after reconstitution. Take extreme care when reconstituting, aliquoting, and freezing. Adding a carrier protein such as BSA can increase its stability.
3. Polyvinyl alcohol does not dissolve readily. We recommend autoclaving 2 L of ddH_2O with a magnetic stir bar. While hot, place on a hot plate with a magnetic stir bar and slowly mix in 5 g of PVA. Wait until dissolved before adding another 5 g. Repeat until all 40 g is added. Continue to stir overnight at room temperature. Autoclave again to sterilize. Do not sterile filter.
4. Be careful not to add more than 1 μM Y-27632 as it will inhibit differentiation.
5. Shake plate front-to-back and side-to-side three times to ensure the cells are distributed evenly.
6. Do not let cells reach over 80% confluency as they will either spontaneously differentiate or senesce.
7. The edges of the colonies will start to curl up and the colonies themselves may lift off the plate with too much agitation.
8. If the media is not pipetted hard enough, it may not lift off the cells. Once the E8 is added, the cells will reattach immediately if not broken off.
9. When carrying out the differentiation protocol through Day 9 without HE enrichment on Day 4 or Day 5, the plating density is absolutely critical to observe hematopoiesis. If the plating density is too low, the cells will not survive the epithelial-to-mesenchymal transition that occurs during the first 2 days when the hPSCs differentiate into primitive mesoderm. However, if the density is too high, the cells will overcrowd by Day 5 and the HE will not undergo EHT to produce hematopoietic progenitors. Thus, it is recommended that initially 6 different densities be tested from 2500 cells/cm^2 up to 15,000 cells/cm^2 in 2500 cells/cm^2 increments. Different hESC and hiPSC lines also have their own optimal densities

as well. Once the optimal density has been determined, the protocol is consistently reproducible. On Day 0, the hPSCs should have attached and either still be single cells or in doublets.

10. The cells are only very gently attached at this point. Pipetting too hard or agitating the plate may dislodge the cells.
11. During the next 2 days, do not open the incubator or agitate the base. The cells undergo an epithelial-to-mesenchymal-like transition at which point they may detach from the plate. Exposure to normoxic conditions may decrease viability during this period.
12. At this point, the cells can be kept in a normoxic incubator.
13. At this point, the cells can also be used for T-lymphoid potential assay. Collect and process 50,000 cells for flow cytometry to analyze differentiation efficiency.
14. After day 4+1, do not aspirate media into waste, as the hematopoietic progenitor cells begin to float and may be lost.
15. Day 4 HE cells are fragile: The addition of low levels of Rock inhibitor for 24 h significantly improved viability of the cells, though higher levels also inhibited EHT. It is critical to achieve plating density of the purified HE within an operational window of 20,000–40,000 cells/cm^2. Too low of a plating density caused the cells to perish, while a plating density that was too high inhibited EHT. Thus, it is recommended that initially 6 different densities be tested from 15,000 cells/cm^2 up to 40,000 cells/cm^2 in 5000 cells/cm^2 increments. Finally, we also found that PHE cells required the addition of IGF-I, IGF-II, and EGF on top of the hematopoietic cytokine combination for their continued survival and growth (Fig. 2a ii).
16. For NOTCH inhibition condition, add DAPT to a final concentration of 20 μM, and for control and NOTCH activation conditions, add the same amount of DMSO to OP9 Differentiation media before distributing 200 μL to each well. **CAUTION**: Do not use the wells on the outer wells, as these wells are susceptible to evaporation, changing the salt concentration and decreasing viability of plated cells.
17. PHE do not survive well in single-cell suspension. It is critical that cells are prepared and single-cell sorted into individual wells of the prepared 96-well plates as quickly as possible. Expect to see hematoendothelial colonies in 50% of the wells.
18. For NOTCH inhibition condition, add DAPT to a final concentration of 20 μM, and for control and NOTCH activation conditions, add the same amount of DMSO to OP9 Differentiation media.

19. AHE do not survive well in single-cell suspension. It is critical that cells are prepared and sorted as quickly as possible.
20. Stroma density is critical for AHE and non-AHE to undergo EHT and produce hematopoietic progeny, particularly for AHE on OP9-DLL4 due to sensitivity to NOTCH activation levels. Thus, it is recommended that initially 6 different densities be tested from 500 cells/cm^2 up to 3000 cells/cm^2 in 500 cells/cm^2 increments.
21. At this point, the cells can also be used for T-lymphoid potential assay. Collect and process 50,000 cells for flow cytometry to analyze differentiation efficiency.

Acknowledgments

We thank Irwin Bernstein for providing DLL1-Fc, and Toru Nakano for providing OP9 cell line. This work was supported by funds from the National Institute of Health (R01HL142665, U01HL134655, and P51OD011106).

References

1. Slukvin II (2016) Generating human hematopoietic stem cells *in vitro* -exploring endothelial to hematopoietic transition as a portal for stemness acquisition. FEBS Lett 590: 4126–4143. https://doi.org/10.1002/1873-3468.12283
2. Gritz E, Hirschi KK (2016) Specification and function of hemogenic endothelium during embryogenesis. Cell Mol Life Sci 73: 1547–1567. https://doi.org/10.1007/s00018-016-2134-0
3. Vodyanik MA, Thomson JA, Slukvin II (2006) Leukosialin (CD43) defines hematopoietic progenitors in human embryonic stem cell differentiation cultures. Blood 108:2095–2105. https://doi.org/10.1182/blood-2006-02-003327
4. Choi K-D, Vodyanik MA, Togarrati PP et al (2012) Identification of the Hemogenic endothelial progenitor and its direct precursor in human pluripotent stem cell differentiation cultures. Cell Rep 2:553–567. https://doi.org/10.1016/j.celrep.2012.08.002
5. Ditadi A, Sturgeon CM, Tober J et al (2015) Human definitive haemogenic endothelium and arterial vascular endothelium represent distinct lineages. Nat Cell Biol 17:580–591. https://doi.org/10.1038/ncb3161
6. Uenishi GI, Jung HS, Kumar A et al (2018) NOTCH signaling specifies arterial-type definitive hemogenic endothelium from human pluripotent stem cells. Nat Commun 9:1828. https://doi.org/10.1038/s41467-018-04134-7
7. Uenishi G, Theisen D, Lee J-H et al (2014) Tenascin C promotes Hematoendothelial development and T lymphoid commitment from human pluripotent stem cells in chemically defined conditions. Stem Cell Rep 3: 1073–1084. https://doi.org/10.1016/j.stemcr.2014.09.014
8. Mesquitta W-T, Wandsnider M, Kang H et al (2019) UM171 expands distinct types of myeloid and NK progenitors from human pluripotent stem cells. Sci Rep 9:6622. https://doi.org/10.1038/s41598-019-43054-4
9. Kennedy M, Awong G, Sturgeon CM et al (2012) T lymphocyte potential Marks the emergence of definitive hematopoietic progenitors in human pluripotent stem cell differentiation cultures. Cell Rep 2:1722–1735. https://doi.org/10.1016/j.celrep.2012.11.003
10. Sturgeon CM, Ditadi A, Awong G et al (2014) Wnt signaling controls the specification of definitive and primitive hematopoiesis from human pluripotent stem cells. Nat Biotechnol 32:554–561. https://doi.org/10.1038/nbt.2915

11. Kang H, Mesquitta W-T, Jung HS et al (2018) GATA2 is dispensable for specification of Hemogenic endothelium but promotes endothelial-to-hematopoietic transition. Stem Cell Rep 11:197–211. https://doi.org/10.1016/j.stemcr.2018.05.002
12. Park MA, Kumar A, Jung HS et al (2018) Activation of the arterial program drives development of definitive Hemogenic endothelium with lymphoid potential. Cell Rep 23: 2467–2481. https://doi.org/10.1016/j.celrep.2018.04.092
13. Jung HS, Uenishi G, Kumar A et al (2016) A human VE-cadherin-tdTomato and CD43-green fluorescent protein dual reporter cell line for study endothelial to hematopoietic transition. Stem Cell Res 17:401–405. https://doi.org/10.1016/j.scr.2016.09.004

Part II

ES/iPS Derived Tissue Stem Cells

Chapter 9

Homogeneous Differentiation of Functional Hepatocytes from Human Induced Pluripotent Stem Cells

Rong Li, Yang Zhao, Jeffrey J. Yourick, Robert L. Sprando, and Xiugong Gao

Abstract

Hepatocyte-like cells (HLCs) generated from human induced pluripotent stem cells (iPSCs) could provide an unlimited source of liver cells for regenerative medicine, disease modeling, drug screening, and toxicology studies. Here we describe a stepwise improved protocol that enables highly efficient, homogeneous, and reproducible differentiation of human iPSCs into functional hepatocytes through controlling all three stages of hepatocyte differentiation, starting from a single cell (non-colony) culture of iPSCs, through homogeneous definitive endoderm induction and highly efficient hepatic specification, and finally arriving at matured HLCs. The final population of cells exhibits morphology closely resembling that of primary human hepatocytes, and expresses specific hepatic markers as evidenced by immunocytochemical staining. More importantly, these HLCs demonstrate key functional characteristics of mature hepatocytes, including major serum protein (e.g., albumin, fibronectin, and alpha-1 antitrypsin) secretion, urea synthesis, glycogen storage, and inducible cytochrome P450 activity.

Key words Hepatocyte, Induced pluripotent stem cell (iPSC), Directed differentiation, Hepatocyte-like cell (HLC), Definitive endoderm, Liver

1 Introduction

A variety of protocols have been established during the past decade for in vitro differentiation of human induced pluripotent stem cells (iPSCs) into hepatocytes, mostly by recapitulating the major signaling pathways involved in the different stages of embryonic hepatogenesis using growth factors or small molecules [1–9]; however, directed differentiation of hepatocytes remains challenging. Although successful derivation of hepatocyte-like cells (HLCs) has been achieved through these protocols, the differentiation efficiency and reproducibility are not ideal. Oftentimes, the differentiation process needs to be optimized for different iPSC lines through fine-tuning the concentration of various growth factors or small molecules, and adjustment of the treatment timeline

Nagarajan Kannan and Philip Beer (eds.), *Stem Cell Assays: Methods and Protocols*, Methods in Molecular Biology, vol. 2429, https://doi.org/10.1007/978-1-0716-1979-7_9,

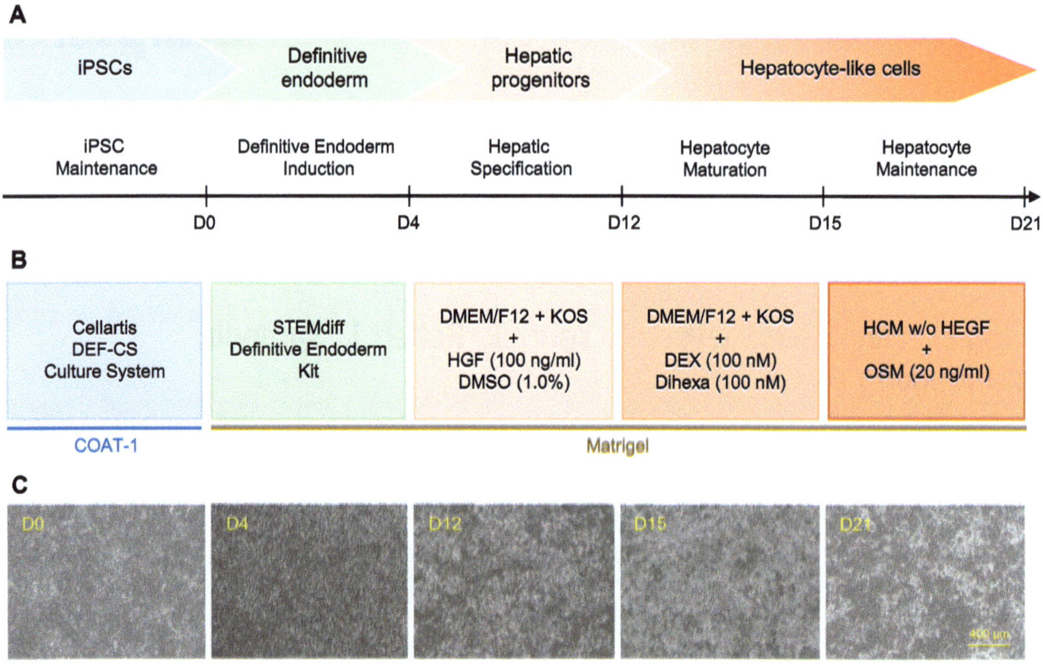

Fig. 1 Schematic illustration of the protocol for the differentiation of HLCs from human iPSCs. (**a**) Cell states, sequential steps, and timeline of the differentiation process. (**b**) Media, growth factors, and small molecules used at each stage of the differentiation. (**c**) Representative phase-contrast micrographs of iPSCs (D0), definitive endoderm cells (D4), hepatoblasts (D12), immature HLCs (D15), and mature HLCs (D21). Scale bar, 400 μm

[5, 10–12]. More importantly, the final differentiated cells have been revealed to be a mixed population of cells with different levels of maturity and sometimes even non-specific cells, which leads to large uncertainty for downstream applications [12–15].

Through evaluating and comparing existing hepatocyte differentiation strategies, we have developed a stepwise improved protocol that enables highly efficient, homogeneous and reproducible differentiation of human iPSCs into functional HLCs (Fig. 1). The protocol starts with growing iPSCs in a medium that supports single cell culture (no colony formation), continues with homogeneous induction of definitive endoderm, further carries on with highly efficient hepatic specification guided by hepatocyte growth factor (HGF) and dimethyl sulfoxide (DMSO), and eventually reaches hepatocyte maturation boosted by *N*-hexanoic-Try-Ile-(6)-amino hexanoic amide (dihexa), dexamethasone (DEX) and oncostatin M (OSM). Using this protocol, we are able to generate highly homogenous functional HLCs at the end of the differentiation process, with a morphology closely resembling that of primary human hepatocyte (PHHs), e.g., distinct cuboidal shape, a high cytoplasm-to-nucleus ratio, sometimes binucleated, intercellular junctions, and apparent lipid droplets in the cytoplasm. These cells express high levels of major hepatic markers as demonstrated by immunofluorescence staining, including hepatocyte nuclear factor 4 alpha (HNF4A), alpha-fetoprotein (AFP), albumin (ALB), cytokeratin 18 (CK18), alpha-1 antitrypsin (A1AT), and cytochrome P450 3A4 (CYP3A4). More importantly, the HLCs possess several key functional characteristics of hepatocytes in vivo, including serum protein (e.g., albumin, fibronectin, and A1AT) secretion, urea synthesis, glycogen storage, and inducible cytochrome P450 (CYP) activity.

2 Materials

2.1 iPSC Line

1. Human iPSC line OARSAi002-A (*see* **Note 1**).

2.2 iPSC Culture Medium

1. Cellartis DEF-CS 500 Culture System (Takara Bio USA): including Basal Medium, COAT-1, and Additives GF-1, GF-2, and GF-3.

2.3 Reagents

1. Dulbecco's Modified Eagle Medium/Nutrient Mixture F-12 (DMEM/F12).
2. Dulbecco's phosphate-buffered saline, with 0.90 mM calcium and 0.50 mM magnesium (DPBS+/+).
3. Dulbecco's phosphate-buffered saline, calcium and magnesium free (DPBS−/−).
4. TrypLE Select Enzyme (1×), no phenol red (Thermo Fisher Scientific).
5. StemPro Accutase Cell Dissociation Reagent (Thermo Fisher Scientific).
6. Rho kinase (ROCK) inhibitor Y-27632 dihydrochloride (Tocris Bioscience).
7. STEMdiff Definitive Endoderm Kit (StemCell Technologies).
8. KnockOut Serum Replacement (KOS) (Thermo Fisher Scientific).
9. MEM Non-Essential Amino Acids (NEAA) Solution (100×) (Thermo Fisher Scientific).
10. Penicillin/Streptomycin (10,000 U/ml).
11. CTS GlutaMAX-I Supplement (Thermo Fisher Scientific).
12. Recombinant hepatocyte growth factor (HGF) (PeproTech).
13. Dimethyl sulfoxide (DMSO).
14. Dexamethasone (DEX) (Sigma-Aldrich).
15. N-hexanoic-Try-Ile-(6)-amino hexanoic amide (dihexa) (MedChemExpress).

16. HCM Hepatocyte Culture Medium BulletKit (Lonza).
17. Recombinant human oncostatin M (OSM) (R&D Systems).
18. Corning Matrigel hESC-Qualified Matrix (Corning).

2.4 Hepatocyte Differentiation and Maturation Media and Solutions

1. Definitive Endoderm Induction Medium 1: STEMdiff Endoderm Basal Medium supplemented with 1× Supplement A and 1× Supplement B (e.g., add 100 μl of Supplement A and 100 μl of Supplement B to 10 ml of Basal Medium).
2. Definitive Endoderm Induction Medium 2: STEMdiff Endoderm Basal Medium supplemented with 1× Supplement B (e.g., add 100 μl of Supplement B to 10 ml of Basal Medium).
3. Hepatocyte Differentiation Basal Medium: 435 ml DMEM/F12, 50 ml KOS, 5 ml NEAA, 5 ml Penicillin/Streptomycin (10,000 U/ml), 5 ml CTS GlutaMAX-I Supplement. Store up to 2 weeks at 4 °C.
4. Hepatic Specification Medium: 99 ml Hepatocyte Differentiation Basal Medium, 100 μl HGF stock solution (*see* **item 7**; 100 ng/ml final), 1.0 ml DMSO (1.0% final). Add HGF and DMSO freshly to the culture medium prior to medium change.
5. Hepatocyte Maturation Medium: 100 ml Hepatocyte Differentiation Basal Medium, 10 μl dihexa stock solution (*see* **item 9**; 100 nM final), 10 μl DEX stock solution (*see* **item 10**; 100 nM final). Add dihexa and DEX freshly to the culture medium prior to medium change.
6. Hepatocyte Maintenance Medium (Modified Lonza HCM Hepatocyte Culture Medium): 500 ml HBM Basal Medium supplemented with the HCM SingleQuots Supplement Pack (including 0.5 ml of transferrin, 0.5 ml of ascorbic acid, 0.5 ml of insulin, 0.5 ml of hydrocortisone, 10.5 ml of BSA (fatty acid free), 0.5 ml of GA-1000, but omitting the human epidermal growth factor (HEGF)), and 0.5 ml OSM stock solution (*see* **item 8**; 20 ng/ml final). Store up to 2 weeks at 4 °C. Add OSM freshly to the culture medium prior to medium change.
7. HGF stock solution (100 μg/ml): Reconstitute recombinant HGF growth factor at 100 μg/ml in sterile DPBS+/+ containing 0.1% bovine serum albumin (BSA). Ensure that all powder in the vial has been thoroughly dissolved. Store aliquots in sterile 1.5 ml microcentrifuge tubes up to 1 year at −80 °C.
8. OSM stock solution (20 μg/ml): Reconstitute OSM at 20 μg/ml in sterile DPBS+/+ containing 0.1% BSA. Ensure that all powder in the vial has been thoroughly dissolved. Store aliquots in sterile 1.5 ml microcentrifuge tubes up to 1 year at −80 °C.
9. Dihexa stock solution (1 mM): Dissolve 10.0 mg dihexa in 19.815 ml DMSO to make 1 mM (10,000×) stock solution.

Ensure that all powder in the vial has been thoroughly dissolved. Stock aliquots in 1.5 ml microcentrifuge tubes up to 1 year at −80 °C.

10. DEX stock solution (1 mM): Dissolve 10.0 mg of DEX in 25.480 ml DMSO to make a 1 mM (10,000×) stock solution. Ensure that all powder in the vial has been thoroughly dissolved. Store aliquots in 1.5 ml microcentrifuge tubes up to 1 year at −80 °C.
11. Y-27632 stock solution (10 mM): Dissolve 10.0 mg of Y-27632 dihydrochloride in 3.123 ml cell culture grade distilled water to make a stock solution of 10 mM (1000×). Ensure that all powder in the vial has been thoroughly dissolved. Store aliquot in 1.5 ml microcentrifuge tubes up to 1 year at −80 °C.
12. Matrigel matrix stock solution: Thaw a 5 ml vial of Corning Matrigel hESC-Qualified Matrix on ice at 4 °C overnight. Keep thawed Matrigel on ice in tissue culture hood and aliquot into sterile, prechilled microcentrifuge tubes. Snap freeze Matrigel aliquots on dry ice and store up to 1 year at −80 °C.

2.5 Equipment

1. Sterile multi-well plates, tissue culture treated polystyrene surface.
2. Class II biological safety cabinet.
3. 37 °C, 5% CO_2 incubator.
4. 37 °C water bath.
5. Liquid nitrogen cryopreserved cell storage tank.
6. Benchtop centrifuge.
7. Phase-contrast microscope.
8. Fluorescence microscope.
9. Microplate reader.
10. Hemocytometer or automated cell counter.

3 Methods

In this protocol, we use 6-well plates for single cell culture of human iPSCs (Subheading 3.1) and definitive endoderm induction (Subheadings 3.2 and 3.3), and 12-well plates for hepatic specification (Subheading 3.4) and hepatocyte maturation (Subheadings 3.5 and 3.6). *See* Table 1 for volumes for other plate formats.

3.1 Single Cell Culture of iPSCs

We highly recommend culturing/adapting iPSCs in a single cell culture system instead of a traditional colony forming culture system before the initiation of hepatocyte differentiation. A

Table 1
Volumes of reagents or media to be used for different plate formats

	Volume per well		
Component	**6-well plate**	**12-well plate**	**24-well plate**
Matrigel	1.5 ml	800 μl	400 μl
Cellartis CEF-CS medium	4.0 ml	2.0 ml	1.0 ml
DPBS	2.0 ml	1.0 ml	500 μl
TrypLE select enzyme	300 μl	120 μl	60 μl
Accutase	1.0 ml	400 μl	200 μl
Definitive Endoderm induction medium 1	2.0 ml	800 μl	400 μl
Definitive Endoderm induction medium 2	2.0 ml	800 μl	400 μl
Hepatic specification medium	2.0 ml	1.0 ml	500 μl
Hepatocyte maturation medium	2.0 ml	1.0 ml	500 μl
Hepatocyte maintenance medium	2.0 ml	1.0 ml	500 μl

homogeneous, synchronized starting population of iPSCs, from which to initiate definitive endoderm differentiation, serves as the key first step to achieve highly efficient, homogeneous, and consistent hepatocyte differentiation. Although this can be achieved through other non-colony type monolayer culture systems developed in individual laboratories [16], we recommend using the commercially available Cellartis DEF-CS Culture System (*see* **Note 2**). In this single cell culture system, iPSC cells are dissociated into single cells using TrypLE Select Enzyme before being plated onto COAT-1 precoated tissue culture plates in Cellartis DEF-CS culture medium. No colonies are formed during the cell proliferation process.

1. Coat tissue culture treated 6-well plates with Cellartis DEF-CS COAT-1 matrix. Dilute required volume of COAT-1 in DPBS +/+ at 1:20 dilution (*see* **Notes 3** and **4**) and add 1.5 ml of diluted COAT-1 solution to each well. Place the plate for ≥20 min in an incubator (37 °C) or 0.5–3 h at room temperature. Aspirate COAT-1 solution from the wells right before cell seeding.
2. Prepare appropriate volume of Cellartis DEF-CS culture medium for cell passaging. Warm the medium to 37 °C and all other reagents to room temperature before use.
3. Check cells under microscope. Aspirate medium from the wells and wash the cells once with DPBS−/−. Add 300 μl of TrypLE Select Enzyme to each well and place the plate in a 37 °C, 5% CO_2 incubator for 4–6 min or until the cell layer has detached.

4. Tap the side of the plate firmly but gently. It is not recommended to tilt or swirl the cell culture plate. Resuspend the cells in Cellartis DEF-CS medium and pipet up and down several times to ensure a single cell suspension.
5. Count the cells using a hemocytometer or automated cell counter. Adjust the cell density using fresh Cellartis DEF-CS medium and add the appropriate volume of cell suspension to a newly coated 6-well plate to obtain the desired density of $4.0–5.0 \times 10^4$ cells/cm^2 (*see* **Note 5**).
6. Shake the plate on an even surface back and forth, then left and right to ensure even distribution of cells over the surface. Place the plate in an incubator at 37 °C, 5% CO_2, and >90% humidity. Change the medium every day until cells reach confluence ($1.5–3.0 \times 10^5$ cells/cm^2), which normally occurs 3–4 days post passage.

3.2 Plating iPSCs for Differentiation

1. Day −1: Examine iPSCs cultured in a 6-well tissue culture plate under the microscope to ensure the cells exhibit typical undifferentiated iPSC morphology (i.e., compact round cells, high nucleus to cytoplasm ratio, and prominent nucleoli) (*see* **Notes 6** and 7).
2. Coat tissue culture plates with Matrigel hESC-Qualified Matrix (*see* **Note 8**). Chill 25 ml DMEM/F12 on ice. Take a 250 μl aliquot of Matrigel from storage at −80 °C and quickly use the cold media to thaw the frozen Matrigel by gently pipetting up and down (*see* **Note 9**). Add 1.5 ml/well of dissolved Matrigel to cover the surface area of the tissue culture plate and place the plate in an incubator at 37 °C, 5% CO_2, and >90% humidity for 1 h or 4 °C overnight. Aspirate Matrigel solution from the wells right before cell seeding.
3. Prepare appropriate volume of Cellartis DEF-CS culture medium. Warm the medium to 37 °C, and warm TrypLE Select Enzyme and DPBS−/− to room temperature.
4. Rinse each well of the 6-well plate of iPSCs ready for passaging with 2 ml of DPBS−/− and remove by aspiration. Add 300 μl of TrypLE Select Enzyme and incubate at 37 °C, 5% CO_2 for 4–6 min.
5. Dislodge cells by pipetting up and down using a pipette with a 1 ml tip. Ensure no cell aggregates remain. Avoid over-pipetting the cells which will cause cell damage.
6. Immediately transfer the cells to a 50 ml conical tube containing 20 ml of Cellartis DEF-CS Basal Medium. Collect the cells by centrifuging for 5 min at $200 \times g$ at room temperature and resuspend the cells into DEF-CS culture medium supplemented with GF-1, GF-2, and GF-3 additives. Count the number of live cells using a hemocytometer or cell counter.

7. Seed cells at a density of 2.1×10^5 cells/cm^2 to a Matrigel precoated 6-well plate (2.0×10^6 cells/well). Adjust cell seeding density if necessary, so that the cells reach ~90–100% confluence after 16–24 h incubation (*see* **Note 10**).
8. Incubate the cells at 37 °C, 5% CO_2, and >90% humidity for 24 h.

3.3 Induction of Definitive Endoderm (Duration: 4 Days; Fig. 1)

1. Day 0: Warm up sufficient volumes of DMEM/F12 at 37 °C. Prepare Definitive Endoderm Induction Medium 1 freshly and warm up to 37 °C.
2. Aspirate medium from the wells and wash with 2 ml/well of DMEM/F12 medium.
3. Remove the wash medium and replace with 2 ml of Definitive Endoderm Induction Medium 1. Incubate at 37 °C, 5% CO_2, and >90% humidity for 24 h.
4. Day 1: Prepare sufficient Definitive Endoderm Induction Medium 2 to be used for Days 1, 2, and 3.
5. Warm up only the amount of Definitive Endoderm Induction Medium 2 required for Day 1 use (2 ml/well). Store the remaining portion at 4 °C. Aspirate medium from the wells and add 2 ml of Definitive Endoderm Induction Medium 2 to each well. Incubate at 37 °C for 24 h.
6. Day 2: Aspirate medium from the wells and replenish with 2 ml of fresh Definitive Endoderm Induction Medium 2. Incubate at 37 °C for 24 h.
7. Day 3: Aspirate medium from the wells and replenish with 2 ml of fresh Definitive Endoderm Induction Medium 2. Incubate at 37 °C for 24 h.
8. Day 4: Cells are ready for assessment of definitive endoderm formation and for downstream hepatic specification (*see* **Note 11**). It is critical that more than 95% of the cells express major markers of definitive endoderm, e.g., SOX17 and FOXA2, at the end of this stage. Also, expression of proteins associated with pluripotency, such as OCT4, should be minimal if detectable at all in the cells.

3.4 Hepatic Specification (Duration: 8 Days)

1. Day 4: Coat 12-well tissue culture plates with Matrigel as described above (Subheading 3.2, **step 2**). Warm sufficient volumes of Hepatocyte Differentiation Basal Medium and StemPro Accutase Cell Dissociation Reagent for definitive endoderm cell dissociation.
2. Aspirate medium from wells of the 6-well plate with induced definitive endoderm cells and rinse with 2 ml of DPBS−/−. Remove the wash medium. Add 1 ml of Accutase to each well and incubate in 37 °C, 5% CO_2 incubator for 5–8 min (*see* **Note 12**).

Table 2
Definitive endoderm cell seeding on different plate formats

Plate format	Culture surface area	Cell numbers per well	Volume
6-well plate	9.5 cm^2	7.84×10^5 cells/well	2.0 ml
12-well plate	3.8 cm^2	3.04×10^5 cells/well	1.0 ml
24-well plate	1.9 cm^2	1.52×10^5 cells/well	0.5 ml

3. Dislodge cells by pipetting up and down using a pipette with a 1 ml tip and immediately transfer cells to a 50 ml conical tube containing 10 ml of Hepatocyte Differentiation Basal Medium (*see* **Note 13**). Collect the cells by centrifuging for 5 min at 200 × *g* at room temperature and resuspend the cells in Hepatic Specification Medium supplemented with 10 μM Y-27632 (*see* **Note 14**).
4. Seed the cells at a density of 8.0×10^4 cells/cm^2 (3.04×10^5 cells/well) to Matrigel precoated plates. Seeding volume depends on the well format chosen (*see* Table 2). Culture the cells at 37 °C, 5% CO_2, and >90% humidity for 24 h (*see* **Notes 15** and **16**).
5. Day 5: Examine the cells under a microscope. Most of the cells should be attached to the plate and cover ~85–95% of the surface area of the well (*see* **Note 17**). Aspirate medium from the wells and replace with fresh Hepatic Specification Medium (1 ml/well) (*see* **Notes 18** and **19**). Incubate the cells at 37 °C for 24 h.
6. Day 6–Day 11: Replenish the wells with fresh Hepatic Specification Medium every day, and continue culturing the cells at 37 °C, 5% CO_2, and >90% humidity.
7. Day 12: Assess the cells for the expression of hepatoblast marker proteins, e.g., HNF4A and AFP, by immunofluorescence staining or flow cytometry.

3.5 Hepatocyte Maturation (Duration: 3 Days)

1. Day 12: Aspirate medium from the wells and replace with Hepatocyte Maturation Medium. Incubate at 37 °C, 5% CO_2, and >90% humidity for 24 h.
2. Day 13–Day 14: Replenish the wells with fresh Hepatocyte Maturation Medium each day, and continue culturing the cells at 37 °C, 5% CO_2, and >90% humidity.
3. Day 15: At this point, cell morphology should be similar to that of PHHs (Fig. 1). Most cells should have a polygonal shape with tight junctions between cells, small round nuclei, and visible lipid droplets in the cytoplasm (*see* **Note 20**).

3.6 Further Maturation and Maintenance of HLCs (Duration: 1–3 Weeks)

1. Day 15: Aspirate medium from the wells and replenish with Hepatocyte Maintenance Medium. Incubate the cells at 37 °C, 5% CO_2, and >90% humidity.
2. Day 16–Day 20: Continue culturing the cells for 5 days with medium change every other day using Hepatocyte Maintenance Medium (*see* **Note 21**). Cell morphology will keep evolving during this time towards more closely resembling that of PHHs.
3. Day 21: At this point, the cells are ready for assessment of hepatocyte functionalities and downstream applications. The final cell population should be highly homogeneous and display a morphology with a distinct cuboidal shape, a high cytoplasm-to-nucleus ratio, sometimes binucleated, obvious intercellular junctions, and apparent lipid droplets in the cytoplasm (Fig. 1). Cells should also express high levels of major hepatic protein markers including HNF4A, AFP, ALB, CK18, A1AT, CYP3A4 (Fig. 2) (*see* **Notes 22** and **23**). Furthermore, cells should possess multiple key hepatocyte functions, including serum protein (ALB, fibronectin, A1AT) secretion (*see* **Note 24**), urea synthesis (*see* **Note 25**), glycogen storage (*see* **Note 26**), and more importantly, inducible CYP activity including CYP1A, CYP 2D6, CYP 2B6, CYP 3A, and CYP 2C9 (Fig. 3) (*see* **Note 27**).
4. Further maintenance of the differentiated HLCs can be carried out in the Hepatocyte Maintenance Medium for another 2 weeks, with medium change every 2–3 days.

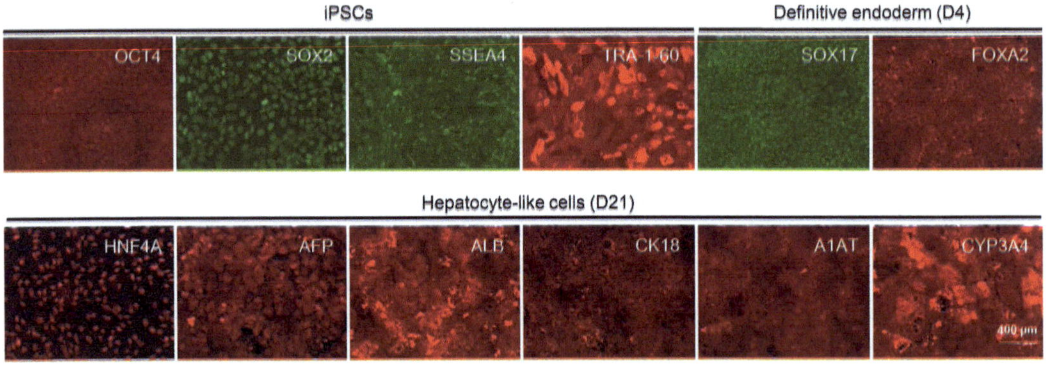

Fig. 2 Cells generated at major stages of the differentiation express characteristic stage-specific protein markers, as demonstrated by immunofluorescence staining of OCT4, SOX2, SSEA4, and TRA-1-60 for iPSCs, SOX17, and FOXA2 for definitive endoderm, and HNF4A, AFP, ALB, CK18, A1AT, and CYP3A4 for HLCs. Scale bar, 400 μm

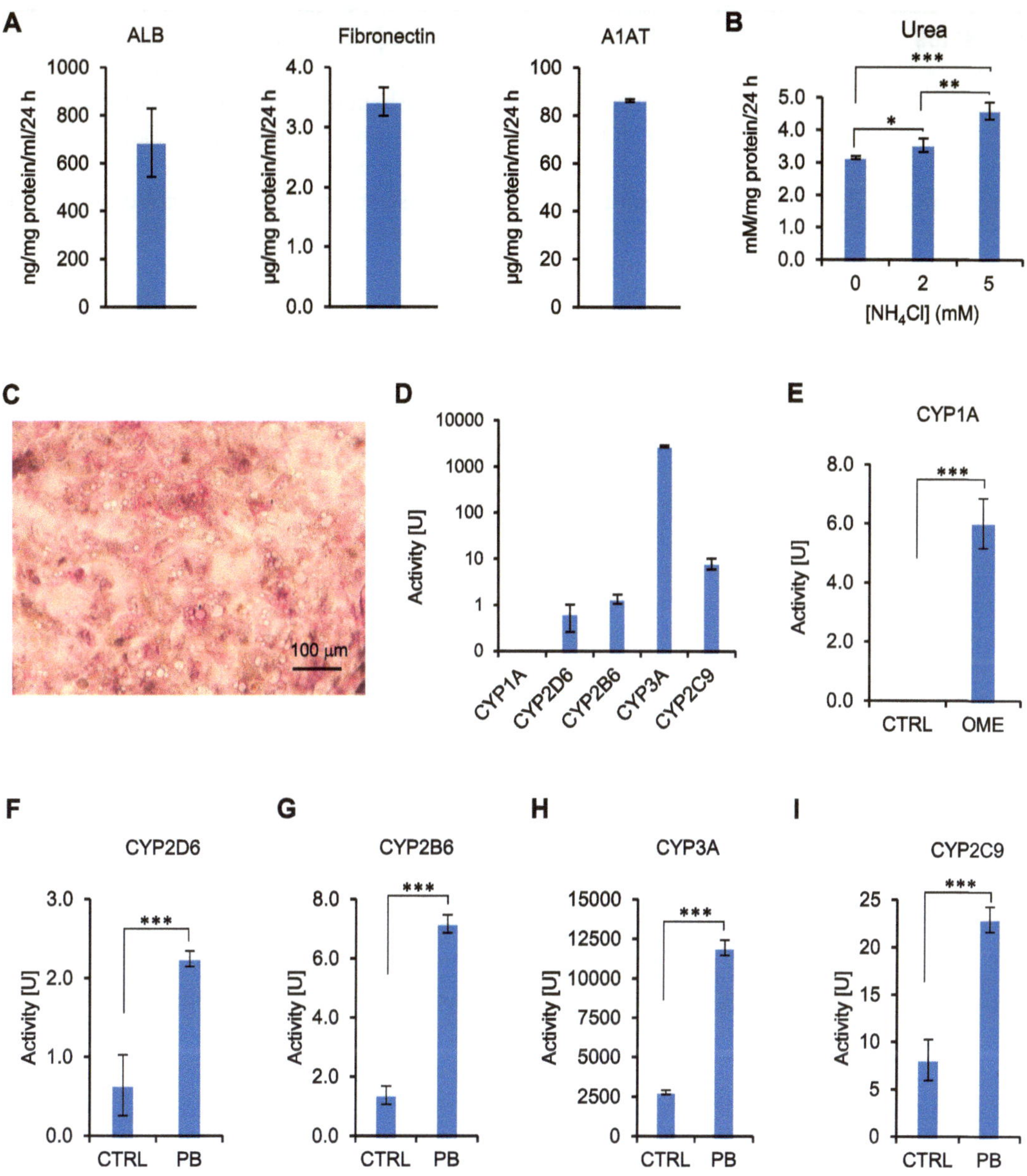

Fig. 3 Functional characterization of the final HLCs. (**a**) Secretion of serum proteins ALB, fibronectin, and A1AT. (**b**) Urea synthesis upon incubation with different concentrations of NH_4Cl. (**c**) PAS staining showing glycogen storage. Scale bar, 100 μm. (**d**) Basal activities of CYP1A, CYP2D6, CYP2B6, CYP3A, and CYP2C9. (**e–i**) Induction of CYP1A, CYP2D6, CYP2B6, CYP3A, and CYP2C9, respectively, after treatment with omeprazole (OME) or phenobarbital (PB). For CYP activity, 1 unit (U) = 1 pmol metabolite/mg protein/h, and the metabolites are acetaminophen, 1′-hydroxy bufuralol, hydroxy bupropion, 1′-hydroxy midazolam, and 4′-hydroxy diclofenac for CYP1A, CYP2D6, CYP2B6, CYP3A, and CYP2C9, respectively. All values are presented as mean ± standard deviation. $n = 3$ for serum protein secretion and for CYP activity, and $n = 4$ for urea synthesis. ***$p < 0.001$ by unpaired two-tailed t test

4 Notes

1. This iPSC line was previously generated in our laboratory from the cord blood of a healthy non-Hispanic white male using self-replicative RNA (srRNA) reprogramming technology [17]. We are using it for the demonstration of the protocol described in this chapter. However, this protocol is generally applicable to all human iPSC lines. We recommend comprehensively characterizing the iPSC line to be differentiated to ensure that the cell line is fully pluripotent and has the correct karyotype before initiating the differentiation [18].
2. iPSC lines maintained in other stem cell culture systems can be readily transferred to the Cellartis DEF-CS Culture System. Fresh cultures can be transferred at passage and cryopreserved cells can be thawed and seeded directly using the Cellartis DEF-CS Culture System following the instructions of the manufacturer. It generally takes 2–5 passages to adapt a cell line to the Cellartis DEF-CS Culture System. When initially transferring iPSCs to the Cellartis DEF-CS Culture System, some cell characteristics might be different from previously used culture systems. The Cellartis DEF-CS Culture System utilizes a single cell passaging method instead of the EDTA-based cell passaging for traditional colony type culture of stem cells, which deposits small cell aggregates for re-plating. The newly passaged cells in the Cellartis DEF-CS Culture System tend to spread out on the tissue culture plate. The cells will get denser with proliferation and exhibit the typical undifferentiated stem cell morphology eventually (e.g., small round compact cells with large nuclei and notable nucleoli).
3. Cellartis DEF-CS COAT-1 should be stored at 2–8 °C. Make sure to dilute COAT-1 using DPBS+/+. Cells do not adhere to the plate if COAT-1 has been diluted in DPBS−/−.
4. When initially transferring iPSCs to the Cellartis DEF-CS Culture System from other stem cell culture systems, it is recommended to use a lower dilution (1:5 or 1:10) of COAT-1 during the first few passages to provide extra support during the adaptation process.
5. Ensure that the seeding density of iPSCs is at least 4.0×10^4 cells/cm^2. If the cells are too sparse or seem to differentiate, consider using higher seeding density. Also, when transferring iPSCs from other culture systems to the Cellartis DEF-CS Culture System, the cells may benefit from a higher seeding density for the first few passages.
6. High quality starting iPSCs serve as the key first step to achieve efficient and consistent hepatocyte differentiation. We recommend routinely monitoring the cell morphology and the

expression of major pluripotency protein markers (e.g., OCT4, SOX2, SSEA4, and TRA-1-60). Low quality starting cells will result in non-uniform differentiation characterized by the presence of heterogeneous cell populations at the end of the differentiation process.

7. Passaged iPSCs generally reach optimal cell density around 1.5–2.0 × 10^5 cells/cm^2 on Day 3 after passaging following this protocol and the cells exhibit typical undifferentiated stem cell morphology; this indicates a good starting point for definitive endoderm differentiation. Decreased cell viability will be observed if cell density reaches 3.0 × 10^5 cells/cm^2 or higher before cell dissociation, resulting in sparse cells after plating and eventually heterogeneous differentiation of definitive endoderm cells.
8. Matrigel provides a good cell culture surface for definitive endoderm and hepatocyte differentiation. Both Corning Matrigel hESC-Qualified Matrix (Corning) and Geltrex hESC-Qualified Matrix (Thermo Fisher Scientific) have been tested and validated for this purpose.
9. Do not allow the Matrigel aliquots to warm to room temperature. It is crucial that all items remain cold to prevent gel forming prior to distribution onto tissue culture plates.
10. The seeding density might need to be adjusted for different iPSC lines so that a 90–100% confluence can be reached after 16–24 h incubation. Longer time of cell culture before initiation of definitive endoderm differentiation is not recommended.
11. Immunofluorescence staining of SOX17 (e.g., Abcam, Cat # ab84990) and FOXA2 (e.g., Abcam, Cat # ab108422) can be used to assess the purity of definitive endoderm. It can also be assessed by flow cytometry analysis following the protocol of STEMdiff Definitive Endoderm Kit using fluorophore-conjugated anti-CXCR4, anti-c-Kit (CD117), and anti-SOX17 antibodies.
12. Determine the optimum duration of Accutase treatment necessary for definitive endoderm cell dissociation. Do not overtreat the cells as it may lead to low cell viability and low rate of cell attachment after re-plating.
13. Prompt but gentle handling of dissociated definitive endoderm cells is important to ensure cell viability for subsequent re-plating.
14. It is essential to add ROCK inhibitor Y-27632 (10 μM final) to the cell suspension before re-plating.
15. The optimal seeding density of definitive endoderm cells for hepatic specification can vary from line to line. Adjust cell

density for different cell lines to achieve 85–95% confluence the day after definitive endoderm cell re-plating. We recommend a starting density of 8.0×10^4 cells/cm^2.

16. Follow the guideline of the manufacturer to thaw and aliquot the Matrigel matrix at low temperature and coat the tissue culture plates following the current protocol. Improper handling of Matrigel may lead to a low rate of definitive endoderm cell attachment.
17. Cell death occurring after definitive endoderm cell dissociation and re-plating may result in sub-optimal cell confluence rate at the beginning of hepatic specification.
18. HGF represents a critical component to guide hepatic specification at this stage. Keep track of the lot number to ensure the quality of HGF. Add HGF stock solution (100 μg/ml) freshly to the Hepatic Specification Medium (1:1000 dilution) prior to medium change.
19. Ensure correct concentration of DMSO (1%) in the Hepatic Specification Medium. Higher concentrations of DMSO may exert cytotoxic effect on the cells.
20. At this stage, cells should express major hepatocyte marker proteins such as ALB and A1AT, which can be evaluated by immunofluorescence staining. Furthermore, cells should secrete certain levels of serum proteins such as ALB and fibronectin, which can be measured by ELISA in cell culture supernatant.
21. Medium change should be carried out gently but quickly. Improper handling of the cells may lead to cell detachment or even peeling off of the entire cell monolayer during the hepatocyte maturation and maintenance stage.
22. Immunofluorescence staining can be carried out to assess the expression of major hepatocyte protein markers. The antibodies in our experiments include anti-HNF4A (Sigma-Aldrich, Cat # HPA004712), anti-AFP (R&D systems, Cat # MAB1368), anti-ALB (R&D systems, Cat # MAB1455), anti-CK18 (R&D systems, Cat # MAB7619), anti-A1AT (R&D systems, Cat # MAB1268), and anti-CYP3A4 (Proteintech, Cat # 67110). Similar antibodies from other sources may also be tested.
23. AFP, which is a marker protein of fetal hepatocytes, remains expressed in HLCs of Day 21, suggesting an immature phenotype. However, the signal of AFP immunofluorescence staining has decreased in comparison with that of Day 15. Moreover, it is expected that AFP will further decrease with prolonged maintenance of HLCs beyond Day 21.

24. ELISA can be used to determine the serum protein secretion in the cell culture supernatant. We used Human Albumin ELISA Kit, Human Fibronectin CatchPoint SimpleStep ELISA Kit, and Human alpha 1 Antitrypsin (SERPINA1) CatchPoint SimpleStep ELISA Kit (all from Abcam) in our experiments.
25. Urea synthesis can be analyzed through measuring urea levels in the cell culture supernatant. We use the Urea Assay Kit from Abcam in our experiments.
26. The Periodic Acid-Schiff (PAS) Kit from Sigma-Aldrich is used in our laboratory to evaluate glycogen storage in HLCs.
27. CYP metabolic activity of HLCs was analyzed following a protocol adapted from [8]. Inducible activities of CYP1A, CYP2D6, CYP2B6, CYP3A, CYP2C9, but not CYP2C19 are detectable in the final HLCs.

Acknowledgments

The authors would like to thank Dr. Marianne Miliotis-Solomotis and Dr. Mary E. Torrence for critical review of the manuscripts as well as for their support of this work. The authors would also like to thank Dr. Seung Bum Park at the National Institute of Diabetes and Digestive and Kidney Diseases for helpful discussions during the development of this protocol. This work was supported by internal funds of the U.S. Food and Drug Administration. The statements made in this article are those of the authors and do not necessarily represent views, opinions, or policies of the U.S. Food and Drug Administration.

References

1. Song Z, Cai J, Liu Y, Zhao D, Yong J, Duo S, Song X, Guo Y, Zhao Y, Qin H, Yin X, Wu C, Che J, Lu S, Ding M, Deng H (2009) Efficient generation of hepatocyte-like cells from human induced pluripotent stem cells. Cell Res 19 (11):1233–1242. https://doi.org/10.1038/cr.2009.107
2. Chen YF, Tseng CY, Wang HW, Kuo HC, Yang VW, Lee OK (2012) Rapid generation of mature hepatocyte-like cells from human induced pluripotent stem cells by an efficient three-step protocol. Hepatology 55 (4):1193–1203. https://doi.org/10.1002/hep.24790
3. Hannan NR, Segeritz CP, Touboul T, Vallier L (2013) Production of hepatocyte-like cells from human pluripotent stem cells. Nat Protoc 8(2):430–437. https://doi.org/10.1038/nprot.2012.153
4. Ma X, Duan Y, Tschudy-Seney B, Roll G, Behbahan IS, Ahuja TP, Tolstikov V, Wang C, McGee J, Khoobyari S, Nolta JA, Willenbring H, Zern MA (2013) Highly efficient differentiation of functional hepatocytes from human induced pluripotent stem cells. Stem Cells Transl Med 2(6):409–419. https://doi.org/10.5966/sctm.2012-0160
5. Mallanna SK, Duncan SA (2013) Differentiation of hepatocytes from pluripotent stem cells. Curr Protoc Stem Cell Biol 26:1G 4 1–1G 4 13. https://doi.org/10.1002/9780470151808.sc01g04s26
6. Siller R, Greenhough S, Naumovska E, Sullivan GJ (2015) Small-molecule-driven hepatocyte differentiation of human pluripotent stem cells. Stem Cell Rep 4(5):939–952. https://doi.org/10.1016/j.stemcr.2015.04.001

7. Carpentier A, Nimgaonkar I, Chu V, Xia Y, Hu Z, Liang TJ (2016) Hepatic differentiation of human pluripotent stem cells in miniaturized format suitable for high-throughput screen. Stem Cell Res 16(3):640–650. https://doi.org/10.1016/j.scr.2016.03.009
8. Asplund A, Pradip A, van Giezen M, Aspegren A, Choukair H, Rehnstrom M, Jacobsson S, Ghosheh N, El Hajjam D, Holmgren S, Larsson S, Benecke J, Butron M, Wigander A, Noaksson K, Sartipy P, Bjorquist P, Edsbagge J, Kuppers-Munther B (2016) One standardized differentiation procedure robustly generates homogenous hepatocyte cultures displaying metabolic diversity from a large panel of human pluripotent stem cells. Stem Cell Rev Rep 12 (1):90–104. https://doi.org/10.1007/s12015-015-9621-9
9. Wang Y, Alhaque S, Cameron K, Meseguer-Ripolles J, Lucendo-Villarin B, Rashidi H, Hay DC (2017) Defined and scalable generation of hepatocyte-like cells from human pluripotent stem cells. J Vis Exp 121:55355. https://doi.org/10.3791/55355
10. Varghese DS, Alawathugoda TT, Ansari SA (2019) Fine tuning of hepatocyte differentiation from human embryonic stem cells: growth factor vs. Small Molecule-Based Approaches. Stem Cells Int 2019:5968236. https://doi.org/10.1155/2019/5968236
11. Siller R, Sullivan GJ (2017) Rapid screening of the endodermal differentiation potential of human pluripotent stem cells. Curr Protoc Stem Cell Biol 43:1G 7 1–1G 7 23. https://doi.org/10.1002/cpsc.36
12. Corbett JL, Duncan SA (2019) iPSC-derived hepatocytes as a platform for disease modeling and drug discovery. Front Med (Lausanne) 6:265. https://doi.org/10.3389/fmed.2019.00265
13. Takayama K, Mizuguchi H (2017) Generation of human pluripotent stem cell-derived hepatocyte-like cells for drug toxicity screening. Drug Metab Pharmacokinet 32(1):12–20. https://doi.org/10.1016/j.dmpk.2016.10.408
14. Roy-Chowdhury N, Wang X, Guha C, Roy-Chowdhury J (2017) Hepatocyte-like cells derived from induced pluripotent stem cells. Hepatol Int 11(1):54–69. https://doi.org/10.1007/s12072-016-9757-y
15. Ang LT, Tan AKY, Autio MI, Goh SH, Choo SH, Lee KL, Tan J, Pan B, Lee JJH, Lum JJ, Lim CYY, Yeo IKX, Wong CJY, Liu M, Oh JLL, Chia CPL, Loh CH, Chen A, Chen Q, Weissman IL, Loh KM, Lim B (2018) A roadmap for human liver differentiation from pluripotent stem cells. Cell Rep 22 (8):2190–2205. https://doi.org/10.1016/j.celrep.2018.01.087
16. Chen KG, Mallon BS, Hamilton RS, Kozhich OA, Park K, Hoeppner DJ, Robey PG, McKay RD (2012) Non-colony type monolayer culture of human embryonic stem cells. Stem Cell Res 9(3):237–248. https://doi.org/10.1016/j.scr.2012.06.003
17. Gao X, Yourick JJ, Sprando RL (2018) Generation of nine induced pluripotent stem cell lines as an ethnic diversity panel. Stem Cell Res 31:193–196. https://doi.org/10.1016/j.scr.2018.07.013
18. Asprer JS, Lakshmipathy U (2015) Current methods and challenges in the comprehensive characterization of human pluripotent stem cells. Stem Cell Rev Rep 11(2):357–372. https://doi.org/10.1007/s12015-014-9580-6

Chapter 10

Differentiation of Human Induced Pluripotent Stem Cells into Cortical Neurons to Advance Precision Medicine

M. Catarina Silva, Ghata Nandi, and Stephen J. Haggarty

Abstract

A major obstacle in studying human central nervous system (CNS) diseases is inaccessibility to the affected tissue and cells. Even in limited cases where tissue is available through surgical interventions, differentiated neurons cannot be maintained for extended time frames, which is prohibitive for experimental repetition and scalability. Advances in methodologies for reprogramming human somatic cells into induced pluripotent stem cells (iPSC) and directed differentiation of human neurons in culture now allow access to physiological and disease relevant cell types. In particular, patient iPSC-derived neurons represent unique ex vivo neuronal networks that allow investigating disease genetic and molecular pathways in physiologically accurate cellular microenvironments, importantly recapitulating molecular and cellular phenotypic aspects of disease. Generation of functional neural cells from iPSCs relies on manipulation of culture formats in the presence of specific factors that promote the conversion of pluripotent stem cells into neurons. To this end, several experimental protocols have been developed. Direct differentiation of stem cells into post-mitotic neurons is usually associated with low throughput, low yield, and high technical variability. Instead, methods relying on expansion of the intermediate neural progenitor cells (NPCs) show incredible potential for posterior generation of suitable neuronal cultures for cellular and biochemical assays, as well as drug screening. NPCs are expandable, self-renewable multipotent cells that can differentiate into astrocytes, oligodendrocytes, and electrically active neurons. Here, we describe a protocol for generating iPSC-derived NPCs via formation of neural aggregates and selection of NPC precursor neural rosettes, followed by a simple and reproducible method for generating a mixed population of cortical-like neurons through growth factor withdrawal. Implementation of this protocol has the potential to advance the goals of precision medicine research for both neurological and psychiatric disorders.

Key words Induced pluripotent stem cells, Neural rosettes, Neural progenitor cells, Neural differentiation, Human neurons, Drug discovery, Precision medicine

1 Introduction

In 1998, Thomson et al. published a methodology for isolating and culturing human embryonic stem cells from blastocysts [1]. However, the need to harvest cells from human embryos raised major ethical concerns and limited broad application of this strategy to the study of diverse human diseases. In 2006, Yamanaka and colleagues

Nagarajan Kannan and Philip Beer (eds.), *Stem Cell Assays: Methods and Protocols*, Methods in Molecular Biology, vol. 2429, https://doi.org/10.1007/978-1-0716-1979-7_10,

demonstrated that human mature somatic cells, such as dermal fibroblasts, could be reprogrammed into induced pluripotent stem cells (iPSCs) through cellular transduction of four key transcription factors: OCT3/4, SOX2, c-MYC, and KLF4, now commonly referred to as the "Yamanaka factors" [2, 3]. Given their pluripotency properties, iPSCs are particularly relevant for the study of tissues and cell types that are of limited availability and cannot be accessed non-invasively, with the central nervous system (CNS) providing a prominent example. Over the ensuing decade and a half, significant technical progress and improved robustness of reprograming strategies [4–10] have allowed iPSCs to become a fundamental tool in the study of cell-replacement therapy, human neuro-development, and generation of genetically accurate model systems of neuronal diseases [11–25]. Moreover, the urgent need for better therapies for CNS diseases has endorsed the use of patient-specific iPSC-derived neuronal models to not only investigate the molecular and cellular mechanisms underlying disease etiology, but also in the drug discovery process, namely in the early testing of efficacy and toxicity of new drug candidates in a human neuronal context, with the goal of increasing later clinical trials success [26, 27]. Genotypic and phenotypic molecular signatures, together with drug response profiles in a physiological and disease relevant context, fundamentally aid in understanding complex disease molecular mechanisms [17, 24, 28].

Several protocols have been developed to convert human iPSCs into differentiated neurons [29–35]. Initial labor-intensive protocols for iPSC direct differentiation lacked scalability and showed significant variability both within a cell line and between iPSC lines, a major challenge for phenotypic studies or drug screening. To circumvent these challenges, methods relying on expansion of an intermediate neural stage were developed [34, 36–41]. With this approach, iPSCs capacity for differentiation is enhanced by promoting the formation of embryoid body-like aggregates (EBs) under non-adherent culture conditions, and subsequent neural rosette structures that are precursors of neural progenitor cells (NPCs). NPCs are expandable, self-renewable multipotent cells under a defined medium with growth factors, and can differentiate into astrocytes, oligodendrocytes, and functional, electrically active neurons [35–37, 40–45].

Here, we will focus on a three-step protocol consisting of (1) culture and propagation of human iPSCs, (2) iPSC neural induction and NPC selection, and (3) neural differentiation (Fig. 1). In part (1), an overview of iPSC feeder-free culture is given, with particular focus on optimal culture conditions conducive to neural induction. In part (2), iPSC formation of EBs is described, based on an optimized strategy using conical, microwell plates designed to promote uniform EB size, maximize neural rosette formation and expansion into monolayer NPCs, which are

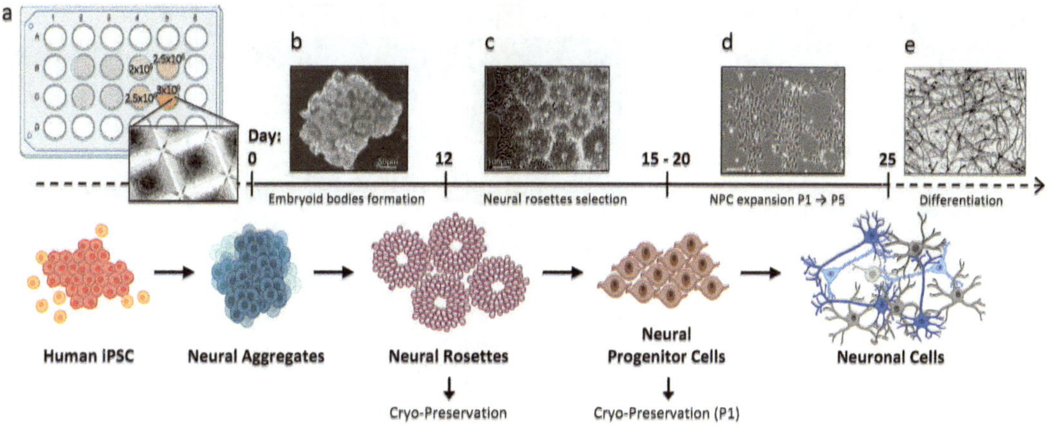

Fig. 1 Overview of the methodology for human iPSC differentiation into neuronal cells. (**a**) iPSCs are plated at multiple densities (for example 2×10^6, 2.5×10^6, 3×10^6 cells/well) in AggreWell-800 microwell plates with a conical-well format conducive to (**b**) formation of embryoid body spheroids (neural aggregates). (**c**) Neural rosettes can then be selected and replated, and (**d**) expanded as a monolayer of neural progenitor cells (NPC). (**e**) Neuronal differentiation is triggered by growth factors withdrawal from the culture medium. Cryopreservation (banking) of neural rosettes and NPCs at passage 1 (P1) allows restarting neural expansion and differentiation from an intermediary state instead of iPSCs. (Created with BioRender.com)

then banked and quality control-assessed [37, 41, 46]. Finally, in part (3), we will focus on NPC differentiation into neurons by growth factor withdrawal and in specific plate formats to accommodate different experimental goals. With this methodology, neural differentiation can be maintained for more than 5 months, without selectivity for a specific neuronal subtype (Fig. 2), giving rise to mature neurons representative of multiple cortical layers (Fig. 2) and with strong action potentials [37, 40, 41, 44]. Some methodology variations will also be addressed in Subheading 4 to suit different experimental needs. Although beyond the scope of this article, it is also crucial that, at each stage, appropriate quality control assessment and cellular characterization are executed to ensure successful generation of human neuronal cultures. This includes proper quality control of iPSCs (Table 1) and characterization of newly generated NPCs (Table 2) [40, 41, 46–48]. By employing the methods described here, we and others have successfully generated multiple neuronal cell models to study the mechanisms of tau protein pathophysiology in frontotemporal dementia [41, 46, 48], as well as other CNS disorders [40, 47, 49–54], and have performed small molecule screens to aid in the development of new therapeutic strategies for neurodegenerative diseases associated with tau pathology [55, 56] as well as for neuropsychiatric disorders [57–60].

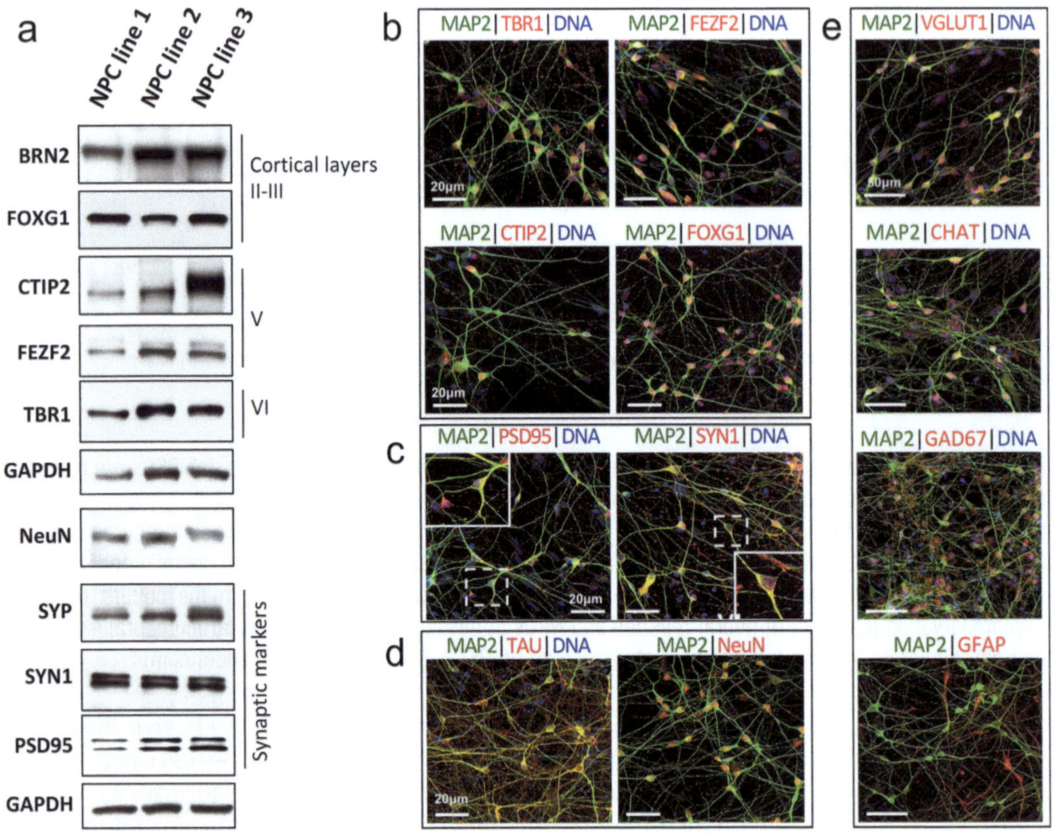

Fig. 2 Human iPSC-derived neurons characterization at the level of synaptic, cortical, and neuro-class markers [68]. (**a**) Expression of cortical layers II–III, V and VI markers and synaptic proteins, in 5-week differentiated neurons from NPC, by western blot analysis. GAPDH served as loading control. (**b–e**) IF analysis of cortical (**b**), synaptic (**c**), neuronal-specific (**d**), and neuronal class (**e**) markers (red), in the background of MAP2 staining (green), in 5-week differentiated neurons. Dotted insets correspond to zoom-in images

2 Materials

All equipment and solutions must be sterile. Tissue culture aseptic practices should be used at all times, including personal protective equipment (PPE) and wiping of all surfaces and materials with 70% ethanol. All materials and reagents should be handled in a sterile hood certified for work with human cells. Most tissue culture medium and reagents are kept at 4 °C or aliquoted and frozen at −20 °C, unless otherwise specified. Carefully follow waste disposal regulations when disposing of biological waste materials.

2.1 Matrigel-Coated Plates

1. Matrigel (for example, Corning 354277), aliquoted according to the Lot number and dilution factor indicated in the respective "Certificate of Analysis." Stored at −20 °C.
2. DMEM/F12 with phenol red, stored at 4 °C.

Table 1
Characterization and quality control of iPSC lines. iPSCs are typically characterized at the level of genomic integrity, morphology, expansion capacity, and expression of pluripotency markers. In vitro differentiation of iPSC-derived embryoid bodies and tri-lineage analysis is employed to assess iPSC capacity for spontaneous differentiation into the three germ layers with expression of β-III tubulin (TUJ1) for ectoderm, smooth muscle actin (SMA) for mesoderm, and alpha-fetoprotein (AFP) for endoderm

	Morphological	Genetic	Pluripotency
Quality control assay	• Light microscopy • Colony formation without differentiation • Expansion capacity	• Silenced reprogramming genes, RT-PCR • Pluripotency markers (*SSEA4, NANOG, OCT4, SOX2, TRA1–60*): immunofluorescence and RT-PCR • Genomic stability by G-band karyotyping • Sanger sequencing	• iPSC-derived embryoid body (EB) formation • iPSC differentiation into 3-germ layers: ectoderm (*TUJ1*), mesoderm (*SMA*), endoderm (*AFP*) markers RT-PCR

3. Tissue culture grade 6-well plates.
4. 50 mL conical tubes, plastic.
5. Keep plates, pipettes, and conical tube where Matrigel solution is prepared at 4 °C and work over ice (*see* **Note 1**).
6. Coating (according to dilution factor of Lot number): In an ice-cold 50 mL conical tube and working over ice, mix an aliquot (e.g., 250 μL) of Matrigel in 25 mL of DMEM/F12. This solution is sufficient to coat 4× 6-well plates (also refrigerated) with 1 mL/well. Make sure the entire surface of the well is covered by the solution, without air bubbles. For other plate formats, adjust volume per well or dish as needed, maintaining a coating ratio of 0.1 mL/cm^2 (*see* **Note 2**). Remove the plates from ice and leave at room temperature (inside the hood) for at least 1 h (3 h is optimal).
7. Plates can be coated in advance and stored at 4 °C for approximately 2–3 weeks. To store, add 1 mL of DMEM/F12 per well, and mix with the Matrigel solution by swirling gently. Seal each plate with parafilm and place at 4 °C.

2.2 Induced Pluripotent Stem Cells (iPSC) Medium

1. Sterile single-use vacuum filter unit 0.2 μm, 500 mL.
2. 15 mL and 50 mL conical tubes, plastic.
3. 10,000 U/mL penicillin–streptomycin (100×), thawed at room temperature and aliquoted in 10 mL volumes into 15 mL conical tubes. Stored at −20 °C.
4. mTeSR Plus medium (StemCell Technology 05825) (*see* **Note 3**): To prepare working medium, thaw the 5× supplement (100 mL) at room temperature, mix with 400 mL mTeSR

Table 2

Characterization of human iPSC-derived neural progenitor cells (NPC) and neurons. NPCs lineage is confirmed by IF of specific markers (NESTIN, SOX2, Musashi, PAX6), and genomic stability can be tested by array-based comparative genomic hybridization (aCGH) analysis to identify sub-karyotype anomalies. Differentiated neurons should also pass quality control before characterization as a model system or drug screening. This includes assessment of expression of specific neuronal markers such as NeuN (neuronal RNA-binding protein, nuclear), microtubule-associated proteins MAP2 and tau, synaptic proteins (synaptophysin, SYN1, PSD95), and cortical layer markers (TBR1, CTIP2, FEZF2, FOXG1, BRN2). Neuronal subtype specific markers can also be tested at this point

Quality control assay	Morphological	Lineage	Cellular
NPC	• Light microscopy	• Lineage markers: NESTIN, SOX2, MUSASHI, PAX6 (immunofluorescence)	• Neural differentiation capacity by growth factors withdrawal
	• Proliferative capacity	• Genomic stability by G-band karyotyping	
	• Passaging survival	• Genomic stability by aCGH (array-based comparative genomic hybridization) for sub-karyotype anomalies	
Neurons	• Processes formation, growth, and branching	• Neuronal proteins expression (western blot) and localization (immunofluorescence): NeuN, MAP2, Tau, PSD95, SYN1	• Examples of markers for subtypes of neurons: Glutamatergic (VGLUT1, GLUR1), cholinergic (VACHT, CHAT), dopaminergic (DRD3, TH), GABAergic (GAD67), serotonergic (5HT1B), astrocytes/glial (GFAP)
	• Light microscopy	• Maturation assessment with cortical layer markers: TBR1, CTIP2, FEZF2, FOXG1, BRN2	• Biochemical and cellular phenotyping: Model-dependent

Plus basal medium, and 5 mL of 100× penicillin-streptomycin (*see* **Note 4**). Filter-sterilize the solution through a 0.2 μm pore-size filter. Store in 25–50 mL aliquots at 4 °C for up to 2 weeks. Store at −20 °C for up to 6 months.

2.3 iPSC Thawing, Maintenance, and Passaging

1. DMEM/F12 with phenol red, stored at 4 °C.
2. Calcium-free PBS.
3. 5 mM Y-27632, 500UG (for example, Millipore 688001), stored at −20 °C.
4. mTeSR Plus medium (Subheading 2.2).
5. ReLeSR Selection and Passaging Reagent (StemCell Technologies 05872), stored at room temperature.

2.4 iPSC Neural Induction

1. 15 mL conical tubes, plastic.
2. StemCell Technologies AggreWell-800 plates (StemCell Technologies 34811).
3. Falcon cell strainers, blue, sterile, 40 μm.
4. Tissue culture grade 6-well plates.
5. Automatic cell counter.
6. Trypan Blue, stored at room temperature.
7. DMEM/F12 with phenol red, stored at 4 °C.
8. 5 mM Y-27632, 500UG (for example, Millipore 688001), stored at −20 °C.
9. Neural Induction Medium/NIM (StemCell Technology 05835): Thaw medium at 4 °C and aliquot 12 mL volumes into 15 mL conical tubes. Re-freeze at −20 °C. Thaw once for use and then keep at 4 °C.
10. Accutase (Sigma A6964): Thaw at room temperature, aliquot into 5 mL volumes, and store at −20 °C. Thaw once for use and then keep at 4 °C.
11. STEMdiff Neural Rosette Selection Reagent (StemCell Technology 05832), stored at room temperature.
12. Neural progenitor cell/NPC medium (Subheading 2.6).
13. Poly-L-ornithine and laminin/POL coated plates (Subheading 2.7).

2.5 Neural Rosettes Cryopreservation

1. 1.5 mL or 2 mL cryovials (sterile).
2. Conical tubes, plastic.
3. 10 mL syringe and 0.2 μm pore-size filter.
4. Cell freezing container.
5. KnockOut Serum Replacement medium/KOSR (Gibco), stored at −20 °C.
6. 2× Neural Freezing medium (5 mL), prepared by mixing 1 mL of DMSO with 4 mL of KOSR, sterilized through a 0.2 μm pore-size filter. It can be stored at 4 °C for up to 4 weeks.

2.6 Neural Progenitor Cell (NPC) Medium

1. Sterile single-use vacuum filter unit 0.2 μm, 500 mL.
2. DMEM, stored at 4 °C.
3. Phosphate-Buffered Saline/PBS without calcium and magnesium.
4. Ham's F12, stored at 4 °C.
5. 10,000 U/mL penicillin–streptomycin (100×), thawed at room temperature and aliquoted in 10 mL volumes into 15 mL conical tubes. Stored at −20 °C.

6. B-27 (Gibco 17504-044), stored at −20 °C.
7. Recombinant human Epidermal Growth Factor/EGF lyophilized powder (for example, Sigma E9644): Prepare 20 μg/mL stock (1000× working concentration) by reconstituting the powder in 1 mL DMEM, and transfer to a 50 mL conical tube. Add 9 mL DMEM and filter-sterilize through a 0.2 μm-size pore filter. Aliquot in 0.5 mL volumes and store at −20 °C. Thaw once and then keep at 4 °C.
8. Recombinant human Fibroblast Growth Factor/FGF (for example, ReproCell 03-0002): Prepare 20 μg/mL stock (1000× working concentration) by adding 1 mL of PBS to 50 μg stock, let dissolve and transfer to a 50 mL conical tube. Make the volume up to 5 mL with PBS and filter-sterilize through a 0.2 μm-size pore filter. Aliquot in 0.5 mL volumes and store at −20 °C. Thaw once and then keep at 4 °C.
9. Heparin, sodium salt (for example, Sigma H3149): Prepare a 5 mg/mL stock (1000× working concentration) by weighing 0.22 g of heparin powder and mixing with 10 mL Ham's F12 in a 50 mL conical tube. Dissolve and add 44 mL of Ham's F12 medium. Filter-sterilize through a 0.2 μm-size pore filter. Aliquot in 1 mL volumes and store at −20 °C. Thaw once and then keep at 4 °C.
10. NPC medium: 350 mL DMEM, 150 mL Ham's F12, 5 mL of penicillin–streptomycin 100×, and 10 mL of B-27 50×. Filter-sterilize using a 0.2 μm filter vacuum bottle (500 mL). Store at 4 °C. Before use in NPC cultures, aliquot and supplement with growth factors EGF, FGF, and heparin (E.F.H.), by adding each at a 1:1000 (v/v) dilution (*see* **Note 5**).

2.7 Poly-L-Ornithine and Laminin (POL) Coated Plates

1. Tissue culture grade 6-well plates.
2. Tissue culture grade plastic bottles.
3. Phosphate-Buffered Saline/PBS without calcium and magnesium.
4. Poly-L-ornithine (for example, Sigma P3655): To a bottle containing 50 mg powder poly-L-ornithine bromide add 5 mL of water, mix and filter-sterilize through a 0.2 μm-size pore filter, making a 10 mg/mL stock solution. Aliquot in 0.5 mL volumes in eppendorf tubes and store at −20 °C. Thaw once and then keep at 4 °C (*see* **Note 6**).
5. Laminin (for example, Sigma L2020-1MG): 1 mg/mL commercial stock solution vials are stored at −20 °C. Thaw only once and then keep at 4 °C.
6. Poly-L-Ornithine and Laminin (POL) plate coating: Label each plate with "POL—date." Dilute the poly-L-ornithine (PO) 1: 500 (v/v) in water to a final concentration of 20 μg/mL. For a

6-well plate, add 2 mL/well and incubate at 37 °C for 2 h. Meanwhile, dilute the laminin 1:200 (v/v) in PBS to a final concentration of 5 μg/mL (*see* **Note 7**). Aspirate the PO solution from each well and replace with 2 mL/well of laminin. Incubate the plates at 37 °C for a minimum of 2 h or overnight (optimal). If using other plate or dish format, keep the coating ratio at 0.2 mL/cm^2. POL-coated plates that are not immediately used can be stored in the laminin solution at 4 °C for up to 4 weeks (*see* **Note 8**).

2.8 NPC Maintenance and Passaging

1. Tissue culture grade 6-well plates.
2. 15 mL or 50 mL conical tubes, plastic.
3. Automatic cell counter.
4. Trypan Blue, stored at room temperature.
5. Phosphate-Buffered Saline/PBS without calcium and magnesium.
6. POL-coated plates (Subheading 2.7).
7. TrypLE, no phenol red (for example, Gibco 12563-029), stored at 4 °C.
8. Neural progenitor cell/NPC medium, and EGF, FGF and heparin (E.F.H.) (Subheading 2.6).

2.9 NPCs Cryopreservation

1. 1.5 mL or 2 mL cryovials (sterile).
2. Conical tubes, plastic.
3. Cell freezing container.
4. DMSO (sterile).
5. NPC medium, and EGF, FGF and heparin (E.F.H.) (Subheading 2.6).
6. NPC Freezing medium: 1 mL of DMSO (sterile) with 9 mL of NPC medium (0.2 μm filter-sterilized) supplemented with E.F.H. This is a 10% DMSO freezing solution (5 mL).

2.10 Thawing Neural Progenitor Cells

1. 15 mL conical tubes, plastic.
2. Tissue culture grade 6-well plates.
3. POL-coated plates (Subheading 2.7).
4. Phosphate-Buffered Saline/PBS without calcium and magnesium.
5. NPC medium, and EGF, FGF and heparin (E.F.H.) (Subheading 2.6).

2.11 NPC Differentiation and Neuronal Cultures

1. Tissue culture grade 6-well plates.
2. Tissue culture grade 24-well plates.

3. Round glass coverslips: RD German Coverslips 12 mm NO1 (Chemglass Life Sciences, Fisher 50-121-5159). High quality borosilicate hydrolytic class 1, ideal for iPSC-derived neuronal cultures.
4. Sterilize the glass coverslips: using 70% ethanol clean tweezers, transfer one coverslip into each well of a 24-well plate, and let the plate(s) stand under the UV light for 30 min.
5. 96-well black polystyrene microplates, flat clear bottom.
6. 15 mL and 50 mL conical tubes, plastic.
7. Cell lifters (for example, Corning CLS3008).
8. Multi-channel (manual or automated) 100–200 μL pipettor.
9. Multi-well vacuum manifold aspirator adaptor.
10. Microplate automated washer (optional).
11. Sterile single-use vacuum filter units 0.2 μm, 500 mL.
12. Automatic cell counter.
13. Trypan Blue, stored at room temperature.
14. DMEM, stored at 4 °C.
15. Ham's F12, stored at 4 °C.
16. 10,000 U/mL penicillin–streptomycin (100×), thawed at room temperature and aliquoted in 10 mL volumes into 15 mL conical tubes. Stored at −20 °C.
17. B-27 (Gibco 17504-044), store at −20 °C.
18. Phosphate-Buffered Saline/PBS without calcium and magnesium.
19. Poly-L-ornithine 10 mg/mL and Laminin 1 mg/mL (Subheading 2.7).
20. POL-coated plates (*see* **Note 9**): For NPC differentiation, plates are simultaneously ("s") coated with poly-L-ornithine and laminin and should be labeled with "POLs—date." Prepare a PBS solution with poly-L-ornithine diluted 1:500 (v/v) to 20 μg/mL, and laminin diluted 1:200 (v/v) to 5 μg/mL. Coat with 0.2 mL/cm^2 of POL per well or dish (2 mL/well is for a 6-well plate). Incubate at 37 °C for a minimum of 4 h or overnight (optimal). If not used immediately, store at 4 °C (*see* **Note 8**).
21. 24-well POL-coated plates: In plates already containing one round glass coverslip per well (sterilized), add 0.5 mL/well of a PBS solution with poly-L-ornithine and laminin (as in **step 20**). Incubate at 37 °C overnight. If not used immediately, store at 4 °C (*see* **Note 8**).
22. 96-well POL-coated plates: Using a multi-channel pipettor, add 100 μL/well of a PBS solution with poly-L-ornithine and

laminin (as in **step 20**). Incubate at 37 °C overnight. If not used immediately, store at 4 °C (*see* **Note 8**).

23. TrypLE, no phenol red (for example, Gibco 12563-029), stored at 4 °C.
24. NPC differentiation medium: 350 mL DMEM, 150 mL Ham's F12, 5 mL of penicillin–streptomycin 100×, and 10 mL of B-27 50× (same as NPC medium, but no E.F.H.). Filter-sterilize using a 0.2 μm filter vacuum bottle (500 mL). Store at 4 °C.

3 Methods

Carry out all procedures under sterile conditions, practicing aseptic techniques at all times. Work in a sterilized hood certified for work with human cells. Cell culture incubations are performed in a humidified 37 °C, 5% CO_2 incubator.

3.1 Induced Pluripotent Stem Cells (iPSC) Thawing

3.1.1 Preparation of Matrigel-Coated Plates

1. Remove the Matrigel plate(s) (Subheading 2.1) from 4 °C and let acclimate at room temperature for approximately 1 h. Warm an aliquot of mTeSR Plus and the DMEM/F12 in a 37 °C water bath.
2. Remove the coating medium from the plate(s) (i.e., Matrigel solution in DMEM/F12) and rinse once with DMEM/F12. Use 0.2 mL/cm^2 of well area, i.e., 2 mL/well for a 6-well plate.
3. Replace the DMEM/F12 with 1 mL/well of mTeSR Plus supplemented with 10 μM Y-27632 (*see* **Note 10**). Place plate(s) in the 37 °C incubator for ≥30 min or until ready to plate cells.
4. Aliquot mTeSR Plus medium into 15 mL conical tubes: for each vial of cryopreserved iPSCs to thaw, prepare a conical tube with 9 mL mTeSR Plus.

3.1.2 Preparation of iPSCs

1. Retrieve the cryovial(s) of iPSCs from liquid nitrogen and keep on dry ice until ready to thaw.
2. Thaw the cryovial(s) in a 37 °C waterbath until only a "pea-sized" ice chunk remains inside the vial. At that point, gently add 1 mL of mTeSR Plus medium from the prepared 9 mL aliquot tube.
3. Transfer the thawed cell mixture to the bottom of the 15 mL conical tube containing the remaining 8 mL medium. This will make a total of 10 mL iPSC suspension in mTeSR Plus. Gently invert the tube once to mix cells and medium (to wash cells from the freezing solution).
4. Centrifuge at 200 × *g* for 2 min to pellet cells.

5. Aspirate the supernatant and add 2 mL of mTeSR Plus supplemented with 10 μM Y-27632 to the pellet. Gently resuspend the cells by pipetting up and down twice without exerting any pressure against the tube (*see* **Note 11**).
6. Transfer the suspension of iPSCs onto one well of the pre-prepared Matrigel plate (Subheading 3.1.1 **steps 1–4**, 6-well plate). The final volume of medium should be 3 mL/well.
7. Place the plate in a 37 °C incubator and gently perform a compass motion (twice) to spread the cells evenly throughout each well. Then leave the plate undisturbed for 48 h, before commencing cell feeding.

3.2 Proliferating iPSC Medium Change ("iPSC Feeding") (See Note 12)

1. Aliquot mTeSR Plus medium into a conical tube, considering that 2 mL of new medium will be needed per well of cells (6-well plate). Warm to 37 °C in a water bath.
2. Using a glass *Pasteur* pipette, aspirate the old medium from each well, aiming at the edge of each well, avoiding touching or aspirating directly from above attached cells. Use a different pipette for each cell line to avoid cross-contamination of cells.
3. Replace with 2 mL of the pre-warmed mTeSR Plus medium. Again, when adding fresh medium, aim to the edge of the well so not to disturb attached cells and growing colonies.
4. In general, iPSCs are ready to be passaged when colonies reach a confluency of 80–90% (*see* **Note 13**).

3.3 iPSC Passage

1. Take a new Matrigel plate(s) from 4 °C and let acclimate to room temperature for approximately 1 h.
2. Pre-warm mTeSR Plus medium (50 mL aliquot(s)) and DMEM/F12 in a 37 °C water bath. The volumes needed will depend on the number of iPSC wells being passaged and dilution(s) used.
3. Inspect cells under the microscope, decide on the dilution for each well of cells or cell line, and determine the number of wells and plates of Matrigel needed (*see* **Note 14**). For example, passage two wells of confluent iPSCs at a 1:3 dilution, into six new Matrigel wells (one 6-well plate).
4. Remove the coating medium (Matrigel solution in DMEM/F12) from the new Matrigel plate(s) and rinse once with DMEM/F12 by adding 2 mL/well.
5. Replace the DMEM/F12 with 2 mL/well of mTeSR Plus and place plate(s) in the 37 °C incubator for ≥30 min or until ready to plate cells.
6. Wash iPSCs by adding 2 mL/well of calcium-free PBS (*see* **Note 15**). Aspirate.

7. Add 1 mL/well of ReLeSR reagent and incubate for 1 min at room temperature. Aspirate.
8. Incubate plate(s) in the 37 °C incubator for 5 min.
9. Add 1 mL of mTeSR Plus per well and firmly tap the side of the plate for approximately 30 s or until cells of interest start to detach, check under the microscope for "chunks" of cells detached. The goal here is to dislodge healthy-looking, dense iPSC colonies, while avoiding detachment of differentiated cells. Adjust the incubation in **step 8** up to 10 min to achieve this goal.
10. Add additional volume of mTeSR Plus to the well with detached cells in order to passage according to the desired dilution and so that the final volume per new well is 4 mL (for example, for a 1:3 dilution, one well-worth of cells will be resuspended in 6 mL mTeSR Plus distributed into three wells of the new plate, with 2 mL cell suspension added per well already containing 2 mL medium). When working with large volumes of medium, transfer cells suspension to a conical tube to perform the dilution and then distribute into the final number of wells.
11. Place the plate(s) in a 37 °C incubator making a compass motion (twice) to ensure even distribution of cells through the well(s).
12. After passaging (day 1), cells should be left undisturbed for 48 h (days 2–3) to promote cell attachment and start of colony formation. There is no need to perform medium change during this period.
13. When utilizing mTeSR Plus, change medium every other day or every 2 days, according to the necessities of each cell line (*see* **Note 12**).

3.4 Derivation of Neural Rosettes from iPSC

This protocol utilizes the StemCell Technologies AggreWell-800 plates (StemCell Technologies 34811, Fig. 1a), which are multi-well culture plates (24-well format) with conical microwells that help promote the formation of spheroid embryoid bodies for neural induction [61–64] (Fig. 1b). Most of this protocol is executed in the absence of antibiotics, therefore the implementation of proper aseptic conditions is crucial.

3.4.1 Day −1 and Before: iPS Cells Expansion

1. Plate iPSCs onto Matrigel-coated plates such that the resulting colonies are medium-to-large in size at confluence, and with less than 10% of cells exhibiting differentiation. If necessary, "manually clean" the culture to scrape off differentiated cells on the day before starting neural differentiation (*see* **Note 13**).
2. The recommended number of iPSCs to setup one well of a 24-well AggreWell-800 plate is 3×10^6 cells, which

corresponds approximately to three wells of a 6-well plate of iPSC at 90% confluency. With this in mind, plan ahead the number of iPSC plates and AggreWell-800 plates needed to initiate neural induction (*see* **Note 16**).

3.4.2 Day 0: Plate Preparation and iPSC Neural Induction

1. Neural induction is initiated a day before iPSCs would normally be passaged.
2. To prepare the AggreWell-800 24-well plate, rinse each well to be used with 1 mL DMEM/F12 at room temperature.
3. Aspirate the DMEM/F12 solution and add 0.5 mL/well of Neural Induction medium (NIM) supplemented with 10 μM Y-27632 (*see* **Note 10**).
4. Centrifuge the AggreWell-800 plate at 2000 × *g* for 5 min to remove air bubbles from the medium in the microwells. Under the microscope, inspect the plate to ensure that no air bubbles remain. Otherwise, repeat plate centrifugation.
5. Place the AggreWell-800 plate in the 37 °C incubator until ready to use. It is recommended to work with only one AggreWell-800 plate at a time.
6. Aspirate medium from the iPSC wells carefully, not to disturb the cell colonies, and wash with 2 mL/well of DMEM/F12.
7. Remove the DMEM/F12 and add 750 μL Accutase per well. Incubate in the 37 °C incubator for 5 min.
8. Take the plate out of the incubator, tap on the sides and gently rock the plate to help dislodge the cells (inspect under the microscope). Return plate to the 37 °C incubator for 2 min.
9. Return plate(s) to the hood and using a 5 mL pipette, pipette the Accutase solution up and down 2–3× to detach cells and break clumps apart. In this case, a single-cell suspension is needed. Check the progress under the microscope if needed.
10. Transfer the DMEM/F12+Accutase+cells suspension to a 15 mL conical tube.
11. Rinse each well with 3 mL DMEM/F12 and transfer to the same conical tube. Repeat. At the end, add DMEM/F12 to the conical tube to a final volume of 10 mL in order to dilute and inactivate the Accutase.
12. Centrifuge the tube with cells at 300 × *g* for 5 min. Aspirate supernatant.
13. Add 1.5 mL of NIM supplemented with 10 μM Y-27632 and resuspend the pellet of cells by pipetting up and down a couple of times.
14. Using a cell counter, determine the concentration of cells in suspension. Calculate the volume of suspension corresponding to 3×10^6 cells, which will be used for one well of the

AggreWell-800 24-well plate (*see* **Note 16** on setting up additional wells per cell line, Fig. 1a).

15. Transfer that volume of iPSCs suspension to the AggreWell. Add additional NIM+Y-27632 medium if needed to reach a final volume of 2 mL/well.
16. Using a P1000 pipette, gently pipette up and down a few times to evenly distribute the cells throughout the microwells.
17. Centrifuge the AggreWell-800 plate at 100 × *g* for 3 min. Inspect under the microscope that the cells are evenly distributed in the microwells (Fig. 1a).
18. Carefully place the plate in the 37 °C incubator without disrupting the cells.

3.4.3 Day 1–4: Daily Medium Change

1. At 24 h after plating iPSCs in the AggreWell-800 plate, a spheroid of cells (aggregate) should be visible in the center of each well (inspect under the microscope, Fig. 1a).
2. Warm an aliquot of NIM in a 37 °C water bath.
3. Remove 1.5 mL of medium from each well very carefully, preferably using a motorized pipettor at a low speed, placing the tip as horizontal as possible against the well-wall and pipetting volume out without creating pressure.
4. In the same fashion, very carefully add 1.5 mL of NIM placing the pipettor as horizontal as possible against the well-wall. This is particularly crucial on the first medium exchange so not to dislodge the forming aggregates of cells at the bottom of each microwell.

3.4.4 Day 5: Neural Aggregates Transfer to POL Plate

1. Use freshly coated POL 6-well plates (Subheading 2.7), pre-warmed to 37 °C. Remove the laminin-PBS solution and wash with 4 mL/well of PBS.
2. Remove PBS and add 1 mL/well of NIM. Return plates to the 37 °C incubator until ready to use.
3. Per well of the AggreWell-800 plate with cells, prepare a 50 mL conical tube with a cell strainer (40 μm) standing on top.
4. On the AggreWell-800 plate, dislodge the neural aggregates from the microwells by firmly pipetting medium (the one already in the wells) up and down into the middle of the well using a P1000 pipettor.
5. Transfer the aggregates to the cell strainer standing on the 50 mL conical tube. If necessary, cut the tip of the P1000-tip to make the transfer easier so that the aggregates do not clog the tip and are not broken apart. Repeat until all aggregates of one well are transferred to the strainer, by repeatedly washing the AggreWell with DMEM/F12 (*see* **Note 17**).

6. Turn the strainer over and onto the destination well of a POL plate already containing 1 mL NIM. Flow 1 mL of NIM through the strainer so that the aggregates are transferred to the POL well. In this step, one AggreWell of aggregates is transferred at a 1:1 ratio onto one POL well (6-well plate).
7. Perform a wash of the cell strainer with 1 mL NIM, so that the final volume of medium is 3 mL/well in the POL plate. Make sure all cell aggregates (visible chunks on the strainer) are transferred, even if a little more medium is used in the wash.
8. Place the POL plate containing the neural aggregates in the 37 °C incubator and perform a compass motion (twice) to ensure distribution of the cells through the entire surface of each well.

3.4.5 Day 6–11: Neural Aggregates Maintenance and Rosette Formation

1. Warm an aliquot of NIM in the 37 °C water bath.
2. Every day at approximately the same time, change medium in each well containing neural aggregates: Aspirate old medium and add 2 mL/well of fresh NIM.
3. Rosette clusters should become visible approximately after 2 days on POL plates or even before (Fig. 1b). In optimal conditions, the majority of the aggregates (>80%) will contain rosette clusters.

3.4.6 Day 12: Selection and Replating Neural Rosettes

1. Neural rosettes are ready to harvest when the initial aggregates have completely flattened, and rosette clusters are clearly visible (Fig. 1c). This is usually the case any time between day three to seven after plating in POL plates but can vary for each cell line. Optimally, at this point there will be, for each cell line, one well of rosette clusters that can be immediately banked by cryopreservation (*see* Subheading 3.5) and at least one well of rosette clusters that can be expanded to generate neural progenitor cells (NPCs) (*see* **Note 16**).
2. Use freshly coated POL 6-well plates (Subheading 2.7), pre-warmed at 37 °C. Remove the laminin-PBS solution and wash with 4 mL/well of PBS.
3. Remove PBS and add 1 mL/well of NIM (*see* **Note 18**). Return plate(s) to the 37 °C incubator until needed.
4. Per well of neural rosette clusters to be transferred, pre-warm 1 mL of Neural Rosette Selection Reagent in a 37 °C water bath (*see* **Note 19**).
5. Aspirate NIM from the neural clusters-containing wells and add 1 mL/well of DMEM/F12 to wash the cells.
6. Aspirate the DMEM/F12 and add 1 mL/well of Neural Rosette Selection Reagent.

7. Incubate in the 37 °C incubator for approximately 1 h. This incubation time may need to be optimized depending on the ease of neural rosettes detachment (check under the microscope).
8. Using a P1000 pipette, carefully remove and discard the Neural Rosette Selection Reagent from each well.
9. With a P1000 pipette and a standard filtered tip, firmly expel 1 mL of DMEM/F12 into the well, aiming specifically at the rosette clusters. This will dislodge the neural rosettes from the clusters and into suspension.
10. Carefully transfer the neural rosette suspension to a 15 mL conical tube, minimizing breakage of the clusters.
11. Repeat **steps 9** and **10** until the majority of neural rosettes have been collected into the 15 mL conical tube, as determined by examination under a microscope (*see* **Note 20**).
12. Centrifuge the rosettes' suspension at 350 × *g* for 5 min.
13. Carefully aspirate the supernatant and add 2 mL of NIM. Using a P1000 pipette, gently resuspend the neural rosettes by pipetting slowly up and down twice. This will break the clusters into 1/4–1/5 of their initial size and will allow expansion of the neural progenitor cells (NPCs) mostly as a monolayer (although rosette-like clumps of cells may still form, Fig. 1d).
14. Transfer the cell suspension (2 mL) to one or multiple wells of a POL 6-well plate (depending on the initial yield and dilution intended). The final volume per well should be 3 mL of medium.
15. Transfer the plate(s) to a 37 °C incubator and perform a compass motion (twice) to distribute the neural progenitor cells across the surface of the well(s).

3.4.7 Day 14

1. Warm an aliquot of NPC medium (Subheading 2.6) in a 37 °C water bath.
2. Change the culture medium from NIM to NPC medium freshly supplemented with growth factors EGF and FGF, and heparin (E.F.H.). Replace NIM with ≥2 mL/well of NPC medium depending on the cell density per well and how quickly the medium turns yellow (due to lowering of medium pH from cell metabolism, pH-sensitive medium).

3.4.8 Day 16, 18

1. Warm an aliquot of NPC medium (Subheading 2.6) in a 37 °C water bath, and supplement with E.F.H.
2. Perform medium change with ≥2 mL/well of NPC medium.

3.4.9 Day ≳ 20: Passage and Cryopreservation of NPC at P1

1. At this point, there should be multiple wells of NPCs per cell line. Allocate a well(s) for passaging and expansion, and a well (s) for cryopreservation (passage no. 1/P1).
2. When NPCs are at more than 80% confluency, can proceed with passage no. 1 (P1) and cryopreservation (Subheading 3.6).
3. For passaging, prepare POL-coated 6-well plates (Subheading 2.7) by washing with 4 mL/well of PBS. Then, add 1.5 mL of freshly prepared NPC medium (Subheading 2.6) supplemented with E.F.H. Return plate(s) to the 37 °C incubator until needed.
4. Warm an aliquot of TrypLE in the 37 °C water bath.
5. Aspirate medium from well(s) with NPCs and add 0.5 mL/well of TrypLE. Swirl the plate to make sure the entire bottom of the well is covered by the solution.
6. Incubate at room temperature for 2–3 min, until cells start to detach. By taping the plate on the side, it should be possible to observe the cells being detached into suspension.
7. To stop the reaction, add 1.5 mL/well of NPC medium +E.F. H. Pipette the suspension up and down a couple of times to wash the entire well and transfer the contents to a 15 mL conical tube.
8. Centrifuge at 200 × *g* for 5 min and carefully aspirate the entire supernatant.
9. To the pellet of cells add 0.5 mL of NPC medium +E.F.H. per well of cells to be plated. That is, at this point, the volume of medium to resuspend cells depends on the dilution intended (*see* **Notes 21** and **22**). For example, for a 1:2 dilution, resuspend the pellet corresponding to one well of cells with 1 mL medium, and for a 1:3 dilution, resuspend the pellet with 1.5 mL medium.
10. Resuspend cells by pipetting up and down and transfer 0.5 mL of NPC suspension per well of the new POL plate (already containing 1.5 mL medium/well), to a final volume of 2 mL medium/well.
11. At this stage, it is important to culture NPCs until confluency, as confluent as possible without excessive detaching or the medium turning yellow (pH-sensitive medium, **Note 21**).
12. Passage and expand NPCs up to P5 by repeating **steps 4–10** (*see* **Note 23**).
13. Make sure to freeze more NPCs at P2 and/or P3.

3.5 Neural Rosettes Cryopreservation

1. Around day 12 of the iPSC Neural Induction procedure (Subheading 3.4), when the neural aggregates have completely

flattened, and rosette clusters are clearly visible, neural rosettes can be cryopreserved (Fig. 1).

2. Label cryovials with cell line name, medium, date, and indication that the vial contains neural rosettes.
3. Pre-chill the freezing container at 4 °C or at −20 °C, according to the type of container.
4. Warm 1 mL of Neural Rosette Selection Reagent in a 37 °C water bath (1 mL per well of rosettes to freeze).
5. Aspirate NIM from the neural clusters-containing wells and add 1 mL/well of DMEM/F12 to wash the cells. Aspirate the DMEM/F12.
6. Add 1 mL/well of Neural Rosette Selection Reagent. Incubate in the 37 °C incubator for approximately 1 h.
7. Using a P1000 pipette, carefully remove the Neural Rosette Selection Reagent from each well.
8. With a P1000 pipette and a standard filtered tip, firmly expel 1 mL of DMEM/F12 into the well, aiming specifically at the rosette clusters. This will dislodge the neural rosettes from the clusters and into suspension. Carefully transfer the neural rosette suspension to a 15 mL conical tube. Repeat until the majority of neural rosettes have been collected.
9. Centrifuge the rosettes suspension at 200 × *g* for 3 min, and discard supernatant.
10. Per number of cryovials to freeze, mix half-volume (mL) of NIM with half-volume (mL) of ice-cold 2× Neural Freezing Medium (Subheading 2.5), and gently resuspend the pellet of neural rosettes. Then, distribute 1 mL suspension into each cryovial (for example, mix 1.5 mL of NIM with 1.5 mL of freezing medium for three cryovials with 1 mL each).
11. Transfer the cryovials into a freezing container and place at −80 °C for at least 48 h. The, transfer to liquid nitrogen for long-term storage.

3.6 Cryopreservation of NPCs

1. When at 80–90% confluency, NPCs can be harvested for cryopreservation.
2. Label the cryovials by including NPC in the name, line ID, passage no., growth medium, and date.
3. Pre-chill freezing container at 4 °C or −20 °C (according to the type of container).
4. Prepare NPC Freezing medium (Subheading 2.9) and keep on ice.
5. Warm an aliquot of TrypLE in the 37 °C water bath.

6. Aspirate medium from well(s) with NPCs and add 0.5 mL/well of TrypLE. Swirl the plate to make sure the entire bottom of the well is covered by the solution.
7. Incubate at room temperature for 2–3 min, until cells start to detach. By taping the plate on the side, it should be visible the cells being detached into suspension.
8. To stop the reaction, add 1.5 mL/well of NPC medium +E.F.H.
9. Pipette the suspension up and down a couple of times to wash the entire well. Transfer the cell suspension into a 15 mL conical tube.
10. Centrifuge at 200 × *g* for 5 min and discard supernatant.
11. Per number of cryovials to freeze, add 1 mL of ice-cold NPC freezing medium to the cell pellet. Gently resuspend the cells by pipetting up and down. Distribute 1 mL cell suspension per cryovial.
12. Transfer the cryovial(s) into a freezing container and place at −80 °C for at least 48 h. Then transfer to liquid nitrogen for long-term storage.

3.7 Thawing Neural Progenitor Cells (See Note 24)

1. Pre-warm new POL-coated 6-well plates in the 37 °C incubator.
2. Pre-warm NPC medium and PBS in a 37 °C water bath.
3. Prepare the POL plate: Aspirate the coating solution and wash with 4 mL/well of PBS. For the number of wells needed (depending on the dilution used for thawing), add 1 mL/well of NPC medium +E.F.H. Transfer the plates to the 37 °C incubator until use.
4. Per cryovial of cells to thaw, prepare a 15 mL conical tube with 9 mL of NPC medium.
5. Retrieve cells from the liquid nitrogen tank and thaw in a 37 °C waterbath until only a "pea-sized" ice chunk remains inside the vial.
6. Back in the hood, gently transfer the thawed cells into the bottom of the conical tube prepared with NPC medium. Wash the cryovial with some of that medium as well, to maximize the number of cells recovered. Pipette up and down twice, to mix cells with medium and wash from freezing medium.
7. Centrifuge the tube(s) at 200 × *g* for 5 min. Aspirate the supernatant.
8. To the cell pellet add 1 mL of NPC medium +E.F.H. per number of wells to be plated (dilution is cell-line dependent). For example, if one cryovial-worth of cells is to be plated onto 4 wells (1:4 dilution), resuspend the pellet in 4 mL of NPC medium +E.F.H.

9. Transfer 1 mL of cell suspension per well of the prepared plate (already containing 1 mL/well of medium) to a final volume of 2 mL/well.
10. Transfer the plate(s) of cells to the 37 °C incubator and perform a compass motion (twice) to evenly distribute cells in the well(s).
11. Culture cells until approximately 90% confluency at which point, NPCs are ready for passaging or to set up neural differentiation (Fig. 1e).

3.8 NPC Differentiation: Neuronal Cultures in 6-Well Plates

1. In general, three wells of a 6-well NPC plate at confluency correspond to approximately 5×10^6 cells. This is a good approximation to help the investigator plan the number of NPC wells needed to initiate differentiation. However, the initial density of cells plated for differentiation might have to be optimized for each cell line for proper differentiation and to obtain a desired neuronal density at a given point in time.
2. Utilize NPC cultures at 90–95% confluency (the medium should not be yellow, which would mean cells are over-confluent, detaching, and stressed).
3. Pre-warm POLs 6-well plates (Subheading 2.11) in the 37 °C incubator.
4. Warm an aliquot of fresh NPC medium in the 37 °C water bath.
5. Prepare POLs plates by washing with 4 mL/well of PBS. Then add 1.5 mL of freshly prepared NPC medium (Subheading 2.6). Return plate(s) to the 37 °C incubator until needed.
6. Warm an aliquot of TrypLE in the 37 °C water bath.
7. Aspirate medium from wells with NPCs and add 0.5 mL/well of TrypLE. Swirl the plate to make sure the entire bottom of the well is covered by the solution.
8. Incubate at room temperature for 2–3 min, until cells start to detach. By taping the plate on the side, it should be possible to observe the cells being detached into suspension.
9. To stop the reaction, add 1.5 mL/well of NPC medium. Pipette the suspension up and down a couple of times to wash the entire well. Transfer all the cells from a single line (multiple wells) into a 15 mL or 50 mL conical tube.
10. Take an aliquot of NPC suspension and perform cell counting to determine the total number and concentration of viable cells in suspension (*see* **Note 22**).
11. Centrifuge at $200 \times g$ for 5 min and discard supernatant.
12. Resuspend the pellet in a specific volume of NPC medium to obtain the desired final concentration so that 800,000 cells are

plated per well in a 0.5 mL volume of NPC medium, i.e., prepare a cell suspension at 1.6×10^6 cells/mL of NPC medium. Usually, for 6-well plates NPCs are plated at density of 40,000–60,000 cells/cm^2 for differentiation, but this needs to be determined and optimized for each cell line and experiment requirements (*see* **Note 25**).

13. Distribute 0.5 mL of cell suspension per well of the POLs plate already containing 1.5 mL/well of NPC medium, to a final volume of 2 mL/well of medium (no growth factors!).
14. Transfer plate(s) to the 37 °C incubator and perform the compass motion (twice) to evenly distribute cells across each well and promote an even neuronal culture.
15. Twice a week (approximately every 3.5 days) replace half-volume of the culture medium with new NPC medium. Take special care to aspirate volumes at low vacuum speed from the edge of the well to avoid detachment of the differentiating neurons.
16. At the time point of interest, these cultures can be utilized for genetic, biochemical (Fig. 2a) or proteomic analysis [41, 49, 56].
17. To collect neurons, first aspirate the NPC medium.
18. (optional) Wash each well of neurons with 2 mL/well of ice-cold PBS.
19. Replace with 1 mL/well of ice-cold PBS and detach neurons into suspension with a cell lifter.
20. Collect the neuronal suspension in PBS into an eppendorf tube.
21. Centrifuge at $3000 \times g$ for 5 min and discard supernatant.
22. Flash-freeze the pellet of neurons on dry ice and store at −80 °C.

3.9 NPC Differentiation: Neuronal Cultures in 24-Well Plates with Glass Slides

1. Utilize NPC cultures at 90–95% confluency.
2. Warm an aliquot of NPC medium in the 37 °C water bath.
3. Pre-warm 24-well glass-slide plates (Subheading 2.11) in the 37 °C incubator.
4. Prepare the plates by washing with 1 mL/well of PBS. Then replace with 250 μL/well of NPC medium (Subheading 2.6). Return plate(s) to the 37 °C incubator until needed.
5. Warm an aliquot of TrypLE in the 37 °C water bath.
6. Aspirate medium from wells with NPCs and add 0.5 mL/well of TrypLE. Swirl the plate to make sure the entire bottom of the well is covered by the solution.

7. Incubate at room temperature for 2–3 min, until cells start to detach. By taping the plate on the side, it should be possible to observe the cells being detached.
8. Add 1.5 mL/well of NPC medium. Pipette the suspension up and down a couple of times to wash the entire well and transfer into a 15 mL conical tube.
9. Take an aliquot of NPC suspension and perform cell counting to determine the total number and concentration of viable cells in suspension (*see* **Note 22**).
10. Centrifuge at 200 × *g* for 5 min and discard supernatant.
11. Resuspend the pellet in a specific volume of NPC medium to obtain the desired final concentration. For 24-well plates NPCs can be plated at a density of 40,000–60,000 cells/cm^2 for differentiation (~90,000 cells/well, **Note 25**).
12. Distribute 250 μL/well of cell suspension (already containing 250 μL/well of NPC medium) to a final volume of 0.5 mL/well of medium (without growth factors!).
13. Transfer plate(s) to the 37 °C incubator and leave undisturbed for 24 h.
14. Twice a week (approximately every 3.5 days) replace half-volume with new NPC medium. Using a P1000 pipette, remove 250 μL of the culture medium and replace with 250 μL of new NPC medium.
15. For these cultures, at the time point of interest, the cover slips can be used for (immuno)-staining and mounted in glass slides for high-resolution microscopy imaging (Fig. 2b–e) [37, 41].

3.10 NPC Differentiation: Neuronal Cultures in 96-Well Plates

1. Start with NPC cultures at 90–95% confluency. It is crucial that for setting up differentiation in 96-well plates, the cells are not over-confluent (stressed) because the change of plate format is more dramatic and only "healthy" cells will do well.
2. Warm an aliquot of NPC medium in the 37 °C water bath.
3. Pre-warm 96-well plates (Subheading 2.11) in the 37 °C incubator.
4. Prepare the plates as follows: using a multi-well vacuum manifold aspirator, remove the coating solution (POL in PBS) and wash the wells with 200 μL/well of PBS. Leave this solution in the wells and place the plate in the 37 °C incubator until needed.
5. Warm an aliquot of TrypLE in the 37 °C water bath.
6. Detach cells with TrypLE as described in Subheadings 3.7 and 3.8, perform cell count and pellet by centrifugation.
7. Resuspend the NPC pellet in a specific volume of NPC medium (no growth factors!) to obtain a final concentration of

70,000–80,000 cells/cm^2 for differentiation, that is, ~35,000 cells/well in 100 μL of medium. This density needs to be optimized per cell line, especially for obtaining optimal neuronal density for imaging.

8. Aspirate the PBS washing solution from the 96-well plate and with a multi-channel pipettor add 100 μL of the NPCs suspension per well.
9. (optional) Quick spin plates up to 150 × *g* to maximize cells immediate contact with the bottom of the well and to promote adherence. Some cell lines will be sensitive to this extra step and, in that case, it is best to skip it.
10. Transfer plate(s) to the 37 °C incubator.
11. On the first day of feeding (2–3 days after plating), simply add 100 μL of NPC medium (no growth factors!) to the wells, to a final volume of 200 μL/well.
12. Twice a week (approximately every 3.5 days) replace half-volume with new NPC medium. To do this, use an automated microplate well washer or a multi-channel pipettor, remove 100 μL/well of the culture medium, and add 100 μL/well of new NPC medium, with the final volume remaining at 200 μL.
13. This culture format can be utilized for microscopy imaging and viability assays, with subsequent collection of neurons for lysis and biochemical analysis. It is particularly suitable for drug screening [37, 55, 56].

4 Notes

1. All materials for handling Matrigel and coating Matrigel plates should be kept at 4 °C, to ensure that the gelatinous protein mixture is kept in liquid form. Matrigel becomes solid at −20 °C and room temperature. So, while coating, keep plates on a bed of ice and use refrigerated pipettes, tubes, and plates.
2. This protocol focuses on the use of 6-well plates but can be adapted to any size multi-well plate or round dish. If working with a new cell line of unknown growth rate, coat only one, two, three, or four wells per 6-well plate, to use well(s)/plate (s) as needed without wasting unused Matrigel-coated wells.
3. Our protocol for culturing human iPSCs in Matrigel-coated plates was initially developed with mTeSR1 medium (StemCell Technology 05850). Then, an "improved" version of mTeSR1 was developed, mTeSR Plus medium (StemCell Technology 05825), with enhanced buffering properties and stabilized FGF2 that according to the manufacturer supports cell

proliferation with fewer medium changes are required. This protocol can be adapted to either medium, only changing frequency of cell feeding.

4. If the 5× supplement solution seems a bit turbid, first ensure that it is at room temperature. If this persists, place in the 37 °C water bath for approximately 5 min, swirling occasionally until the supplement solution becomes clear, that is, until all components are appropriately dissolved prior to adding to the basal medium.
5. The growth factors FGF and EGF, as well as heparin, should always be added fresh to an aliquot of NPC medium, just prior to passaging cells. Avoid re-heating the same aliquot of medium already containing growth factors. Also, aliquots of FGF, EGF, and heparin are only thawed once and kept at 4 °C for the remainder of their use. Do not re-freeze.
6. Poly-L-ornithine is solid at −20 °C, at room temperature and 37 °C, and it is liquid at 4 °C.
7. Make sure to dilute laminin in PBS and not water to ensure optimal solubility and proper coating. If a mistake is made and laminin is diluted in water the probable outcome will be that the cells will not attach properly to the plate.
8. Because POL-coated plates will need to be maneuvered outside of the hood and stored in a 4 °C refrigerator (non-sterile), it is best to wrap the plates in aluminum foil for protection.
9. Plates simultaneously coated with poly-L-ornithine and laminin (in PBS) show increased adherence capacity of the cells to the coating matrix and increased coating longevity, which is advantageous for long-term cell differentiation without replating cells. This may also be a helpful strategy to grow cells that have a high tendency to detach upon thawing or at low density.
10. The cell-permeable, Rho-associated coiled-coil containing protein kinase (ROCK) inhibitor Y-27632 is used here to promote cryopreserved or passaged cells' survival by preventing dissociation-induced (i.e., single-cell associated) apoptosis. It significantly improves cells adherence and proliferation.
11. It is important to resuspend the pellet very gently. "Chunks" of cells should be visible and should not be disrupted or disintegrated, because these are the colony-forming iPSCs.
12. The frequency of medium change, i.e., iPSC feeding schedule, will depend on factors such as rate of cell proliferation, tendency for differentiation and medium used (mTeSR1 or mTeSR1 Plus, **Note 3**). With mTeSR1, daily feeding is usually optimal to promote good cell proliferation with minimal differentiation. With mTeSR1 Plus, thanks to increased buffer potency and stabilized growth factors (according to the

commercial source), it is possible to feed iPS cells every other day or every 2 days. This frequency should be optimized for each cell line. Independently of the days when medium change is executed, iPSCs should be fed approximately at the same time of the day in order to properly document feeding regimen for each cell line, for example every 24 h or every 48 h, etc.

13. If iPSCs start to differentiate before reaching the desired confluency for passaging, the investigator can perform "manual cleaning" of the culture. For this procedure, a hood with an integrated light microscope (sterile) is needed. Place the plate of cells under the microscope, inside the hood. Using an angle fire-polished *Pasteur* pipette (with a rounded tip), score around differentiated cells and cells with abnormal morphology, scrape them and lift cells into the medium. Making sure all the abnormal/differentiated cells are in suspension, aspirate this medium and replace with 2 mL of fresh pre-warmed mTeSR1 Plus medium. Let cell colonies continue to grow until the desired confluency.
14. The dilution factor needs to be optimized for each iPSC line. It is preferable to have cells become dense with confluent colonies earlier, rather than plate cells at an excessive high dilution, which in many cases will delay or even prevent colony formation and possibly promote cellular differentiation. The investigator will get a sense for the optimal dilution factor for each cell line with subsequent passages and continued handling of the cells.
15. It is recommended to use calcium-free PBS, because calcium will interfere with the activity of the reagent ReLeSR used to dislodge the iPSCs.
16. Unless specific information is available a priori, it is recommended to plate two or three cell densities for each cell line in the AggreWell-800 24-well plate. Different cell lines will have different optimal initial densities at which they will form EBs and generate neural rosettes. This is an inherent characteristic of each pluripotent stem cell line. An example of starting iPSC densities would be as follows: one well with 2×10^6 cells, two wells with 2.5×10^6 cells/well, one well with 3×10^6 cells (Fig. 1a). This strategy will ensure that the optimal density for neural rosettes generation is achieved and should also provide sufficient number of rosettes for cryopreservation (neural rosettes banking). That is, with four wells worth of neural aggregates per cell line, one well can be frozen as neural rosettes and the remaining three wells expanded into neural progenitor cells. In this case, if approximately three wells (90% confluence) are needed for one well of the AggreWell-800 plate, 2× 6-well plates of iPSCs will be needed to set up four AggreWell-800 wells.

17. If there are unused wells in the AggreWell-800 24-well plate: aspirate any liquid left in all used wells and cross out the corresponding spots of used wells on the plate lid. Seal the plate with parafilm to keep the new wells sterile and store at room temperature. The unused new wells can be later used for another experiment.
18. From this point on, the NIM could potentially be supplemented with the antibiotic penicillin-streptomycin (at a 1:100 dilution), although originally this protocol was developed without use of antibiotics. It is unclear whether the presence of antibiotics has any influence on the formation of neural rosettes and progenitor cells.
19. The STEMdiff Neural Rosette Selection Reagent is an enzyme-free reagent used for the selective detachment of neural rosette clusters from adherent neural aggregates, in place of manual scoring and scraping under the microscope. Collecting and replating rosette clusters after incubation with Neural Rosette Selection Reagent is expected, in most cases, to yield highly pure populations of neural progenitor cells. Alternatively, (when rosettes have sub-optimal morphology or low frequency per cluster), manual selection can be used to transfer rosettes. To this end, remove medium, wash with 2 mL/well of DMEM/F12 and replace with 1 mL/well of NIM. Under a hood-integrated light microscope and using an angle fire-polished *Pasteur* pipette, score around each rosette cluster and detach the clusters from the surrounding flat cells, lifting them into suspension. Transfer the rosette clusters suspension into a 15 mL conical tube and pipette up and down a few times to break apart the clusters into 1/4 or 1/5 of their original size. Then transfer onto wells of a new POL-coated plate (already containing 1 mL/well of NIM) to a final volume of 2 mL/well of NIM.
20. To avoid contamination of neural rosettes with non-CNS type of progenitor cells, do not over-select. It is preferable to exclude some rosettes, trading yield for higher purity.
21. Early passage NPCs are generally cultured at high density. Due to the embryonic-like nature of these neural progenitor cells, single cells have difficulty in proliferating, whereas small clusters of cells tend to survive and proliferate better. Therefore, when passaging "young" NPCs, maintain a low dilution ration (usually 1:2 or 1:3), without a need for cell counting, and only passage when 90–100% confluency is reached.
22. To ensure a specific number of cells is plated or to uniformize NPC proliferation and passages across cell lines (and measure cell viability), cell counting can also be done as follows: detach cells with TrypLE, transfer the cell suspension into a conical

tube and take an aliquot of 10 μL. Mix with 10 μL of Trypan Blue and load 10 μL of this 1:1 mixture onto a disposable cell counting chamber slide. An automated cell countess can be used to determine initial cell concentration, which multiplied by the initial cell suspension volume will give the total number of cells. This number can then be resuspended in a specific volume of NPC medium +E.F.H. to a final concentration so that 400,000–600,000 cells are plated in a 0.5 mL volume of medium per well.

23. To avoid proliferation of non-CNS neural progenitors, in particular peripheral nervous system neural crest cells that can take over the culture and do not differentiate under these conditions, it is crucial to keep NPCs in proliferative state for no more than an average of approximately five passages. That is, the NPCs can be passaged and expanded from P1 to P5, with P5 corresponding to the differentiation initiation step. This is a limitation of this protocol but, if combined with fluorescence-activated cell sorting (FACS) [37, 41] or magnetic-activated cell sorting (MACS) [65], NPCs can be further enriched for cortical precursors based on selection of specific cell-surface markers (CD184-positive, CD133-positive, and CD271-negative) [35, 37, 41, 42, 66, 67]. The outcome is a CNS-enriched NPC population with high homogeneity and extensive proliferative capacity. Although these methodologies are time-consuming and can delay initiation of differentiation, these NPCs can be stably propagated in culture for up to 1 year or at least 50 passages, without losing differentiation potential or karyotype integrity [37, 41, 42].
24. The method used for thawing NPCs can be applied to any passage number NPC. It can also be applied to neural rosettes or NPCs from a different source (from a different lab or cell bank), even if they have been grown in slightly different conditions of plate coating or medium composition. In the latter case, extra surveillance of the cultures might be necessary to ensure attachment to POL-coated plates and proper proliferation (avoiding culture conditions that are too diluted or overgrown). For cell lines with less-than-optimal attachment or proliferation characteristics, the medium can be initially supplemented with the ROCK inhibitor Y-27632 at 10 μM and cells can be thawed into POLs (PO and laminin simultaneously)-coated plates that provide higher adherence.
25. Alternatively, NPC lines that are sensitive to growth conditions, density and attachment, can also be plated in NPC medium +E.F.H. overnight to promote attachment and correct starting density. After no longer than 18–24 h exchange medium as follows: carefully aspirate all medium and perform one wash with 1.5 mL/well of NPC medium to ensure

complete removal of growth factors. Then, add 2 mL/well of fresh NPC medium-only to initiate differentiation. In this case and for most cell lines, it is advisable to start with a slightly lower density of cells (for example 600,000–700,000 cells/well) because even after growth factors withdrawal, the cells still tend to proliferate a little. This number needs to be optimized for each cell line and the experimental goals.

References

1. Thomson JA, Itskovitz-Eldor J, Shapiro SS et al (1998) Embryonic stem cell lines derived from human blastocysts. Science 282(5391): 1145–1147
2. Takahashi K, Tanabe K, Ohnuki M et al (2007) Induction of pluripotent stem cells from adult human fibroblasts by defined factors. Cell 131(5):861–872. https://doi.org/10.1016/j.cell.2007.11.019
3. Yu J, Vodyanik MA, Smuga-Otto K et al (2007) Induced pluripotent stem cell lines derived from human somatic cells. Science 318(5858):1917–1920. https://doi.org/10.1126/science.1151526
4. Fusaki N, Ban H, Nishiyama A et al (2009) Efficient induction of transgene-free human pluripotent stem cells using a vector based on Sendai virus, an RNA virus that does not integrate into the host genome. Proc Jpn Acad Ser B Phys Biol Sci 85(8):348–362
5. Stadtfeld M, Nagaya M, Utikal J et al (2008) Induced pluripotent stem cells generated without viral integration. Science 322(5903): 945–949. https://doi.org/10.1126/science.1162494
6. Warren L, Manos PD, Ahfeldt T et al (2010) Highly efficient reprogramming to pluripotency and directed differentiation of human cells with synthetic modified mRNA. Cell Stem Cell 7(5):618–630. https://doi.org/10.1016/j.stem.2010.08.012
7. Yu J, Hu K, Smuga-Otto K et al (2009) Human induced pluripotent stem cells free of vector and transgene sequences. Science 324(5928):797–801. https://doi.org/10.1126/science.1172482
8. Zhang Y, Pak C, Han Y et al (2013) Rapid single-step induction of functional neurons from human pluripotent stem cells. Neuron 78(5):785–798. https://doi.org/10.1016/j.neuron.2013.05.029
9. Victor MB, Richner M, Hermanstyne TO et al (2014) Generation of human striatal neurons by microRNA-dependent direct conversion of fibroblasts. Neuron 84(2):311–323. https://doi.org/10.1016/j.neuron.2014.10.016
10. Okita K, Matsumura Y, Sato Y et al (2011) A more efficient method to generate integration-free human iPS cells. Nat Methods 8(5): 409–412. https://doi.org/10.1038/nmeth.1591
11. Lancaster MA, Knoblich JA (2014) Organogenesis in a dish: modeling development and disease using organoid technologies. Science 345(6194):1247125. https://doi.org/10.1126/science.1247125
12. Lancaster MA, Renner M, Martin CA et al (2013) Cerebral organoids model human brain development and microcephaly. Nature 501(7467):373–379. https://doi.org/10.1038/nature12517
13. Haggarty SJ, Silva MC, Cross A et al (2016) Advancing drug discovery for neuropsychiatric disorders using patient-specific stem cell models. Mol Cell Neurosci 73:104–115. https://doi.org/10.1016/j.mcn.2016.01.011
14. Brennand KJ, Gage FH (2012) Modeling psychiatric disorders through reprogramming. Dis Model Mech 5(1):26–32. https://doi.org/10.1242/dmm.008268
15. Brennand KJ, Marchetto MC, Benvenisty N et al (2015) Creating patient-specific neural cells for the in vitro study of brain disorders. Stem Cell Rep 5(6):933–945. https://doi.org/10.1016/j.stemcr.2015.10.011
16. Di Lullo E, Kriegstein AR (2017) The use of brain organoids to investigate neural development and disease. Nat Rev Neurosci 18(10): 573–584. https://doi.org/10.1038/nrn.2017.107
17. Dolmetsch R, Geschwind DH (2011) The human brain in a dish: the promise of iPSC-derived neurons. Cell 145(6):831–834. https://doi.org/10.1016/j.cell.2011.05.034
18. Ebert AD, Liang P, Wu JC (2012) Induced pluripotent stem cells as a disease modeling and drug screening platform. J Cardiovasc

Pharmacol 60(4):408–416. https://doi.org/10.1097/FJC.0b013e318247f642
19. Engle SJ, Puppala D (2013) Integrating human pluripotent stem cells into drug development. Cell Stem Cell 12(6):669–677. https://doi.org/10.1016/j.stem.2013.05.011
20. Ichida JK, Kiskinis E (2015) Probing disorders of the nervous system using reprogramming approaches. EMBO J 34(11):1456–1477. https://doi.org/10.15252/embj.201591267
21. Ko HC, Gelb BD (2014) Concise review: drug discovery in the age of the induced pluripotent stem cell. Stem Cells Transl Med 3(4): 500–509. https://doi.org/10.5966/sctm.2013-0162
22. Marchetto MC, Winner B, Gage FH (2010) Pluripotent stem cells in neurodegenerative and neurodevelopmental diseases. Hum Mol Genet 19(R1):R71–R76. https://doi.org/10.1093/hmg/ddq159
23. Sayed N, Liu C, Wu JC (2016) Translation of human-induced pluripotent stem cells: from clinical trial in a dish to precision medicine. J Am Coll Cardiol 67(18):2161–2176. https://doi.org/10.1016/j.jacc.2016.01.083
24. Silva MC, Cross A, Brandon NJ, Haggarty SJ (2017) Human iPSC models in drug discovery: opportunities and challenges. In: DPRaSEW SC (ed) Comprehensive medicinal chemistry III, vol 1. Elsevier, Oxford, pp 48–73. https://doi.org/10.1016/B978-0-12-409547-2.12284-X
25. Shi Y, Kirwan P, Smith J et al (2012) Human cerebral cortex development from pluripotent stem cells to functional excitatory synapses. Nat Neurosci 15(3):477–486., , S471. https://doi.org/10.1038/nn.3041
26. Desbordes SC, Studer L (2013) Adapting human pluripotent stem cells to high-throughput and high-content screening. Nat Protoc 8(1):111–130. https://doi.org/10.1038/nprot.2012.139
27. Jensen J, Hyllner J, Bjorquist P (2009) Human embryonic stem cell technologies and drug discovery. J Cell Physiol 219(3):513–519. https://doi.org/10.1002/jcp.21732
28. Silva MC, Haggarty SJ (2020) Human pluripotent stem cell-derived models and drug screening in CNS precision medicine. Ann N Y Acad Sci 1471(1):18–56. https://doi.org/10.1111/nyas.14012
29. Pang ZP, Yang N, Vierbuchen T et al (2011) Induction of human neuronal cells by defined transcription factors. Nature 476(7359): 220–223. https://doi.org/10.1038/nature10202
30. Son EY, Ichida JK, Wainger BJ et al (2011) Conversion of mouse and human fibroblasts into functional spinal motor neurons. Cell Stem Cell 9(3):205–218. https://doi.org/10.1016/j.stem.2011.07.014
31. Tsunemoto RK, Eade KT, Blanchard JW et al (2015) Forward engineering neuronal diversity using direct reprogramming. EMBO J 34(11): 1445–1455. https://doi.org/10.15252/embj.201591402
32. Yoo AS, Sun AX, Li L et al (2011) MicroRNA-mediated conversion of human fibroblasts to neurons. Nature 476(7359):228–231. https://doi.org/10.1038/nature10323
33. Pasca SP, Portmann T, Voineagu I et al (2011) Using iPSC-derived neurons to uncover cellular phenotypes associated with Timothy syndrome. Nat Med 17(12):1657–1662. https://doi.org/10.1038/nm.2576
34. Li W, Sun W, Zhang Y et al (2011) Rapid induction and long-term self-renewal of primitive neural precursors from human embryonic stem cells by small molecule inhibitors. Proc Natl Acad Sci U S A 108(20):8299–8304. https://doi.org/10.1073/pnas.1014041108
35. Muratore CR, Srikanth P, Callahan DG et al (2014) Comparison and optimization of hiPSC forebrain cortical differentiation protocols. PLoS One 9(8):e105807. https://doi.org/10.1371/journal.pone.0105807
36. Chambers SM, Fasano CA, Papapetrou EP et al (2009) Highly efficient neural conversion of human ES and iPS cells by dual inhibition of SMAD signaling. Nat Biotechnol 27(3): 275–280. https://doi.org/10.1038/nbt.1529
37. Cheng C, Fass DM, Folz-Donahue K et al (2017) Highly expandable human iPS cell-derived neural progenitor cells (NPC) and neurons for central nervous system disease modeling and high-throughput screening. Curr Protoc Hum Genet 92:21 28 21–21 28 21. https://doi.org/10.1002/cphg.33
38. Odorico JS, Kaufman DS, Thomson JA (2001) Multilineage differentiation from human embryonic stem cell lines. Stem Cells 19(3): 193–204. https://doi.org/10.1634/stemcells.19-3-193
39. Sato C, Barthelemy NR, Mawuenyega KG et al (2018) Tau kinetics in neurons and the human central nervous system. Neuron 98(4): 861–864. https://doi.org/10.1016/j.neuron.2018.04.035
40. Sheridan SD, Theriault KM, Reis SA et al (2011) Epigenetic characterization of the FMR1 gene and aberrant neurodevelopment in human induced pluripotent stem cell models

of fragile X syndrome. PLoS One 6(10): e26203. https://doi.org/10.1371/journal.pone.0026203

41. Silva MC, Cheng C, Mair W, Almeida S, Fong H, Biswas MHU, Zhang Z, Huang Y, Temple S, Coppola G, Geschwind DH, Karydas A, Miller BL, Kosik KS, Gao FB, Steen JA, Haggarty SJ (2016) Human iPSC-derived neuronal model of tau-A152T frontotemporal dementia reveals tau-mediated mechanisms of neuronal vulnerability. Stem Cell Reports 7(3):325–340
42. Nemati S, Hatami M, Kiani S et al (2011) Long-term self-renewable feeder-free human induced pluripotent stem cell-derived neural progenitors. Stem Cells Dev 20(3):503–514. https://doi.org/10.1089/scd.2010.0143
43. Lee S, Huang EJ (2015) Modeling ALS and FTD with iPSC-derived neurons. Brain Res 1656:88–97. https://doi.org/10.1016/j.brainres.2015.10.003
44. Zhao WN, Cheng C, Theriault KM et al (2012) A high-throughput screen for Wnt/beta-catenin signaling pathway modulators in human iPSC-derived neural progenitors. J Biomol Screen 17(9):1252–1263. https://doi.org/10.1177/1087057112456876
45. Koch P, Opitz T, Steinbeck JA et al (2009) A rosette-type, self-renewing human ES cell-derived neural stem cell with potential for in vitro instruction and synaptic integration. Proc Natl Acad Sci U S A 106(9):3225–3230. https://doi.org/10.1073/pnas.0808387106
46. Karch CM, Kao AW, Karydas A et al (2019) A comprehensive resource for induced pluripotent stem cells from patients with primary Tauopathies. Stem Cell Rep 13(5):939–955. https://doi.org/10.1016/j.stemcr.2019.09.006
47. Almeida S, Zhang Z, Coppola G et al (2012) Induced pluripotent stem cell models of progranulin-deficient frontotemporal dementia uncover specific reversible neuronal defects. Cell Rep 2(4):789–798. https://doi.org/10.1016/j.celrep.2012.09.007
48. Fong H, Wang C, Knoferle J et al (2013) Genetic correction of tauopathy phenotypes in neurons derived from human induced pluripotent stem cells. Stem Cell Rep 1(3): 226–234. https://doi.org/10.1016/j.stemcr.2013.08.001
49. She A, Kurtser I, Reis SA et al (2017) Selectivity and kinetic requirements of HDAC inhibitors as Progranulin enhancers for treating frontotemporal dementia. Cell Chem Biol 24(7):892–906. e895. https://doi.org/10.1016/j.chembiol.2017.06.010
50. Bavamian S, Mellios N, Lalonde J et al (2015) Dysregulation of miR-34a links neuronal development to genetic risk factors for bipolar disorder. Mol Psychiatry 20(5):573–584. https://doi.org/10.1038/mp.2014.176
51. Madison JM, Zhou F, Nigam A et al (2015) Characterization of bipolar disorder patient-specific induced pluripotent stem cells from a family reveals neurodevelopmental and mRNA expression abnormalities. Mol Psychiatry 20(6):703–717. https://doi.org/10.1038/mp.2015.7
52. Martin P, Wagh V, Reis SA et al (2020) TSC patient-derived isogenic neural progenitor cells reveal altered early neurodevelopmental phenotypes and rapamycin-induced MNK-eIF4E signaling. Mol Autism 11(1):2. https://doi.org/10.1186/s13229-019-0311-3
53. Mellios N, Feldman DA, Sheridan SD et al (2018) MeCP2-regulated miRNAs control early human neurogenesis through differential effects on ERK and AKT signaling. Mol Psychiatry 23(4):1051–1065. https://doi.org/10.1038/mp.2017.86
54. Tobe BTD, Crain AM, Winquist AM et al (2017) Probing the lithium-response pathway in hiPSCs implicates the phosphoregulatory set-point for a cytoskeletal modulator in bipolar pathogenesis. Proc Natl Acad Sci U S A 114(22):E4462–E4471. https://doi.org/10.1073/pnas.1700111114
55. Silva MC, Ferguson FM, Cai Q et al (2019) Targeted degradation of aberrant tau in frontotemporal dementia patient-derived neuronal cell models. Elife 8:e45457. https://doi.org/10.7554/eLife.45457
56. Silva MC, Nandi GA, Tentarelli S et al (2020) Prolonged tau clearance and stress vulnerability rescue by pharmacological activation of autophagy in tauopathy neurons. Nat Commun 11(1):3258. https://doi.org/10.1038/s41467-020-16984-1
57. Bernard-Gauthier V, Mossine AV, Knight A et al (2019) Structural basis for achieving GSK-3beta inhibition with high potency, selectivity, and brain exposure for positron emission tomography imaging and drug discovery. J Med Chem 62(21):9600–9617. https://doi.org/10.1021/acs.jmedchem.9b01030
58. Zhao WN, Ghosh B, Tyler M et al (2018) Class I histone deacetylase inhibition by Tianeptinaline modulates neuroplasticity and enhances memory. ACS Chem Neurosci 9(9): 2262–2273. https://doi.org/10.1021/acschemneuro.8b00116
59. Zhao WN, Hylton NK, Wang J et al (2019) Activation of WNT and CREB signaling

pathways in human neuronal cells in response to the Omega-3 fatty acid docosahexaenoic acid (DHA). Mol Cell Neurosci 99:103386. https://doi.org/10.1016/j.mcn.2019.06.006

60. Zhao WN, Tobe BTD, Udeshi ND et al (2020) Discovery of suppressors of CRMP2 phosphorylation reveals compounds that mimic the behavioral effects of lithium on amphetamine-induced hyperlocomotion. Transl Psychiatry 10(1):76. https://doi.org/10.1038/s41398-020-0753-6

61. Bauwens CL, Peerani R, Niebruegge S et al (2008) Control of human embryonic stem cell colony and aggregate size heterogeneity influences differentiation trajectories. Stem Cells 26(9):2300–2310. https://doi.org/10.1634/stemcells.2008-0183

62. Elkabetz Y, Studer L (2008) Human ESC-derived neural rosettes and neural stem cell progression. Cold Spring Harb Symp Quant Biol 73:377–387. https://doi.org/10.1101/sqb.2008.73.052

63. Pettinato G, Wen X, Zhang N (2015) Engineering strategies for the formation of Embryoid bodies from human pluripotent stem cells. Stem Cells Dev 24(14):1595–1609. https://doi.org/10.1089/scd.2014.0427

64. Werner A, Iwasaki S, McGourty CA et al (2015) Cell-fate determination by ubiquitin-dependent regulation of translation. Nature 525(7570):523–527. https://doi.org/10.1038/nature14978

65. Bowles KR, Tcw J, Qian L et al (2019) Reduced variability of neural progenitor cells and improved purity of neuronal cultures using magnetic activated cell sorting. PLoS One 14(3):e0213374. https://doi.org/10.1371/journal.pone.0213374

66. Kim DS, Lee DR, Kim HS et al (2012) Highly pure and expandable PSA-NCAM-positive neural precursors from human ESC and iPSC-derived neural rosettes. PLoS One 7(7): e39715. https://doi.org/10.1371/journal.pone.0039715

67. Yuan SH, Martin J, Elia J et al (2011) Cell-surface marker signatures for the isolation of neural stem cells, glia and neurons derived from human pluripotent stem cells. PLoS One 6(3): e17540. https://doi.org/10.1371/journal.pone.0017540

68. Silva MC, Cheng C, Mair W et al (2016) Human iPSC-derived neuronal model of tau-A152T frontotemporal dementia reveals tau-mediated mechanisms of neuronal vulnerability. Stem Cell Rep 7(3):325–340. https://doi.org/10.1016/j.stemcr.2016.08.001

Chapter 11

Differentiation of iPS-Cells into Peripheral Sensory Neurons

Anika Neureiter, Esther Eberhardt, and Angelika Lampert

Abstract

Induced pluripotent stem cells (iPS-cells) have significantly expanded our experimental possibilities, by creating new strategies for the molecular study of human disease and drug development. Treatment of pain has not seen much improvement over the past decade, likely due to species differences in preclinical models. Thus, iPS-cell derived sensory neurons offer a highly welcome translational approach for research and drug development. Although central neuronal differentiation is relatively straightforward, the successful and reliable generation of peripheral neurons requires more complex measures. Here, we describe a small molecule-based protocol for the differentiation of human sensory neurons from iPS-cells which renders functional nociceptor-like cells within several weeks.

Key words Induced pluripotent stem cells, iPS-cells, Sensory neurons, Differentiation, Neural crest cells, Peripheral neurons, Disease modeling, Pain, Sensory neuropathy

1 Introduction

Pain research has been hampered in recent years because translation from preclinical models to humans has severe shortcomings. As such, highly promising drugs which passed many preclinical models have failed when introduced to patients; one highly visible example is the effort to develop a blocker of the sodium channel Nav1.7, a human validated target, which to date has not delivered clinical application [1].

In order to bridge this translational gap, human cellular models are needed. As human sensory neurons are scarce, induced pluripotent stem cell (iPS-cell) derived peripheral neurons offer an attractive alternative. With the publication of the first protocol in 2012 [2] the field moved significantly forward as this protocol was applied and adapted [3–7]. Other protocols revealing mechanosensors [8] or proprioceptors were reported [9]. Using dual SMAD inhibition differentiating iPS-cells [2] pass a state of SOX10 positive neuronal precursor cells, followed by a phase of about 40 days of maturation in the presence of growth factors. The resulting cells

Nagarajan Kannan and Philip Beer (eds.), *Stem Cell Assays: Methods and Protocols*, Methods in Molecular Biology, vol. 2429, https://doi.org/10.1007/978-1-0716-1979-7_11,

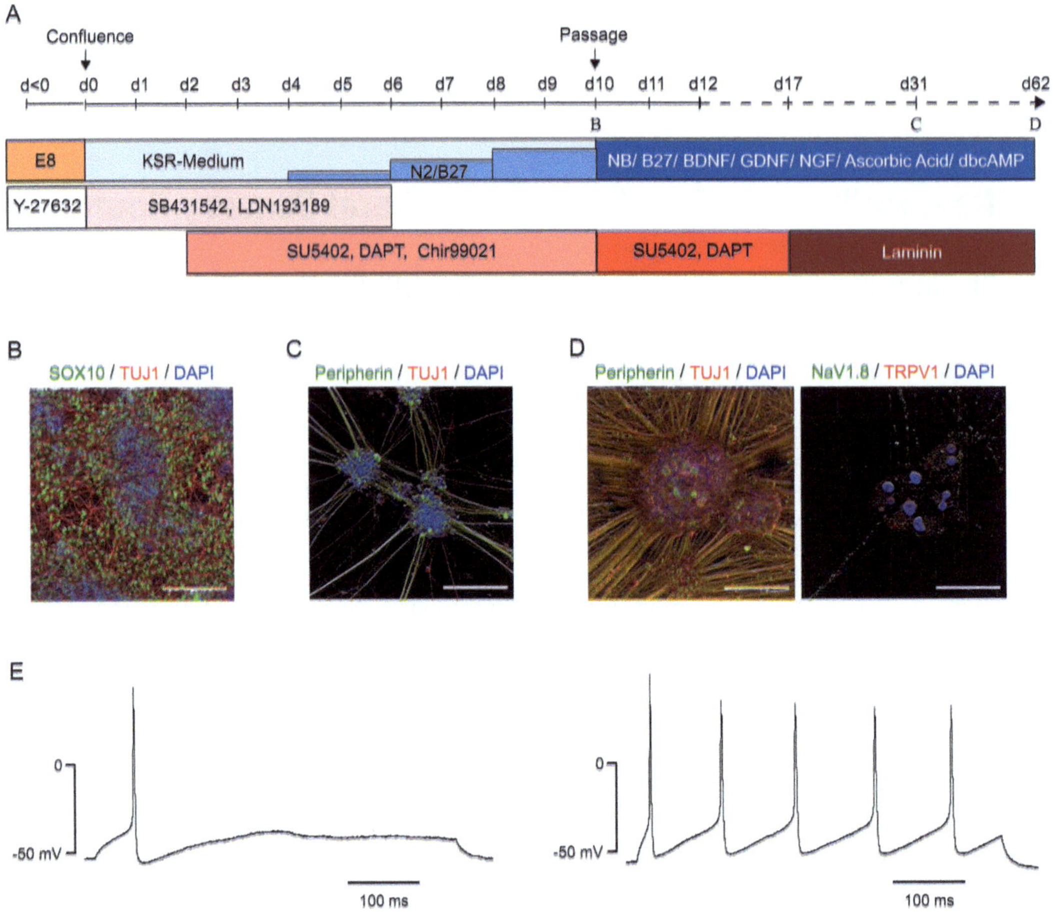

Fig. 1 Timeline of sensory neuron differentiation. (**a**) Schematic overview of the protocol for sensory neuron differentiation. (**b**) Cells were stained at d10 of differentiation for expression of SOX10 (Neural Crest Cells and sensory neuron precursors) and Tuj1 (pan-neuronal marker). Nuclei were counterstained with DAPI. Scale bar represents 200 μm. (**c**) Immunofluorescence staining at d31 of differentiation. Peripherin (PNS-neurons) and Tuj1 (pan-neuron marker) were stained. Nuclei are stained with DAPI. Scale bar represents 200 μm. (**d**) Co-staining of Peripherin and Tuj1 (left) or NaV1.8 and TRPV1 (right, marking nociceptors) at d62 of differentiation. Nuclei are counterstained with DAPI. Scale bar represents 200 μm (right) and 50 μm (left). (**e**) Representative whole cell current-clamp recording of a mature sensory neuron-like neuron after 8 weeks in culture showing tonic firing behavior with a pronounced overshoot. Evoked activity in response to a current injection of 50 pA (rheobase, left) and 100 pA (2× rheobase, right). Membrane potential was set to -55 mV by injecting -30 pA current; pipettes (2.5–3.5 mΩ) were filled with 4 mM NaCl, 135 mM K-gluconate, 3 mM $MgCl_2$, 5 mM EGTA, 5 mM HEPES, 2 mM Na_2-ATP, and 0.3 mM Na_3-GTP (pH 7.25). The bath solution contained 140 mM NaCl, 3 mM KCl, 1 mM $CaCl_2$, 1 mM $MgCl_2$, and 10 mM HEPES (pH 7.4)

should show expression of peripherin (in addition to the neuronal marker Tuj1) and sensory/nociceptive markers, such as the ion channels TrpV1, Nav1.7, and Nav1.8.

In this chapter we describe a protocol based on [2] with modifications which improve the protocol in our hands (Fig. 1). iPS-cells are differentiated into neuronal precursor cells (*see* Subheading 3.1), which are then transferred to triple coated coverslips

[8] (*see* Subheading 3.2). These SOX10 positive neuronal precursors are matured for several weeks before they can be used for experiments (*see* Subheading 3.3). Elimination of non-neuronal cells can be achieved by treatment with cytostatic drugs (*see* Subheading 3.4). In this protocol, we count days of differentiation continuously, although some labs (including ours [4]) also sometimes separate a differentiation phase (d0–d10) from a maturation phase (d11 and following, restart counting with m0 and following). This results from the possibility to freeze cells on d10 when they are split. This protocol does not include freezing; therefore, we continue counting throughout the differentiation and maturation phase.

It is important to note that this long protocol (about 50 days and longer) shows high clone-to-clone variability and not every iPS-cell clone can be successfully differentiated into sensory neurons. Selection of clones that are predisposed to peripheral neuronal differentiation also helps in the investigation of patient-derived sensory neurons, offering a useful tool to study pathological mechanisms [4, 10] and pave the road for the identification of patient-specific individual treatment [7].

2 Materials

2.1 Cells

hiPS-cells: Feeder-independent human induced pluripotent stem cells reprogrammed from peripheral blood mononuclear cells, fibroblasts, or mesenchymal stromal cells of healthy donors or patients. iPS-cells are cultured in E8-Medium on Vitronectin coated 6-well plates. Coating is performed according to the manufacturer's recommendations (*see* **Note 1**). Cells are passaged every 5–7 days in a ratio of 1:5–1:10 using 0.5 mM EDTA in PBS for 3 min (*see* **Note 2**). hiPS-cells require a daily medium change (*see* **Note 3**) until cultures reach approximately 70% confluence. At this time point, the differentiation of neural crest-like cells and sensory neurons is started.

2.2 Reagents and Supplements

1. Accutase: Thaw at 4 °C. Prepare 10 ml aliquots and store at −20 °C.
2. B27 supplement + vitamin A: B27 supplement is thawed at 4 °C, aliquoted and aliquots are stored at −20 °C (*see* **Note 4**).
3. BDNF (200 μg/ml): Dilute 100 μg BDNF (e.g., Peprotech, cat. no. 450-02) in 500 μl 0.1% BSA in PBS; 25 μl aliquots are stored at −20 to −80 °C for long-term storage or at 4 °C for immediate use.
4. 10 mM CHIR99021: Dilute 10 mg CHIR99021 (e.g., Tocris Bioscience, cat. No. 4423) in 7.3 ml sterile DMSO; 50 μl aliquots are stored at −20 °C for long-term storage or at 4 °C for immediate use.

5. 2 mM Ara-C (Cytosine β-D-arabinofuranoside hydrochloride): Prepare as 2 mM stock in sterile water, (1000× stock) final concentration 2 μM.
6. 100 mM DAPT: Dilute 10 mg DAPT (e.g., Tocris Bioscience, cat. no. 2634) in 231.2 μl sterile DMSO; 50 μl aliquots are stored at −20 °C for long-term storage or at 4 °C for immediate use.
7. 500 mM dbcAMP: Dilute 500 μg dbcAMP (e.g., Stem Cell Technologies, cat. no. 73886) in 2.04 ml sterile water; 50 μl aliquots are stored at −20 to −80 °C for long-term storage or at 4 °C for immediate use.
8. DMEM F12 or DMEM-F12 + GlutaMAX (*see* **Note 5**).
9. E8-Supplement (100×): DMEM F12 (adjust volume), 221.1 mM L-Ascorbic Acid 2-Phosphate, 8.18 mM Sodium Selenite, 0.107 mg/ml Transferrin 2 mg/ml Insulin, 10 μg/ml hbFGF (e.g., Peprotech, cat. no. 100-18B), 0.2 μg/ml TGFß1 (e.g., Peprotech, cat. no 100-21). Prepare aliquots of 5 ml and store at −20 °C.
10. 0.5 mM EDTA: Dilute 50 μl of 0.5 M, pH 8.0 EDTA in 50 ml sterile PBS and store at 4 °C.
11. 0.5 mg/ml Fibronectin bovine protein: Dilute 1 mg Fibronectin in 2 ml cell culture grade water and store aliquots at −20 °C.
12. GDNF (200 μg/ml): Dilute 100 μg GDNF (e.g., Peprotech, cat. no. 450-10) in 500 μl 0.1% BSA in PBS, giving 200 μg/ml. 25 μl aliquots are stored at −20 to −80 °C for long-term storage or at 4 °C for immediate use.
13. Geltrex: Thaw Geltrex overnight at 4C° on ice. Aliquot Geltrex with prechilled pipette tips and microcentrifuge tubes, while keeping Geltrex on ice to avoid polymerization; do not warm Geltrex above 15 °C. Prepare 250 μl aliquots on ice and store at −20 °C.
14. 1 M HEPES: Dilute 23.8 g HEPES in 100 ml cell culture grade, sterile water and filter through a 0.2 μm filter for sterilization.
15. Knock-Out Serum replacement (KOSR): KOSR is thawed at 4 °C, aliquoted and stored at −20 °C.
16. Laminin: Laminin-concentration varies between 1–2 mg/ml depending on Lot. Thaw laminin at 4 °C, prepare aliquots and store at −20 °C.
17. 200 mM L-Ascorbic Acid: Dilute 252.2 mg L-Ascorbic Acid in 10 ml sterile water and filter sterilize through 0.2 μm filter. 25 μl aliquots are stored at −20 to −80 °C for long-term storage or at 4 °C for immediate use.

18. 1 mM LDN-193189: Dilute 2 mg LDN-193189 (e.g., Miltenyi Biotec, cat. no. 130-106-925) in 4.92 ml sterile DMSO; aliquots are stored at −20 °C for long-term storage or at 4 °C for immediate use.
19. 200 mM L-glutamine: L-Glutamine is thawed at 37 °C in the water bath until fully dissolved, aliquoted and aliquots are stored at −20 °C.
20. MEM-NEAA (100×).
21. N2 supplement: N2 supplement is thawed at 4 °C, aliquoted and aliquots are stored at −20 °C.
22. Neurobasal medium.
23. NGF (200 μg/ml): Dilute 100 μg NGF (e.g., R&D-Systems, cat. no. 256-GF-100) in 500 μl 0.1% BSA in PBS, 25 μl aliquots are stored at −20 to −80 °C for long-term storage or at 4 °C for immediate use.
24. Neurotrophin 3 (hNT3): Prepare 20 μg/ml stock solution in PBS with 0.1% BSA, (1000× stock) final concentration in medium is 20 ng/ml (e.g., Peprotech, cat no. 450-03).
25. PBS (1×).
26. Penicillin/Streptomycin (100×): Penicillin/Streptomycin is thawed at 4 °C, aliquoted and aliquots are stored at −20 °C.
27. Poly-L-Ornithine (10 mg/ml): Dilute 10 mg Poly-L-Ornithine in 1 ml cell culture grade water and store aliquots at −20 °C.
28. 10 mM SB-431542: Dilute 10 mg SB-431542 (e.g., Miltenyi Biotec, cat. no 130-106-543) in 2.6 ml sterile DMSO, 50 μl aliquots are stored at −20 °C for long-term storage or at 4 °C for immediate use.
29. 10 mM SU 5402: Dilute 1 mg SU-5402 (e.g., Tocris Bioscience, cat. no 3300) in 337 μl sterile DMSO; 50 μl aliquots are stored at −20 °C for long-term storage or at 4 °C for immediate use.
30. Vitronectin: Thaw Vitronectin at room temperature and prepare aliquots according to the desired volume. Store aliquots at −80 °C.
31. 10 mM Y-27632: Dilute 10 mg Y-27632 in 3.1 ml 1×PBS. 50 μl aliquots are stored at −20 °C for long-term storage or at 4 °C for immediate use.
32. 50 mM β-mercaptoethanol.

2.3 Cell Culture Media

1. E8-Medium: For 500 ml E8-Medium combine 482.5 ml DMEM-F12, 5 ml E8-Supplement (final concentration: 1×), 7.5 ml HEPES (final concentration: 15 mM), and 5 ml L-glutamine (final concentration: 2 mM). For long-term-

Table 1
Schedule for cell culture media and small molecules. KSR-Medium and N2/B27-Medium are mixed in the indicated percentages and prepared for 2 days (e.g., for d0 and d1). The small molecules are also added for both days. During the 10 days of differentiation, coat glass coverslips or the desired cell culture ware with Poly-L-Ornithine, Laminin, and Fibronectin between days 7 and 10 (*see* Subheading 3.2)

Day	Medium	Small molecules	
d0 and d1	100% KSR-medium	LDN193189 SB431542	1 μM 10 μM
d2 and d3	100% KSR-medium	LDN193189 SB431542 SU5402 DAPT CHIR99021	1 μM 10 μM 10 μM 10 μM 3 μM
d4 and d5	75% KSR-medium +25% N2/B27	LDN193189 SB431542 SU5402 DAPT CHIR99021	1 μM 10 μM 10 μM 10 μM 3 μM
d6 and d7	50% KSR-medium +50% N2/B27	SU5402 DAPT CHIR99021	10 μM 10 μM 3 μM
d8 and d9	25% KSR-medium +75% N2/B27	SU5402 DAPT CHIR99021	10 μM 10 μM 3 μM

storage, prepare 50 ml aliquots and store at −20 °C or for immediate use the medium is stored at 4 °C. Prewarm at room temperature prior to use.

2. KSR-Medium: For 100 ml KSR-Medium combine DMEM F12 + GlutaMAX, 15 ml Knock-Out Serum Replacement, 1 ml MEM-NEAA (final concentration: 1×), and 200 μl ß-mercaptoethanol (final concentration: 0.1 mM). The Medium can be stored for 1 week at 4 °C. Small molecules are added freshly prior to use according to Table 1. Prewarm at room temperature prior to use.
3. N2/B27-Medium: For 100 ml N2/B27-Medium combine 47.5 ml Neurobasal medium, 47.5 ml DMEM-F12 + GlutaMAX, 0.5 ml L-glutamine (final concentration: 1 mM), 1 ml MEM-NEAA (final concentration: 1×), 1 ml N2 (final concentration: 1×), 2 ml B27 + vitamin A (final concentration: 1×) and optional 1 ml penicillin/streptomycin. The medium can be stored for 2 weeks at 4 °C. Small molecules are added freshly according to Table 1. Prewarm at room temperature prior to use.

4. B27-Medium: For 100 ml B27-Medium combine 96 ml Neurobasal, 2 ml B27 supplement + vitamin A (final concentration: 1×), 1 ml L-glutamine (final concentration: 2 mM) and optional 1 ml penicillin/streptomycin. The medium can be stored for 2 weeks at 4 °C. Small molecules are added freshly according to Table 1 prior to use. Prewarm at room temperature prior to use.

3 Methods

3.1 Differentiation of Neural Crest-like Cells (d0–d10, Fig. 2)

1. Maintain iPS-cells in a 6-well plate format until they reach a confluence of approximately 70% and are ready for passaging.
2. Coat tissue culture treated cell culture dishes with Geltrex according to the manufacturer's recommendations. Briefly, thaw one aliquot of Geltrex at 4 °C and dilute Geltrex 1:100 in cold DMEM F12 (*see* **Note 6**). Use 1 ml of the cold Geltrex solution to coat one 6-well (*see* **Note 7**). Incubate at least for 1 h at 37 °C (*see* **Note 8**). For seeding cells, aspirate Geltrex solution and add 1 ml of E8-medium containing 10 μM Y-27632.
3. If necessary, mark all differentiated areas in the iPS-cell culture with a pen under the microscope (*see* **Note 9**).
4. Aspirate E8-Medium and all marked areas of differentiated cells. Add 1 ml Accutase per 6-well and incubate at 37 °C for 5 min or until colonies start to break up and go into suspension when tapping the plate.
5. Add 1 ml DMEM F12 and pipette up and down to generate a single cell suspension. Check under the microscope for remaining aggregates.

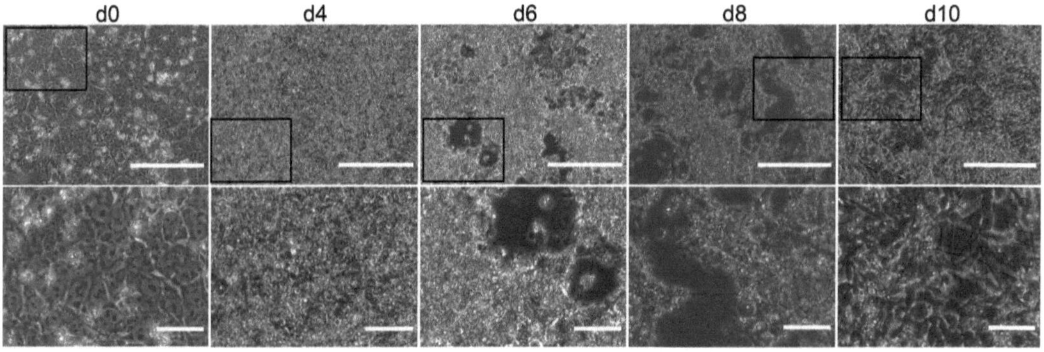

Fig. 2 Differentiation of neural crest-like cells from d0 to d10. Phase contrast images of pluripotent stem cells seeded at higher density for differentiation (d0, step 11), and during the differentiation towards neural crest-like cells and sensory neuron progenitors (d10). The upper panel shows an overview of the cultures (scale bar: 200 μm). The area in the box is depicted in the lower panel at higher magnification (scale bar: 50 μm)

6. Transfer the cell suspension to a conical tube and wash the well with 1 ml DMEM F12. Add the wash solution to the conical tube.
7. Centrifuge the single cell suspension at 200 × *g* for 5 min. Resuspend the cell pellet in E8-Medium containing 10 μM Y-27632.
8. Count live cells. Take 20 μl of the cell suspension (mix well before) and mix with an equal amount of Trypan blue. Pipette the cell suspension/Trypan blue onto a hemocytometer and count all living cells (*see* **Note 10**).
9. Seed 1.5–2 × 10^5 cells/cm^2 (*see* **Note 11**) and place the plate back in the incubator (37 °C, 5% CO_2).
10. After 24 h, cultures should be 80–90% confluent. In case of lower confluence change the medium: Use E8 medium without Y- 27632 and start the differentiation at the following day (*see* **Note 12**). When cells are dense enough differentiation is started (=d0, Fig. 2).
11. To start the differentiation (=d0) medium is changed to differentiation medium according to Table 1 (*see* **Note 13**). Medium has to be changed daily. 2 ml of medium is used per well of a 6-well plate. If using other well sizes, reduce volume relative to the surface area of the well.

3.2 Coating of Glass Coverslips with Poly-L-Ornithine, Laminin, and Fibronectin

1. Starting at d7 of neural crest cell differentiation, pretreat coverslips with 1 M HCl by putting them in a conical tube filled with 40 ml 1 M HCl overnight on a rotator.
2. After 24 h, discard HCl and wash coverslips 3 times with water, followed by 70% Ethanol for at least 30 min.
3. Separate the coverslips on filter paper in the cell culture hood and let them dry under UV-light for 1 h.
4. Transfer the coverslips to a coating chamber prepared with parafilm with sterile forceps and place all coverslips on top of the parafilm (*see* **Note 14**).
5. Add 100 μl of 15 μg/ml Poly-L-Ornithine diluted in cell culture grade sterile water onto each coverslip. Perform coating overnight at 37 °C.
6. On the next day, wash the Poly-L-Ornithine coated surfaces three times with water by directly pipetting 100–200 μl water on each coverslip and aspirating it.
7. Let the coverslips dry under the cell culture hood under UV-light.
8. Coated coverslips can now be stored at 4 °C (*see* **Note 15**).

9. Coating of Poly-L-Ornithine treated coverslips with Laminin and Fibronectin is performed simultaneously. Transfer coverslips in 24-well plates for coating.
10. 10 μg/ml Laminin and 10 μg/ml Fibronectin are diluted in water and 100 μl of the dilution is added on top of each coverslip without coating the surface of the 24-well plate.
11. The Laminin and Fibronectin coating is performed at least 16 h before use (e.g., overnight) at 37 °C. Coverslips should not dry out before use (*see* **Note 16**).
12. For seeding cells, aspirate the coating solution immediately before adding medium or plating the cells.

3.3 Differentiation and Maturation of Sensory Neurons (d10-End, Fig. 3)

1. Aspirate the medium from the neural crest-like cells and add 1 ml Accutase/6-well. Incubate for 1 h at 37 °C (*see* **Note 17**).
2. Add 1 ml medium (use the remaining KSR or N2/B27-Medium without small molecules, DMEM F12 or Neurobasal) and generate a single cell suspension by pipetting with a 1 ml pipette. Transfer the cell suspension to a conical tube and wash the well with medium. Collect all cells in one conical tube. Centrifuge at 200 × *g* for 5 min.
3. Resuspend cell pellet in B27-Medium without small molecules.
4. Determine the cell number by using a hemocytometer and live-dead-staining with Trypan blue as described in Subheading 3.1.
5. Adjust cell number to 5 × 10^4 cells/100 μl.

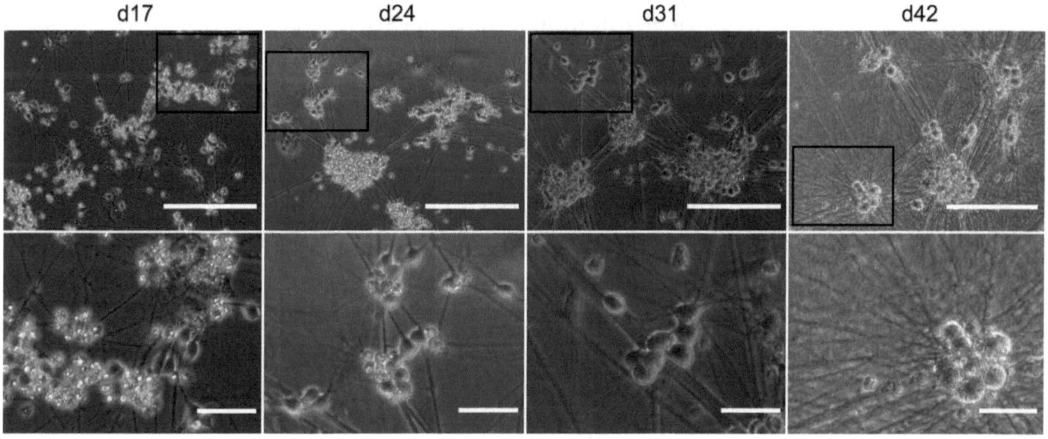

Fig. 3 Differentiation of sensory neuron-like cells from d10 onwards. Phase contrast images of sensory neuron progenitors and sensory neurons (d17) during the differentiation towards mature sensory neurons (d42). The upper panel shows an overview of the cultures (scale bar: 200 μm). The area in the box is depicted in the lower panel at higher magnification (scale bar: 50 μm)

6. Aspirate Laminin/Fibronectin-solution from the coverslips. Avoid touching the surface of the coverslip and seed 100 μl of the single cell suspension (2.5×10^4 cells/cm^2) directly on top of the coverslips in a single centered droplet.
7. Let the cells adhere to the coverslip for at least 20–30 min at 37 °C.
8. Combine B27-Medium with 10 μM Y-27632, 0.5 μg/ml Laminin, 10 μM DAPT, 10 μM SU5402, 20 ng/ml BDNF, 20 ng/ml GDNF, 20 ng/ml NGF, 0.2 mM L-Ascorbic Acid, and 0.5 mM dbcAMP (*see* **Note 18**).
9. Check if the cells are fully attached to the coverslips. Carefully add 400 μl/24-well (final volume is 500 μl/well) B27-Medium containing all small molecules, proteins, and Y-27632 without detaching the cells.
10. Change half of the B27-medium containing all small molecules except Y-27632 after 3–4 days. Aspirate half of the medium and add 350–400 μl B27-Medium with freshly added small molecules and proteins. Do not tilt the plate. The cells should always be covered with liquid (*see* **Note 19**).
11. At d17–d18, change the medium as described in **step 10** but do not continue DAPT and SU5402.
12. From now on, aspirate half of the medium every 3–4 days and replace with 350–400 μl B27-Medium. BDNF, NGF, GDNF, dbcAMP, L-Ascorbic Acid, and Laminin are added immediately before use (*see* **Notes 20** and **21**).

3.4 Selection of Neurons by Ara-C Treatment

If many non-neuronal cells are present in the culture, cytotoxic agents can be included in the maturation medium in the first week of maturation (~day 14) to reduce the risk of overgrowth (*see* **Note 22**):

1. Apply the 2 mM stock solution of Ara-C directly to the maturation medium in a 1:1000 ratio (final concentration of 2 μM in the dish) as soon as non-neuronal cells start to proliferate. Place the culture dish back in the incubator.
2. Carefully aspirate the whole medium 18–24 h after incubation and add fresh medium including small molecules to remove all traces of Ara-C (*see* **Note 23**). If no media change is scheduled that day in the protocol, perform an additional one.
3. Proceed with the protocol as indicated above (*see* Subheading 3.3).

4 Notes

1. Besides Vitronectin-coating, iPS-cells can also be maintained on other coatings, e.g., Matrigel (growth factor reduced) or Geltrex (growth factor reduced) following the manufacturers' recommendations.
2. The success of sensory neuron differentiation is greatly dependent on the iPSC lines used. We recommend testing the differentiation-bias of each cell line beforehand by spontaneous 3-germ layer differentiation or short-term sensory neuron differentiation as described. Suitable cell lines are expected to efficiently form ectodermal cells, especially NCLCs, during spontaneous differentiation. They also yield high numbers of NCLCs after 10 days of directed sensory neuron differentiation quantified by flow cytometry for CD271 or immunostaining for SOX10. Successfully differentiating iPSC lines form high numbers of CD271 and SOX10 expressing NCCs at d10.
3. iPS-cells generally require daily medium changes for maintaining pluripotency. However, it is possible to replace 1 day per week by performing a double feeding (double the volume) if cultures are at a low density.
4. B27 supplement containing Retinoic Acid can be replaced by B27 supplement without Retinoic acid. In our hands, B27 with vitamin A yields purer neuron cultures during sensory neuron differentiation.
5. DMEM F12 + GlutaMAX can be replaced by DMEM-F12/KO and L-Glutamine.
6. Geltrex polymerizes at room temperature. It is therefore important to work quickly and with precooled equipment.
7. 1 ml of the coating solution is sufficient to coat one well of a tissue culture treated 6-well plate. For other sized wells or dishes, adjust volume relative to surface area of the well. Make sure that the whole surface is covered and does not dry out.
8. When coating glass coverslips, the coating time should be prolonged for better attachment of the cells.
9. Only "high-quality" iPS-cultures without any or minimal number of differentiated cells (less than 5%) should be used for neural crest cell and sensory neuron differentiation. The iPS-cells should display all features of human pluripotent cells, e.g., marker expression, functionality, and genetic characteristics for the specific donor.
10. Alternatively, count with an automated cell counter according to your protocol.

11. The optimal cell number needs to be tested for each researcher and cell line individually.
12. The differentiation should be started 24–48 h after seeding. Waiting for too many days to reach the appropriate cell density could negatively impact the quality of cells and, hence, the outcome of the differentiation.
13. We recommend the use of 1 μM LDN. In our experience, some cell lines require lower concentrations for successful NCC induction. Therefore, it is advisable to test different LDN concentrations, e.g., 100 nM, 250 nM, 500 nM, and no LDN if the differentiation results in low numbers of neural crest cells.
14. The coating chamber consists of a 15 cm cell culture dish. The lid of a 10 cm dish is placed inside and parafilm is put on top of the lid (*see* **Note 15**). The coverslips are placed on top of the parafilm. This way, the Poly-Ornithine solution stays on top of the coverslip and does not coat the back of the coverslip or any other material that will be in contact with the cell suspension or cultured cells later on. This prevents the attachment and growth of cells on all non-desired surfaces except on top of the coverslip. When taking out the coverslip for subsequent experiments, the neuronal network will be restricted to the coverslip and the risk of detachment is minimized due to strong attachment to the coverslip surface.
15. The parafilm in the coating chamber changes its stickiness when stretched or stored for too long at 37 °C or at 4 °C. Therefore, it is important not to stretch it when preparing the coating chamber. The coverslips should be transferred to a 10 cm dish containing sterile filter paper after Poly-Ornithine coating. The 10 cm dish is then closed with parafilm to keep the coverslips sterile when stored at 4 °C.
16. Due to time constraints, Laminin and Fibronectin coating can be shortened to several hours but cells may tend to detach in long-term differentiation setting. When using glass surfaces, overnight coating is strongly recommended.
17. The exact incubation time for treatment with Accutase needs to be determined for every passaging individually. In general, Accutase treatment is stopped when a single cell suspension appears and remaining clumps can be easily triturated by pipetting. Do not incubate for too short, since insufficient disaggregation causes clumps and too harsh triturating can sometimes cause cell death due to damage of the cells and their processes. In some cases, a 40 μm cell strainer might be helpful to get rid of remaining clumps.
18. Add all small molecules, cytokines and proteins immediately before using the medium and only to the volume needed.

19. Performing full media changes or exposing the cells to air will favor detachment even when pipetting carefully and slowly.
20. Some groups have described the use of NT3 during maturation [5, 8, 11, 12]. In our experience, NT3 can optionally be included in the maturation medium (final concentration 20 ng/ml) but its impact and benefit during maturation remains to be determined.
21. The exact duration of the differentiation needs to be determined for each experiment individually. We perform the differentiation at least until d30 but neurons can be kept in culture for up to 90 days (or even longer) [4, 11, 13]. For most experiments, a maturation period of 40–70 days seems to be sufficient. However, as known for iPS-cell-derived central nervous neurons, differentiation of sensory neuron-like neurons can yield different functional states [14]. It is therefore desirable to validate the functionality, e.g., by patch-clamp or multi-electrode array recordings (MEA) or calcium imaging to confirm efficacy of sensory neuron differentiation and optimize timing for other experiments. A majority of sensory neurons should obtain a functionally mature state during the first weeks of maturation. A lack of mature functional states after more than 5 weeks of maturation usually does not resolve by prolonged maturation but is indicative for a more fundamental problem during the differentiation process.
22. Ara-C effectively decreases the amount of non-neuronal cells and does not seem to be as harmful as an alternative treatment with Mitomycin C [11, 15].
23. Longer incubation times are not recommended to reduce harm to sensory neuron like cells although the effect of Ara-C might not be visible in optical microscopy after 24 h but usually evolves during the following days.

References

1. Kingwell K (2019) Nav1.7 withholds its pain potential. Nat Rev Drug Discov. https://doi.org/10.1038/d41573-019-00065-0
2. Chambers SM, Qi Y, Mica Y, Lee G, Zhang XJ, Niu L, Bilsland J, Cao L, Stevens E, Whiting P, Shi SH, Studer L (2012) Combined small-molecule inhibition accelerates developmental timing and converts human pluripotent stem cells into nociceptors. Nat Biotechnol 30(7): 715–720. https://doi.org/10.1038/nbt.2249
3. Eberhardt E, Havlicek S, Schmidt D, Link AS, Neacsu C, Kohl Z, Hampl M, Kist AM, Klinger A, Nau C, Schüttler J, Alzheimer C, Winkler J, Namer B, Winner B, Lampert A (2015) Pattern of functional TTX-resistant sodium channels reveals a developmental stage of human iPSC- and ESC-derived nociceptors. Stem Cell Reports 5(3):305–313. https://doi.org/10.1016/j.stemcr.2015.07.010
4. Meents JE, Bressan E, Sontag S, Foerster A, Hautvast P, Rosseler C, Hampl M, Schuler H, Goetzke R, Le TKC, Kleggetveit IP, Le Cann K, Kerth C, Rush AM, Rogers M, Kohl Z, Schmelz M, Wagner W, Jorum E, Namer B, Winner B, Zenke M, Lampert A (2019) The role of Nav1.7 in human nociceptors: insights from human induced pluripotent stem cell-derived sensory neurons of erythromelalgia patients. Pain 160(6):1327–1341.

https://doi.org/10.1097/j.pain.0000000000001511

5. Young GT, Gutteridge A, Fox H, Wilbrey AL, Cao L, Cho LT, Brown AR, Benn CL, Kammonen LR, Friedman JH, Bictash M, Whiting P, Bilsland JG, Stevens EB (2014) Characterizing human stem cell-derived sensory neurons at the single-cell level reveals their ion channel expression and utility in pain research. Mol Ther 22(8):1530–1543. https://doi.org/10.1038/mt.2014.86
6. Cao L, McDonnell A, Nitzsche A, Alexandrou A, Saintot P-P, Loucif AJC, Brown AR, Young G, Mis M, Randall A, Waxman SG, Stanley P, Kirby S, Tarabar S, Gutteridge A, Butt R, McKernan RM, Whiting P, Ali Z, Bilsland J, Stevens EB (2016) Pharmacological reversal of a pain phenotype in iPSC-derived sensory neurons and patients with inherited erythromelalgia. Sci Transl Med 8(335):335ra56
7. Namer B, Schmidt D, Eberhardt E, Maroni M, Dorfmeister E, Kleggetveit IP, Kaluza L, Meents J, Gerlach A, Lin Z, Winterpacht A, Dragicevic E, Kohl Z, Schuttler J, Kurth I, Warncke T, Jorum E, Winner B, Lampert A (2019) Pain relief in a neuropathy patient by lacosamide: proof of principle of clinical translation from patient-specific iPS cell-derived nociceptors. EBioMedicine 39:401–408. https://doi.org/10.1016/j.ebiom.2018.11.042
8. Schrenk-Siemens K, Wende H, Prato V, Song K, Rostock C, Loewer A, Utikal J, Lewin GR, Lechner SG, Siemens J (2015) PIEZO2 is required for mechanotransduction in human stem cell-derived touch receptors. Nat Neurosci 18(1):10–16. https://doi.org/10.1038/nn.3894
9. Dionisi C, Rai M, Chazalon M, Schiffmann SN, Pandolfo M (2020) Primary proprioceptive neurons from human induced pluripotent stem cells: a cell model for afferent ataxias. Sci Rep 10(1):7752. https://doi.org/10.1038/s41598-020-64831-6
10. Mis MA, Yang Y, Tanaka BS, Gomis-Perez C, Liu S, Dib-Hajj F, Adi T, Garcia-Milian R, Schulman BR, Dib-Hajj SD, Waxman SG (2019) Resilience to pain: a peripheral component identified using induced pluripotent stem cells and dynamic clamp. J Neurosci 39(3):382–392. https://doi.org/10.1523/JNEUROSCI.2433-18.2018
11. Clark AJ, Kaller MS, Galino J, Willison HJ, Rinaldi S, Bennett DLH (2017) Co-cultures with stem cell-derived human sensory neurons reveal regulators of peripheral myelination. Brain 140(4):898–913. https://doi.org/10.1093/brain/awx012
12. McDermott LA, Weir GA, Themistocleous AC, Segerdahl AR, Blesneac I, Baskozos G, Clark AJ, Millar V, Peck LJ, Ebner D, Tracey I, Serra J, Bennett DL (2019) Defining the functional role of Na(V)1.7 in human nociception. Neuron 101(5):905–919.e908. https://doi.org/10.1016/j.neuron.2019.01.047
13. Bennett DL, Clark AJ, Huang J, Waxman SG, Dib-Hajj SD (2019) The role of voltage-gated sodium channels in pain signaling. Physiol Rev 99(2):1079–1151. https://doi.org/10.1152/physrev.00052.2017
14. Bardy C, van den Hurk M, Kakaradov B, Erwin JA, Jaeger BN, Hernandez RV, Eames T, Paucar AA, Gorris M, Marchand C, Jappelli R, Barron J, Bryant AK, Kellogg M, Lasken RS, Rutten BP, Steinbusch HW, Yeo GW, Gage FH (2016) Predicting the functional states of human iPSC-derived neurons with single-cell RNA-seq and electrophysiology. Mol Psychiatry 21(11):1573–1588. https://doi.org/10.1038/mp.2016.158
15. Chambers SM, Mica Y, Lee G, Studer L, Tomishima MJ (2016) Dual-SMAD inhibition/WNT activation-based methods to induce neural crest and derivatives from human pluripotent stem cells. Methods Mol Biol 1307:329–343. https://doi.org/10.1007/7651_2013_59

Chapter 12

Culture of Human iPSC-Derived Motoneurons in Compartmentalized Microfluidic Devices and Quantitative Assays for Studying Axonal Phenotypes

Maria Giovanna Garone, Chiara D'Antoni, and Alessandro Rosa

Abstract

In order to use induced Pluripotent Stem Cells (iPSCs) to model neurodegenerative diseases, efficient and homogeneous generation of neurons in vitro represents a key step. Here we describe a method to obtain and characterize functional human spinal and cranial motoneurons using a combined approach of microfluidic chips and programs designed for scientific multidimensional imaging. We have used this approach to analyze axonal phenotypes. These tools are useful to investigate the cellular and molecular bases of neuromuscular diseases, including amyotrophic lateral sclerosis and spinal muscular atrophy.

Key words iPSC, piggyBac, Spinal motoneuron, Cranial motoneuron, Differentiation protocol, Microfluidic device, Axon, Axotomy, Amyotrophic lateral sclerosis

1 Introduction

Human induced Pluripotent Stem Cells (hiPSCs) provide a valuable resource for modeling diseases of the nervous system. Several neuronal and glial subtypes can be obtained from hiPSCs by differentiation or by transcription factor(s)-mediated conversion [1–5]. For instance, we and others have shown that inducible expression of specific transcription factors can effectively convert pluripotent cells, including hiPSCs, into distinct populations of motoneurons (MNs) [6–8]. hiPSC-derived MNs are a useful in vitro tool for studying the molecular and cellular basis of degeneration occurring in neuromuscular diseases, such as amyotrophic lateral sclerosis (ALS) and spinal muscular atrophy (SMA) [9–11]. In these diseases, analysis of post-mortem samples does not provide information on the early events leading to MN death and animal models might fail to recapitulate unique key features of the human nervous system. In particular, aberrant axon branching and growth have recently emerged as early phenotypes in ALS,

Nagarajan Kannan and Philip Beer (eds.), *Stem Cell Assays: Methods and Protocols*, Methods in Molecular Biology, vol. 2429, https://doi.org/10.1007/978-1-0716-1979-7_12,

impairing physiological signal transduction functions in MNs and thus playing a potential important role in ALS pathophysiology [12]. Here we show a method to obtain in a short time, and with high efficiency, spinal or cranial MNs from hiPSCs, with ALS-mutant or control genetic background, and their culture in multi-chambered microfluidic chips. Such an experimental setup facilitates analyzing the axonal compartment with the aim of identifying and characterizing early disease phenotypes. Individual MN fates are induced by piggyBac-mediated expression of "programming modules" of three transcription factors: Ngn2-Isl1-Lhx3 (NIL) for spinal MNs or Ngn2-Isl1-Phox2a (NIP) for cranial MNs. After reaching the MN progenitor stage in 5 days, cells are cultured in compartmentalized devices in which the cell bodies are separated from the axon chamber by 450–500 μm microchannels. Analyses of axonal phenotypes can be performed after 7 days, or at subsequent time points as desired following subsequent maturation on the chip. During this time, cells can be treated with compounds or transfected. Axotomy is performed with chemical (trypsin or accutase treatment) or mechanical (vacuum application) methods and regeneration dynamics can be studied in real time, by time-lapse microscopy, or at an end-point, by immunostaining of neurofilament markers. With this method, we have shown that spinal MNs carrying a severe ALS mutation in the FUS gene show altered axon branching and growth upon injury [13].

2 Materials

2.1 Transfection Reagents

1. Neon Transfection System (Thermo Fisher Scientific). Components: Resuspension Buffer R, Resuspension Buffer T, Electrolytic Buffer E2, 100 μL Tips, Electroporation Tubes (*see* **Note 1**).
2. Electroporator (Thermo Fisher Scientific) (*see* **Note 1**).
3. Plasmids: epB-Bsd-TT-NIL, epB-Bsd-TT-NIP and the piggyBac transposase (*see* **Note 2**).
4. Blasticidin (Sigma-Aldrich): Resuspend in water. Store at 4 °C.

2.2 Cell Reagents

1. Human iPSC medium supplemented with 0.1× Penicillin/Streptomycin. Store at 4 °C (*see* **Note 3**).
2. ROCK inhibitor (Y-27632): Resuspend in PBS (pH 7.2) or water or DMSO. Store at −20 °C (*see* **Note 4**).
3. Cell dissociation reagent: Accutase (for example Sigma-Aldrich) (*see* **Note 5**).
4. Differentiation medium: DMEM/F-12, supplemented with 1× stable L-glutamine analog, 1× non-essential amino acid (NEAA), and 0.5 × penicillin/streptomycin. Store at 4 °C.

5. Doxycycline: Resuspend in water. Store at 4 °C (*see* **Note 6**).
6. Neurobasal/B-27 medium: Neurobasal Medium supplemented with 1× B-27, 1× stable L-glutamine analog, 1× non-essential amino acid (NEAA), and 0.5 × penicillin/streptomycin. Store at 4 °C (*see* **Note 7**).
7. DAPT: Prepare a 25 mM stock solution in DMSO. Store at −20 °C (*see* **Note 5**).
8. SU5402: Prepare a 20 mM stock solution in DMSO. Store at −20 °C (*see* **Note 5**).
9. Neuronal medium: Neurobasal Medium supplemented with 1× B-27, 1× stable L-glutamine analog, 1× non-essential amino acid (NEAA), and 0.5 × penicillin/streptomycin. Store at 4 °C (*see* **Note 7**).
10. Brain-derived neurotrophic factor (BDNF): Prepare a 20 mM stock solution in water. Store at 4 °C.
11. Glial cell line-derived neurotrophic factor (GDNF): Prepare a 20 mM stock solution in water. Store at 4 °C.
12. L-ascorbic acid: Prepare a 20 mM stock solution in water. Store at 4 °C.
13. Poly-ornithine: Prepared in cell culture grade water. Store at 4 °C.
14. Laminin: Prepare a 2 mM stock solution in EDTA. Store at 4 °C.
15. Matrix (for example Corning Matrigel hESC-qualified Matrix): Dispense matrix into aliquots in pre-chilled cryotubes on ice. Freeze unused aliquots at −20 °C (*see* **Note 8**).
16. Phosphate-buffered saline (PBS) (Ca^{2+}/Mg^{2+} free).
17. Cell incubator: temperature of 37 °C, and a relative humidity of about 95 percent. The CO_2 concentration is about 5%.

2.3 Microfluidic Chambers

1. PDMS Sylgard 184: It is comprised of Base/Curing Agent to be mixed in a 10 (base): 1 (curing agent) ratio by weight for manual mixing (*see* **Note 9**).
2. Vacuum and desiccator.
3. Soft lithography: It allows to obtain a sealed microfluidic device and the microchannels are imprinted in the PDMS layer closed with a slide (*see* **Note 9**).
4. Structure on a master in a soft elastomer: It allows to create features with geometries defined by the mold's relief structure. After fabrication, the master is filled with PDMS (*see* **Note 10**).
5. Oven.
6. Arc eyelet hole punch cutter tool.
7. Plasma cleaner (*see* **Note 15**).

8. Glass-bottom dishes (*see* **Note 15**).
9. Poly-D-Lysine: Resuspend in water. Store at 4 °C (*see* **Note 11**).
10. Bovine Serum Albumin (BSA): Dilute in PBS (Ca^{2+}/Mg^{2+} free).
11. Complete Embryonic Stem Cell medium: Ready to use medium for culture of pluripotent stem (ES and iPS) cells and feeders. Store at 4 °C.

2.4 MN Seeding in Microfluidic Chambers (MFCs)

1. Cell incubator: temperature of 37 °C, and a relative humidity of about 95%. CO_2 concentration is about 5%.
2. Neuronal medium: Follow the instructions Subheading 2.2, **item 9**.

2.5 Axon Branching and Branch point Analysis

1. Paraformaldehyde (PFA): Dilute in PBS (Ca^{2+}/Mg^{2+}).
2. Horse serum: Store at 4 °C.
3. Triton X-100: 10% in H_2O.
4. Anti-TUJ1 (for example Sigma-Aldrich): concentration 1: 1000.
5. DAPI.
6. Mounting medium (for example Invitrogen): ProLong Diamond Antifade Mountant.
7. Confocal Microscope.
8. PBS (Ca^{2+}/Mg^{2+}).
9. ImageJ software: Analyze Skeleton plugin.

2.6 Axotomy Assay

1. PBS (Ca2+/Mg2+ free).
2. Cell incubator: temperature of 37 °C, and a relative humidity of about 95 percent. The CO_2 concentration is about 5%.
3. Trypsin-EDTA 0.25%.
4. Accutase.
5. Vacuum hood.
6. Confocal Microscope.
7. Cell incubator: temperature of 37 °C, and a relative humidity of about 95%. CO_2 concentration is about 5%.

3 Methods

3.1 Generation of NIL and NIP Inducible iPSC Lines

1. Place one aliquot of Matrigel on ice for about 2 h to thaw.
2. Dilute the aliquot of matrix with 20 mL of cold DMEM/F-12 in a 50 mL conical tube.

3. Mix well and dispense 1 mL of diluted matrix into 35 mm dishes (equivalent amounts per surface area of other dishes).
4. Keep the dishes containing diluted matrix for 1 h at room temperature to allow coating (*see* **Note 13**).
5. Aspirate iPSC culture medium.
6. Rinse iPSCs with PBS (Ca^{2+}/Mg^{2+} free).
7. Add cell dissociation reagent (0.35 mL for a 35 mm dish) and incubate at 37 °C until single cells are separated (5–10 min).
8. Gently complete cell separation by pipetting up and down with a P1000 pipettor 3–4 times.
9. Collect in a 15 mL tube and add PBS (Ca^{2+}/Mg^{2+} free) to 10 mL. Count the cells.
10. Pellet 10^6 cells and resuspend in 100 μL of Buffer R.
11. Add plasmid DNA for transfection: 4.5 μg of transposable vector (epB-Bsd-TT-NIL or epB-Bsd-TT-NIP) and 0.5 μg of the piggyBac transposase plasmid.
12. Transfect with the cell electroporation system following parameters: 1200 V voltage, 30 ms width, 1 pulse. Seed the cells in human iPSC medium supplemented with 10 μM ROCK inhibitor in a matrix-coated dish.
13. The day after transfection, add 5 μg/mL blasticidin to the culture medium.
14. Keep iPSCs in selection conditions for 2 weeks.

3.2 Motoneuron Differentiation (Fig. 1)

1. Aspirate iPSC culture medium.
2. Add cell dissociation reagent (0.35 mL for a 35 mm dish) and incubate at 37 °C until single cells are separated (5–10 min).
3. Gently complete cell separation by pipetting up and down with a P1000 pipettor 3–4 times.

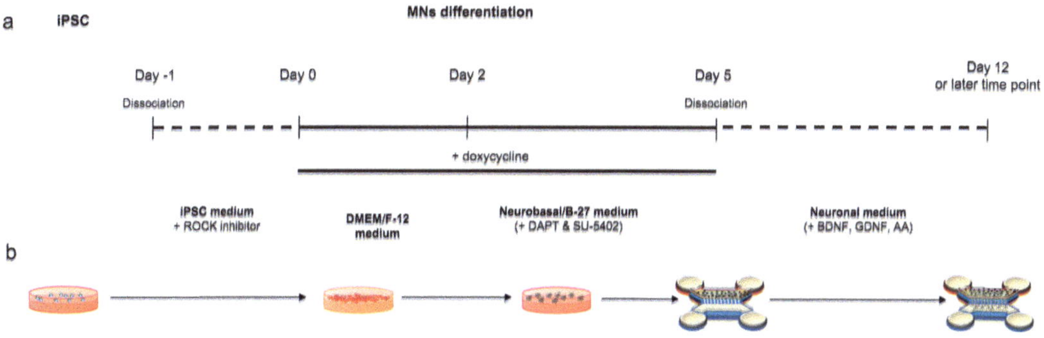

Fig. 1 Motoneuron differentiation protocol. (**a**) Schematic representation of the differentiation protocol, starting with stably transfected iPSCs. (**b**) Representation of main steps to derive iPSC-MNs for axon analysis in microfluidic devices

4. Collect dissociated cells in a 15 mL tube and dilute with 5 volumes of DMEM/F-12. Pellet the cells and resuspend in human iPSC medium supplemented with 10 μM ROCK inhibitor.
5. Count the cells and seed on matrix-coated dishes at a density of 62,500 cells/cm^2.
6. The following day, replace the medium with DMEM/F-12 medium containing 1 μg/mL doxycycline. This is considered as day 0 of differentiation. On day 1, refresh the medium and doxycycline.
7. On day 2, change the medium to Neurobasal/B-27 medium containing 5 μM DAPT, 4 μM SU5402, and 1 μg/mL doxycycline.
8. Refresh the medium and doxycycline every day until day 5.
9. Day 5: Dissociate the neuron-like cells with cell dissociation reagent as described (**steps 1–3**).
10. Pellet the cells and resuspend in Neuronal medium supplemented with 20 ng/mL BDNF, 10 ng/mL GDNF, 20 ng/mL L-ascorbic acid, and 10 μM ROCK inhibitor.
11. Seed the cells on poly-ornithine/laminin- or alternatively on matrix-coated supports at a density of 100,000 cells/cm^2 (*see* **Note 15**).
12. Refresh the medium, without ROCK inhibitor, every 2 days.
13. The MNs are electrophysiologically active on day 12 of differentiation (*see* **Note 2**).

3.3 Fabrication of Microfluidic Chambers (MFCs) in PDMS

Follow the instructions below if you have soft lithography (*see* **Note 14**).

1. Mix vigorously the polymeric base with curing agent in a plastic cup with a wooden tongue depressor (proportion is 10 parts of base for 1 part of curing agent). Bubbles will be formed; this is not a problem.
2. Put the mix under vacuum in a desiccator for approximately 30 min. This will remove the bubbles.
3. Make sure that epoxy masters are clean and free of dust.
4. When PDMS is ready, pour the liquid mix on the masters and bring to oven. Bake at 65–70 °C for 1 h.
5. Peel the PDMS piece from the epoxy master inserting a razorblade in the edge of the PDMS part. Gently detach from master. Be careful not to scratch the epoxy resin as any scratches will be copied in subsequent PDMS pieces.
6. Cut the PDMS part to the desired shape, cut also the four wells.

7. Clean the PDMS pieces of dust with magic tape (an adhesive tape that is suitable for temporary application and can be removed or repositioned).
8. Mount the MFCs onto glass-bottom dishes.
9. Pre-coat MFCs with 1:200 100× poly-D-Lysine for 2 h (*see* **Note 15**).
10. Rinse five times in water.
11. Block MFCs with 0.8% BSA in ES overnight.
12. Aspirate ES medium.
13. Proceed with the matrigel coating (follow the instructions Subheading 3.1, **steps 1–4**).

3.4 MN Seeding in MFCs

1. Follow the instructions Subheading 3.2, **steps 9** and **10**.
2. Resuspend MNs pellet at the density of 1000 cells/μL.
3. Seed MNs in "Soma Chamber" (*see* Fig. 2).
4. Put the MFCs in the incubator and wait 20 min.
5. Add 250 μL of Neuronal medium for each "Soma Chamber" reservoir, and 150 μL for each "Axon Chamber" reservoir (*see* Fig. 2).
6. Refresh the medium without ROCK inhibitor, every 2 days.
7. The MNs are electrophysiologically active on day 12 of differentiation.

3.5 Axon Branching and Branch point Analysis (Fig. 3)

1. Fix the cells in 4% PFA in PBS for 10 min at room temperature.
2. Wash with PBS for 3 times.
3. Permeabilize and block for 15 min using a solution of 0.5% BSA, 10% HRS, 0.2% Triton X-100 in PBS. Incubate sample with primary antibody Anti-TUJ1 (for TUBB3 detection) in 0.5% BSA, 10% HRS in PBS for 1 h at room temperature. Dispense the mix in "Soma chamber" reservoirs and then in "Axon chamber" reservoirs.
4. Rinse three times in PBS for 5 min each.
5. Incubate sample with fluorochrome-conjugated secondary antibody at appropriate dilution in 0.5% BSA, 10% HRS in PBS for 1 h at room temperature in dark. Dispense the mix in "Soma Chamber" reservoirs and then in "Axon Chamber" reservoirs.
6. Rinse three times in PBS for 5 min each in dark.
7. Incubate samples with 1 μg/mL DAPI.
8. Mount samples with a drop of mounting medium. Dispense medium in "Soma Chamber" reservoirs and then in "Axon Chamber" reservoirs.

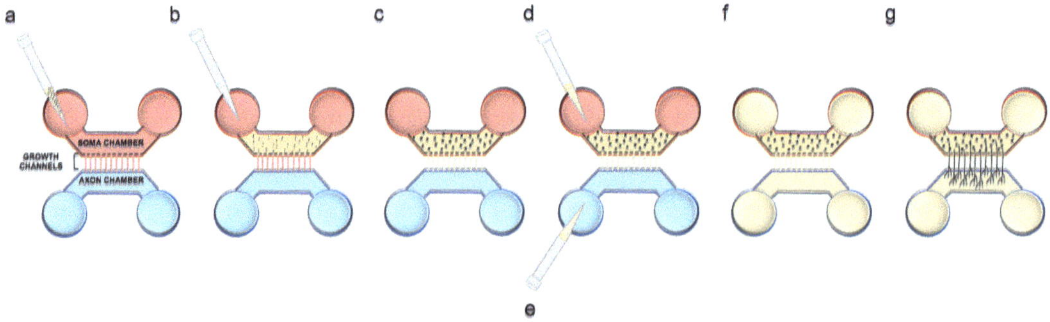

Fig. 2 Schematic representation of motoneurons seeding in MFCs. After MNs dissociation (**a**) the cell pellet is resuspended at the density of 10,000 cells/μL. (**b**) MNs are seeded in the first chamber: "Soma Chamber." (**c**) MFC is left in incubator for 20 min and the cells acquire the star shape: MNs are stuck to the MFC surface and the neural medium fill the growth channels. (**d**) Neural medium is added in "Soma Chamber" and subsequently (**e**) it is added in "Axon Chamber." (**f**) MFCs are left in incubator for 7 days. Reservoirs are refilled every 2 days. (**g**) At day 12, MNs are mature and the terminal axons are visible in "Axon Chamber"

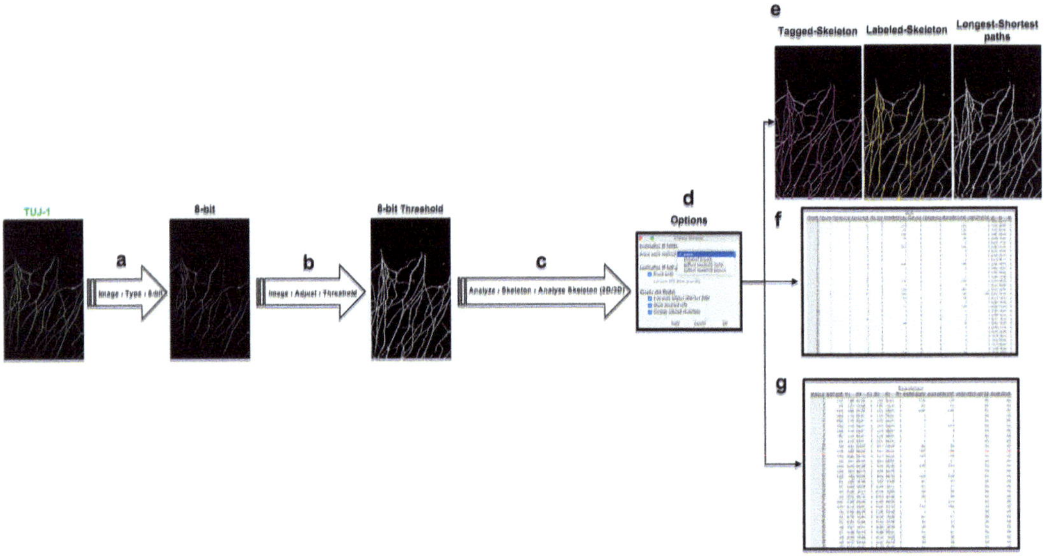

Fig. 3 Steps to follow for use of Analyze Skeleton (ImageJ plugin). (**a**) Upload the TUJ1 image and convert it in 8-bit image. Follow these instructions: Image › Type › 8-bit. (**b**) Adjust the image threshold. Follow these instructions: Image › Adjust › Threshold. (**c**) Use ImageJ plugin. Follow these instructions: Analyze › Skeleton › Analyze Skeleton (2D/3D). (**d**) Select some options in the main dialog of the plugin: prune the possible loops, prune any branch that ends in an end-point, calculate the largest shortest path, show detailed info, display labeled skeletons. (**e**) In this case, the shortest path will be displayed in a new window containing the skeleton in white, the shortest path in magenta, and skeleton labeled with its corresponding skeleton ID in yellow. (**f**) "Results" window is displayed showing for each skeleton in the image: number of branches, number of voxels of every type, number of actual branch points, number of triple and quadruple points, average and maximum length of branches. (**g**) "Branch information" box is displayed showing all branches information: skeleton ID, calibrated branch length, 3D coordinates of the extremes of the branch, and the Euclidean distance between those extreme points

9. Acquire the image.
10. Use Analyze Skeleton (ImageJ plugin) to investigate a skeleton image.
11. Follow these instructions: Analyze › Skeleton › Analyze Skeleton (2D/3D).
12. Skeletonize (2D/3D) works with 8-bit 2D images and stacks, expecting the image to be binary. If not, Skeletonize3D considers all pixel values above 0 to be white (255).
13. Select some options in the main dialog of the plugin: Prune the possible loops, Prune any branch that ends in an end-point, Calculate the largest shortest path, Show detailed info, Display labeled skeletons.
14. Then, a "Results" window is displayed showing for each skeleton in the image:
    - The number of branches.
    - The number of voxels of every type. The number of actual branch points with an arbitrary number of projecting branches.
    - The number of triple points (branch points with exactly 3 branches) and quadruple points (4 branches).
    - The average and maximum length of branches, in the corresponding units.

3.6 Axotomy Assay

At day 12, axons sit in "axon chamber."

1. Wash samples two times with PBS (Ca^{2+}/Mg^{2+} free).
2. Treat "Axon Chamber" with Trypsin-EDTA 0.25% for 15 min (*see* **Note 16**).
3. Reperfuse the axon chamber with PBS (Ca^{2+}/Mg^{2+} free) until effective removal of the damaged axons, without disturbing the cell bodies in the "Soma Chamber."
4. Change the medium in each reservoir: Add 250 μL of Neuronal medium for each "soma chamber" reservoir, and 150 μL for each "Axon Chamber" reservoir (*see* Fig. 2).
5. After 30 h, perform immunofluorescence staining with an anti-TUJ1 antibody as "Axon Branching and Junction Analysis" section.
6. Acquire the image.

4 Notes

1. Follow manufacturer's instructions.
2. References: De Santis et al., 2018; Garone et al., 2019.

3. Storage condition is −20 °C, upon thawing, the medium may be stored at 2–8 °C for 14 days. Medium can be aliquoted into smaller working volumes to avoid repeated freeze/thaw cycles. Avoid exposure to light.
4. Storage condition is −20 °C as supplied. Protect from prolonged exposure to light. This product is soluble water/PBS (pH 7.2) ≤30 mM, DMSO ≤90 mM.
5. Protect from the light.
6. This product is soluble in water (50 mg/mL), yielding a clear, yellow-green solution. The solution should be used within 48 h of preparation and protected from direct sunlight.
7. Protect from the light. After filtering the medium add the B-27.
8. Aliquots are made according to the dilution factor indicated on the datasheet, specific for the individual lot.
9. It is a silicone-based elastomeric kit that is a two-component system with a polymeric base and a curing agent which cross-links with the polymeric matrix. The resulting composite formed is a polydimethylsiloxane (PDMS) with tensile strength (UTS) of ~5.2 MPa and shore hardness of ~44 at room temperature. The UTS, hardness, and the young's modulus (E) increase at a higher curing temperature.
10. The microfluidic channels are designed in a CAD program and printed onto a high-resolution transparency (~5000 dpi) (or converted into a conventional chrome mask). This transparency is used as a photomask in 1:1 contact photolithography (usually using SU-8 or PMMA as photoresist) to produce a master. This master consists of a positive bas-relief of photoresist on a silicon wafer and serves as a mold for PDMS.
11. Excellent optical properties for high-resolution microscopy. Very low autofluorescence.
12. Dishes, sealed with parafilm, can be stored at 4 °C for up to 2 weeks.
13. Follow manufacturer's instructions for poly-ornithine/laminin coated dish.
14. It is possible to buy a pre-assembled microfluidic device.
15. For best results, you can stick MFC on the glass dish using the plasma cleaner.
16. It is possible to treat the axons with Accutase or Vacuum hood.

Acknowledgments

The authors are grateful to Nicol Birsa and Pietro Fratta (UCL, London) and Maria Rosito (IIT, Rome) for helpful discussion. This work was partially supported by Sapienza University, Fondazione Istituto Italiano di Tecnologia and a grant from Istituto Pasteur Italia - Fondazione Cenci Bolognetti to A.R.

Authors' contributions: Conceptualization, M.G.G.; Supervision, A.R.; Writing—original draft, M.G.G, C.D., and A.R.

References

1. Zhang Y, Pak C, Han Y, Ahlenius H, Zhang Z, Chanda S, Marro S, Patzke C, Acuna C, Covy J, Xu W, Yang N, Danko T, Chen L, Wernig M, Südhof TC (2013) Rapid single-step induction of functional neurons from human pluripotent stem cells. Neuron 78(5): 785–798. https://doi.org/10.1016/j.neuron.2013.05.029
2. Davis-Dusenbery BN, Williams LA, Klim JR, Eggan K (2014) How to make spinal motor neurons. Development 141:491–501. https://doi.org/10.1242/dev.097410
3. Yu DX, Di Giorgio FP, Yao J, Marchetto MC, Brennand K, Wright R, Mei A, Mchenry L, Lisuk D, Grasmick JM, Silberman P, Silberman G, Jappelli R, Gage FH (2014) Modeling hippocampal neurogenesis using human pluripotent stem cells. Stem Cell Reports 2:295–310. https://doi.org/10.1016/j.stemcr.2014.01.009
4. Suzuki IK, Vanderhaeghen P (2015) Is this a brain which I see before me? Modeling human neural development with pluripotent stem cells. Development 142:3138–3150. https://doi.org/10.1242/dev.120568
5. Arenas E, Denham M, Villaescusa JC (2015) How to make a midbrain dopaminergic neuron. Development 142:1918–1936. https://doi.org/10.1242/dev.097394
6. Mazzoni EO, Mahony S, Closser M, Morrison CA, Nedelec S, Williams DJ, An D, Gifford DK, Wichterle H (2013) Synergistic binding of transcription factors to cell- specific enhancers programs motor neuron identity. Nat Neurosci 16:1219–1227. https://doi.org/10.1038/nn.3467
7. De Santis R, Garone MG, Pagani F, de Turris V, Di Angelantonio S, Rosa A (2018) Direct conversion of human pluripotent stem cells into cranial motor neurons using a piggyBac vector. Stem Cell Res 29:189–196. https://doi.org/10.1016/j.scr.2018.04.012
8. Garone MG, de Turris V, Soloperto A, Brighi C, De Santis R, Pagani F, Di Angelantonio S, Rosa A (2019) Conversion of human induced pluripotent stem cells (iPSCs) into functional spinal and cranial motor neurons using PiggyBac vectors. J Vis Exp (147). https://doi.org/10.3791/59321
9. Dimos JT, Rodolfa KT, Niakan KK, Weisenthal LM, Mitsumoto H, Chung W, Croft GF, Saphier G, Leibel R, Goland R, Wichterle H, Henderson CE, Eggan K (2008) Induced pluripotent stem cells generated from patients with ALS can be differentiated into motor neurons. Science 321(5893):1218–1221. https://doi.org/10.1126/science.1158799
10. Ebert AD, Yu J, Rose FF Jr, Mattis VB, Lorson CL, Thomson JA, Svendsen CN (2009) Induced pluripotent stem cells from a spinal muscular atrophy patient. Nature 457(7227): 277–280. https://doi.org/10.1038/nature07677
11. Lenzi J, De Santis R, de Turris V, Morlando M, Laneve P, Calvo A, Caliendo V, Chiò A, Rosa A, Bozzoni I (2015) ALS mutant FUS proteins are recruited into stress granules in induced pluripotent stem cells (iPSCs) derived motoneurons. Dis Model Mech 8:755–766. https://doi.org/10.1242/dmm.020099
12. Suzuki N, Akiyama T, Warita H, Aoki M (2020) Omics approach to axonal dysfunction of motor neurons in amyotrophic lateral sclerosis (ALS). Front Neurosci 14:194. https://doi.org/10.3389/fnins.2020.00194
13. Garone MG, Birsa N, Rosito M, Salaris F, Mochi M, de Turris V, Nair RR, Cunningham TJ, Fisher EMC, Morlando M, Fratta P, Rosa A. ALS-related FUS mutations alter axon growth in motoneurons and affect HuD/ELAVL4 and FMRP activity. Commun Biol. 2021 Sep 1;4(1):1025. https://doi.org/10.1038/s42003-021-02538-8

Chapter 13

iPS Cell Differentiation into Brain Microvascular Endothelial Cells

Angelica Medina and Hengli Tang

Abstract

The blood–brain barrier is a tissue structure that modulates the selective entry of molecules into the brain compartment. This barrier offers protection to the brain microenvironment from toxins or any fluctuations in the composition of the blood plasma via a layer of endothelial cells connected by tight junctions and supported by pericytes and astrocytes. Disruption of the barrier can be either a cause or a consequence of central nervous system pathogenesis. Therefore, research based on understanding the structure, function, and the mechanisms of breaching the blood–brain barrier is of primary interest for diverse disciplines including drug discovery, brain pathology, and infectious disease. The following protocol describes a detailed differentiation method that uses defined serum components during stem cell culture to deliver cellular cues in order to drive the cells towards brain endothelial cell lineage. This method can be used to obtain reproducible and scalable cultures of brain microvascular endothelial cells with barrier characteristics and functionality. These endothelial cells can also be stored long term or shipped frozen.

Key words Stem cells, Blood–brain barrier, Endothelial cells, Barrier integrity, TEER

1 Introduction

The brain tissue is separated from the blood flow by the blood–brain barrier (BBB). The cells that compose this barrier strictly maintain and regulate the exchange of materials such as nutrients and toxins between the systemic blood circulation and the central nervous system (CNS) [1–4]. This structure acts primarily as a physical barrier due to the presence of complex tight junction (TJ) proteins in the brain endothelial cells that restrict the passage of most molecules across the BBB into the brain parenchyma [3, 5]. The main component of the BBB is the endothelial cell lining the blood capillaries in the brain, which are surrounded by other brain cell types such as neurons and glial cells. The organized interaction of these different cell types is referred to as the neurovascular unit (Fig. 1) [3, 6].

Nagarajan Kannan and Philip Beer (eds.), *Stem Cell Assays: Methods and Protocols*, Methods in Molecular Biology, vol. 2429, https://doi.org/10.1007/978-1-0716-1979-7_13,

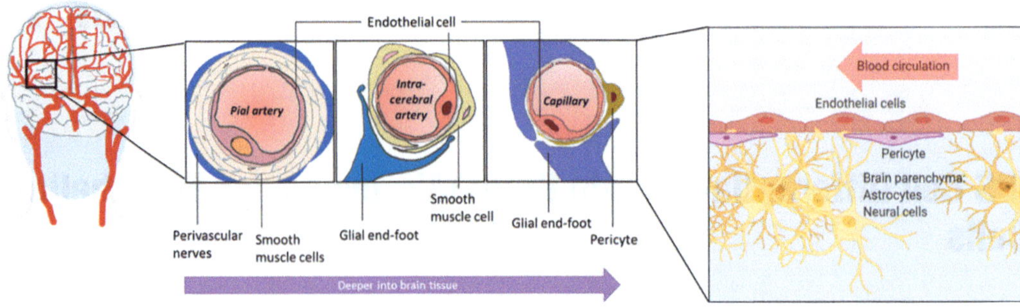

Fig. 1 Schematic representation of cellular components forming the blood–brain barrier. Oxygen in the blood circulation is supplied to the brain by the internal carotid and vertebral arteries that, as they deepen into the brain tissue, branch into pial arteries, arterioles, and capillaries that run along and throughout the brain. The capillaries are the highest in density and responsible for most of the interactions between the brain parenchyma and the blood components

Researchers modeling the BBB have traditionally used animal models, cultured primary brain microvascular endothelial cells (BMECs), and a few cell lines derived from human BMECs (Table 1); however, all these have limitations, including species differences in the case of animal models and low barrier tightness due to the extended subculture of BMECs and immortalized BMECs cell lines [12, 13]. The accessibility of primary human BMECs also presents a challenge. Hence, there is a great need of in vitro models that are reliable and can fully replicate the complex relationships existing within the BBB neurovascular unit in vivo. Desired characteristics include both physiological functions and disease phenotypes such as barrier breakdown and the ensuing immune response [14, 15]. Here, we provide a detailed step-by-step protocol for a stem cell differentiation scheme with proven applications in drug permeability prediction [16], evaluation of nutrient uptake by the BBB [12], and barrier dysfunction in the context of neurological disorders [17]. The method consists of differentiation of pluripotent stem cells guided by serum and growth factor addition to culture medium and subsequent selection of BBB microvascular endothelial cells via capture by basement membrane proteins (Fig. 2). The endothelial cells can then either be directly used for experimental manipulations or frozen down using conventional cell culture techniques and shipped to other labs, facilitating experimental design and implementation of this model. This procedure allows for consistent production of cultures with desired barrier phenotypes such as high barrier tightness as measured by their transendothelial electrical resistance (TEER) values that average between 2000 and 4000 $\Omega.cm^2$.

Table 1
Reported TEER values for BBB culture in different models

	Reported TEER values (Ω.cm²)
In vivo	
Rat BBB	30–5900 [7]
Transwell culture	
iPSC monolayer co-culture with astrocytes	1450 [8]
iPSC monolayer co-culture with astrocytes	2000–8000 [9]
Primary BMECs	70 [10]
hCMEC/D3	60 [10]
Dynamic BBB model	
Primary BMECs	1200 [10]
hCMEC/D3	1200 [10]
b.End3	20 [11]
Microfluidic based model	
b.End3	150 [11]
b.End3 co-culture with astrocytes	300 [11]

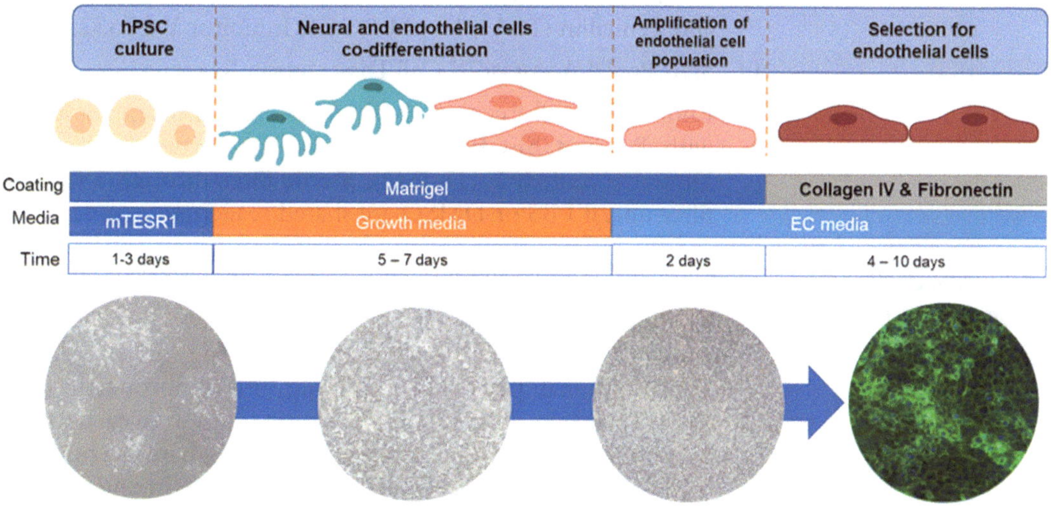

Fig. 2 Differentiation time-course diagram. Stem cells are cultured on Matrigel-coated plates for 1–3 days, then medium is changed to growth medium for a minimum of 5 days until culture flasks are confluent and morphological changes are observed on the cell monolayer. After this, the cells are cultured for 2 days on EC medium with supplements to favor endothelial cell growth. Finally, the cells are selected for and subcultured using collagen IV and fibronectin coated culture vessels

2 Materials

2.1 Stem Cell Culture

1. Stem cells: hESCs (H9)11 or hiPSCs (iPS(IMR90)-4, iPSC DF19-9-11T.H).
2. Corning® Matrigel® Growth Factor Reduced (GFR) Basement Membrane Matrix.
3. T-75 cell culture flasks.
4. Stem cell culture medium (suggested: mTESR1, STEMCELL Technologies, 85857).
5. Dissociation solution (suggested: ReLeSR or Versene solution).
6. Astrocyte Medium.
7. Trypsin/EDTA Solution.
8. Dulbecco's phosphate-buffered saline (DPBS).
9. Growth medium: DMEM/Ham's F12, 20% Knockout Serum Replacer, 1× MEM nonessential amino acids, 1 mM l-glutamine, 0.1 mM β-mercaptoethanol, 100 ng/mL Human basic fibroblast growth factor (bFGF), 10 μM Retinoic Acid (RA).
10. Endothelial cell (EC) medium: Human Endothelial Serum-Free Medium, 1× B27 supplement or 1× N2 supplement, 1× Antibiotic-Antimycotic: Streptomycin, Amphotericin B and Penicillin (PSA), 10 μM ROCK Inhibitor Y-27632.
11. StemPro™ Accutase™ Cell Dissociation Reagent.
12. Endothelial cell freezing medium: EC medium supplemented with RA and bFGF, 10% Fetal Bovine Serum (FBS), 10% Dimethyl Sulfoxide (DMSO). Sterile filter through a 0.2 μm filter unit if FBS or DMSO are shared reagents.

2.2 Plate Coating for Endothelial Cell Seeding

1. Corning® Transwell® polyester membrane cell culture inserts.
2. Coating solution: Mix in sterile distilled water the protein stocks to reach a final concentration of 400 μg/mL of collagen IV from human placenta and 100 μg/mL of fibronectin from human placenta. Add 250 μL of coating solution to each well of a 24-well plate or a 1.12 cm^2 transwell. Incubate overnight at 37 °C, 5% CO_2 for best results in cell attachment and growth.

2.3 TEER Measurements

1. EVOM2 voltohmmeter (World Precision Instruments).
2. STX2 electrode (World Precision Instruments).

2.4 Sodium Fluorescein Assay

1. Fluorescein salt solution: Dilute Fluorescein Sodium salt in fresh and sterile filtered EC medium to achieve a concentration of 100 ng/mL.

2. Spectrophotometer.
3. 96-well plates compatible with spectrophotometer plate reader.

2.5 Poly-D-Lysine Coating for T-75 Flasks

Coating solution: Use 10 mL of sterile water to dilute poly-D-lysine stock to a concentration of 0.1 mg/mL. For a T-75 flask use a coating volume of 10 mL and incubate for minimum of 1 h or preferably overnight at 37 °C, 5% CO_2.

2.6 Fixing and Permeabilization of Cells

1. Fixing solution: 4% paraformaldehyde.
2. PBT buffer for permeabilization of cells: Phosphate-Buffered Saline (PBS) with 0.1% Tween 20.

2.7 PBTG Blocking Solution

1. PBTG: Add 5% Normal Goat Serum to PBT solution and sterile filter through a 0.2 μm filter unit. For long-term storage, place in 4 °C fridge.

3 Methods

3.1 General Recommendations to Prevent Contamination During the Differentiation Process

1. During the process of stem cell culture and differentiation, it is crucial to maintain the cells free of bacterial and fungal contamination. Use strict aseptic technique in the cell culture hood. Routinely screen stem cell stocks for hard to detect bacteria such as Mycoplasma, as contamination with this microorganism can lead to decreased cell proliferation.
2. Before and after working in the sterile hood, use 70% ethanol to decontaminate reagents containers, gloves, and the general work surface. The work area should contain only the reagents required for the specific procedure that is about to be performed.
3. Date and label the reagents pertinent for cell culture and differentiation procedures to track and ensure sterility. Avoid leaving reagents uncapped and unattended. Sterile filter media stocks and divide in ready to use aliquots using sterile pipettes.

3.2 Maintenance of Stem Cell Culture

3.2.1 Preparation of Matrigel-Coated Flasks

1. Thaw the aliquots of Matrigel in cold (4 °C) DMEM/Ham's F12 to achieve a concentration of 0.2 mg/mL of Matrigel. For a 24-well plate, use a volume of 200–400 μL to coat the wells.
2. Coat flasks or plates the day before using. Incubate at 37 °C, 5% CO_2 overnight (*see* **Notes 1–2**).

3.2.2 Thawing Stem Cell Stocks

1. Before thawing the cells, prepare the flasks that will be used for culture. Retrieve the flasks from the incubator and aspirate the medium used for coating the flask. If using a T-75 flask, add 5 mL of mTESR1 medium and set aside with appropriate labels.

2. Prepare 15 mL falcon tubes to centrifuge the stem cell stocks by adding 5 mL of mTESR1 medium into each falcon tube. Remove cryovial from liquid nitrogen and use 70% ethanol to clean the outside of the cryovial, then thaw the cells by gently swirling the frozen cryovial in the 37 °C water bath until a small crystal remains inside.
3. Dry excess water and use 70% ethanol to sterilize the outside of the cryovial, especially around the cap, before placing it inside the cell culture hood. Briefly let the cryovial air dry for 20 s, then transfer the contents to the pre-prepared 15 mL falcon tube with 5 mL of culture medium. Centrifuge for 5 min at 200 × *g*. Aspirate medium and resuspend the cells gently in 12 mL mTESR1 supplemented with 10 μM ROCK inhibitor (*see* **Notes 3–6**).
4. After seeding the cells on the flask, incubate overnight at 37 °C, 5% CO_2 without disturbing the cells in the incubator for at least 10 h. Allow the cells to get to 70–80% confluency before splitting to seed into flasks to start the differentiation process.

3.2.3 Maintenance and Subculture of Stem Cells

1. During stem cell culture, perform daily media changes adding 12 mL of fresh mTESR1 to the flask. After the flasks have reached the desired confluency (between 70% and 80%) to passage the cells, aspirate the culture medium and add 3–5 mL of ReLeSR to the T-75 flask, incubate for 5 min at 37 °C.
2. After the incubation period, resuspend the cells with mTESR1 without ROCK inhibitor to obtain a final volume of 6 mL. Count the cells to seed approximately 1×10^6 to 1.5×10^6 cells in a new T-75 flask with a final volume of 12 mL of mTESR1.
3. Incubate overnight at 37 °C, 5% CO_2 and check for attachment and cell viability the next day before starting the differentiation process.

3.3 Differentiation Protocol

3.3.1 Differentiation Protocol Setup

1. The day after seeding the appropriate number of stem cells and culturing for 1–2 days in mTeSR1, switch the stem cell culture medium to growth medium containing knockout serum replacer to initiate differentiation [8] (*see* **Notes 7–8**). Cells will proliferate quickly and go through morphological changes from a stem cell state to a mixed culture of endothelial and neural cells by day 4 of medium exposure. Replace medium daily with fresh growth medium during this phase.
2. After day 5 (Fig. 2), switch the growth medium to EC medium supplemented with 20 ng/mL bFGF and 10 μM RA [9].
3. After 48 h of incubation at 37 °C, dissociate the cell monolayer with Accutase (*see* **Note 9**). Use an appropriate volume to cover

the cell monolayer and incubate for 30 min to 1 h at 37 °C. Use 5 mL of Accutase for a T-75 flask.

4. Resuspend the dissociated monolayer by dispensing fresh EC medium lacking RA and bFGF onto the flask. Use a cell scraper to gently resuspend sections that are hard to reach with the serological pipettes if culturing in flasks. Collect in 15 mL falcon tubes and centrifuge the cells for 5 min at 200 × *g*. Aspirate the medium and resuspend the cells in EC medium without RA and bFGF but supplemented with 10 μM of ROCK inhibitor (*see* **Note 10**).
5. Use Trypan blue to evaluate viability after centrifugation and resuspension, use the live cell count for seeding calculations.
6. Plate cells onto 24-well tissue culture plates or 1.12 cm^2 Transwell-Clear permeable inserts (0.4–8 μm pore size) coated with 250 μL of a 4:1 mixture of human collagen IV and fibronectin. Seed between 5×10^5–8×10^5 cells/well. Incubate the seeded cells overnight and change the medium (without RA and bFGF) the next day.
7. The next day after seeding onto collagen IV and fibronectin coated plates, measure the TEER values on the Transwell plates using a STX2 chopstick electrode and an EVOM2 voltohmmeter. TEER measurements can be monitored in a timely manner relevant to the experiment but for maintenance culture purposes, measure every 24 h.

3.4 Freezing Down Differentiated Endothelial Cells

1. Freeze the endothelial cell stocks before the final selection on collagen IV and fibronectin coated culture vessels. After 30 min–1-h incubation with Accutase on day 8 (Fig. 2), collect and centrifuge the cells for 5 min at 200 × *g*.
2. Aspirate medium and resuspend the cells with the endothelial cell freezing medium. Make stocks of 8×10^6 cells/mL making sure to break down the cell aggregates into small or single cell masses.
3. Aliquot in prelabeled cryovials and store at −80 °C overnight inside an isopropanol freezer container to allow the cells to freeze slowly. The next day, for extended storage, store cell vials in liquid nitrogen. Frozen cell stocks can be used within a month from the date of storage and yield a barrier with similar characteristics as non-frozen endothelial cell monolayers.

3.5 Co-culture of Stem Cell Derived Brain Microvascular Endothelial Cells with Primary Astrocytes

1. To set up the co-culture during the experiments, seed primary human astrocytes on poly-D-lysine coated plates at a final concentration of 2 μg/cm^2.
2. A day after the establishment of the astrocyte culture from cryopreserved cell stocks, replace medium with fresh astrocyte complete medium. For regular maintenance, change medium

every 3 days and monitor to passage the cells when the culture reaches 80–90% confluency.

3. To passage and seed the astrocyte cells for the co-culture experiments, aspirate culture medium, wash cell monolayer using 5 mL of DPBS, and incubate the cells in 1 mL of 0.05% trypsin/EDTA solution for 5 min at 37 °C. After incubation, confirm under a microscope the change in cell morphology and detachment from the culture vessel.
4. Gently tap the side of the flask to help release the cells from the culture surface and transfer the trypsin and cell solution into a pre-prepared 15 mL falcon tubes containing 5 mL astrocyte complete medium. Rinse the flask with an additional 5 mL of astrocyte complete medium to collect any remaining cells. Centrifuge the falcon tubes containing the cells for 5 min at 200 × *g*.
5. After centrifugation resuspend the cell pellet in astrocyte complete medium for plating into new coated culture plates. For maintenance culture, seed approximately 5000 cells/cm^2. When setting up the co-culture transwell experiments, seed 20,000 cells per well in a 12-well plate. When the astrocyte culture has reached 70% confluency in the wells, start the co-culture by transferring transwell inserts already seeded with endothelial cells into the astrocyte plate, one insert per well.

3.6 Immunofluorescence Staining of Brain Microvascular Endothelial Cells Grown on Transwell Membrane

1. To perform immunostaining of the endothelial cells grown on the transwell membrane inserts, fix cells for 15 min with appropriate volume of 4% Paraformaldehyde (PFA) at room temperature. For a 1.12cm^2 transwell size use 200–250 μL of the fixing solution. After the incubation period, wash once with 1× PBS, aspirate, add 200 μL 1× PBT, and incubate for another 5 min at room temperature.
2. After incubation, aspirate 1× PBT and wash once with 1× PBS, aspirate, add 200 μL of PBTG per well, and incubate for 1 h at room temperature.
3. Make primary antibody solution diluting primary antibodies in PBTG according to the concentration recommended by the supplier. After incubation period, aspirate PBTG and add 200 μL of primary antibody solution per well. Incubate the primary antibody 1 h at room temperature or overnight at 4 °C. After incubation, wash 3 times with 1× PBS for 5 min at room temperature.
4. Aspirate 1× PBS and add 200 μL of secondary antibody solution (secondary antibodies diluted in PBTG according to the supplier instructions) and incubate at room temperature for 1 h. Aspirate and wash 3 times with 1× PBS.

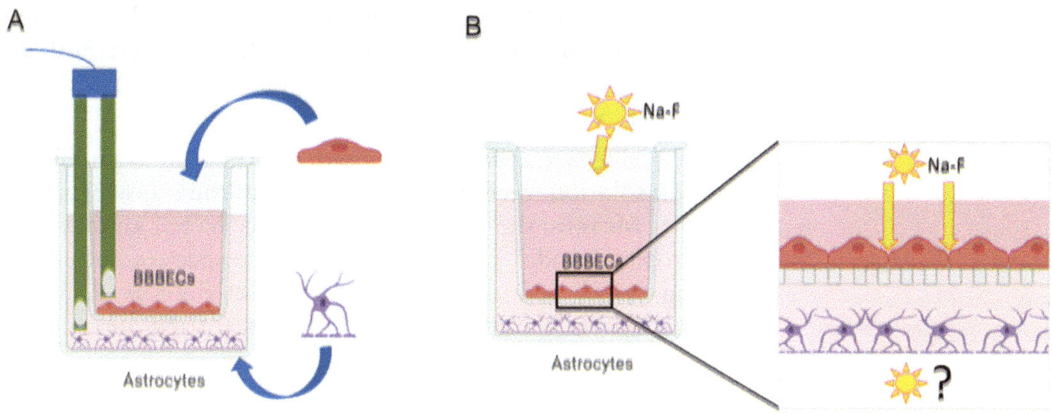

Fig. 3 Barrier integrity evaluation methods. (**a**) Transendothelial electrical resistance measurement of the endothelial cell monolayer with chopstick electrodes. Electrode placement between the transwell containing endothelial cells and plate well containing astrocytes. The electrical resistance is measured in Ohms (Ω) and adjusted to the growth surface area of the transwell (cm^2). (**b**) Sodium fluorescein assay schematics. Permeability to sodium fluorescein is measured based on the salt concentration that diffuses through the monolayer of endothelial cells into the basolateral side of the transwell chamber over a time

5. After the final wash, aspirate 1× PBS and cut the transwell membrane carefully using a scalpel or razor blade. Place scalpel on the border of the membrane from the basolateral side of the transwell and turn the transwell to release the membrane. Use tweezers to hold the membrane to avoid damaging the cell monolayer, dab excess 1× PBS on the membrane using a paper towel and mount the cells on coverslips then place on microscope slides. Make sure to mount the membrane cell-size down onto the coverslip with mounting medium. After placing the coverslips on the microscope slides, seal the edges with clear nail polish to preserve the slide.

3.7 Measurement of TEER for Barrier Tightness Assessment

1. Take TEER measurements using STX2 electrodes attached to an EVOM voltohmmeter. Place the STX2 electrodes within the transwell and the containing well (*see* **Note 11**, Fig. 3A).
2. Measure the resistance value in each transwell 5 times, including a cell-free transwell to use as control for membrane resistance.
3. Adjust the average electrical resistance value (Ω) with the surface growth area (cm^2) of the transwell to obtain the barrier TEER ($\Omega\ .cm^2$) value of each transwell.

3.8 Sodium Fluorescein Assay

1. Aspirate medium from the apical chamber of the transwell plate and replace it with sodium fluorescein (Na-F) in fresh EC medium (Fig. 3b). Every 30 min, during the following 2-h window, remove 500 μL of medium from the basolateral chamber of each transwell. Immediately, replace the volume on the

basolateral chamber with 500 μL of fresh medium. In a 96-well plate, transfer 100 μL of the previously removed 500 μL medium to measure the concentration of Na- F flowing through the membrane.

2. Use a cell-free transwell coated with 4:1 collagen IV and fibronectin solution to control for the effect of the transwell membrane. Measure the fluorescence of the collected samples in a microplate reader compatible with your spectrophotometer using Ex(λ) 485 ± 10 nm and Em(λ) 530 ± 12.5 nm [18, 19]. Perform this assay with technical triplicates for the fluorescent label reading.
3. Calculate the permeability coefficient (Pe) value by obtaining the clearance of Na-F flowing from the apical to the basolateral side of the chamber. Calculate clearance (μL) from the initial 100 ng/mL sodium fluorescein (Na-F) diluted in fresh EC medium added to the apical side of the chamber and the final concentration of Na-F in the basolateral side as: Clearance (μL) = $C_A \times V_A / C_i$. In this case, C_A is the basolateral Na-F concentration, V_A is the volume of the basolateral chamber, and C_i is the initial Na-F concentration used (100 ng/mL).
4. Plot the average clearance vs time for each timepoint (every 30-min measurement of the removed 500 μL from the basolateral side of the transwell set up). Use linear regression analysis to obtain the slope of the clearance curve. The slope (P) of the clearance curve for the monolayer is the permeability by the surface area of the transwell (μL/min). Denote the permeability of the control cell-free transwell membrane as Pc and the permeability for the experimental transwell culture as Pt. Then calculate the value for Pt using the equation: $1/P = 1/Pc + 1/Pt$. Divide Pt by the surface area of the transwell to obtain Pe of each transwell in cm/min [18, 20].

4 Notes

1. It is recommended to coat the Matrigel plates prior to the day of seeding the stem cells. If planning to thaw and monitor multiple batches of stem cells in the same week, it is possible to coat multiple flasks with Matrigel. Be sure to use these flasks within 5–7 days after coating to avoid the DMEM/Ham's F12 medium from evaporating. Monitor humidity inside the incubator to prevent the flasks or Matrigel-coated plates from drying out faster than expected. If there are sections of evaporated medium on the Matrigel plates, do not use them as the dried patches compromise the coating quality and limits the attachment of the stem cells in the culture vessels.

2. To aliquot Matrigel stocks and to dissolve into DMEM/Ham's F12 for coating flasks, it is recommended to use chilled equipment (for example, tips and medium) to slowly dissolve the Matrigel into the medium and immediately apply the coating solution to the culture vessel. Avoid applying body heat from handler's fingertips directly on the Matrigel aliquot by holding the centrifuge tube away from the frozen Matrigel pellet.
3. When thawing the stem cells, do not release the contents of the cryovial against the wall of the falcon tubes, aim to slowly release the cells directly into the 5 mL medium to avoid extra shear stress applied to the cells.
4. To resuspend the stem cells, Accutase can also be used; however, it tends to break the stem cell colonies into small colonies or even single cell cultures. This can cause spontaneous differentiation or lower the attachment efficiency of the stem cells. We recommend using ReLeSR instead, as incubation period is short (~5 min) and the culture is maintained in small to medium colonies that attach well when re-seeded in new Matrigel-coated flasks, maintaining a high yield of differentiated endothelial cells.
5. It is recommended to resuspend stem cells so that they form small to medium sized colonies when plated. To achieve this, resuspend the cells in mTESR1 medium with a 5 mL pipette 2–4 times. Do not force the cells against the wall of the falcon tube or flask as the excess shear force will reduce attachment and viability.
6. Ideally, keep stem cell cultures that will be used to differentiate into endothelial cells under passage 40. Thaw and culture the stem cells until the cells reach splitting confluency (70–80% confluency on a T-75 flask). Split stem cells in a 1:6 ratio into new T-75 flasks to start the differentiation process. Check for spontaneous differentiation in the stem cell culture: if the flask contains over 20% differentiated colonies, it is not recommended to use the flask for differentiation.
7. Change growth medium every day and increase medium volume from 12 to 15 mL during the growth phase as the cell number increases and the flask gets confluent (day 4 and 5). It is recommended to prepare fresh batches of medium stocks that will be used within 1 week and keep them stored at 4 °C.
8. During day 6–8 it is recommended not to change the endothelial cell culture medium; however, cell numbers might be high, and nutrients could be used up quickly during this 48 h incubation period. To avoid nutrient starvation and drastic low pH in the culture, increase culture medium volume to 20 mL and monitor the pH at 24 h. If the medium pH reaches 6.5–6.7, it is preferable to change medium to freshly made endothelial cell medium than inducing further stress in the culture.

9. After day 8 of differentiation, to resuspend the cells for seeding on coated Transwells, plates, or slides, incubate at 37 °C for 30 min to 1 h with Accutase. Do not let the incubation period go longer than 1 h as it will decrease cell viability. After 30 min of incubation, monitor the cell monolayer looking for signs of lifted cells in the culture flask. Rock the flasks back and forth to avoid drying out the cells while the Accutase is incubating. After the cells have lifted, use 5 mL of endothelial cell medium to resuspend the cells and break up into small clumps. A cell scraper could be used to lift any remaining cells and help collect all of the differentiated cells.
10. After centrifugation of the differentiated cells at day 8, it is recommended to break the pellet into single cells by pipetting up and down several times in small volume using a p1000 micropipette. This is recommended for either seeding or freezing down the endothelial cell stocks.
11. When measuring the TEER values, place the electrodes straight over the transwell culture (Fig. 3), avoid placing the electrodes with inclination or touching the monolayer on the apical side of the transwell so the monolayer is not disturbed.

References

1. He Y, Yao Y, Tsirka SE, Cao Y (2014) Cell-culture models of the blood–brain barrier. Stroke 45:2514–2526. https://doi.org/10.1161/STROKEAHA.114.005427
2. Huber JD, Egleton RD, Davis TP (2001) Molecular physiology and pathophysiology of tight junctions in the blood-brain barrier. Trends Neurosci 24:719–725
3. Abbott NJ, Rönnbäck L, Hansson E (2006) Astrocyte-endothelial interactions at the blood-brain barrier. Nat Rev Neurosci 7:41–53
4. Wilhelm I, Fazakas C, Krizbai IA (2011) In vitro models of the blood-brain barrier. Acta Neurobiol Exp (Wars) 71:113–128
5. Ballabh P, Braun A, Nedergaard M (2004) The blood – brain barrier : an overview Structure , regulation , and clinical implications. Neurobiol Dis 16:1–13
6. Cho H et al (2015) Three-dimensional blood-brain barrier model for in vitro studies of neurovascular pathology. Sci Rep 5:15222. https://doi.org/10.1038/srep15222
7. Butt BYAM, Jones HC, Abbott NJ (1990) Electrical resistance across the blood-brain barrier in anaesthetized rats: a developmental study. J Physiol 429:47–62
8. Lippmann ES et al (2012) Derivation of blood-brain barrier endothelial cells from human pluripotent stem cells. Nat Biotechnol 30:783
9. Neal EH et al (2019) A simplified, fully defined differentiation scheme for producing blood-brain barrier endothelial cells from human iPSCs. Stem Cell Reports 12:1380–1388
10. Cucullo L et al (2008) Immortalized human brain endothelial cells and flow-based vascular modeling : a marriage of convenience for rational neurovascular studies. J Cereb Blood Flow Metab 28:312–328. https://doi.org/10.1038/sj.jcbfm.9600525
11. Booth R, Kim H (2012) Miniaturisation for chemistry, physics, biology, materials science and bioengineering. Lab Chip 12(19)
12. Al-Ahmad AJ (2017) Comparative study of expression and activity of glucose transporters between stem cell-derived brain microvascular endothelial cells and hCMEC/D3 cells. Am J Physiol Cell Physiol 313:C421–C429
13. Lauschke K, Frederiksen L, Hall VJ (2017) Paving the way toward complex blood-brain barrier models using pluripotent stem cells. Stem Cells Dev 26:857–874

14. Jeffrey P, Summerfield S (2010) Assessment of the blood-brain barrier in CNS drug discovery. Neurobiol Dis 37:33–37
15. Banks WA (2016) From blood – brain barrier to blood – brain interface : new opportunities for CNS drug delivery. Nat Rev Drug Discov 15:275–292
16. Ohshima M et al (2019) Prediction of drug permeability using in vitro blood-brain barrier models with human induced pluripotent stem cell-derived brain microvascular endothelial cells. Biores Open Access 8:200–209
17. Lim RG et al (2017) Huntington's disease iPSC-derived brain microvascular endothelial cells reveal WNT-mediated angiogenic and blood-brain barrier deficits. Cell Rep 19: 1365–1377
18. Dohgu S, Takata F, Yamauchi A, Nakagawa S (2005) Brain pericytes contribute to the induction and up-regulation of blood – brain barrier functions through transforming growth factor-h production. Brain Res 1038:208–215
19. Dohgu S et al (2007) Adverse effect of cyclosporin A on barrier functions of cerebral microvascular endothelial cells after hypoxia-reoxygenation damage in vitro. Cell Mol Neurobiol 27:889–899
20. Dehouck M-P et al (1992) Drug transfer across the blood-brain barrier: correlation between in vitro and in vivo models. J Neurochem 58: 1790–1797

Chapter 14

Chromatin Immunoprecipitation in Human Pluripotent Stem Cell-Derived 3D Organoids to Analyze DNA–Protein Interactions

Wei Xuan Tan, Chek Mei Bok, Natasha Hui Jin Ng, and Adrian Kee Keong Teo

Abstract

Chromatin immunoprecipitation (ChIP) is a technique that has been widely used to interrogate DNA--protein interactions in cells. In recent years, human pluripotent stem cell (hPSC)-derived 3D organoids have emerged as a powerful model to understand human development and diseases. Performing ChIP in hPSC-derived 3D organoids is a useful approach to dissect the roles of transcription factors or co-factors and to understand the epigenetic landscape in human development and diseases. However, performing ChIP in 3D organoids is more challenging than monolayer cultures, and an optimized protocol is needed for interpretable data. Hence, in this chapter, we describe in detail a protocol for performing ChIP in hPSC-derived islet-like cells as an example, from organoid harvest to ChIP-qPCR data analysis. This chapter also highlights potential pitfalls and provides recommendations for troubleshooting.

Key words Chromatin immunoprecipitation, ChIP, Pluripotent, Stem cell, Differentiation, 3D, Organoids, Islet, Beta cell, Transcription factor

1 Introduction

Chromatin immunoprecipitation (ChIP) is a useful technique to study the interaction(s) between specific proteins and genomic DNA in cells. It is commonly used to identify target genes of transcription factors or co-factors, or to map the location of various modified forms of histone proteins. In principle, live cells are first fixed with crosslinking agents to stabilize protein–DNA interactions. After harvesting the nuclear material, the protein-bound chromatin is released and then fragmented into shorter lengths of DNA, typically through sonication. A specific antibody targeting the protein of interest is added to pull down the protein-DNA complex. Following the reversal of crosslinking and subsequent DNA purification, the immunoprecipitated DNA can be identified

Nagarajan Kannan and Philip Beer (eds.), *Stem Cell Assays: Methods and Protocols*, Methods in Molecular Biology, vol. 2429, https://doi.org/10.1007/978-1-0716-1979-7_14,

and quantified using various methods, such as qPCR (ChIP-qPCR) and high-throughput DNA sequencing (ChIP-Seq). While DNA microarrays (ChIP-on-chip) can also be used to analyze immunoprecipitated DNA, it is now less commonly used due to the predominance of high-throughput sequencing technologies.

With the recent rapid advancement of organoid technology, human pluripotent stem cell (hPSC)-derived organoids have been very useful in understanding embryonic development and as a model to study human diseases [1]. ChIP can be performed on hPSC-derived organoids to understand the epigenetic regulation or the role of transcription factors or co-factors in disease, embryonic development or cellular function. However, the standard protocol for ChIP is highly laborious and many of the steps can contribute to its variability, making data interpretation challenging [2]. Moreover, performing ChIP on 3D organoids is more challenging than on 2D cultures due to difficulties in estimating cell numbers and inaccessibility of the cells, which may hinder crosslinking, nuclear lysis and/or sonication. Given that 3D organoids mimic in vivo tissues more closely than monolayer cultures, the ability to perform ChIP reproducibly on 3D organoids is highly important. Therefore, in this chapter, we describe in detail a protocol for performing ChIP [3–5] in hPSC-derived 3D organoids, using hPSC-derived islet-like cells (obtained using a published differentiation protocol [6]) as an example.

2 Materials

Prepare all buffers using ultrapure water and analytical grade reagents.

2.1 Organoid Dissociation and Crosslinking

1. Phosphate-buffered saline (PBS).
2. TrypLE™ Express Enzyme (1×), phenol red (Gibco™).
3. Fetal bovine serum (FBS).
4. Dimethyl sulfoxide (DMSO).
5. (Optional) Dimethyl 3,3′-dithiopropionimidate dihydrochloride (DTBP) (*see* **Note 1**).
6. (Optional) 3,3′-Dithiodipropionic acid di (N-hydroxysuccinimide ester) (DSP) (*see* **Note 1**).
7. 37% Formaldehyde.
8. 2 M glycine. Store at room temperature.
9. Water bath.
10. Centrifuge.
11. Rocking shaker platform.

2.2 Chromatin Extraction and Sonication

1. Leupeptin hemisulfate. Prepare stock aliquots and store them at −20 °C.
2. Pepstatin A. Prepare stock aliquots and store them at −20 °C.
3. Phenylmethylsulfonyl fluoride (PMSF). Prepare stock aliquots and store them at −20 °C.
4. Cell lysis buffer: 10 mM Tris–HCl pH 8, 10 mM NaCl, and 0.2% NP-40. Store at 4 °C.
5. Nuclear lysis buffer: 50 mM Tris–HCl pH 8, 10 mM EDTA, and 1% SDS. Store at 4 °C (*see* **Note 2**).
6. Immunoprecipitation (IP) dilution buffer: 20 mM Tris–HCl pH 8, 2 mM EDTA, 150 mM NaCl, 0.01% SDS, and 1% Triton X-100. Store at 4 °C.
7. 70% ethanol (cleaning).
8. Non-abrasive cleaning wipes.
9. Distilled water.
10. Misonix Q500–220 ultrasonic probe sonicator (500 W, 220 V), with 1.6 mm probe (0.2–5 mL) (*see* **Note 3**).

2.3 Chromatin Immunoprecipitation

1. Mock IgG (corresponding to the species of the antibody of interest).
2. Antibody for protein of interest (Ab) (*see* **Note 4**).
3. Protein A/G agarose beads (25% volume/volume in PBS) (*see* **Note 5**).
4. IP Wash Buffer 1: 20 mM Tris–HCl pH 8, 2 mM EDTA, 50 mM NaCl, 0.1% SDS, and 1% TritonX-100. Store at 4 °C.
5. IP Wash Buffer 2: 10 mM Tris–HCl pH 8, 1 mM EDTA, 0.25 M LiCl, 1% NP-40, and 1% Sodium deoxycholate. Store at 4 °C.
6. Tris-EDTA (TE) buffer.
7. Elution Buffer: 100 mM $NaHCO_3$ and 1% SDS. Store at 4 °C (*see* **Note 2**).
8. (Optional) Dithiothreitol (DTT). Prepare 1 M stock aliquots and store them in −20 °C (*see* **Note 6**).
9. Tube rotator.

2.4 DNA Purification

1. RNase A (10 μg/μL).
2. 5 M NaCl. Store at room temperature.
3. Proteinase K (20 μg/μL).
4. Phenol:Chloroform:Isoamyl Alcohol 25:24:1.
5. 3 M NaAc pH 5.2. Store at room temperature.
6. GlycoBlue™ Coprecipitant (15 mg/mL) (Invitrogen™).

7. 100% ethanol.
8. 70% ethanol in nuclease-free water.
9. Nuclease-free water.

2.5 ChIP-qPCR

1. SYBR Green qPCR Master Mix.
2. Primers targeting regions of interest (*see* **Note 7**).
3. Primers targeting a negative control region (*see* **Note 8**).
4. 384-well PCR plate.
5. qPCR machine.

3 Methods

Figure 1a sums up the workflow of a ChIP experiment in hPSC-derived 3D organoids, starting from cell harvest to downstream analyses. The protocol described below is meant for one ChIP reaction (pulldown of one protein of interest and one Mock IgG) and can be scaled up accordingly for more samples. It is important to note that the amount of starting material is crucial for a successful ChIP experiment. However, it is extremely challenging to accurately control the number of cells in 3D organoid cultures. Moreover, counting the number of organoids can be too tedious and inaccurate as organoids vary in size. Hence, we gauge the starting amount of material for ChIP by its pellet size. Based on our experience, we estimate that a pellet size as shown in Fig. 1b would be sufficient for one ChIP reaction (*see* **Note 9**). Using hPSC-derived islet-like cells as an example, this pellet size is equivalent to ~400 organoids, that are typically 300–500 μm in diameter each. Each organoid is estimated to contain ~1000 cells, although this is highly dependent on the size of the organoid.

3.1 Reagent Preparation

Prepare the following reagents just before the start of the experiment.

1. Prepare 6 mL of 5% FBS in PBS.
2. Add protease inhibitors (for final concentration of 100 μM Leupeptin hemisulfate, 1 μM Pepstatin A, and 1 mM PMSF) to 2 mL of cell lysis buffer, 1.25 mL of nuclear lysis buffer, and 4.25 mL of IP dilution buffer. Keep the buffers on ice.
3. (Optional) Prepare crosslinkers (10.8 mM DTBP and 2.5 mM DSP). First, dissolve 3.79 mg of DSP in 0.375 mL of DMSO. Next, dissolve 12.5 mg of DTBP in 3.375 mL of PBS. Finally, mix the two together to obtain 3.75 mL (*see* **Note 10**).

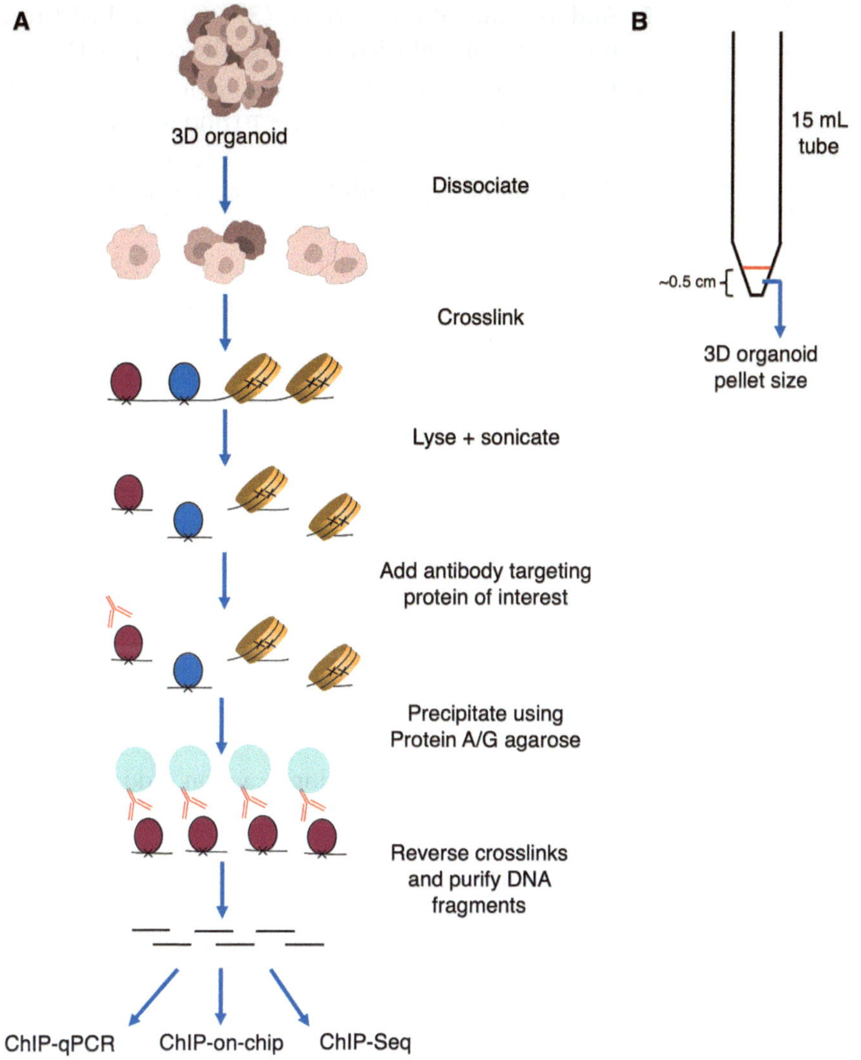

Fig. 1 Schematic illustration of ChIP in hPSC-derived 3D organoids. (**a**) Typical workflow for ChIP. (**b**) Cell pellet size in a 15 mL tube that is recommended (minimum) for one ChIP reaction. The pellet size shown is roughly 0.5 cm in height in a typical 15 mL tube. This is approximately equivalent to 400 organoids that are 300–500 μm in diameter each

3.2 Organoid Dissociation and Crosslinking

All procedures and centrifugation described in this sub-section are performed at room temperature, unless otherwise stated.

1. Transfer organoids into a 15 mL tube.
2. Let organoids sink to the bottom of the tube by gravity, which may take a few minutes.
3. Aspirate the supernatant.
4. Wash the organoids twice with 5 mL PBS, by repeating **steps 2** and **3**.

5. Add 1–2 mL of pre-warmed (37 °C) TrypLE™ to the organoids (volume sufficient to cover the cell pellet).
6. Place the tube in a 37 °C water bath for 3–4 min. While the tube is in the water bath, use a P1000 micropipette (with a cut tip to increase the pore size for organoids to pass) to gently dissociate the organoids by continuous pipetting. Avoid vigorous pipetting to prevent cell lysis (*see* **Note 11**).
7. To dilute TrypLE™, add 3× volume of PBS with 5% FBS to the cell suspension and pipette up and down gently for 5–6 times to mix thoroughly (*see* **Note 12**).
8. Centrifuge the cells at 200 × *g* for 3 min and aspirate the supernatant.
9. Wash the cells with 5 mL of PBS and repeat **step 8**.
10. (Optional) Add 3.75 mL of crosslinkers (10.8 mM DTBP and 2.5 mM DSP) to the cells in 15 mL tube, tighten the tube cap, and place the tube horizontally on a rocking shaker platform for 15 min at 30 rpm.
11. Add 101.25 μL of 37% formaldehyde (1% final concentration) and incubate for another 15 min at 30 rpm (*see* **Note 13**). If **step 10** is not performed, add 3.75 mL of PBS prior to **step 11**.
12. To quench the crosslinking reaction, add 234.4 μL of 2 M Glycine (final concentration 0.125 M) and incubate for 5 min at 30 rpm.

3.3 Chromatin Extraction and Sonication

From this step onwards, keep the samples on ice at all times and perform all centrifugation steps at 4 °C.

1. Spin the cells down at 800 × *g* for 5 min and aspirate the supernatant.
2. Wash the cells twice with 3 mL of ice-cold PBS and repeat **step 1**.
3. Resuspend the cell pellet in 2 mL of cell lysis buffer (with protease inhibitors) and incubate for 10 min on ice. Resuspend 2–3 times during the 10 min incubation as the cells tend to sink.
4. Spin down the cells at 3200 × *g* for 5 min and aspirate the supernatant.
5. Resuspend the cell pellet in 1.25 mL of nuclear lysis buffer (with protease inhibitors) and incubate for 10 min on ice.
6. Add 0.75 mL of IP dilution buffer so that the total volume is 2 mL and proceed to sonication using an ultrasonic probe sonicator.

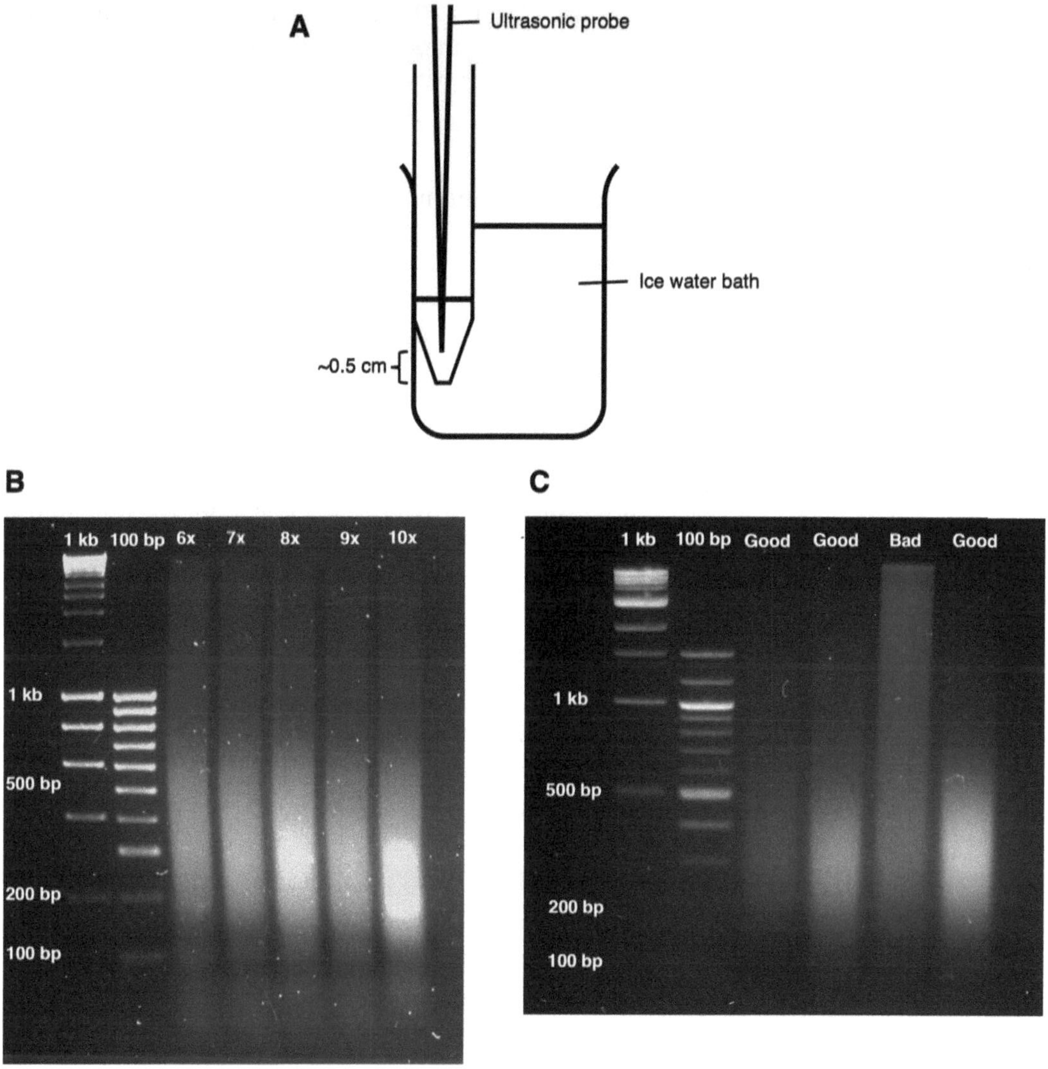

Fig. 2 Sonication of chromatin samples using an ultrasonic probe. (**a**) Schematic illustration for the setup of sonication. The 15 mL tube containing 2 mL of chromatin sample is held stable and immobilized to the side of a beaker filled with ice water. The ultrasonic probe is lowered into the tube until ~0.5 cm above the bottom of the tube. The probe should be in the center of the tube without touching the walls of the tube (*see* **Note 14**). (**b**) The effect of the number of sonication cycles (6–10) on DNA fragment size. The number of sonication cycles should be optimized to get most of the DNA fragments within 200–500 bp. (**c**) An example of badly versus well-sonicated gDNA samples

7. Clean the probe with 70% ethanol and non-abrasive cleaning wipes.
8. Clean the probe again with distilled water and non-abrasive cleaning wipes.
9. Refer to Fig. 2a for the sonication setup. Place the tube upright in an ice water bath (*see* **Note 14**). Sonicate cells in a cold room

using Misonix Q500–220 ultrasonic probe sonicator at 30% power, 30 s on for a duration of 6 min (12 cycles of sonication with 45 s rest intervals between each cycle).

10. Spin down the cells at 3200 × *g* for 5 min at 4 °C (*see* **Note 15**).
11. Collect the supernatant in a new 15 mL tube.

3.4 Chromatin Immunoprecipitation

1. Add 3.5 mL of IP dilution buffer (with protease inhibitors) to make up a total volume of 5.5 mL.
2. Add 10 μg of Mock IgG for pre-clearing (*see* **Note 16**).
3. Incubate the samples on a tube rotator (10 rpm) for 1 h at 4 °C.
4. Add 60 μL of Protein A/G agarose beads.
5. Incubate the samples on a tube rotator (10 rpm) for 2 h at 4 °C.
6. Spin down Protein A/G agarose beads at 3200 × *g* for 2 min at 4 °C.
7. Collect 550 μL of supernatant into a new 1.5 mL tube. This is the 10% input of the sonicated gDNA. Store at −20 °C.
8. Collect the remaining 5 mL of supernatant into two tubes of 2.5 mL each.
9. Add 10 μg of Mock IgG into one of the sample tubes and 10 μg of antibody into the other (*see* **Note 17**).
10. Incubate the samples on a tube rotator (10 rpm) overnight at 4 °C.
11. The next day, add 60 μL of Protein A/G agarose beads.
12. Incubate the samples on a tube rotator (10 rpm) for 1.5 h at 4 °C (*see* **Note 18**). From this step onwards, samples can be handled at room temperature and all centrifugation steps can be done at room temperature as well.
13. Spin down Protein A/G agarose beads at 3200 × *g* for 2 min. The beads now contain the chromatin regions of interest.
14. Aspirate the supernatant. Be careful not to aspirate any beads while removing the supernatant.
15. Add 500 μL of IP wash buffer 1 and transfer all the Protein A/G agarose beads into a new 1.5 mL tube. Keep the pipette tip as it contains residual agarose beads.
16. Using a new pipette tip, add another 500 μL of IP wash buffer 1 into the same 15 mL tube (*see* **Note 19**). Discard this pipette tip.
17. Re-attach the old pipette tip from **step 15** and transfer the 500 μL from the 15 mL tube to the same 1.5 mL tube, which should now contain 1 mL of IP wash buffer 1 with the Protein A/G agarose beads (*see* **Note 20**).

18. Incubate tubes on a tube rotator (10 rpm) for 4 min at room temperature.
19. Spin down the Protein A/G agarose beads at 1800 × *g* for 2 min.
20. Remove the supernatant.
21. Wash the Protein A/G agarose beads with another 500 μL of IP wash buffer 1 and repeat **steps 18–20** (*see* **Note 21**).
22. Wash the Protein A/G agarose beads with 500 μL of IP wash buffer 2 and repeat **steps 18–20**.
23. Wash the Protein A/G agarose beads with 500 μL of TE buffer twice. Repeat **steps 18–20**.
24. After aspirating the supernatant of the second TE buffer wash, add 250 μL of elution buffer with 100 mM DTT (optional) (*see* **Note 6**).
25. Agitate the tubes on a heat block shaker at 1400 rpm for 10 min at room temperature. The exact speed of the shaker is not crucial as long as the beads are sufficiently mixed around or agitated. Alternatively, the tubes can also be placed on a vortex for 10 min.
26. Spin down the Protein A/G agarose beads at 1800 × *g* for 2 min.
27. Collect the supernatant into a new 1.5 mL tube. The supernatant now contains the immunoprecipitated DNA. Add another 250 μL of elution buffer with 100 mM DTT (optional) into the original tube with the Protein A/G agarose beads for a second elution. Repeat **steps 25** and **26**.
28. Combine the supernatant in the same tube, which now contains 500 μL of eluted chromatin samples.

3.5 DNA Purification

1. Add 30 μL of 5 M NaCl into the eluted samples and input sample.
2. Add 0.1 μL of RNase A (10 μg/uL) to every tube and incubate them at 67 °C for at least 4 h (*see* **Note 22**).
3. Add 3 μL of proteinase K (20 μg/uL) into every tube and incubate them at 45 °C for at least 3 h (*see* **Note 23**).
4. Add 500 μL of phenol-chloroform-isoamyl alcohol into every tube.
5. Vortex and centrifuge at 18,000 × *g* for 5 min at room temperature.
6. Transfer the top aqueous layer into a new 1.5 mL tube.
7. Add another 500 μL of phenol-chloroform-isoamyl alcohol to the top aqueous layer (*see* **Note 24**). Repeat **steps 5** and **6**. There should be ~500 μL of each sample now.

8. Add 50 μL of 3 M NaAc pH 5.2, 30 μg of GlycoBlue™ Coprecipitant, and 750 μL of 100% ethanol to every sample.
9. Incubate at −80 °C for at least 30 min (*see* **Note 25**).
10. Spin down the DNA pellet at 18,000 × *g* for 30 min at 4 °C. A blue pellet should be visible.
11. Aspirate the supernatant.
12. Add 750 μL of 70% ethanol (in nuclease-free water) to the tubes for washing.
13. Spin down the DNA pellet at 18,000 × *g* for 5 min at 4 °C.
14. Aspirate the supernatant. Be very careful not to touch the DNA pellet (*see* **Note 26**).
15. Air dry the DNA pellet by placing the tubes on the bench at room temperature for 10–15 min (*see* **Note 27**).
16. Resuspend the DNA pellet in 35 μL of nuclease-free water for Mock IgG and pulldown samples. Resuspend the input DNA pellet in 70 μL of nuclease-free water for the 10% input sample (*see* **Note 28**). Keep the samples on ice at all times and store them at −20 °C. Minimize free-thaw cycles.
17. Run 1–2 μg of input DNA on a 2% agarose gel to evaluate the extent and quality of sonication (Fig. 2b, c). The smear should appear below 1 kb, with the majority of fragments falling within a size range of 200–500 bp (*see* **Note 29**).
18. Proceed to analyze immunoprecipitated DNA via ChIP-qPCR and/or ChIP-Seq.

3.6 ChIP-qPCR

ChIP-qPCR is typically performed to experimentally confirm that a protein of interest is indeed bound to a specific genomic region of interest, indicating that the ChIP experiment is successful. This sub-section briefly describes the setup of ChIP-qPCR (similar to a routine qPCR setup) and two common methods of presenting ChIP-qPCR data.

1. Before using a set of primers for the first time, perform a validation test for each pair of primers using sonicated input DNA. Plot a standard curve of Ct value against amount of input DNA (25 ng, 5 ng, 1 ng, 0.2 ng) (*see* **Note 30**).
2. Dilute the final 5% input by 50× and, Mock IgG and pulldown samples by 5×, using nuclease-free water (*see* **Note 31**).
3. Set up qPCR reactions in a 384-well plate for input, Mock IgG, and pulldown samples. Evaluate protein-bound target region (s) of interest and a non-binding negative control for each sample. In a 10 μL qPCR reaction, add 2.5 μL of either diluted input, Mock IgG or pulldown sample, SYBR green PCR master mix, 0.3 μM of forward and reverse primers, and top it up with nuclease-free water. Perform at least a technical duplicate for each sample.

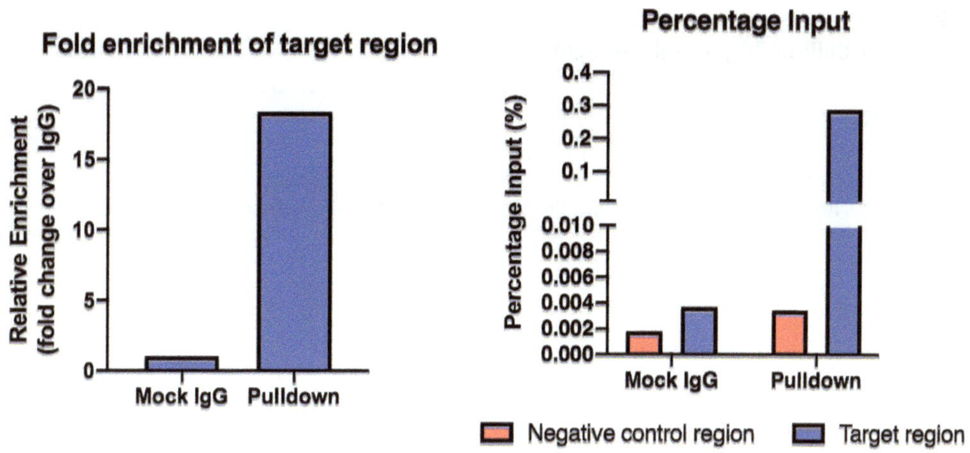

Fig. 3 Presentation of ChIP-qPCR data by fold enrichment (left) or percentage input (right). Refer to Tables 1 and 2 for detailed calculations

Table 1
Calculation for ChIP-qPCR fold enrichment

	Average Ct value	Amount of DNA (Interpolate Ct values on the standard curve)	Fold enrichment (normalized to Mock IgG)	Fold enrichment (normalized to the control region)
Mock IgG (control region)	32.370	0.018	1	–
Pulldown (control region)	31.438	0.036	2.025	–
Mock IgG (target region)	34.127	0.411	1	1
Pulldown (target region)	27.848	15.223	37.077	18.307

Generally, there are two ways of representing ChIP-qPCR results (Fig. 3). The pulldown of target gDNA can be represented as (a) a fold enrichment over that of the IgG control, or as (b) a percentage of total input chromatin. Detailed calculation(s) for fold enrichment and percentage input can be found in Tables 1 and 2, respectively.

3.7 Preparation for ChIP-Seq

Typically, the input and the pulldown samples can be sent for ChIP-Seq, without the Mock IgG, to identify genome-wide binding targets of the protein of interest. This is because the amount of

Table 2
Calculation for ChIP-qPCR percentage input

Input % of starting chromatin			0.5		
Dilution factor			100/0.5 = 200		
Dilution factor in cycles			Log (200, 2) = 7.644		
	Average Ct value	Adjust Ct of 0.5% input to 100%	Normalize Ct to 100% input	Fold enrichment in relative to 100% input	Percentage input
0.5% input (control region)	24.261	24.261–7.644 = 16.617	–	–	–
Mock IgG (control region)	32.370	–	16.617–32.370 = −15.753	2^(−15.753) = 0.0000181	100 × 0.0000181 = 0.00181
Pulldown (control region)	31.438	–	16.617–31.438 = −14.822	2^(−14.822) = 0.0000345	100 × 0.0000345 = 0.00345
0.5% input (target region)	27.050	27.050–7.644 = 19.406	–		–
Mock IgG (target region)	34.127	–	19.406–34.127 = −14.721	2^(−14.721) = 0.0000370	100 × 0.0000370 = 0.00370
Pulldown (target region)	27.848	–	19.406–27.848 = −8.442	2^(−8.442) = 0.00287	100 × 0.00287 = 0.287

immunoprecipitated DNA from Mock IgG is usually low and of insufficient quality for ChIP-Seq. One of the most commonly encountered issues when it comes to ChIP-Seq is low yield of pulldown gDNA to be sequenced. For ChIP-Seq, a minimum of 5 ng of immunoprecipitated DNA is ideal. Samples must be quantitated before sequencing via a sensitive assay such as the Qubit® dsDNA HS Assay Kit (Invitrogen™) (*see* **Note 32**).

Given that organoids are highly heterogenous, performing ChIP-Seq in organoids can produce variable results. To identify targets with high confidence, it is recommended to repeat the ChIP-Seq for at least three times or to compare with ChIP-Seq data in other similar models.

4 Notes

1. The use of additional crosslinkers DTBP and DSP is optional as formaldehyde is typically sufficient as a crosslinking agent. Formaldehyde links macromolecules that are about 2 Å apart together [7]. On the other hand, DTBP and DSP have a spacer arm length of 11.9 Å [8] and 12 Å [9] respectively. Therefore,

DTBP and DSP can be used to enhance chromatin pulldown by stabilizing the interactions over a greater distance.

2. The nuclear lysis buffer and elution buffer tend to precipitate in the cold due to the high concentration of SDS. Before use, equilibrate the buffers to room temperature until all precipitate has dissolved. Alternatively, the buffers can be warmed briefly in a 37 °C water bath. Using the buffers with precipitates may result in inconsistencies in concentration.

3. Apart from using an ultrasonic probe, sonication can also be performed using an ultrasonic water bath.

4. The success of a ChIP experiment is heavily dependent on the quality of the antibody. Where possible, use a ChIP-grade antibody for the pulldown. If there are no well-validated antibodies for the protein of interest, it is recommended to perform antibody characterization. The ENCODE and modENCODE consortia have also published a set of detailed guidelines to validate antibodies for ChIP [2].

   Briefly, the primary characterization of antibodies for ChIP involves performing western blot or immunofluorescence assay. For western blot validation, whole-cell lysates, nuclear extracts, or immunoprecipitated materials can be used for analysis. The band of interest must be within 20% of the expected protein size and represents at least 50% of the signals in whole-cell lysates or nuclear extracts. For immunofluorescence, nuclear staining must be observed in relevant cell types and in appropriate conditions. The antibody can be further characterized by performing at least one of the following four tests: (a) performing gene knockout and detecting the loss of signal via western blot or immunofluorescence assay, (b) performing immunoprecipitation followed by protein sequencing, (c) performing ChIP using antibodies that target different proteins of the same complex, or (d) performing ChIP with an epitope-tagged version of the protein [2].

5. Magnetic beads can be used in place of agarose beads. Although the use of magnetic beads is generally faster, easier (possibility of automation) and provides higher yield, it is more expensive than agarose beads. It is also important to note that Protein A and Protein G bind to antibodies from different species with different affinity [10]. Hence, a mixture of these two would ensure a comprehensive coverage of most species.

6. The purpose of DTT is to reverse the DNA-protein crosslink induced by DTBP and DSP. If DTBP and DSP are not used, DTT is not required during the elution step.

7. For transcription factor ChIP, primers should be designed to flank any known/predicted DNA-binding motif. The intended ChIP-qPCR product should range from 70 bp to 200 bp, or

shorter than the majority of the DNA fragments after sonication. If the exact binding region of the transcription factor is not known, primer walking may be performed on the target region.

8. A negative control region for ChIP-qPCR is simply a genomic locus that is unlikely to be a binding site for the protein of interest. In the case of a transcription factor, a negative control region can be anywhere within a gene body that is far away from the promoter. However, if there are no established negative control regions for a particular protein, a newly designed negative control region needs to be validated to ensure there is no binding across multiple ChIP runs.
9. The pellet size shown in Fig. 1b serves as an estimate, which has worked for ChIP of various proteins in our laboratory. However, it is important to note that the amount of starting material varies on a case-by-case basis. Some considerations include the expression level of the protein of interest and the efficiency of the antibody. It is recommended to optimize the amount of starting material, which can be evaluated by ChIP-qPCR of known binding regions and quantifying the total amount of pulldown DNA.
10. Only perform this step if the use of DTBP and DSP as additional crosslinkers is desired. Allow DTBP and DSP to reach room temperature before opening the vials. This is to prevent condensation as they are sensitive to moisture [8, 9]. Minimize air and light exposure when weighing the DTBP and DSP powder. Use the crosslinkers DTBP and DSP as soon as possible after reconstitution as they are relatively unstable when in solution. Do not keep any leftover crosslinkers after reconstitution.
11. It is not necessary to obtain a single cell suspension upon dissociation of the organoids, as crosslinking agents are able to access the cells in small clusters. However, if the dissociated organoids are still too large after 3–4 min of TrypLE™ treatment, the duration can be extended, without compromising on cell integrity. Optimization for your organoids is required.
12. Addition of 5% FBS helps to reduce the appearance of strand-like material.
13. It is important not to go beyond the recommended duration of crosslinking as excessive crosslinking can alter epitope accessibility and hinder sonication [11].
14. During sonication, the tube has to be immersed in an ice-cold water bath and the sonicator should be placed in a cold room to prevent overheating. The probe should go as deep as possible, leaving ~0.5 cm gap from the bottom of the tube to ensure a thorough sonication of the sample without foaming

(generation of many microbubbles). The tube must be upright and parallel to the probe as much as possible, such that the probe does not touch the walls of the tubes. Observe at least the first two cycles of sonication for any signs of foaming, although a small amount of bubbles is normal. If large amounts of bubbles are being generated, stop the cycle, readjust the tube, and allow the foam to dissipate before restarting the sonication. Preventing the samples from foaming during sonication is key as foaming will interfere with sonication efficiency.

15. After sonication followed by centrifugation, a grayish-black pellet should be observed to indicate that the sonication may be successful. Collect the supernatant without touching the pellet. If no grayish-black pellet is observed, the sonication step needs to be further optimized.
16. The purpose of pre-clearing with a Mock IgG of the same species as the antibody of interest is to reduce non-specific binding of DNA onto the IgG molecule. The amount of Mock IgG to be used should be the same as the amount of antibody of interest.
17. Although the required amount of antibody targeting the protein of interest may vary depending on the expression level of the protein of interest and the efficiency of the antibody, 10 μg of pulldown antibody is usually sufficient in our experience. As ChIP is an inefficient assay, antibody saturation is seldom an issue and increasing the amount of antibody is typically done to increase the enrichment. However, one notable exception is histone ChIP as histone proteins are expressed in abundance and crosslink very efficiently with DNA. It has been shown that high amounts of antibody can saturate the signal and instead reduce the enrichment [12]. In this case, titrating the amount of antibody may be required to obtain the optimal enrichment of target loci.
18. The length of incubation in this step must be shorter than the pre-clearing step.
19. The Protein A/G agarose beads tend to stick to the walls of the tube, resulting in the loss of chromatin fragments. Addition of another 500 μL of IP wash buffer 1 helps to rinse down any agarose beads stuck on the wall of the tube.
20. The Protein A/G agarose beads tend to stick to the walls of the pipette tip. Re-using the old pipette tip to transfer the buffer helps to reduce the loss of chromatin samples.
21. Thorough washing is critical in ensuring a successful ChIP. However, do take extra care to minimize the loss of agarose beads during washing. When adding wash buffers to the Protein A/G agarose beads, be sure not to pipette the beads as the beads will stick to the walls of the pipette tip.

22. The purpose of adding RNase A is to remove any immunoprecipitated RNA as it may interfere with downstream DNA analysis, while the purpose of the 67 °C incubation is to reverse the crosslinks. The length of incubation is minimally 4 h and can go up to overnight. It is also possible to pause at this step by keeping the samples at −20 °C.
23. The purpose of the proteinase K digestion is to digest the antibody and other proteins in the chromatin as they will affect downstream DNA analysis. The length of incubation is minimally 3 h and can go up to overnight. It is also possible to pause at this step by keeping the samples at −20 °C.
24. Performing a second round of phenol-chloroform extraction enhances the DNA purity.
25. The minimum length of incubation is 30 min and there is no maximum incubation time for this step, to our knowledge. This can be a potential stop point.
26. After most of the supernatant has been removed, using a P10 micropipette tip will help to remove the residual supernatant more cleanly.
27. Avoid over-drying the DNA pellets as they will be more difficult to be resuspended in water. Once the DNA pellet appears to be gelatinous or transparent, it should be ready for resuspension.
28. Note that the final input is 5% instead of 10% as the resuspension volume of the input sample is twice that of pulldown or IgG samples.
29. Having a tight smear within 200–500 bp is key to a successful ChIP experiment. If majority of the DNA fragments are too large, the power of the sonicator and the number of cycles of sonication can be increased and vice versa. This should be optimized for each cell type. Our recommendations serve as a starting point for consideration.
30. Primers to be used in ChIP-qPCR should also be validated beforehand by performing a standard curve. This can be achieved by performing a qPCR with varying amounts of sonicated input samples (i.e., 25 ng, 5 ng, 1 ng, 0.2 ng) and plotting a standard curve of Ct value against amount of DNA. Using the standard curve, the primer efficiency can be calculated using various online software. It is also important to check that there is only one melt peak for each pair of primers to ensure specificity. This validation test only needs to be done once and the standard curve can be applied to other ChIP experiments using the same set of primers.
31. The purpose of diluting the samples by 5× is to save on the samples which are typically very low in abundance. The

dilution for Mock IgG and pulldown samples should be kept the same. As the 5% input is diluted 10× more than the Mock IgG and pulldown samples, the final percentage of input for ChIP-qPCR calculation will be 0.5%.

32. To increase yield, one can either perform multiple ChIP experiments and pool the pulldown samples or scale up one ChIP experiment.

Acknowledgments

The authors thank members of the Teo laboratory for the critical reading of this manuscript. W.X.T. is supported by the NUS Research Scholarship. N.H.J.N. is supported by NMRC OFYIRG18may040 and the first A*STAR Career Development Award (CDA). A.K.K.T. is supported by the Institute of Molecular and Cell Biology (IMCB), A*STAR, NMRC Open Fund-Young Individual Research Grant (OF-YIRG) OFYIRG16may014, A*STAR ETPL Gap Funding ETPL/18-GAP005-R20H, Lee Foundation Grant SHTX/LFG/002/2018, Skin Innovation Grant SIG18011, NMRC OF-LCG/DYNAMO, FY2019 SingHealth Duke-NUS Surgery Academic Clinical Programme Research Support Programme Grant, Precision Medicine and Personalised Therapeutics Joint Research Grant 2019, Industry Alignment Fund – Industry Collaboration Project (IAF-ICP) I1901E0049 and the second A*STAR-AMED Joint Grant Call 192B9002.

Declaration of interests: N.H.J.N. and A.K.K.T. are founders and shareholders of BetaLife Pte Ltd.

References

1. Dutta D, Heo I, Clevers H (2017) Disease modeling in stem cell-derived 3D organoid systems. Trends Mol Med 23(5):393–410
2. Landt SG, Marinov GK, Kundaje A, Kheradpour P, Pauli F, Batzoglou S, Bernstein BE, Bickel P, Brown JB, Cayting P (2012) ChIP-seq guidelines and practices of the ENCODE and modENCODE consortia. Genome Res 22(9):1813–1831
3. Teo AKK, Tsuneyoshi N, Hoon S, Tan EK, Stanton LW, Wright CV, Dunn NR (2015) PDX1 binds and represses hepatic genes to ensure robust pancreatic commitment in differentiating human embryonic stem cells. Stem Cell Reports 4(4):578–590
4. Ng NHJ, Jasmen JB, Lim CS, Lau HH, Krishnan VG, Kadiwala J, Kulkarni RN, Ræder H, Vallier L, Hoon S (2019) HNF4A haploinsufficiency in MODY1 abrogates liver and pancreas differentiation from patient-derived induced pluripotent stem cells. iScience 16:192–205
5. Low BSJ, Lim CS, Ding SSL, Tan YS, Ng NHJ, Krishnan VG, Ang SF, Neo CWY, Verma CS, Hoon S, Lim SC, Tai ES, Teo AKK (2021) Decreased GLUT2 and glucose uptake contribute to insulin secretion defects in MODY3/HNF1A hiPSC-derived mutant β cells. Nat Commun 12(1):3133. https://doi.org/10.1038/s41467-021-22843-4. PMID: 34035238 Free PMC article
6. Pagliuca FW, Millman JR, Gürtler M, Segel M, Van Dervort A, Ryu JH, Peterson QP, Greiner D, Melton DA (2014) Generation of functional human pancreatic β cells in vitro. Cell 159(2):428–439

7. Hoffman EA, Frey BL, Smith LM, Auble DT (2015) Formaldehyde crosslinking: a tool for the study of chromatin complexes. J Biol Chem 290(44):26404–26411
8. Thermo Fisher Scientific Inc. (2012) Imidoester Crosslinkers: DMA, DMP, DMS, DTBP. https://assets.fishersci.com/TFS-Assets/LSG/manuals/MAN0011314_ImidoesterCrsLnk_DMA_DMP_DMS_DTBP_UG.pdf?_ga=2.143383925.1951848112.1596835662-1324288756.1595448388. Accessed 6 August 2020
9. Thermo Fisher Scientific Inc. (2012) DTSSP DSP. https://www.thermofisher.com/document-connect/document-connect.html?url=https%3A%2F%2Fassets.thermofisher.com%2FTFS-Assets%2FLSG%2Fmanuals%2FMAN0011280_DTSSP_DSP_UG.pdf&title=VXNlciBHdWlkZTogIERUU1NQIERTUA. Accessed 6 August 2020
10. Merck (2016) Protein G and protein A bind to different IgG. https://www.sigmaaldrich.com/technical-documents/articles/biology/affinity-chromatography-antibodies/protein-a-g-binding.html?gclid=Cj0KCQjwvIT5BRCqARIsAAwwD-Qh3Z19-nhKd_UhDWUUTzUVyiLFAnHJHyHB6NK2MAvSZTrHF50IpREaAjHfEALw_wcB. Accessed 6 August 2020
11. Orlando V, Strutt H, Paro R (1997) Analysis of chromatin structure by in vivo formaldehyde cross-linking. Methods 11(2):205–214
12. Asp P (2018) How to combine ChIP with qPCR. In: Neus Visa AJ-P (ed) Chromatin immunuoprecipitation: methods and protocol, Methods in molecular biology, vol 1689. Humana Press, New York, NY, pp 29–42

Chapter 15

Generation of Embryonic Origin-Specific Vascular Smooth Muscle Cells from Human Induced Pluripotent Stem Cells

Mengcheng Shen, Chun Liu, and Joseph C. Wu

Abstract

Vascular smooth muscle cells (VSMCs), a highly mosaic tissue, arise from multiple distinct embryonic origins and populate different regions of our vascular network with defined boundaries. Accumulating evidence has revealed that the heterogeneity of VSMC origins contributes to region-specific vascular diseases such as atherosclerosis and aortic aneurysm. These findings highlight the necessity of taking into account lineage-dependent responses of VSMCs to common vascular risk factors when studying vascular diseases. This chapter describes a reproducible, stepwise protocol for the generation of isogenic VSMC subtypes originated from proepicardium, second heart field, cardiac neural crest, and ventral somite using human induced pluripotent stem cells. By leveraging this robust induction protocol, patient-derived VSMC subtypes of desired embryonic origins can be generated for disease modeling as well as drug screening and development for vasculopathies with regional susceptibility.

Key words Induced pluripotent stem cells, Smooth muscle cell, Embryonic origin, Vascular disease, Regional susceptibility

1 Introduction

Vascular smooth muscle cells (VSMCs), the predominant cell type in medium and large blood vessels, are essential for providing vasomotor control of circulation to regulate blood pressure and for maintaining cardiovascular health through cell–cell and cell–matrix interactions [1, 2]. However, unlike other myogenic cell types, VSMCs retain a high degree of plasticity and can undergo phenotypic switching from a quiescent and contractile state to a proliferative, proinflammatory, and synthetic state in response to factors such as mechanical stretch, growth factors, and inflammatory mediators [2, 3]. Phenotypically modified VSMCs play a key role in arterial remodeling, which can consequently lead to a variety of region-specific cardiovascular complications such as atherosclerosis, myocardial infarction, stroke, in-stent restenosis, calcification, and aortic aneurysm [1].

Nagarajan Kannan and Philip Beer (eds.), *Stem Cell Assays: Methods and Protocols*, Methods in Molecular Biology, vol. 2429, https://doi.org/10.1007/978-1-0716-1979-7_15,

Fate mapping studies on vertebrate embryos have identified that VSMCs with distinct embryonic origins exhibit a highly mosaic distribution along the aorta and its distal branches [4]. Specifically, VSMCs are derived from proepicardium (PE)/epicardium (EPI) in coronary arteries [5, 6]; from second heart field (SHF) in the aortic root and the outer layer of the ascending aorta [7]; from cardiac neural crest (CNC) in the inner layer of the ascending aorta, the aortic arch and branches, and the pulmonary trunk [8]; from ventral somite (VS) in the descending aorta [9, 10]; and from tissue-specific mesothelium (progenies of splanchnic mesoderm) in most coelomic organs such as gut, liver, and kidneys [11–13].

The heterogeneity of VSMC embryonic origins, in addition to other factors such as variable hemodynamics and vascular structures, contributes to a propensity for vascular diseases to exhibit regional differences. For example, a recent study showed that the development of aortic root aneurysm in a mouse model of Loeys–Dietz syndrome was the result of opposing Smad signaling activities in CNC- versus SHF-derived VSMC subtypes residing in the aortic root [14]. Likewise, single-cell transcriptomics of mouse aortas showed that the expression levels of genes associated with inflammation, adhesion, and migration are much higher in the aortic arch (the athero-prone region) SMCs than in the descending aorta (the athero-resistant region) SMCs [15].

The signaling pathways involved in diversifying embryonic origin-specific SMCs during vascular development can recur in arterial remodeling postnatally [16–18]. Therefore, a better understanding of the key molecular events that govern the diverse developmental trajectories of VSMCs under the same genetic background will help reveal the heterogeneous adaptive or pathogenic responses of anatomically distinct arteries to common risk factors. In this regard, human induced pluripotent stem cell (iPSC)-derived isogenic embryonic origin-specific VSMC subtypes represent a promising in vitro platform for understanding and modeling region-specific vascular diseases, because they can overcome the challenges of procuring arterial tissues from different segments of the same patients for comparative studies. Moreover, the scalable nature of deriving VSMCs from human iPSCs also opens up an avenue for the discovery of therapeutic targets to prevent or treat vascular diseases using high-throughput drug screening [19, 20].

Here, we detail a stepwise differentiation protocol to derive embryonic origin-specific VSMC subtypes from human iPSCs via four intermediate lineages, namely: PE/EPI, SHF, CNC, and VS. These four intermediate lineages are generated in a chemically defined medium (CDM) using different combinations of growth factors and small molecules. We temporally control differentiation of these lineages by mirroring their respective sequential signals activated at each lineage step during embryo development

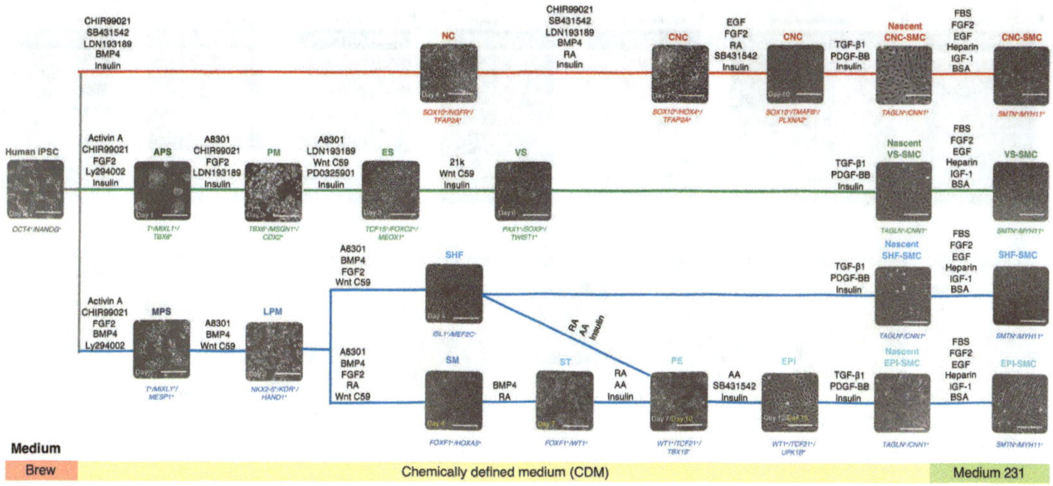

Fig. 1 Schematic diagram of the conditions for deriving embryonic origin-specific VSMCs from human iPSCs. Lineage-specific intermediate cell types that give rise to each VSMC subtype are differentiated in a stepwise fashion. Each intermediate cell population is subjected to TGF-β1 and PDGF-BB treatment for 6 days before switching to a smooth muscle cell growth medium (Medium 231) for another 14–21 days. Phenotypic markers for specific cell types are represented in *italics* at the bottom of each bright-field image. *NC* neural crest, *CNC* cardiac neural crest, *APS* anterior primitive streak, *PM* paraxial mesoderm, *ES* early somite, *VS* ventral somite, *MPS* mid-primitive streak, *LPM* lateral plate mesoderm, *SHF* second heart field, *SM* splanchnic mesoderm, *ST* septum transversum, *PE* proepicardium, *EPI* epicardium. Scale bars represent 50 μm

(Fig. 1). This stepwise differentiation approach yields highly pure, lineage-specific intermediate cell types and VSMC subtypes without the need for cell sorting to enrich target cell populations (Fig. 2). We further show that treating proliferating human iPSC-derived VSMC subtypes with a MEK inhibitor for 6 days can induce a more mature and contractile phenotype (smooth muscle myosin heavy chain 11, MYH11^{+}) of these cells (Fig. 3). This robust and developmental trajectory-defined differentiation protocol can be deployed for the mass production of isogenic embryonic origin-specific VSMC subtypes from patient iPSCs for vascular disease modeling as well as drug screening and development.

2 Materials

2.1 Cell Culture Reagents

1. Stem cell growth medium (for example, Stemmacs™ iPS-Brew XF, Miltenyi Biotech, cat # 130-104-368).
2. DMEM/F-12, HEPES.
3. Ham's F-12 Nutrient Mix.
4. IMDM.
5. Smooth muscle cell growth basal medium (for example, Medium 231, Gibco, cat # M231500).

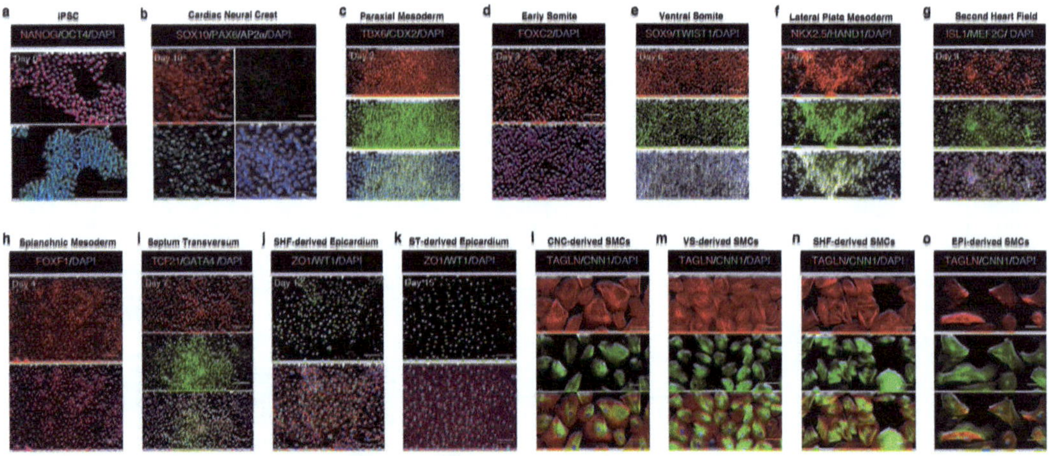

Fig. 2 Representative immunofluorescence images showing phenotypic markers for specific cell types during embryonic origin-specific VSMC differentiation. (**a–k**) The time point labeled on each set of cell marker-positive fluorescence images is consistent with that shown on the bright-field image of the same cell type in Fig. 1. (**l–o**) Embryonic origin-specific VSMCs are positive for TAGLN and CNN1 after being cultured in the SMC growth medium for 14 days. Scale bars represent 100 μm

6. Smooth muscle growth supplement (for example, SMGS, Gibco, cat # S00725).
7. Gentamicin.
8. Distilled H_2O.
9. Dimethyl sulfoxide (DMSO).
10. DPBS without calcium and magnesium.
11. Chemically defined lipid concentrate.
12. Glutamax.
13. Insulin.
14. Polyvinyl alcohol (PVA) (*see* **Note 1**).
15. Transferrin (30 mg/ml).
16. Bovine serum albumin (BSA, 0.1% wt/vol).
17. Monothioglycerol.
18. Gelatin solution (0.1% wt/vol).
19. Matrigel Growth Factor Reduced Basement Membrane Matrix, phenol red-free, LDEV-free (0.4% wt/vol, Corning, cat # 356231).
20. Accutase solution.
21. Gentle Cell Dissociation Reagent.
22. Bambanker (Wako Chemicals USA, cat # 30214681).

2.2 Small Molecules

1. ALK5 inhibitor A83-01 (1 mM).
2. ALK5 inhibitor SB431542 (10 mM).

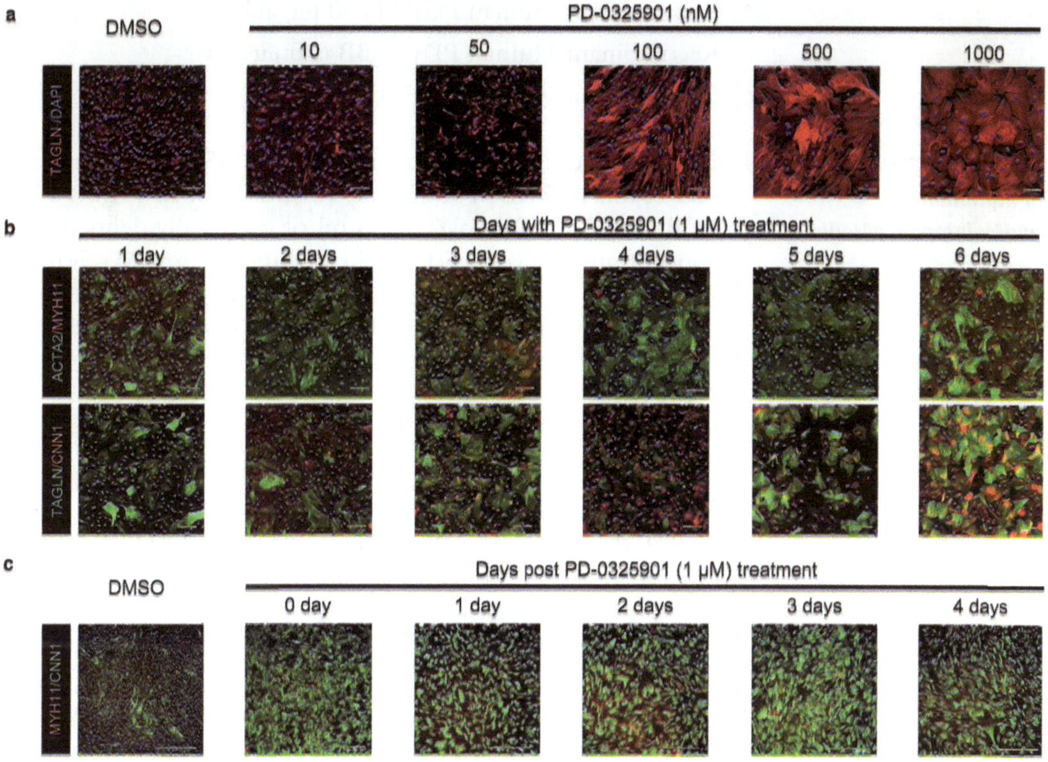

Fig. 3 PD0325901, a MEK inhibitor, can upregulate the expression of SMC contractile proteins. (**a**) Human iPSC-derived VSMCs show a dose-dependent response to PD0325901-mediated upregulation of SMC contractile proteins. (**b**) Human iPSC-derived VSMCs show a time-dependent upregulation of SMC contractile proteins with the treatment of PD0325901 at a concentration of 1 μM. (**c**) Human iPSC-derived VSMCs can preserve a contractile phenotype for at least 4 days after the withdrawal of PD0325901. Because PD0325901 application demonstrates the same effect on different VSMC subtypes, only representative images generated from CNC-specific VSMCs are shown in this figure. Scale bars represent 100 μm

3. L-Ascorbic acid 2-phosphate sesquimagnesium salt hydrate (25 mg/ml).
4. BMP inhibitor LDN193189 2HCl (1 mM).
5. GSK3 inhibitor CHIR99021 (10 mM).
6. Hedgehog activator SAG 21k (5 μM).
7. MEK inhibitor PD0325901 (1 mM).
8. PI3K-AKT inhibitor LY294002 (10 mM).
9. Retinoic acid (10 mM).
10. ROCK inhibitor Y27632 2HCl (10 mM).
11. Wnt inhibitor C59 (1 mM).

2.3 Growth Factors

1. Recombinant Human/Murine/Rat Activin A (10 μg/ml).
2. Recombinant Human BMP4 (100 μg/ml).
3. Recombinant Human EGF (100 μg/ml).

4. Recombinant Human FGF2 (100 μg/ml).
5. Recombinant Human PDGF-BB (10 μg/ml).
6. Recombinant Human TGF-β1 (2 μg/ml).

2.4 Preparation of Induction and Maintenance Media

1. Basal chemically defined medium (CDM), 500 ml: consisting of 240 ml of IMDM (50% vol/vol), 240 ml of Ham's F-12 Nutrient Mix (50% vol/vol), 5 ml of chemically defined lipid concentrate (1% vol/vol); 5 ml of Glutamax (2 mM), 5 ml of PVA (1 mg/ml), 250 μl of transferrin (15 μg/ml), and 20 μl of monothioglycerol (450 μM). Basal CDM is sterile filtered after preparation and can be stored for up to 4 weeks at 4 °C (*see* **Note 2**).
2. Neural crest (NC) induction medium, 50 ml: consisting of 50 ml of CDM, 5 μl of CHIR99021 (1 μM), 10 μl of SB431542 (2 μM), 12.5 μl of LDN193189 (250 nM), 7.5 μl of BMP4 (15 ng/ml), and 35 μl of insulin (7 μg/ml).
3. Neural crest (NC) maintenance medium, 50 ml: consisting of 50 ml of CDM, 5 μl of FGF2 (10 ng/ml), 5 μl of EGF (10 ng/ml), 10 μl of SB431542 (2 μM), and 35 μl of insulin (7 μg/ml).
4. Anterior primitive streak (APS) induction medium, 50 ml: consisting of 50 ml of CDM, 150 μl of Activin A (30 ng/ml), 20 μl of CHIR99021 (4 μM), 10 μl of FGF2 (20 ng/ml), 10 μl of LY294002 (2 μM), and 35 μl of insulin (7 μg/ml).
5. Paraxial mesoderm (PM) induction medium, 50 ml: consisting of 50 ml of CDM, 50 μl of A83-01 (1 μM), 15 μl of CHIR99021 (3 μM), 12.5 μl of LDN192189 (250 nM), 10 μl of FGF2 (20 ng/ml), and 35 μl of insulin (7 μg/ml).
6. Early somite (ES) induction medium, 50 ml: consisting of 50 ml of CDM, 50 μl of A83-01 (1 μM), 12.5 μl of LDN193189 (250 nM), 50 μl of C59 (1 μM), 25 μl of PD0325901 (500 nM), and 35 μl of insulin (7 μg/ml).
7. Ventral somite (VS) induction medium, 50 ml: consisting of 50 ml of CDM, 50 μl of SAG 21 k (5 nM), 50 μl of C59 (1 μM), and 35 μl of insulin (7 μg/ml).
8. Mid-primitive streak (MPS) induction medium, 50 ml: consisting of 50 ml of CDM, 50 μl of Activin A (10 ng/ml), 30 μl of CHIR99021 (6 μM), 25 μl of BMP4 (50 ng/ml), 10 μl of FGF2 (20 ng/ml), and 10 μl of LY294002 (2 μM).
9. Lateral plate mesoderm (LPM) induction medium, 50 ml: consisting of 50 ml of CDM, 50 μl of A83-01 (1 μM), 15 μl of BMP4 (30 ng/ml), and 50 μl of C59 (1 μM).
10. Second heart field (SHF) induction medium (*see* **Note 3**), 50 ml: consisting of 50 ml of CDM, 50 μl of A83–01 (1 μM), 15 μl of BMP4 (30 ng/ml), 50 μl of C59 (1 μM), and 10 μl of FGF2 (20 ng/ml).

11. Splanchnic mesoderm (SM) induction medium, 50 ml: consisting of 50 ml of CDM, 50 μl of A83-01 (1 μM), 15 μl of BMP4 (30 ng/ml), 50 μl of C59 (1 μM), 10 μl of FGF2 (20 ng/ml), and 10 μl of retinoic acid (2 μM).
12. Septum transversum (ST) induction medium, 50 ml: consisting of 50 ml of CDM, 10 μl of retinoic acid (2 μM), and 20 μl of BMP4 (40 ng/ml).
13. Proepicardial cell (PE) induction medium, 50 ml: consisting of 50 ml of CDM, 200 μl of ascorbic acid (100 μg/ml), 10 μl of retinoic acid (2 μM), and 35 μl of insulin (7 μg/ml).
14. Epicardial cell (EPI) maintenance medium, 50 ml: consisting of 50 ml of CDM, 200 μl of ascorbic acid (100 μg/ml), 10 μl of SB431542 (2 μM), and 35 μl of insulin (7 μg/ml).
15. Smooth muscle cell (SMC) induction medium, 50 ml: consisting of 50 ml of CDM, 50 μl of TGF-β1 (2 ng/ml), 50 μl of PDGF-BB (10 ng/ml), and 35 μl of insulin (7 μg/ml).
16. Smooth muscle cell (SMC) maturation medium, 50 ml: consisting of 50 ml of Medium 231, 2.5 ml of SMGS, and 50 μl of PD0325901 (1 μM).

3 Methods

3.1 Maintenance of Human iPSCs

Human iPSCs are routinely maintained under feeder-free conditions in Stemmacs™ iPS-Brew XF (Brew) medium. Karyotype analysis should be performed periodically to ensure high quality of iPSCs. Mycoplasma contamination tests should be performed on a weekly basis, and mycoplasma positive cells should be discarded immediately. All intermediate progenitor cell types are differentiated in 6-well plates.

1. Incubate Matrigel-coated 6-well plates (1.5 ml per well) at 37 °C for at least 30 min before seeding iPSCs.
2. When iPSCs reach ~80% confluency, add 1 ml of Gentle Cell Dissociation Reagent to each well and incubate the plates in an incubator for 6 min.
3. Check the morphology changes of iPSCs under the microscope. Carefully aspirate the dissociation reagent when iPSCs become granular and show clear boundaries. Otherwise, put the plates back to the incubator for an additional 1–2 min.
4. Add 1 ml of Brew medium +10 μM Y27632 perpendicular to each well and triturate iPSCs to small clumps (~10 cells per clump).
5. Remove Matrigel solution from coated plates and briefly wash each well with DPBS. Add 2 ml of Brew medium +10 μM Y27632 to each well of the new plates, and then add 100 μl of iPSC clumps to each well.

6. Gently shake the plates in three quick back-and-forth and left-and-right motions to distribute iPSCs evenly across the wells. Cells are cultured at 37 °C in an atmosphere of 5% CO_2/ 95% air.
7. On the next day, aspirate the medium from each well and add 3 ml of fresh room temperature Brew medium. Refresh the Brew medium every other day. When the cells reach ~80% confluency, passage them as described above, or proceed to differentiating lineage-specific intermediate progenitor cell types and respective VSMC subtypes.

3.2 Generation of Lineage-Specific SMC Intermediate Populations

The induction protocols for generating lineage-specific SMC precursors have been optimized based on previous studies [21–23]. Figure 1 shows a detailed schematic diagram of the differentiation steps.

3.2.1 Cardiac Neural Crest (CNC) Cell Differentiation

1. Day 0: Dissociate ~80% confluent iPSCs into small clumps as described in Subheading 3.1, and sparsely seed cells at a density of 5×10^3 cells/cm^2 onto new Matrigel-coated 6-well plates in Brew medium +10 μM Y27632. Make sure iPSCs express high levels of pluripotency markers (Fig. 2a) before differentiation is initiated.
2. Day 1 ~ 4: Wash each well with DPBS and add 2 ml of NC induction medium (*see* Subheading 2.4, **item 2**). Change the NC induction medium daily.
3. Day 5 ~ 6: Change the NC induction medium +1 μM retinoic acid daily to facilitate the specification of cardiac NC (CNC) cells (*see* **Note 4**).
4. Day 7: Aspirate the spent NC induction medium and wash the wells with DPBS. Add 1 ml of Accutase solution into each well and incubate the cells at 37 °C for 5 min. Add 4 ml of CDM basal medium (*see* Subheading 2.4, **item 1**) to 15 ml conical tubes, and then collect triturated cells from each well of the 6-well plates. Centrifuge the cells at 200 × *g* for 5 min at room temperature. Decant the supernatant and resuspend the cell pellets at a dilution ratio of 1:6 in the NC maintenance medium (*see* Subheading 2.4, **item 3**) + 1 μM retinoic acid. Cells are seeded on Matrigel-coated 6-well plates, and 10 μM of Y27632 can be added to the medium to improve cell survival and attachment.
5. Day 8 ~ 9: Refresh NC maintenance medium +1 μM retinoic acid daily.
6. Day 10: Dissociate the cells with Accutase solution (1 ml per well) at 37 °C for 5 min. The cell pellets are resuspended in the NC maintenance medium and seeded at a split ratio of 1:3 to 1: 6 on gelatin-coated 6-well plates. Replated CNC cells on Day

10 are labeled as passage 1. The majority of the cells on day 10 are positive for SOX10/AP2α and negative for PAX6 (Fig. 2b). Change the NC maintenance medium every day and passage the cells when they reach ~90% confluency. CNC cells at passage 2 can be cryopreserved for future expansion. Otherwise, proceed to Subheading 3.3 for CNC-specific (ascending and aortic arch) SMC differentiation by seeding CNCs at a density of 2×10^4 cells/cm^2.

3.2.2 Ventral Somite (VS) Differentiation

1. Day 0: Dissociate ~80% confluent iPSCs into small clumps as described in Subheading 3.1, and sparsely seed cells at a density of 1×10^4 cells/cm^2 onto new Matrigel-coated 6-well plates in Brew medium +10 μM Y27632.
2. Day 1: Wash each well with DPBS and add 2 ml of APS induction medium (*see* Subheading 2.4, **item 4**).
3. Day 2: Wash each well with DPBS and add 2 ml of PM induction medium (*see* Subheading 2.4, **item 5**). The majority of the cells on day 2 are positive for TBX6 and CDX2 (Fig. 2c).
4. Day 3: Wash each well with DPBS and add 2 ml of ES induction medium (*see* Subheading 2.4, **item 6**). The majority of the cells on day 3 are positive for FOXC2 (Fig. 2d).
5. Day 4 ~ 6: Wash each well with DPBS and add 2 ml of VS induction medium (Subheading 2.4, **item** 7). Change the medium daily. The majority of the cells on day 6 are positive for SOX9 and TWIST1 (Fig. 2e).
6. Day 7: Dissociate the cells with Accutase solution (1 ml per well) at 37 °C for 5 min. Resuspend the cell pellets in CDM + 10 μM Y27632, and seed 2×10^4 cells/cm^2 on gelatin-coated 6-well plates. Proceed to Subheading 3.3 for VS-specific (descending aortic) SMC differentiation, as the maintenance conditions for VS cells have not been optimized.

3.2.3 Second Heart Field (SHF) Differentiation

1. Day 0: Dissociate ~80% confluent iPSCs into small clumps as described in Subheading 3.1, and sparsely seed cells at a density of 1.5×10^4 cells/cm^2 onto new Matrigel-coated 6-well plates in Brew medium +10 μM Y27632.
2. Day 1: Wash each well with DPBS and add 2 ml of MPS induction medium (*see* Subheading 2.4, **item 8**).
3. Day 2: Wash each well with DPBS and add 2 ml of LPM induction medium (*see* Subheading 2.4, **item 9**). The majority of the cells on day 2 are positive for NKX2.5 and HAND1 (Fig. 2f).
4. Day 3 ~ 4: Wash each well with DPBS and add 2 ml of SHF cell induction medium (*see* Subheading 2.4, **item 10**). Change the medium daily.

5. Day 5: Dissociate the cells with Accutase solution (1 ml per well) at 37 °C for 5 min. Resuspend the cell pellets in CDM + 10 μM Y27632. The majority of cells on day 4 are positive for ISL1 and MEF2C (Fig. 2g). Seed 1 × 10^3 cells/cm^2 (*see* **Note 5**) or 3 × 10^4 cells/cm^2 on gelatin-coated 6-well plates, and proceed to Subheading 3.2.5 for PE/EPI differentiation or Subheading 3.3 for SHF-specific (aortic root and ascending aortic) SMC differentiation, respectively. The maintenance conditions for SHF cells have not been optimized.

3.2.4 Septum Transversum (ST) Differentiation

1. Day 0 ~ 2: Generate LPM cells as described in Subheading 3.2.3, **steps 1–3**.
2. Day 3 ~ 4: Wash each well with DPBS and add 2 ml of SM induction medium (*see* Subheading 2.4, **item 11**). Change the medium daily. The majority of the cells on day 4 are positive for FOXF1 (Fig. 2h).
3. Day 5 ~ 7: Wash each well with DPBS and add 2 ml of ST induction medium (*see* Subheading 2.4, **item 12**). Change the medium daily. The majority of the cells on day 7 are positive for GATA4 and TCF21 (Fig. 2i).
4. Day 8: Dissociate the cells with Accutase solution (1 ml per well) at 37 °C for 5 min. Resuspend the cell pellets in PE induction medium +10 μM Y27632. Seed 1 × 10^3 cells/cm^2 on gelatin-coated 6-well plates and proceed to Subheading 3.2.5 for PE/EPI differentiation. The maintenance conditions for ST cells have not been optimized.

3.2.5 Proepicardium (PE)/Epicardium (EPI) Differentiation

As the PE has been shown to be derived from SHF and ST cells [24], the PE/EPI differentiation process continues after day 5 and day 8 in Subheadings 3.2.3 and 3.2.4, respectively.

1. Day 6 ~ 7 (for the SHF lineage, the same as below) or 9 ~ 10 (for the ST lineage, the same as below): Wash each well with DPBS and add 2 ml of PE induction medium (*see* Subheading 2.2, **item 13**). Maintain cells in the medium without change for 2 days.
2. Day 8 ~ 9 or 11 ~ 12: Wash each well with DPBS and 2 ml of CDM + 100 μg/ml ascorbic acid + 7 μg/ml insulin. Maintain cells in the medium without change for 2 days. Most SHF- and ST-derived EPI cells are positive for WT1 and ZO1 (Fig. 2j, k).
3. Day 10 ~ 12 or 12 ~ 15: Dissociate the cells with Accutase solution (1 ml per well) at 37 °C for 5 min. Resuspend the cell pellets and seed 1.5 × 10^4 cells/cm^2 on gelatin-coated 6-well plates in EPI maintenance medium (*see* Subheading 2.4, **item 14**). Change the medium every 2 days until the cells become confluent. EPI cells can be maintained in this medium for over 15 passages without losing their cell type-specific markers. EPI

cells can be cryopreserved after passage 2 for future expansion. Otherwise, proceed to Subheading 3.3 for coronary VSMC differentiation by seeding EPI cells at a density of 3×10^4 cells/cm^2.

3.3 Differentiation and Maturation of Embryonic Origin-Specific VSMC Subtypes

1. Day 0 ~ 5: Wash lineage-specific intermediate progenitor cells from Subheadings 3.2.1, **step** 7 (CNC); 3.2.2, **step 6** (VS); 3.2.3, **step 5** (SHF); and 3.2.5, **step 3** (SHF-EPI and ST-EPI) with DPBS and add 2 ml of SMC induction medium (*see* Subheading 2.4, **item 15**). Change the medium daily.
2. Day 6 ~ 20: Wash cells with DPBS and maintain in Medium 231 + SMGS for 2 weeks or longer as desired. Change the medium every 2 days and split the VSMCs at a 1:2 ratio (*see* **Note 6**) when the cells become confluent. Embryonic origin-specific VSMC subtypes express similar levels of VSMC contractile proteins, such as TAGLN and CNN1 (Fig. 2l–o).
3. Day 21 ~ 27: Wash cells with DPBS and maintain in SMC maturation medium (*see* Subheading 2.4, **item 16**) for 6 days (*see* **Note** 7). Change the medium every 2 days. We observe that treating iPSC-derived VSMC subtypes with PD0325901 at a concentration of 1 μM (Fig. 3a) for 6 days (Fig. 3b) can result in high expression levels of MYH11, a mature marker of VSMCs. This mature and contractile phenotype can be preserved for at least 4 days after the withdrawal of PD0325901 (Fig. 3c).

4 Notes

1. To prepare 100 mg/ml PVA solution, weigh 10 g of PVA and add the powder slowly into a sterile flask containing 100 ml of distilled H_2O with constant gentle agitation until the PVA is completely dissolved. The solution can be stored at 4 °C for 2 months.
2. To avoid the risk of protein retention in some filters, we recommend adding growth factors (and small molecules) directly to filtered basal CDM. Make sure the stock growth factors and small molecules are prepared in sterile conditions. The complete induction and maintenance media can be stored up to 1 week at 4 °C.
3. Because a definitive surface marker for SHF cells is lacking, here we define SHF as *ISL1*$^+$/*MEF2C*$^+$/*GATA4*$^+$ cells [25].
4. The cardiac neural crest is the most caudal of the cranial neural crest located postotically in the neural tube during development [26]. To skew the differentiation more toward a posterior axis to get CNC cells (*PLXNA2*$^+$/*HOX4*$^+$) [27], retinoic acid

is added to the NC induction medium on day 5 as a caudalizing factor. In contrast, a cranial-dominant NC cell population (*OTX2*$^{+}$) was reported to be successfully derived using retinoic acid-free NC induction medium [21].

5. We observe that it is critical to seed cardiac progenitor cells (i.e., SHF cells and ST cells) at a very low density to promote the differentiation of PE/EPI cells.
6. Cell–cell interaction and paracrine effects are critical to the growth of VSMCs. Therefore, we recommend using a low split ratio (<1:3) to passage human iPSC-derived VSMCs of any embryonic origins. We observe that both SHF- and EPI-VSMC subtypes are more quiescent than CNC- and VS-VSMCs in the SMC growth medium. Therefore, different seeding densities of lineage-specific progenitors are required at the VSMC induction stage to ensure adequate end-point numbers of each VSMC subtype for downstream experiments.
7. Although MEK inhibition can effectively upregulate the expression of VSMC contractile proteins by blocking the p38 signaling pathway, caution should be taken when highly active ERK signaling plays a pathogenic role in the development of vascular diseases such as Marfan syndrome [28]. Repsox, a TGF-β inhibitor, was reported to significantly improve the expression of MYH11 in iPSC-derived VSMCs [29]. However, this small molecule does not work even in healthy iPSC-derived VSMC subtypes in our hands. Therefore, an alternative but less effective way to promote the maturation of Marfan syndrome iPSC-derived VSMCs could be maintaining the cells in an SMC growth medium supplemented with heparin and TGF-β1 for at least 6 days [30].

Acknowledgments

This work was supported in part by research grants from NIH R01 HL1265276, R01 HL133272, R01 HL146690, R01 HL141371 (JCW), Tobacco-Related Disease Research Program (TRDRP) 30FT0852 (MS) and 27IR-0012 (JCW), and AHA 17MERIT33610009 (JCW) and AHA Career Development Award 19CDA34760019 (CL).

References

1. Owens GK, Kumar MS, Wamhoff BR (2004) Molecular regulation of vascular smooth muscle cell differentiation in development and disease. Physiol Rev 84(3):767–801. https://doi.org/10.1152/physrev.00041.2003
2. Berk BC (2001) Vascular smooth muscle growth: autocrine growth mechanisms. Physiol Rev 81(3):999–1030. https://doi.org/10.1152/physrev.2001.81.3.999

3. Basatemur GL, Jorgensen HF, Clarke MCH, Bennett MR, Mallat Z (2019) Vascular smooth muscle cells in atherosclerosis. Nat Rev Cardiol 16(12):727–744. https://doi.org/10.1038/s41569-019-0227-9
4. Majesky MW (2007) Developmental basis of vascular smooth muscle diversity. Arterioscler Thromb Vasc Biol 27(6):1248–1258. https://doi.org/10.1161/ATVBAHA.107.141069
5. Cai CL, Martin JC, Sun Y, Cui L, Wang L, Ouyang K, Yang L, Bu L, Liang X, Zhang X, Stallcup WB, Denton CP, McCulloch A, Chen J, Evans SM (2008) A myocardial lineage derives from Tbx18 epicardial cells. Nature 454(7200):104–108. https://doi.org/10.1038/nature06969
6. Zhou B, Ma Q, Rajagopal S, Wu SM, Domian I, Rivera-Feliciano J, Jiang D, von Gise A, Ikeda S, Chien KR, Pu WT (2008) Epicardial progenitors contribute to the cardiomyocyte lineage in the developing heart. Nature 454(7200):109–113. https://doi.org/10.1038/nature07060
7. Sawada H, Rateri DL, Moorleghen JJ, Majesky MW, Daugherty A (2017) Smooth muscle cells derived from second heart field and cardiac neural crest reside in spatially distinct domains in the media of the ascending aorta. Arterioscler Thromb Vasc Biol 37(9):1722–1726. https://doi.org/10.1161/ATVBAHA.117.309599
8. Jiang X, Rowitch DH, Soriano P, McMahon AP, Sucov HM (2000) Fate of the mammalian cardiac neural crest. Development 127(8):1607–1616
9. Pouget C, Gautier R, Teillet MA, Jaffredo T (2006) Somite-derived cells replace ventral aortic hemangioblasts and provide aortic smooth muscle cells of the trunk. Development 133(6):1013–1022. https://doi.org/10.1242/dev.02269
10. Wasteson P, Johansson BR, Jukkola T, Breuer S, Akyurek LM, Partanen J, Lindahl P (2008) Developmental origin of smooth muscle cells in the descending aorta in mice. Development 135(10):1823–1832. https://doi.org/10.1242/dev.020958
11. Wilm B, Ipenberg A, Hastie ND, Burch JB, Bader DM (2005) The serosal mesothelium is a major source of smooth muscle cells of the gut vasculature. Development 132(23):5317–5328. https://doi.org/10.1242/dev.02141
12. Asahina K, Zhou B, Pu WT, Tsukamoto H (2011) Septum transversum-derived mesothelium gives rise to hepatic stellate cells and perivascular mesenchymal cells in developing mouse liver. Hepatology 53(3):983–995. https://doi.org/10.1002/hep.24119
13. Que J, Wilm B, Hasegawa H, Wang F, Bader D, Hogan BL (2008) Mesothelium contributes to vascular smooth muscle and mesenchyme during lung development. Proc Natl Acad Sci U S A 105(43):16626–16630. https://doi.org/10.1073/pnas.0808649105
14. MacFarlane EG, Parker SJ, Shin JY, Kang BE, Ziegler SG, Creamer TJ, Bagirzadeh R, Bedja D, Chen Y, Calderon JF, Weissler K, Frischmeyer-Guerrerio PA, Lindsay ME, Habashi JP, Dietz HC (2019) Lineage-specific events underlie aortic root aneurysm pathogenesis in Loeys-Dietz syndrome. J Clin Invest 129(2):659–675. https://doi.org/10.1172/JCI123547
15. Dobnikar L, Taylor AL, Chappell J, Oldach P, Harman JL, Oerton E, Dzierzak E, Bennett MR, Spivakov M, Jorgensen HF (2018) Disease-relevant transcriptional signatures identified in individual smooth muscle cells from healthy mouse vessels. Nat Commun 9(1):4567. https://doi.org/10.1038/s41467-018-06891-x
16. Sinha S, Santoro MM (2018) New models to study vascular mural cell embryonic origin: implications in vascular diseases. Cardiovasc Res 114(4):481–491. https://doi.org/10.1093/cvr/cvy005
17. Klein D (2018) iPSCs-based generation of vascular cells: reprogramming approaches and applications. Cell Mol Life Sci 75(8):1411–1433. https://doi.org/10.1007/s00018-017-2730-7
18. Roostalu U, Wong JK (2018) Arterial smooth muscle dynamics in development and repair. Dev Biol 435(2):109–121. https://doi.org/10.1016/j.ydbio.2018.01.018
19. Matsa E, Ahrens JH, Wu JC (2016) Human induced pluripotent stem cells as a platform for personalized and precision cardiovascular medicine. Physiol Rev 96(3):1093–1126. https://doi.org/10.1152/physrev.00036.2015
20. Paik DT, Chandy M, Wu JC (2020) Patient and disease-specific induced pluripotent stem cells for discovery of personalized cardiovascular drugs and therapeutics. Pharmacol Rev 72(1):320–342. https://doi.org/10.1124/pr.116.013003
21. Fukuta M, Nakai Y, Kirino K, Nakagawa M, Sekiguchi K, Nagata S, Matsumoto Y, Yamamoto T, Umeda K, Heike T, Okumura N, Koizumi N, Sato T, Nakahata T, Saito M, Otsuka T, Kinoshita S, Ueno M, Ikeya M, Toguchida J (2014) Derivation of mesenchymal stromal cells from pluripotent

stem cells through a neural crest lineage using small molecule compounds with defined media. PLoS One 9(12):e112291. https://doi.org/10.1371/journal.pone.0112291

22. Hackland JOS, Frith TJR, Thompson O, Marin Navarro A, Garcia-Castro MI, Unger C, Andrews PW (2017) Top-down inhibition of BMP signaling enables robust induction of hPSCs into neural crest in fully defined, xeno-free conditions. Stem Cell Reports 9(4): 1043–1052. https://doi.org/10.1016/j.stemcr.2017.08.008
23. Loh KM, Chen A, Koh PW, Deng TZ, Sinha R, Tsai JM, Barkal AA, Shen KY, Jain R, Morganti RM, Shyh-Chang N, Fernhoff NB, George BM, Wernig G, Salomon REA, Chen Z, Vogel H, Epstein JA, Kundaje A, Talbot WS, Beachy PA, Ang LT, Weissman IL (2016) Mapping the pairwise choices leading from pluripotency to human bone, heart, and other mesoderm cell types. Cell 166(2):451–467. https://doi.org/10.1016/j.cell.2016.06.011
24. Brade T, Pane LS, Moretti A, Chien KR, Laugwitz KL (2013) Embryonic heart progenitors and cardiogenesis. Cold Spring Harb Perspect Med 3(10):a013847. https://doi.org/10.1101/cshperspect.a013847
25. Buckingham M, Meilhac S, Zaffran S (2005) Building the mammalian heart from two sources of myocardial cells. Nat Rev Genet 6(11):826–835. https://doi.org/10.1038/nrg1710
26. Thattaliyath B, Hutson M (2016) Neural crest. In: Congenital heart diseases: the broken heart. Springer, Vienna, pp 41–53
27. Brown CB, Feiner L, Lu MM, Li J, Ma X, Webber AL, Jia L, Raper JA, Epstein JA (2001) PlexinA2 and semaphorin signaling during cardiac neural crest development. Development 128(16):3071–3080
28. Holm TM, Habashi JP, Doyle JJ, Bedja D, Chen Y, van Erp C, Lindsay ME, Kim D, Schoenhoff F, Cohn RD, Loeys BL, Thomas CJ, Patnaik S, Marugan JJ, Judge DP, Dietz HC (2011) Noncanonical TGFbeta signaling contributes to aortic aneurysm progression in Marfan syndrome mice. Science 332(6027): 358–361. https://doi.org/10.1126/science.1192149
29. Zhang J, McIntosh BE, Wang B, Brown ME, Probasco MD, Webster S, Duffin B, Zhou Y, Guo LW, Burlingham WJ, Kent C, Ferris M, Thomson JA (2019) A human pluripotent stem cell-based screen for smooth muscle cell differentiation and maturation identifies inhibitors of intimal hyperplasia. Stem Cell Reports 12(6):1269–1281. https://doi.org/10.1016/j.stemcr.2019.04.013
30. Patsch C, Challet-Meylan L, Thoma EC, Urich E, Heckel T, O'Sullivan JF, Grainger SJ, Kapp FG, Sun L, Christensen K, Xia Y, Florido MH, He W, Pan W, Prummer M, Warren CR, Jakob-Roetne R, Certa U, Jagasia R, Freskgard PO, Adatto I, Kling D, Huang P, Zon LI, Chaikof EL, Gerszten RE, Graf M, Iacone R, Cowan CA (2015) Generation of vascular endothelial and smooth muscle cells from human pluripotent stem cells. Nat Cell Biol 17(8):994–1003. https://doi.org/10.1038/ncb3205

Chapter 16

Generation of Salivary Gland Organoids from Mouse Embryonic Stem Cells

Junichi Tanaka and Kenji Mishima

Abstract

Salivary glands are exocrine glands composed of several cell types, including the ductal, acinar, and basal/myoepithelial cells. They play important roles in maintaining oral homeostasis and health. During early murine development, the salivary glands, which arise as epithelial buds, are produced from primitive oral epithelia through an interaction between the oral epithelium and mesenchyme.

We recently reported that salivary gland organoids can be generated from mouse embryonic stem cells (ESCs). We recapitulated the process of embryonic salivary gland development using an organoid culture system. The mouse ESC-derived salivary gland organoids consisted of acinar-, ductal-, and myoepithelial-like cells. In this chapter, we describe a protocol for differentiating salivary gland organoids from ESCs.

Key words Salivary gland, Embryonic stem cells, Organoids, Ectoderm, SFEBq culture

1 Introduction

There are three major salivary glands: the parotid gland, the sublingual gland, and the submandibular gland. The salivary gland is an exocrine gland that secretes saliva into the oral cavity. Saliva has several important antibacterial, mucosal-protective, and digestive functions. Salivary gland hypofunction is caused by the autoimmune disease Sjögren's syndrome and by irradiation treatments of head and neck cancer [1]. Salivary gland hypofunction results in xerostomia, which can cause various oral diseases including mucosal infections and dysphagia. Current therapies for xerostomia include the administration of artificial saliva substitutes and parasympathomimetic drugs [2, 3]. However, such therapies are not effective in cases where most acinar cells are already lost.

Several culture systems that use embryonic stem cells (ESCs) and induced pluripotent stem cells (iPSCs) have been developed to generate organoids of various tissues including brain, pituitary, optic cup, inner ear, thyroid, intestine, liver, and kidney [4–13]. These studies illustrate how ESCs can be used to recapitulate

Nagarajan Kannan and Philip Beer (eds.), *Stem Cell Assays: Methods and Protocols*, Methods in Molecular Biology, vol. 2429, https://doi.org/10.1007/978-1-0716-1979-7_16,

in vivo organogenesis. These organoids are composed of multiple cell types and have a complex structure that mimics in vivo organs. Additionally, these studies suggest that mouse and human iPSCs can differentiate into ectodermal organoids when cultured as floating aggregates under serum-free conditions [4, 9, 14]. Subsequently, we reported on the generation of salivary gland organoids using mouse ESCs [15]. We identified Sox9 and Foxc1 as critical genes for organ-inductive signals. These genes are involved in the commitment of the primitive oral epithelium to differentiation into salivary glands in self-organized ESCs. We recapitulated the process of embryonic salivary gland development using an organoid culture system (Fig. 1a, b). In this chapter, we describe a protocol for differentiating salivary gland organoids from mouse ESCs. Our model is a powerful tool that will enhance our understanding of salivary gland development and disease.

2 Materials

1. 0.1% gelatin solution: Dissolve 0.5 g of gelatin in 500 mL of PBS (−) by autoclaving. The solution can be stored at 4 °C.
2. Growth factor-reduced Matrigel: Thaw Matrigel overnight at 4 °C. Keep Matrigel on ice, make 400 μL aliquots (*see* **Note 1**), and store the aliquots at −20 °C until use.
3. FGF2 stock solution: Reconstitute 100 μg of FGF2 with 500 μL of 5 mM Tris–HCL (pH 7.4) to a concentration of 200 ng/mL. Aliquot the solution (20 μL/tube) and store the aliquots at −20 °C until use.
4. BMP4 stock solution: Reconstitute 10 μg of BMP4 with 100 μL of 0.1% (w/v) bovine serum albumin (BSA)/PBS to a concentration of 100 ng/mL. Aliquot the solution (20 μL/tube) and store the aliquots at −20 °C until use.
5. DNase I stock solution: Reconstitute 10 μg of DNase I with 100 μL of distilled water. Aliquot the solution (20 μL/tube) and store the aliquots at −20 °C until use.
6. Mouse ESC maintenance medium: Glasgow's minimum essential medium (MEM) supplemented with 10% (v/v) knockout serum replacement, 0.1 mM nonessential amino acids, 0.1 mM 2-mercaptoethanol, 1 mM sodium pyruvate, and 1% (v/v) penicillin-streptomycin. The solution can be stored at 4 °C for 1 month. Add 2000 U/mL leukemia inhibitory factor (LIF) just before use.
7. Growth factor-free medium: Iscove's MEM/F-12 supplemented with 1% (v/v) chemically defined lipid concentrate, 450 μM monothioglycerol, 5 mg/mL BSA, and 1% (v/v) penicillin-streptomycin. The solution can be stored at 4 °C for 2 weeks.

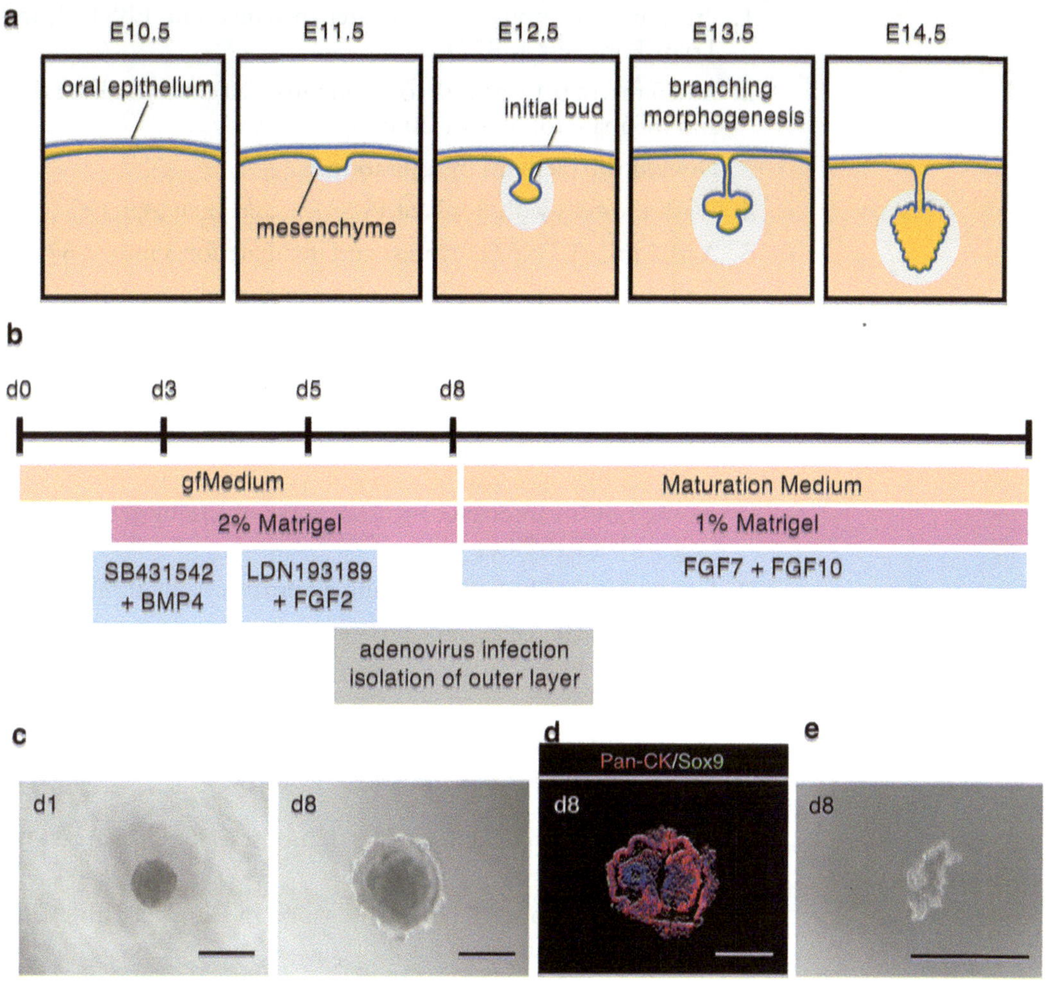

Fig. 1 Generation of salivary gland organoids in floating mouse ESC culture. (**a**) Schematic diagram of mouse salivary gland development. (**b**) Timetable of salivary gland organoid induction from mouse ESCs. (**c**) Morphological changes of the ESCs aggregates during day 1 (left) to day 8 (right). (**d**) Pan-cytokeratin (Pan-CK, red) was detected via immunofluorescence analysis. (**e**) Separated outer layer from the ESCs aggregates on day 8. Scale bars, 300 μm. (**c–e**) Modified from Tanaka et al. *Nat Commun* 2018.

8. Maturation medium: Advanced Dulbecco's MEM/F-12 supplemented with 1% (v/v) N2 supplement, 1% (v/v) Glutamax, 1% (v/v) penicillin-streptomycin, 1% (v/v) growth factor-reduced Matrigel, 100 ng/mL FGF7, and 200 ng/mL FGF10. Prepare the solution just before use (*see* **Note 2**).

3 Methods

3.1 Maintenance of Mouse ESCs

Mouse ESCs are cultured without feeder cells in mouse ESC maintenance medium on gelatin-coated dishes. Cells are cultured at 37 °C and 5% CO_2 (*see* **Note 3**).

1. Pre-warm the mouse ESC maintenance medium, PBS (−), and TrypLE express at 37 °C.
2. Add 5 mL of 0.1% gelatin solution into a 10 cm cell culture dish and incubate for at least 20 min at 37 °C.
3. Discard the medium by aspiration.
4. Wash the cells with 5 mL of PBS (−), and then aspirate.
5. Add 2 mL of TrypLE express and incubate for 3 min at 37 °C.
6. Add 8 mL of mouse ESC maintenance medium without LIF and detach the cells by pipetting up and down with a 10 mL pipette.
7. Transfer the cell suspension into a 15 mL conical tube and centrifuge at 300 × *g* for 3 min at room temperature (RT).
8. Remove the supernatant and resuspend the cells in 1 mL of mouse ESC maintenance medium without LIF by gently pipetting with a P1000 tip.
9. Count the number of cells using a hematocytometer.
10. Plate 5×10^5 cells into a 10 cm gelatin-coated dish (**steps 2** and **3**).
11. Incubate the plate at 37 °C under 5% CO_2. Starting on the next day, change the mouse ESC maintenance medium (10 mL) once daily and passage the cells every 3 days.

3.2 Oral Ectoderm Differentiation from Mouse ESCs

3.2.1 Day 0

1. Prepare one 10 cm culture dish of mouse ESCs grown to 70% confluency.
2. Change the mouse ESC maintenance medium (10 mL) and incubate the cells again for 3 h.
3. Discard the medium by aspiration.
4. Wash twice with 5 mL of PBS (−), and then aspirate.
5. Add 2 mL of TrypLE express and incubate for 3 min at 37 °C (*see* **Note 4**).
6. Add 8 mL of mouse ESC maintenance medium without LIF and detach the cells by pipetting up and down with a 10 mL pipette.
7. Transfer the cell suspension into a 15 mL conical tube and centrifuge at 300 × *g* for 3 min at RT.
8. Remove the supernatant and resuspend the cells in 1 mL of growth factor-free medium by gently pipetting up and down with a P1000 tip.
9. Add 2 mL of growth factor-free medium and pour the cell suspension into a 15 mL conical tube with a 40 μm cell strainer.
10. Count the number of cells using a hematocytometer.

11. Adjust the number of cells to 3×10^4 cells/mL with growth factor-free medium.
12. Seed dissociated mouse ESCs into a 96-well U-bottomed culture plate at a density of 3000 cells/well in a volume of 100 μL.
13. Incubate the plate at 37 °C under 5% CO_2.

3.2.2 Day 1

1. Add growth factor-reduced Matrigel to each well to a final concentration of 2% (v/v): Prepare a 4% Matrigel solution with growth factor-free medium, discard 50 μL of medium from each well using a P100 tip and then add 50 μL of 4% Matrigel solution to each well. Mix well by pipetting up and down two to three times using a P100 tip.

3.2.3 Day 2

1. Add BMP4 and SB-431542 to each well to final concentrations of 10 ng/mL and 1 μM, respectively: Prepare 50 ng/mL BMP4 and 5 μM SB-431542 with growth factor-free medium, and add 25 μL of growth factor-free medium containing BMP4 and SB-431542 to each well. Mix well by pipetting up and down two to three times using a P100 tip. Each well should now contain 125 μL of medium.

3.2.4 Day 5

1. Add FGF2 and LDN-193189 to each well to final concentrations of 25 ng/mL and 100 nM, respectively: Prepare 150 ng/mL FGF2 and 600 nM LDN-193189 with growth factor-free medium, add 25 μL of growth factor-free medium containing FGF2 and LDN-193189 to each well, and mix well by pipetting up and down two to three times using a P100 tip. Each well should now contain 150 μL of medium. Continue to incubate cells for another 3 days (Fig. 1c, d).

3.3 Salivary Gland Organoid Differentiation from Mouse ESC-Derived Oral Ectoderm

3.3.1 Day 8

1. On day 8 of differentiation (*see* **Note 5**), perform transcription factor induction: remove 70 μL of medium from each well, and add recombinant adenoviruses (3×10^6 pfu Ad-Sox9 and 3×10^6 pfu Ad-Foxc1) in 20 μL growth factor-free medium. Incubate for 60 min at 37 °C (*see* **Note 6**).
2. After 60 min of incubation, transfer the infected aggregates into a 15 mL conical tube and wash twice with 5 mL of maturation medium without Matrigel and FGF, then discard the supernatant.
3. Add 2 mL of dispase and incubate for 2 min at 37 °C (*see* **Note 7**).
4. Wash twice with 5 mL of maturation medium without Matrigel and FGF.
5. Add 5 mL of DNase I solution to the aggregates in 5 mL of maturation medium and incubate for 5 min at RT (*see* **Note 8**).

6. Wash twice with 5 mL of maturation medium without Matrigel and FGF. Transfer the infected aggregates to a 10 cm bacterial dish.
7. Cut the outer layer of the aggregates with a 29 G needle syringe under an inverted microscope (*see* **Note 9**). Transfer the isolated outer layers of the aggregates to 24-well low-cell-adhesion plates in maturation medium containing FGF7, FGF10, and Matrigel (Fig. 1e) (*see* **Note 10**). Incubate the plate at 37 °C under 5% CO_2.

3.3.2 Day 8–30

1. Change the maturation medium (500 mL) containing FGF7, FGF10, and Matrigel once every other day using a P1000 tip (*see* **Note 11**). The dissected regions of ESC aggregates gradually increased in size after 15 days in culture and then show extensive protrusion from the aggregates towards the outer space, as well as branching, during days 20–28 of differentiation (Fig. 2).

3.4 Immunostaining of Salivary Gland Organoids

1. Transfer the organoids to a 1.5 mL tube and wash the organoids with PBS (−).
2. Discard the PBS (−), add 1.5 mL of 4% (w/v) paraformaldehyde in PBS (−), and incubate overnight at 4 °C.
3. Discard the paraformaldehyde and wash with PBS (−).
4. Discard the PBS (−), add 1 mL of 10% (w/v) sucrose in PBS (−), and incubate for 60 min at 4 °C.
5. Discard the 10% sucrose, add 1 mL of 20% (w/v) sucrose in PBS (−), and incubate for 60 min at 4 °C.
6. Discard the 20% sucrose, add 1 mL of 30% (w/v) sucrose in PBS (−), and incubate overnight at 4 °C.
7. Remove the excess 30% sucrose using a P200 tip and add the optimal cutting temperature (OCT) compound.
8. Transfer the organoids with OCT compound to a cryomold using a wide-opening P200 tip. Place the organoids in the bottom of the cryomold.
9. Freeze the organoids with liquid nitrogen-cooled isopentane.
10. Make frozen organoid sections using a cryostat (*see* **Note 12**).
11. Carry out immunostaining using the antibodies listed in Table 1.

4 Notes

1. The Matrigel, 1.5 mL tubes, and P1000 tips should be ice-cold.

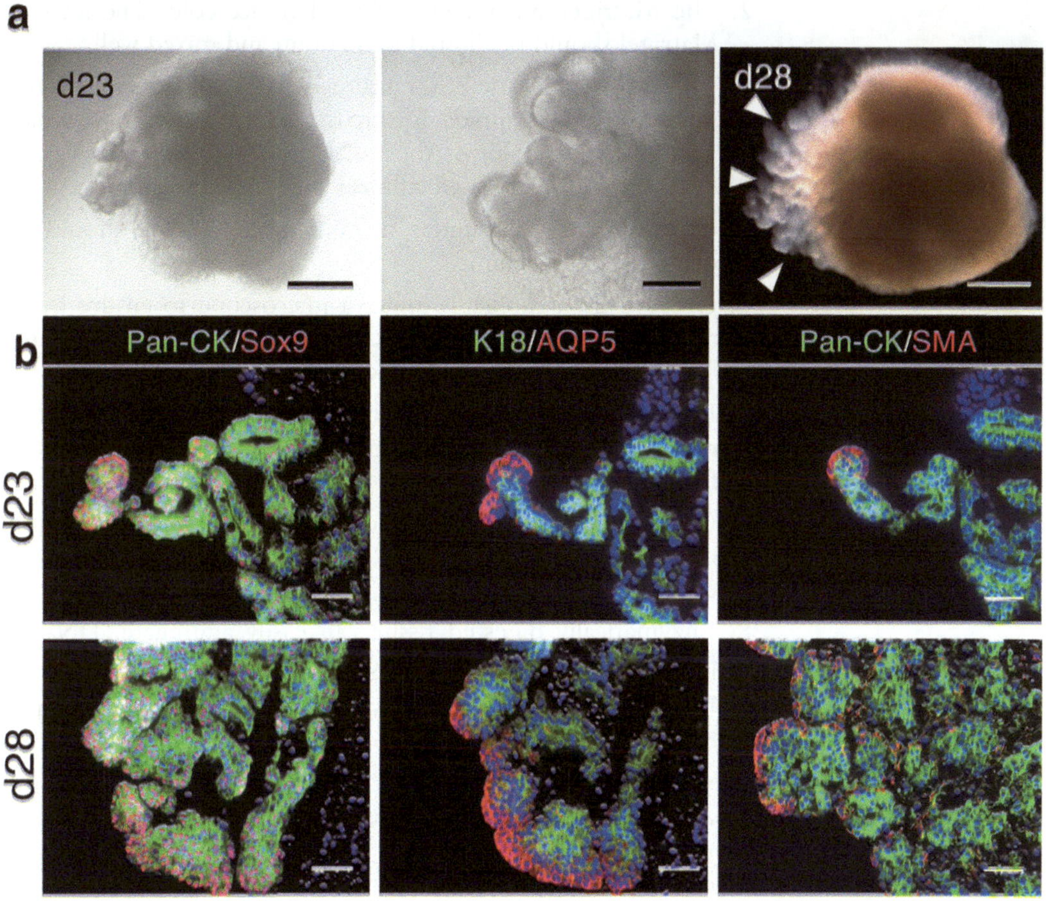

Fig. 2 Analysis of the salivary gland organoids by immunofluorescence. (**a**) Phase-contrast representative images showing the morphological changes in the aggregates at d23. The square outline indicates epithelial bud formation (left and middle). Bright-field view of the aggregate at d28. Arrowheads indicate the epithelial branching structure (right). Scale bars, 300 μm (left and right), 50 μm (middle). (**b**) Immunofluorescence images of the salivary gland organoid at d23 and d28. The epithelial marker Pan-CK (green), the ductal marker K18 (green), the acinar cell marker AQP5 (red), and the myoepithelial cell marker a-SMA (red) were detected. Scale bars 50 μm. Modified from Tanaka et al. *Nat Commun* 2018

Table 1
Antibodies required

Antibody	Host	Supplier	Cat. No.	Dilution
Pan-cytokeratin	Mouse	Sigma-Aldrich	C2562	1:200
Sox9	Rabbit	Chemicon	AB5535	1:500
K18	Mouse	Progen	61028	1:200
AQP5	Rabbit	Alomone Labs	AQP-005	1:200
α-SMA	Rabbit	Abcam	Ab5694	1:200

2. The Matrigel and medium should be ice-cold. The ice-cold Matrigel should be diluted in medium and mixed well.
3. The quality of mouse ESC culture is critical for the induction of salivary gland organoids. Ensure the cultured ESCs have typical ESC-like morphology. Abnormal morphology is the sign of differentiation, and such cultures should be discarded.
4. Observe the cells under a microscope. ESCs should be detached and in clumps.
5. Observe the aggregates under a microscope to ensure that a translucent structure is formed on the outer layer of aggregates. This structure is the oral ectoderm and should be isolated.
6. HEK293 packaging cells are used for amplification of recombinant adenoviruses. Adenoviral titer is measured by a plaque formation assay.
7. Carefully follow the suggested reaction time and temperature of dispase treatment. Rapid and precise washing steps are essential.
8. Enzymatically digested tissue can easily aggregate. DNase I treatment can prevent this aggregation.
9. Dissection of the outer layer should be performed using a pair of 29 G needles. Be careful not to press the outer layer to avoid destroying the tissue.
10. A 20 μL pipette tip is suitable to transfer the isolated outer layer to a 24-well dish. We suggest using a 24-well dish and placing one aggregate per well to avoid the aggregate adhesion.
11. Salivary gland organoids arise as protrusions from the aggregates towards the outer space or branching during differentiation (days 20–28).
12. We suggest generating 4 μm serial sections for immunostaining.

References

1. Atkinson JC, Grisius M, Massey W (2005) Salivary hypofunction and xerostomia: diagnosis and treatment. Dent Clin N Am 49:309–326
2. Guggenheimer J, Moore PA (2003) Xerostomia: etiology, recognition and treatment. J Am Dent Assoc 134:61–119
3. Rocchi C, Emmerson E (2020) Mouth-watering results: clinical need, current approaches, and future directions for salivary gland regeneration. Trends Mol Med 26(7): 649–669. https://doi.org/10.1016/j.molmed.2020.03.009
4. Koehler KR, Mikosz AM, Molosh AI et al (2013) Generation of inner ear sensory epithelia from pluripotent stem cells in 3D culture. Nature 500:217–221
5. Eiraku M, Watanabe K, Matsuo-Takasaki M et al (2008) Self-organized formation of polarized cortical tissues from ESCs and its active manipulation by extrinsic signals. Cell Stem Cell 3:519–532
6. Suga H, Kadoshima T, Minaguchi M et al (2011) Self-formation of functional adenohypophysis in three-dimensional culture. Nature 480:57–62
7. Antonica F, Kasprzyk DF, Opitz R et al (2012) Generation of functional thyroid from embryonic stem cells. Nature 491:66–71

8. Lancaster MA, Knoblich JA (2014) Organogenesis in a dish: modeling development and disease using organoid technologies. Science 345:1247125
9. Eiraku M, Takata N, Ishibashi H et al (2011) Self-organizing optic-cup morphogenesis in three-dimensional culture. Nature 472:51–56
10. Spence JR, Mayhew CN, Rankin SA et al (2011) Directed differentiation of human pluripotent stem cells into intestinal tissue in vitro. Nature 470:105–109
11. Takebe T, Sekine K, Enomura M et al (2013) Vascularized and functional human liver from an iPSC-derived organ bud transplant. Nature 499:481–484
12. Taguchi A, Nishinakamura R (2017) Higher-order kidney organogenesis from pluripotent stem cells. Cell Stem Cell 21:730–746.e6
13. Takasato M, Er PX, Chiu HS et al (2016) Kidney organoids from human iPS cells contain multiple lineages and model human nephrogenesis. Nature 536:238
14. Ozone C, Suga H, Eiraku M et al (2016) Functional anterior pituitary generated in self-organizing culture of human embryonic stem cells. Nat Commun 7:1–10
15. Tanaka J, Ogawa M, Hojo H et al (2018) Generation of orthotopically functional salivary gland from embryonic stem cells. Nat Commun 9:4216

Chapter 17

In Vitro Generation of Heart Field Specific Cardiomyocytes

Arash Pezhouman, Ngoc B. Nguyen, Allison Shevtsov, Rong Qiao, and Reza Ardehali

Abstract

Myocardial infarction (MI) can lead to irreversible loss of cardiomyocytes (CMs), primarily localized to the left ventricle (LV) of the heart. The CMs of the LV are predominantly derived from first heart field (FHF) progenitors, whereas the majority of CMs within the right ventricle originate from the second heart field (SHF) during early cardiogenesis. Human embryonic stem cells (hESCs) serve as a valuable source of CMs for understanding early cardiac development and lineage commitment of CMs within these two heart fields that ultimately enable the development of more effective candidates for cell therapy. An ideal candidate may be FHF CMs that share the same ontogeny with the LV CMs that die after MI. We previously generated a double reporter hESC line that utilizes two important cardiac transcription factors, TBX5 and NKX2-5. TBX5 marks FHF progenitors and CMs, while NKX2-5 is expressed in nearly all myocytes of the developing heart. Here, we describe a step-by-step approach to efficiently generate FHF and SHF CMs using this double reporter hESC line. In addition, this approach can be applied to any non-genetically modified hESC lines to enrich FHF and SHF CMs. Obtaining enriched populations of these two CM subtypes provides a platform for downstream comparative analyses and in vitro studies to facilitate a deeper understanding of cardiovascular lineage commitment and the development of more effective candidates for cell therapy to treat diseases or defects that affect specific regions of the heart.

Key words Human embryonic stem cell, First heart field cardiomyocyte, Second heart field cardiomyocyte, Human heart development, Cardiac differentiation

1 Introduction

Human embryonic stem cells (hESCs) can self-renew indefinitely while maintaining the capacity to differentiate into other cell types, including cardiomyocytes (CMs) [1]. As such, hESCs represent an unlimited source of committed CMs of potential use in understanding heart development, cell-based cardiac therapies, and disease modeling [2–4].

Previous studies have shown significant molecular differences between left and right ventricular CMs originating from first and second heart fields (FHF and SHF) progenitors, respectively [5]. Given these differences, it seems that an ideal candidate for

Nagarajan Kannan and Philip Beer (eds.), *Stem Cell Assays: Methods and Protocols*, Methods in Molecular Biology, vol. 2429, https://doi.org/10.1007/978-1-0716-1979-7_17,

regenerative therapy could be FHF CMs which share the same ontogeny with the lost CMs after a myocardial infarction (MI) affecting the LV. Another important prerequisite for successful chamber-specific regenerative therapies is the ability to generate millions of functional hESC-derived FHF CMs. Although a number of studies [6–8] have previously described efficient cardiac differentiation protocols, none has generated the high throughput heart field-specific protocol.

Therefore, the ability to generate pure and high-yield FHF and SHF populations not only circumvents some of the obstacles in regenerative therapy but also provides an in vitro model to investigate chamber-specific congenital heart diseases such as hypoplastic left heart syndrome or arrhythmogenic right ventricular dysplasia [9, 10].

Here, we describe a simple and scalable method to generate FHF- and SHF-like progenitors and CMs in vitro using our previously generated double reporter (*TBX5-TdTomato*$^{+/W}$/ *NKX2–5*$^{eGFP/W}$) hESC line [11]. This protocol relies on temporal and dosedependent modulation of canonical Wnt signaling pathway in feeder- and serum-free method. Precise modulation of hESC seeding density and glycogen synthase kinase 3 (GSK3) inhibitor concentration is sufficient to generate high-yield FHF and SHF CMs in 12 days. Functional heart field specific CMs, differentiated via these protocols may constitute a potential cell source for cell-based regenerative medicine, congenital heart disease modeling, and drug screening. Considering the limited accessibility of the human heart during early cardiogenesis, in vitro generation of heart field specific CMs may also be used to decode some aspects of early human heart development. Despite its advantages, it is important to note that this is an in vitro system and may not precisely mimic in vivo human heart development.

In summary, this protocol details the necessary steps to generate highly enriched FHF (70–80%) and SHF (35–45%) cardiac progenitors from differentiating hESCs. Additionally, we provide simple approaches for validation of first and second heart field CMs prior to downstream analyses.

2 Materials

2.1 Double Reporter hESC Line

HES3 (*TBX5-TdTomato*$^{+/W}$/*NKX2–5*$^{eGFP/W}$) hESC line (Ardehali Lab, UCLA).

2.2 hESC Line Maintenance

1. ReLeSR (STEMCELL Technologies, 05873).
2. *mTeSR 1 Media Preparation*: Thaw mTeSR™ 5× Supplement (Stem Cell™ Technologies, 85852) at room temperature (15–25 °C) or overnight at 2–8 °C and mix thoroughly (*see* **Note 1**). Add 100 mL of mTeSR™ 1 5× Supplement and

5 mL Penicillin/Streptomycin (10,000 U/mL, Thermo Fisher Scientific, 15140122) to 400 mL of mTeSR™ 1 Basal Medium (Stem Cell™ Technologies, 85850) and mix thoroughly.

3. *Geltrex™ Matrix Preparation*: Thaw Geltrex™ LDEV-Free Reduced Growth Factor Basement Membrane Matrix (Gibco™, A1413202) overnight at 2–8 °C. Dilute 1 mL of Geltrex™ Matrix solution into 99 mL of pre-chilled (4 °C) DMEM medium (Gibco™, 21068028, 1% final concentration) (*see* **Note 2**).

2.3 Cardiac Differentiation

1. RPMI 1640 Medium (Gibco™, 11875093).
2. B-27™ supplement, minus Insulin (B-27™ minus insulin) (Gibco™, A1895601).
3. B-27™ supplement (B-27™ plus insulin) (50×) (Gibco™, A3582801).
4. TrypLE Express (Thermo Fisher, 12604021).
5. *CHIR99021 Preparation*: Molecular weights of CHIR99021 (Tocris, 4423) vary batch to batch, which will affect the solvent volumes required to prepare stock solutions. To calculate the solvent volume (sterile DMSO), click on Reconstitution Calculator button from the ref. 12. Add calculated volume of DMSO (Max concentration of 20 mM) in CHIR99021 and mix thoroughly. Aliquot and store in −20 °C.
6. *IWP-2 Preparation*: Molecular weights of IWP-2 (Tocris, 3533) vary batch to batch, which will affect the solvent volumes required to prepare stock solutions. To calculate the solvent volume (sterile DMSO), click on Reconstitution Calculator button from the ref. 13. Add calculated volume of DMSO (Max concentration of 5 mM) in IWP2 and mix thoroughly. Aliquot and store in −20 °C.
7. *Y-27632 Preparation*: Molecular weights of Y-27632 (Tocris, 1254) vary batch to batch, which will affect the solvent volumes required to prepare stock solutions. To calculate the solvent volume (sterile water), click on Reconstitution Calculator button from the ref. 14. Aliquot and keep them in −20 °C.

2.4 Validation of hESC-Derived CM Populations

1. Flow cytometric analysis.
   (a) *FACS Buffer Preparation*: To prepare 50 mL of FACS buffer, combine 49.8 mL HBSS (Life Technologies, PN 14175-095] with 0.2 mL 500 mM EDTA (Sigma, PN 03690) and 0.250 g BSA Fraction V (Sigma, PN A9418) with 5 ul DAPI (Biolegend, 422,801).
2. Antibodies for Immunocytochemistry.
   (a) Human anti-TBX5 (Abcam, ab123665).
   (b) Human anti-Islet-1 (R&D Systems, AF1837).

   (c) Human anti-Cardiac Troponin T (R&D System, MAB1874).
3. Primers for qPCR analysis.
   (a) *TNNT2* F: 5′- AGCGGAAAAGTGGGAAGAG-3′;
   R: 5′- TCCAAGTTATAGATGCTCTGCC -3′.
   (b) *ACTN2* F: 5′- CAAACCTGACCGGGGAAAAAT -3′;
   R: 5′- CTGAATAGCAAAGCGAAGGATGA -3′.
   (c) *TBX5* F: 5′-GACTTCCTACCAGAACCACAAG-3′;
   R: 5′-GGGACCACGGGATATTCTTTAC-3′.
   (d) *ISL1* F: 5′-ACCTTGGAAAGTACTGAGCG-3′;
   R: 5′-GGTGTATCTGGAAGTTGAGAGG-3′.
   (e) *GAPDH* F: 5′- ACATCGCTCAGACACCATG -3′;
   R: 5′-TGTAGTTGAGGTCAATGAAGGG -3′.

3 Methods

3.1 hESC Line Maintenance

3.1.1 Thawing the hESCs

1. Add 1 mL diluted Geltrex™ Matrix solution to cover entire growth surface area of one well of 6-well plate.
2. Incubate the coated plates at 37 °C for a minimum of 30 min.
3. Remove a frozen cell vial from liquid N_2 and immerse the vial in a 37 °C water bath without submerging the cap. Swirl the vial gently for 2–3 min until the contents are completely thawed.
4. Spray the vial with 70% ethanol and place it in tissue culture hood. Use a sterile 1 mL pipette to gently transfer the cells into a sterile 15 mL conical tube containing 5 mL of room temperature mTeSR1 medium.
5. Centrifuge the cells at 200 × *g* for 3 min. Aspirate and discard the supernatant with a sterilized Pasteur pipette.
6. Resuspend the cell pellet in 4 mL mTeSR1 medium +4 μL of (5 μM Y27632).
7. Aspirate diluted Geltrex™ Matrix solution from the wells and add 2 mL the cell suspension into each well of the Geltrex-coated 6-well plate (1 vial to 2 wells). Move the plate in two quick, short, back-and-forth and side-to-side motions to disperse cells across the surface of the wells. Put the plate back into the 37 °C, 5% CO_2 incubator (*see* **Note 3**).
8. The next day, aspirate the medium in each well and replace with 2 mL fresh room temperature mTeSR1 medium daily.

3.1.2 Passaging hESCs Using ReleSR

1. When the cells are 85–90% confluent (usually 2–3 days after passaging) aspirate the old medium then add 1 mL room temperature sterile DPBS and wash gently. Aspirate DPBS and add 1 mL ReLeSR to each well and wait for 1 min then aspirate the ReLeSR.
2. Incubate the plate at 37 °C, 5% CO_2 and wait for 2 min.
3. Add 2 mL mTeSR1 perpendicularly to the 1 well from 6-well plate and dispense the medium over the surface of the well until all the cells are detached (*see* **Note 4**).
4. Pool the contents of the well into a sterile conical tube containing 10 mL mTeSR1 and gently mix 3 times using 5 mL pipette. Seed 2 mL of the cell suspension into each well of a Geltrex-coated 6-well plate (split ratio of 1:6 is performed here, *see* **Note 5**).
5. Move the plate in two quick, short, back-and-forth and side-to-side motions to disperse cells across the surface of the wells. Return the plate to the incubator after plating the cells.
6. Each day aspirate the medium and add 2 mL of fresh room temperature mTeSR1 per well.

3.2 Heart Field Specific Cardiac Differentiation

To generate heart field specific CMs from differentiating hESCs, a standard protocol (Fig. 1a) was modified to determine the optimized cell density and CHIR99021 concentration to achieve high yield of FHF or SHF CMs. Sequential cellular morphological

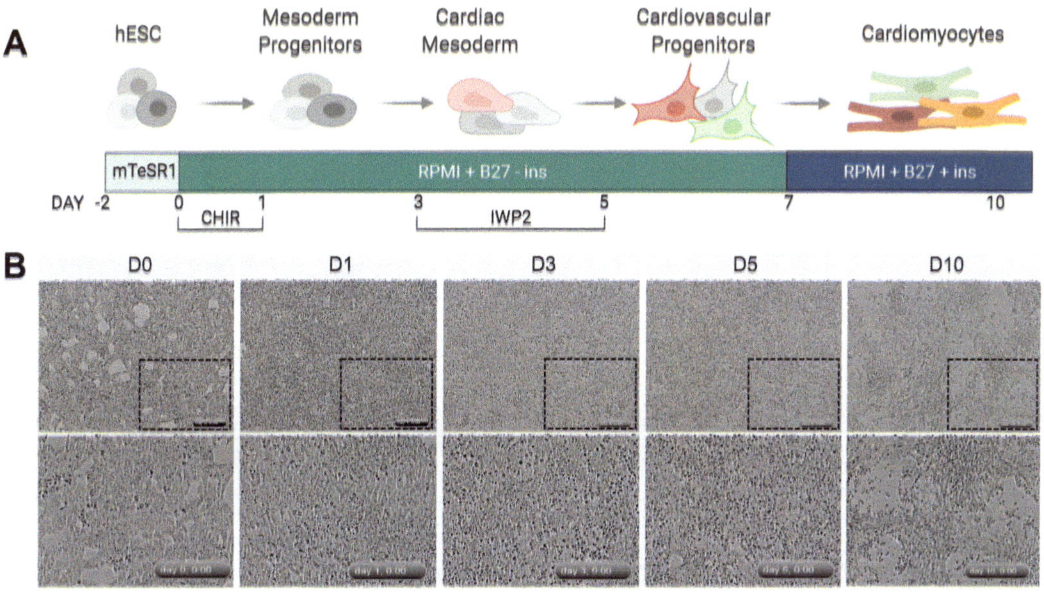

Fig. 1 (**a**) Schematic representation of monolayer cardiac differentiation protocol with stepwise lineage commitment of hESCs into cardiomyocytes. (**b**) Representative phase-contrast images showing the morphological change of day 0 (hESCs) to day 10 (beating cardiomyocytes)

changes during cardiac differentiation are shown in Fig. 1b for reference.

3.2.1 Basal Media Preparation

1. Thaw B-27™ minus Insulin supplement overnight at 2–8 °C.
2. Add 10 mL of B-27™ minus Insulin supplement into 500 mL RPMI and mix thoroughly.

3.2.2 First Heart Field Specific Cardiomyocyte Differentiation

1. Culture the hESCs on Geltrex-coated 6-well plate in mTeSR1 medium to 85–90% confluence.
2. Aspirate the medium and add 1 mL of room temperature TrypLE to each well. Incubate the cells in a 37 °C, 5% CO_2 incubator and wait for 3 min.
3. Add 2 mL of DPBS into each well of the 6-well plate and pool all of the cells into a 50 mL conical tube. Centrifuge the cells at 200 × *g* for 3 min. Aspirate the supernatant and resuspend the cells in mTeSR1 + 5 μM Y27632. Count the total cell number, for example by using an automate cell counter (such as Countess II Automated Cell Counter, Thermo Fisher).
4. Re-plate the cells onto Geltrex-coated 6-well plates at 1.3×10^5 cells/cm^2 density (Fig. 2a, top) and add mTeSR1 + 5 μM Y27632 medium to each well to make a final volume of 2 mL (Day −2 of differentiation, *see* **Note 6**).
5. On day −1, aspirate the medium and replace with 2 mL room temperature mTeSR1™ per well of the 6-well plate and Incubate the cells in a 37 °C, 5% CO_2 incubator.
6. At day 0, aspirate the medium and add 2 mL DPBS and wash the cells gently.
7. Aspirate DPBS and replace with 2 mL room temperature RPMI + B-27™ minus Insulin containing CHIR99021

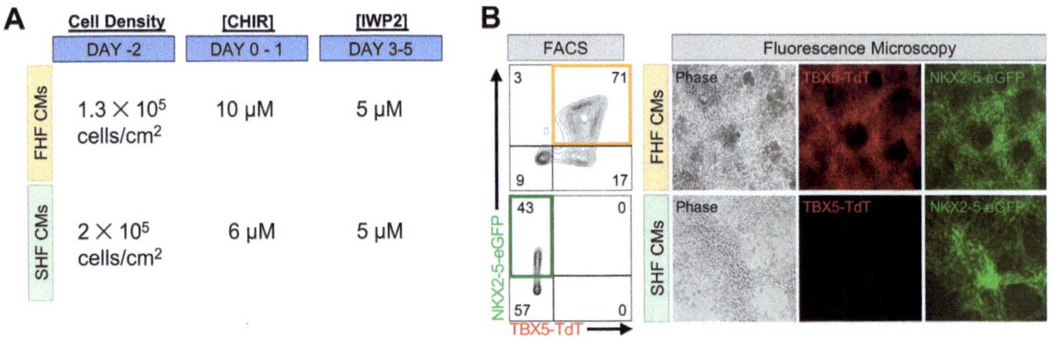

Fig. 2 (**a**) Optimized cardiac differentiation protocols to generate FHF and SHF cardiomyocytes via small-molecule modulation of Wnt signaling and cell seeding densities. (**b**) Flow cytometric analysis and fluorescence microscopy of optimized differentiations of FHF ($TBX5^+$ /$NKX2\text{-}5^+$) and SHF (TBX5/$NKX2\text{-}5^+$) cardiomyocytes

(10 μM) and incubate the cells in a 37 °C, 5% CO_2 incubator for 24 h (*see* **Note 7**).

8. At day 1, aspirate the medium and replace with 2 mL room temperature RPMI + B-27™ minus insulin and place back into incubator for 48 h.
9. At day 3, aspirate the medium and replace with 2 mL room temperature RPMI + B-27™ minus insulin containing IWP2 (5 μM) and place back into incubator for 48 h.
10. At day 5, aspirate the medium and replace with 2 mL room temperature RPMI + B-27™ minus Insulin and place back into incubator for 48 h.
11. At day 7, aspirate the medium and replace with 2 mL room temperature RPMI + B-27™ Plus Insulin and place back into incubator for 48 h.
12. Maintain the cells in RPMI + B-27™ Plus Insulin and replace media every 3 days thereafter.

3.2.3 Second Heart Field Specific Cardiomyocyte Differentiation

1. Culture the hESCs on Geltrex-coated 6-well plate in mTeSR1 medium to 85–90% confluence.
2. Aspirate the medium and add 1 mL of room temperature TrypLE to each well. Incubate the cells in a 37 °C, 5% CO_2 incubator and wait for 3 min.
3. Add 2 mL of DPBS into each well of the 6-well plate and pool all of the cells into a 50 mL conical tube. Centrifuge the cells at $200 \times g$ for 3 min. Aspirate the supernatant and resuspend the cells in mTeSR1 + 5 μM Y27632. Determine cell count.
4. Re-plate the cells onto Geltrex-coated 6-well plates at 2×10^5 cells/cm^2 density (Fig. 2a, bottom) and add mTeSR1 + 5 μM Y27632 medium to each well to make a final volume of 2 mL (Day −2 of differentiation, *see* **Note 6**).
5. On day −1, aspirate the medium and replace with 2 mL room temperature mTeSR1™ per well of the 6-well plate and place back into incubator.
6. At day 0, aspirate the medium and add 2 mL DPBS and wash the cells gently.
7. Aspirate DPBS and replace with 2 mL room temperature RPMI + B-27™ minus Insulin containing CHIR99021 (6 μM) and place back into incubator for 24 h.
8. At day 1, aspirate the medium and replace with 2 mL room temperature RPMI + B-27™ minus Insulin and place back into incubator for 48 h.
9. At day 3, aspirate the medium and replace with 2 mL room temperature RPMI + B-27™ minus Insulin containing IWP2 (5 μM) and place back into incubator for 48 h.

10. At day 5, aspirate the medium and replace with 2 mL room temperature RPMI + B-27™ minus insulin and place back into incubator for 48 h.
11. At day 7, aspirate the medium and replace with 2 mL room temperature RPMI + B-27™ Plus Insulin and place back into incubator for 48 h.
12. Maintain the cells in RPMI + B-27™ Plus Insulin and replace media every 3 days thereafter.

3.3 Validation of hESC-Derived CM Populations

3.3.1 Flow Cytometry Analysis

To determine the percentage of FHF or SHF CMs generated from the cardiac differentiation protocols, isolate cells into a single cell suspension for flow cytometric analysis on Day 10 of differentiation.

1. Aspirate the medium and add 1 mL of room temperature TrypLE to each well. Incubate the cells and wait for 3 min (*see* **Note 8**).
2. Add 2 mL of DPBS and resuspend cells into a 15 mL conical tube and add additional DPBS until the 10 mL mark. Centrifuge at 200 × *g* for 3 min. Aspirate the supernatant and resuspend the cells in 300 μL per well FACS buffer.
3. Strain cells through a 70 μm cell strainer and add 0.3 μL DAPI for each 300 μL FACS buffer added.

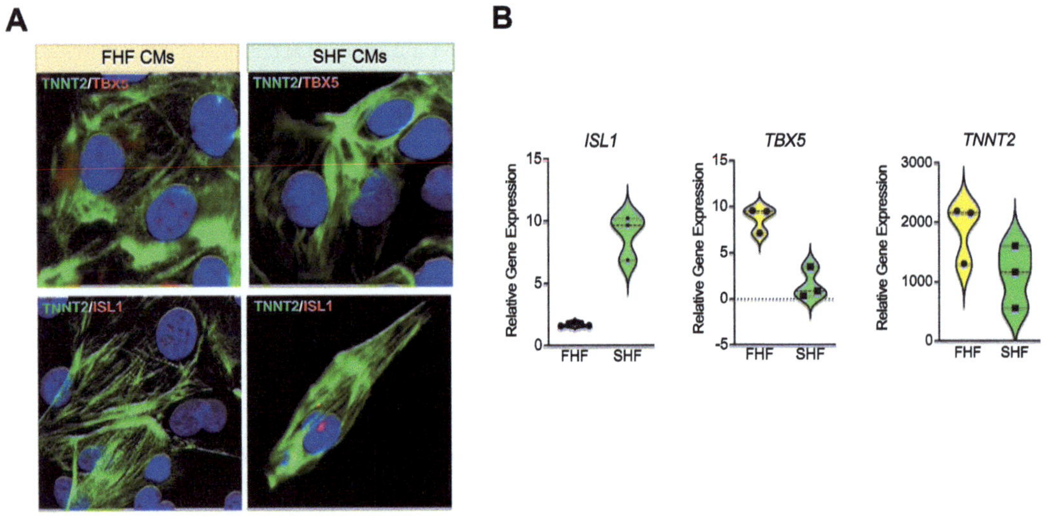

Fig. 3 (**a**) Immunocytochemical analysis of Day 10 sorted FHF- and SHF-like cardiomyocytes for specific markers of FHF (TBX5) and SHF (ISL1). The presence of cardiomyocytes is indicated by the expression of TNNT2 and all nuclei were visualized with DAPI. (**b**) Gene expression analysis of *ISL1*, *TBX5*, and *TNNT2* of FHF and SHF populations

4. Perform flow cytometry analysis of TdTomato (TBX5) and GFP (NKX2-5) (Fig. 2b).

3.3.2 Immunocytochemistry and qPCR

After optimization of the above protocols for generation of first or second heart field CMs, we recommend performing immunocytochemistry (Fig. 3a) and quantitative RT-PCR (Fig. 3b) for known CM (ACTN2, TNNT2) and heart field markers (TBX5, ISL1) before downstream assays with the generated cells (*see* **Note 9**).

4 Notes

1. mTeSR™1 5× Supplement may appear slightly cloudy after thawing. If this is noted, ensure that the supplement is at room temperature. If cloudiness persists, place in a 37 °C water bath for approximately 5 min, swirling occasionally until supplement becomes clear. Supplement must be free of cloudiness before adding to basal medium. mTeSR™1 medium can be stored at 2–8 °C for up to 2 weeks. Alternatively, aliquot and store at −20 °C for up to 6 months. After thawing the aliquoted complete medium, use immediately or store at 2–8 °C for up to 2 weeks. Do not refreeze.
2. Geltrex™ Matrix solution gels in 5–10 min when kept above 15 °C. When working from a full 5 mL vial, it is unnecessary to keep the vial on ice if it is used within 5 min and the environmental temperature does not exceed 25 °C. Because smaller volumes warm more quickly, keep partial tubes and aliquots on ice to prevent premature gelling. Store remaining stock solution at −80 °C to −20 °C. Avoid multiple freeze/thaw cycles.
3. Avoid circular motions as cells will accumulate toward the center of the well, leading to inconsistent densities across the well and affects efficiency of differentiation.
4. Do not attempt to forcefully detach cells that firmly remain attached as they are likely to have spontaneously differentiated.
5. The split ratio is variable, though generally between 1:3 and 1:6 is appropriate. A general rule is to observe the last split ratio and adjust the ratio according to the appearance of the colonies. If the cells have enough space (more than 10 μm) between them to expand, split using the same ratio; if they are overly dense and crowded (colony diameter > 500 μm, spacing between colonies <10 μm), increase the ratio; and if the cells are sparse (colony diameter < 50 μm, spacing between colonies >500 μm), decrease the ratio.
6. We determined that out of all the variables involved in the cardiac differentiation process, seeding density and the

concentration of CHIR99021 had the most effect on directing the cardiac differentiation toward specific lineage commitment. The initial plating density and CHIR concentration may also need to be optimized for different cell lines. To determine the optimal parameters, we recommend using a 12-well plate containing a certain cell density and varying CHIR99021 concentration for each column. On Day 10 of differentiation, isolate cells into a single cell suspension and use flow cytometry to determine the maximum percentage of TBX5 + NKX2–5+ (FHF) or TBX5-NKX2–5+ cells (SHF). The combination of cell density and CHIR concentration that yields the highest percentage of the respective populations are the optimal conditions to use for future differentiations.

7. It is important to note the starting time of the addition of CHIR99021 to ensure that media will be changed no later than 24 h to avoid toxicity from prolonged exposure.
8. To analyze the differentiated cells at a later time point, it may be necessary to incubate the cells with TrypLE for a longer period of time up to 5 min as cells may be more firmly attached to each other and the well surface. We recommend monitoring the cellular attachment every minute (after 3 min) under a microscope with gentle shaking of the plate to avoid under- or over-digestion. This process is important as it affects downstream cellular viability.
9. Our general ICC protocol is as follows. After fixation with 4% paraformaldehyde in PBS for 10 min, cells are prmeabilized with 0.2% TritonX for 20 min and then blocked for 30 min with 1% BSA/PBS-T. We have found the following dilutions for primary antibodies to work well for hESC-derived CMs: anti-TBX5 (1:50), anti-Islet-1 (1:40, 5 μg/mL), anti-Cardiac Troponin T (1:100, 5 μg/mL).

Acknowledgments

This work was supported in part by grants from the Eli & Edythe Broad Center of Regenerative Medicine and Stem Cell Research at UCLA Postdoctoral Fellowship (A.P.), Department of Defense Discovery Award (W81XWH-19-1-0244) (A.P.), Ruth L. Kirschstein Predoctoral Fellowship (HL144057) (N.B.N), California Institute for Regenerative Medicine (CIRM) (RN3-06378) (R.A.) and Eli & Edythe Broad Center of Regenerative Medicine and Stem Cell Research at UCLA Research Award (R.A.).

References

1. Ilic D, Ogilvie C (2017) Concise review: human embryonic stem cells-what have we done? What are we doing? Where are we going? Stem Cells 35(1):17–25. https://doi.org/10.1002/stem.2450
2. Romagnuolo R, Masoudpour H, Porta-Sánchez A et al (2019) Human embryonic stem cell-derived cardiomyocytes regenerate the infarcted pig heart but induce ventricular tachyarrhythmias. Stem Cell Reports 12(5): 967–981. https://doi.org/10.1016/j.stemcr.2019.04.005
3. Kim D, Kim SB, Ryu JL et al (2020) Human embryonic stem cell-derived Wilson's disease model for screening drug efficacy. Cells 9(4): 872. https://doi.org/10.3390/cells9040872
4. Crespo M, Vilar E, Tsai SY et al (2017) Colonic organoids derived from human induced pluripotent stem cells for modeling colorectal cancer and drug testing. Nat Med 23(7):878–884. https://doi.org/10.1038/nm.4355
5. DeLaughter DM, Bick AG, Wakimoto H et al (2016) Single-cell resolution of temporal gene expression during heart development. Dev Cell 39(4):480–490. https://doi.org/10.1016/j.devcel.2016.10.001
6. Kattman SJ, Witty AD, Gagliardi M et al (2011) Stage-specific optimization of activin/nodal and BMP signaling promotes cardiac differentiation of mouse and human pluripotent stem cell lines. Cell Stem Cell 8(2): 228–240. https://doi.org/10.1016/j.stem.2010.12.008
7. Lian X, Zhang J, Azarin SM et al (2013) Directed cardiomyocyte differentiation from human pluripotent stem cells by modulating Wnt/β-catenin signaling under fully defined conditions. Nat Protoc 8(1):162–175. https://doi.org/10.1038/nprot.2012.150
8. Yang L, Soonpaa MH, Adler ED et al (2008) Human cardiovascular progenitor cells develop from a KDR+ embryonic-stem-cell-derived population. Nature 453(7194):524–528. https://doi.org/10.1038/nature06894
9. Liu X, Yagi H, Saeed S et al (2017) The complex genetics of hypoplastic left heart syndrome. Nat Genet 49(7):1152–1159. https://doi.org/10.1038/ng.3870
10. Corrado D, Link MS, Calkins H (2017) Arrhythmogenic right ventricular cardiomyopathy. N Engl J Med 376(15):1489–1490. https://doi.org/10.1056/NEJMc1701400
11. Pezhouman A, Engel JL, Nguyen NB et al (2021) Isolation and characterization of hESC-derived heart field-specific cardiomyocytes unravels new insights into their transcriptional and electrophysiological profiles. Cardiovasc Res. https://doi.org/10.1093/cvr/cvab102
12. CHIR 99021. https://www.tocris.com/products/chir-99021_4423?gclid=CjwKCAjwm_P5BRAhEiwAwRzSO587e4PfNDdgpk4MfnlHGmMHzxLzmr1D8tGbiSPKg-9ATGBFLqVn9BoCvUYQAvD_BwE#ds_technical_data
13. IWP2. https://www.tocris.com/products/iwp-2_3533
14. Y-27632 dihydrochloride. https://www.tocris.com/products/y-27632-dihydrochloride_1254?gclid=CjwKCAjwm_P5BRAhEiwAwRzSO4qVIcMe6q8JJZPJ981Ad_mG07y4RMghUf5GSx5AaXzSRBqKHGHODhoCcmoQAvD_BwE

Part III

Tissue Stem Cells

Chapter 18

Isolation and Characterization of Extracellular Vesicles Derived from Human Umbilical Cord Mesenchymal Stem Cells

Noridzzaida Ridzuan, Darius Widera, and Badrul Hisham Yahaya

Abstract

The safety and efficacy of mesenchymal stem cells/marrow stromal cells (MSC) have been widely studied. Since they are hypoimmunogenic, MSC can escape immune recognition, thus making them an attractive tool in clinical settings beyond autologous cell-based therapy. Paracrine factors including extracellular vesicles (EVs) released by MSC play a significant role in exerting therapeutic effects of MSC. Since their first discovery, MSC-EVs have been widely studied in an attempt to tackle the mechanisms of their therapeutic effects in various disease models. However, currently there are no standard methods to isolate EVs. Here, we describe a differential centrifugation-based protocol for isolation of EVs derived from human umbilical cord MSC (huc-MSC). In addition, the protocol describes methods for characterization of the EVs using transmission electron microscope, Western blot, and nanoparticle tracking analysis.

Key words Extracellular vesicles, Human umbilical cord, Mesenchymal stem cells

1 Introduction

Mesenchymal stem cells have been the subject of interest in cellular therapy since Freidenstein first isolated the cells from the bone marrow of guinea pigs [1–3]. MSC are not only able to self-renew but also to differentiate into various types of cells making them an attractive tool for regenerative medicine [4]. Another intriguing feature of MSC is their ability to escape immune recognition and modulate immune cells in a wide range of diseases such as inflammatory bowel disease, traumatic brain injury, asthma, and chronic obstructive pulmonary disease [5–8]. This immunomodulation is mediated by cell-to-cell contacts and via the release of paracrine factors such as cytokines, growth factors, and extracellular vesicles (EVs) [9]. The paracrine effects of MSC were first observed in MSC cultured under different conditions. MSC cultured in normal growth medium, dexamethasone supplemented medium, and IL-1α supplemented medium were shown to express different

Nagarajan Kannan and Philip Beer (eds.), *Stem Cell Assays: Methods and Protocols*, Methods in Molecular Biology, vol. 2429, https://doi.org/10.1007/978-1-0716-1979-7_18,

cytokines profiles [10]. It was not until 2009 when Chen and colleague discovered EVs enriched with pre-microRNA in conditioned medium from embryonic stem cell-derived MSC [11].

EVs are heterogeneous particles involved in cellular communication that are released into the microenvironment by all cell types. Different terms have been used for EVs, including ectosomes, microparticles, prostasomes, oncosomes, membrane particles, migrasomes, and epididimosomes. To standardize the nomenclature of EVs, the International Society for Extracellular Vesicles (ISEV) suggests using the term "extracellular vesicle" unless the subcellular origin of the vesicle is demonstrated [12]. The most widely studied types of EVs are exosomes and microvesicles. Exosomes are formed from inward budding of endosome that forms multivesicular bodies (MVBs) and are released when MVBs fuse with the cell membrane [13]. Exosomes are heterogeneous in size ranging from 40 to 100 nm in diameter. Released exosomes can be taken up by other cells via endocytosis, fusion, or phagocytosis [14]. Exosomes contain several specific markers such as tetraspanins (CD63, CD9, CD81, and CD82), flotillin, TSG101, and heat shock proteins (HSP60, HSP70, HSPA5, CCT2, and HSP90) [15]. In contrast, microvesicles are formed by outward budding of the cell membrane and subsequently released to the extracellular space. The size of microvesicles ranges from 50 to 1000 nm with integrins, selectins, and CD40 being the main protein markers of MVs [15]. Recently, an increasing number of studies have been conducted to decipher the therapeutic effects of MSC-EVs in various diseases and conditions including, but not limited to, multiple sclerosis, diabetic nephropathy, myocardial infarction, and brain injury [16–19]. These effects have been found to be mediated by various biomolecules including miRNA, protein, lipid, and RNA [20, 21].

Ultracentrifugation remains the most utilized method to isolate EVs as the protocol is simple and can produce high yield of EVs [22]. Characterization of EVs is usually conducted by a combination of morphological observation, protein marker analysis as well as by analysis of the size distribution [23]. Our study has found that huc-MSC-derived EV reduced the inflammatory responses in animal model of chronic obstructive pulmonary disease (COPD) thus EVs could serve as a future cell-free therapy for the treatment of chronic lung diseases [24]. This chapter describes the isolation of huc-MSC-EVs using differential centrifugation followed by their characterization using three independent methods. Morphology of huc-MSC-EVs was conducted using the transmission electron microscope (TEM); meanwhile, protein marker characterization and size distribution were conducted using western blot and nanoparticle tracking analysis (NTA) respectively. The methods applied in this study can also be used to isolate and characterize EVs from other sources.

2 Materials

Reagents

1. Dulbecco's Modified Eagle Medium/F12 (DMEM/F12).
2. Fetal bovine serum (FBS).
3. L-glutamine.
4. Antibiotic antimycotic.
5. Basic fibroblast growth factor.
6. 1.5 M Tris/HCL.
7. Ammonium Persulfate.
8. Sodium Dodecyl Sulfate (SDS).
9. Acrylamide/Bis solution 37.5:1.
10. *N,N,N′,N′*-Tetramethylethylene-1,2-diamine (Temed).
11. Tween-20.
12. Methanol.
13. Rabbit monoclonal antibody β-actin (13E5).
14. Rabbit monoclonal antibody CD63 (EPR5702).
15. Goat polyclonal anti-rabbit IgG FITC.
16. Bovine serum albumin.
17. Phosphate-Buffered Saline.

Equipment

- Biological safety cabinet Class II.
- 15 mL polypropylene conical centrifuge tube.
- 50 mL polypropylene conical centrifuge tube.
- 75cm^2 flask.
- NU-5510/E Air-Jacketed DHD Autoflow CO_2 Incubator.
- 26.3 mL polycarbonate aluminium bottle with cap assembly.
- Allegra X-15R Ultracentrifuge.
- Eppendorf centrifuge 5810R.
- Type 50.2Ti fixed-angle rotor, Optima L-100 K Ultracentrifuge.
- Carbon-coated copper grids.
- Uranyl Acetate.
- Energy filtered transmission electron microscopy.
- Nanosight NS300.
- Polyvinylidine difluoride (PVDF) membrane.
- Mini-PROTEAN Short Plate.

- Mini Trans-Blot Cell.
- Li-Cor Odyssey Fc Dual Mode Imaging System.

2.1 EV-Free FBS Medium

Supplement Dulbecco's Modified Eagle Medium/F12 (DMEM/F12) with 20% fetal bovine serum (FBS) and centrifuge using Type 50.2Ti fixed-angle rotor, Optima L-100 K Ultracentrifuge for 18 h at 100,000 × *g* (*see* **Note 1**). Collect the supernatant from the ultracentrifuge tube and discard the pellet. Prepare 10% FBS-EV free DMEM/F12 by diluting with DMEM/F12 and supplement with 1% antibiotic antimycotic and 1% L-glutamine in EV-free FBS medium.

2.2 Cell Culture Medium

Supplement DMEM/F12 with 10% FBS, 1% antibiotic antimycotic, and 1% L-glutamine.

2.3 SDS Polyacrylamide Gel

2.3.1 12% Separating Gel 10 mL

Mix 2.5 mL of 1.5 M Tris/HCL pH 8.8 with 100 μL of 10% SDS, 100 μL of 10% ammonium persulfate (*see* **Note 2**), 3 mL of 40% acrylamide/Bis solution 37.5:1, 4.3 mL of ddH_2O, and 2.5 μL of *N,N,N′,N′*-Tetramethylethylene-1,2-diamine (Temed) (*see* **Note 3**).

2.3.2 3% Stacking Gel 5 mL

Mix 1.25 mL of 0.5 M Tris/HCL pH 6.8, 50 μL of 10% SDS, 50 μL of 10% Ammonium Persulfate, 375 μL of 40% acrylamide/Bis solution 37.5:1, 3.275 mL of ddH_2O, and 2.5 μL of Temed (*see* **Note 3**).

2.4 Running Buffer (1×)

Add 100 mL of 10× Running Buffer and 10 mL of 10% SDS in 890 mL ddH_2O. Prepare 1 L of 10× Running Buffer 1 L by mixing 30.3 g of Tris/HCL (trizma Base) Mr. 121.14, pH 8.3, and 188 g of Glycine in ddH_2O.

2.5 Transfer Buffer (1×)

Add 50 mL of 20× Transfer buffer, 200 mL of methanol, and 750 mL of ddH_2O. Prepare 1 L of 20× Transfer Buffer by mixing 116.3 g of Tris/HCL (trizma Base) Mr. 121.14, 58.6 g of Glycine, 7.4 g of SDS in ddH_2O.

2.6 Antibodies

Mix rabbit monoclonal antibody β-actin in 5 mL of 1× PBS with 2% BSA (*see* **Note 4**).

Mix rabbit monoclonal antibody CD63 in 5 mL of 1× PBS with 2% BSA.

Mix goat polyclonal anti-rabbit IgG in 5 mL of 1× PBS with 2% BSA.

2.7 Blocking Buffer

Mix 10 mL of 1× PBS and 0.2 g of BSA to prepare 2% BSA. Blocking buffer should be prepared fresh for every use.

2.8 2% Uranyl Acetate

Add 1 g of uranyl acetate in 40 mL of distilled water in a 50 mL amber bottle with cap. Stir the solution slowly until fully dissolve. Top up the solution with distilled water to 50 mL. Keep in 4 °C.

3 Methods

3.1 Culture of huc-MSC

1. Cultivate 4000 cells/cm^2 of huc-MSC in 10 mL cell culture medium with 20 ng/mL basic fibroblast growth factor (bFGF) added. Incubate the flask in a humidified incubator at 37 °C supplied with 5% carbon dioxide.
2. Maintain the cells in 75cm^2 flask. After 48 h, change the medium to EV-free FBS medium to collect huc-MSC-EVs. Add 20 mL of EV-free FBS medium into the flask and incubate in the humidified incubator at 37 °C supplied with 5% carbon dioxide.

3.2 Isolation of huc-MSC-EVs

1. Collect the conditioned medium from the 75cm^2 flask after 72 h and transfer to a 50 mL polypropylene conical centrifuge tube.
2. Centrifuge the conditioned medium at 300 × *g* for 10 min to remove dead cells.
3. Collect the conditioned medium into a new 50 mL polypropylene conical centrifuge tube and centrifuge at 10,000 × *g* for 30 min in an ultracentrifuge to remove apoptotic bodies.
4. Collect the supernatant into a polycarbonate aluminium bottle and centrifuge using ultracentrifuge at 100,000 × *g* for 2 h using Type 50.2Ti fixed-angle rotor to concentrate huc-MSC-EVs.
5. Discard the conditioned medium. Add 23 mL of 1× PBS in the polycarbonate aluminium bottle containing huc-MSC-EVs and centrifuge again using an ultracentrifuge at 100,000 × *g* for 2 h using Type 50.2Ti fixed-angle rotor to wash the EVs.
6. Discard the supernatant and add 150 μL of 1× PBS in the tube. Gently resuspend the huc-MSC-EVs and collect the huc-MSC-EVs in 0.5 mL microcentrifuge tube.

3.3 Transmission Electron Microscope Analysis

1. Load the freshly isolated huc-MSC-EVs in 1× PBS (*see* **Note 5**) onto the carbon-coated copper grids and incubate for 10 min.
2. Blot the grid with filter paper to remove the excess liquid. Stain the EVs with 2% uranyl acetate for 1 min.
3. Remove the excessive uranyl acetate and let the grid dry for 15 min before viewing using an appropriate electron microscope (*see* Fig. 1 for example image [24]).

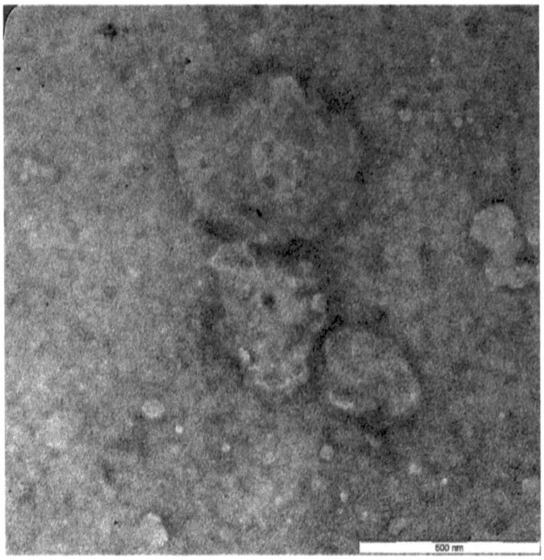

Fig. 1 Morphological observation of huc-MSC-EVs using TEM

3.4 Nanoparticle Tracking Analysis

1. Dilute the huc-MSC-EVs with 1× PBS to the concentration of 20 to 60 particles per field (*see* **Note 6**).
2. Load the sample using a 1 mL syringe into the nanoparticle tracking analysis. Be careful not to form bubble as bubbles will interfere with the reading (*see* Fig. 2 and Table 1 for example readouts).
3. Record five readings for each huc-MSC-EV sample.

3.5 Western Blot

1. The gel should be prepared a day before running the western blot.
2. Add 7.5 mL to the casting system and gently layer 250 μL of water-saturated isobutanol (*see* **Note 7**).
3. When the gel has set, drain the water-saturated isobutanol and gently dry with filter paper.
4. Layer with 3% Stacking gel to the top and add in the comb. Be careful not to form bubbles.
5. Keep at 4 °C overnight.

3.6 Gel Electrophoresis

1. Mix sample with 8 μL of loading dye and denature the sample at 100 °C for 3 min.
2. Cool the sample and centrifuge at 27,548 × *g* for 1 min.
3. Load the sample in the gel and fill the tank with 1× running buffer and run at 20 mA for 3 h.

3.7 Transfer Protocol

1. Cut the PVDF membrane to the size of 9 × 7.5 cm.
2. Soak it in methanol for 15 s and ddH_2O for 2 min.

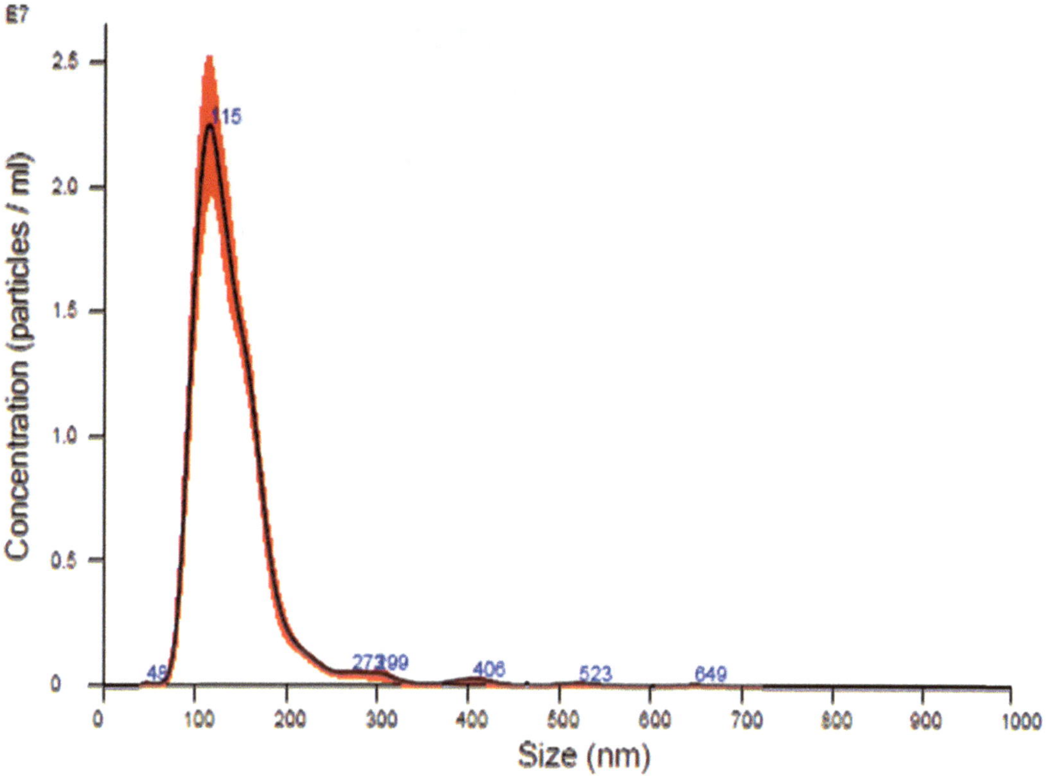

Fig. 2 Size distribution analysis of hUCMSC-EV. Particle distribution by Nanosight NS300 reported an average MSC-EV diameter of 153 nm. The graph represents 3 independent experiments

Table 1
Analysis of hUCMSC-EV size distribution

Sample	Mean (nm)	Mode (nm)	SD (nm)	Range (nm)
1	141.2	115.0	51.4	36–737
2	156.5	116.9	68.4	64–795
3	163.0	123.1	68.3	25–740

3. Equilibrate in 1× transfer buffer for 10 min.
4. All sponges, filter paper must be soaked in 1× transfer buffer.
5. Layer the sandwich on the cassette and place the cassette in the mini-trans blot cell.
6. Fill the cooling unit with dry-ice (*see* **Note 8**), fill with transfer buffer, and a stirring bar and place on stirring unit.
7. Run at constant current, 1 amp per cell, for 1 h.
8. Layer the PVDF membrane on the tissue and dry it at 37 °C for 10 min to make sure the protein sticks to the membrane properly.

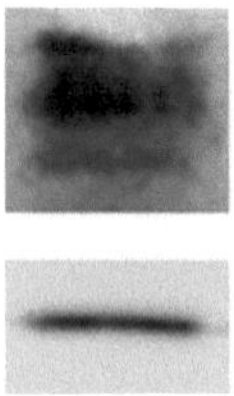

Fig. 3 Expression of protein marker using western blot. β-actin and CD63 were assessed on UCMSC-EV

9. Rehydrate the membrane in methanol for 15 s, and ddH_2O for 5 s. Transfer the membrane to 1× PBS for 1 min.
10. Add blocking buffer and incubate for 1 h at room temperature on a rocker.
11. Add primary antibody and incubate overnight at 4 °C on a rocker.
12. Wash the membrane 6 times with 1× PBST (*see* **Note 9**) 5 min each washing.
13. Incubate with secondary antibody for 1 h at room temperature on a rocker.
14. Wash the membrane 6 times with 1× PBST 5 min each washing.
15. Ready to view with suitable imaging system (*see* Fig. 3 for example blot [24]).

4 Notes

1. Centrifugation of medium containing EVs for 18 h will remove 95% of the EVs from the medium.
2. 10% ammonium persulfate should be prepared fresh in dH_2O.
3. Add Temed prior to loading the gel in the gel cast to avoid solidification of the mixture before gel loading.
4. Dilute the antibody as per manufacturer's suggestion.
5. Use freshly prepared 1× PBS to avoid precipitation that will interfere with the viewing.
6. The dilution can be done gradually. When too many particles are seen in the field, stop the machine, and dilute the sample until the desired number of particles in the field are achieved.
7. Mix equal parts of isobutanol and dh_2O and keep in room temperature for further use.
8. Dry-ice is used to cool off the mini trans-blot cell as running at 1 amp will produce heat.
9. Add 1 mL of tween-20 in 999 mL of 1× PBS. Mix well and keep in 4 °C.

Acknowledgements

This study was supported by the Universiti Sains Malaysia Research University Grant Scheme (1001/CIPPT/8012203). The authors also acknowledged the CryoCord Sdn Bhd (Malaysia) for providing the human umbilical cord-derived mesenchymal stem cells to be used in our study. The authors declare that there is no potential conflict of interest.

References

1. Friedenstein A, Chailakhjan R, Lalykina K (1970) The development of fibroblast colonies in monolayer cultures of guinea-pig bone marrow and spleen cells. Cell Prolif 3(4):393–403
2. Zeng SL, Wang LH, Li P, Wang W, Yang J (2015) Mesenchymal stem cells abrogate experimental asthma by altering dendritic cell function. Mol Med Rep 12(2):2511–2520
3. Wang N, Chen C, Yang D, Liao Q, Luo H, Wang X, Zhou F, Yang X, Yang J, Zeng C (2017) Mesenchymal stem cells-derived extracellular vesicles, via miR-210, improve infarcted cardiac function by promotion of angiogenesis. Biochim Biophys Acta Mol Basis Dis 1863(8):2085–2092
4. Samsonraj RM, Raghunath M, Nurcombe V, Hui JH, van Wijnen AJ, Cool SM (2017) Concise review: multifaceted characterization of human mesenchymal stem cells for use in regenerative medicine. Stem Cells Transl Med 6(12):173–2185
5. Mao F, Tu Q, Wang L, Chu F, Li X, Li HS, Xu W (2017) Mesenchymal stem cells and their therapeutic applications in inflammatory bowel disease. Oncotarget 8(23):38008
6. Peruzzaro S, Andrews MM, Al-Gharaibeh A, Pupiec O, Resk M, Story D, Maiti P, Rossignol J, Dunbar G (2019) Transplantation of mesenchymal stem cells genetically engineered to overexpress interleukin-10 promotes alternative inflammatory response in rat model of traumatic brain injury. J Neuroinflammation 16(1):1–15
7. Dai R, Liu J, Cai S, Zheng C, Zhou X (2017) Delivery of adipose-derived mesenchymal stem cells attenuates airway responsiveness and inflammation in a mouse model of ovalbumin-induced asthma. Am J Transl Res 9(5):2421
8. H-m L, Y-t L, Zhang J, Ma L-j (2017) Bone marrow mesenchymal stem cells ameliorate lung injury through anti-inflammatory and antibacterial effect in COPD mice. J Huazhong Univ Sci Technolog Med Sci 37(4):496–504
9. Bruno S, Kholia S, Deregibus MC, Camussi G (2019) The role of extracellular vesicles as paracrine effectors in stem cell-based therapies. Adv Exp Med Biol 1201:175–193
10. Haynesworth SE, Baber MA, Caplan AI (1996) Cytokine expression by human marrow-derived mesenchymal progenitor cells in vitro: effects of dexamethasone and IL-1α. J Cell Physiol 166(3):585–592
11. Chen TS, Lai RC, Lee MM, Choo ABH, Lee CN, Lim SK (2009) Mesenchymal stem cell secretes microparticles enriched in pre-microRNAs. Nucleic Acids Res 38 (1):215–224
12. Théry C, Witwer KW, Aikawa E, Alcaraz MJ, Anderson JD, Andriantsitohaina R, Antoniou A, Arab T, Archer F, Atkin-Smith GK (2018) Minimal information for studies of extracellular vesicles 2018 (MISEV2018): a position statement of the International Society for Extracellular Vesicles and update of the MISEV2014 guidelines. J Extracell Vesicles 7 (1):1535750
13. Sarko DK, McKinney CE (2017) Exosomes: origins and therapeutic potential for neurodegenerative disease. Front Neurosci 11:82
14. Rani S, Ryan AE, Griffin MD, Ritter T (2015) Mesenchymal stem cell-derived extracellular vesicles: toward cell-free therapeutic applications. Mol Ther 23(5):812–823
15. Konoshenko MY, Lekchnov EA, Vlassov AV, Laktionov PP (2018) Isolation of extracellular vesicles: general methodologies and latest trends. Biomed Res Int 2018:8545347
16. Clark K, Zhang S, Barthe S, Kumar P, Pivetti C, Kreutzberg N, Reed C, Wang Y, Paxton Z, Farmer D (2019) Placental mesenchymal stem cell-derived extracellular vesicles promote myelin regeneration in an animal model of multiple sclerosis. Cell 8(12):1497
17. Grange C, Tritta S, Tapparo M, Cedrino M, Tetta C, Camussi G, Brizzi MF (2019) Stem cell-derived extracellular vesicles inhibit and

revert fibrosis progression in a mouse model of diabetic nephropathy. Sci Rep 9(1):1–13
18. Yang L, Zhu J, Zhang C, Wang J, Yue F, Jia X, Liu H (2019) Stem cell-derived extracellular vesicles for myocardial infarction: a meta-analysis of controlled animal studies. Aging (Albany NY) 11(4):1129
19. Dabrowska S, Andrzejewska A, Strzemecki D, Muraca M, Janowski M, Lukomska B (2019) Human bone marrow mesenchymal stem cell-derived extracellular vesicles attenuate neuroinflammation evoked by focal brain injury in rats. J Neuroinflammation 16(1):216
20. Wang LT, Ting CH, Yen ML, Liu KJ, Sytwu HK, Wu KK, Yen BL (2016) Human mesenchymal stem cells (MSCs) for treatment towards immune- and inflammation-mediated diseases: review of current clinical trials. J Biomed Sci 23(1):76
21. Penfornis P, Vallabhaneni KC, Whitt J, Pochampally R (2016) Extracellular vesicles as carriers of microRNA, proteins and lipids in tumor microenvironment. Int J Cancer 138(1):14–21
22. Paganini C, Capasso Palmiero U, Pocsfalvi G, Touzet N, Bongiovanni A, Arosio P (2019) Scalable production and isolation of extracellular vesicles: available sources and lessons from current industrial bioprocesses. Biotechnol J 14(10):1800528
23. Borosch S, Dahmen E, Beckers C, Stoppe C, Buhl EM, Denecke B, Goetzenich A, Kraemer S (2017) Characterization of extracellular vesicles derived from cardiac cells in an in vitro model of preconditioning. J Extracell Vesicles 6(1):1390391
24. Ridzuan N, Zakaria N, Widera D, Sheard J, Morimoto M, Kiyokawa H, Isa SA, Singh GK, Then KY, Ooi GC, Yahaya BH (2021) Human umbilical cord mesenchymal stem cell-derived extracellular vesicles ameliorate airway inflammation in a rat model of chronic obstructive pulmonary disease (COPD). Stem Cell Res Ther 12(1):1–21

Chapter 19

Identification and Validation of CRISPR/Cas9 Off-Target Activity in Hematopoietic Stem and Progenitor Cells

So Hyun Park, Ciaran M. Lee, and Gang Bao

Abstract

Targeted genome editing in hematopoietic stem and progenitor cells (HSPCs) using CRISPR/Cas9 can potentially provide a permanent cure for hematologic diseases. However, the utility of CRISPR/Cas9 systems for therapeutic genome editing can be compromised by their off-target effects. In this chapter, we outline the procedures for CRISPR/Cas9 off-target identification and validation in HSPCs. This method is broadly applicable to diverse CRISPR/Cas9 systems and cell types. Using this protocol, researchers can perform computational prediction and experimental identification of potential off-target sites followed by off-target activity quantification by next-generation sequencing.

Key words Genome editing, CRISPR/Cas9, Off-target, Next-generation sequencing (NGS), Hematopoietic stem and progenitor cells (HSPCs), GUIDE-seq

1 Introduction

Targeted genome editing using CRISPR/Cas9 can potentially provide a permanent cure for genetic disorders through targeted deletion, precise sequence replacement, or site specific insertion of exogenous DNA. The hematologic diseases such as sickle cell disease, β-thalassemia, and primary immunodeficiencies are ideal candidates for ex vivo hematopoietic stem cell (HSC) gene therapy followed by autologous transplantation. However, the utility of CRISPR-Cas9 systems for therapeutic genome editing may be compromised by their off-target toxicity due to the tolerance for nucleotide mismatches between the genome and the targeting region of the guide RNA (gRNA) [1–3]. The off-target activity of Cas9 nuclease can disrupt normal gene function and induce genome instability via large chromosomal rearrangements between two simultaneous DNA breaks [4], which is of serious concern in human gene therapies, potentially leading to difficult-to-predict side effects. As CRISPR/Cas9 moves towards the clinic, it is reasonable to assume that CRISPR/Cas9 treatments will have some

Nagarajan Kannan and Philip Beer (eds.), *Stem Cell Assays: Methods and Protocols*, Methods in Molecular Biology, vol. 2429, https://doi.org/10.1007/978-1-0716-1979-7_19,

degree of off-target edits that will require careful monitoring over time to ensure that off-target edits in hematopoietic stem cells (HSCs) do not have a proliferative effect in vivo. Therefore, potential off-target effects need to be carefully analyzed, and significant challenges exist with both accurately predicting potential off-target sites and performing genome-wide unbiased searches. Herein, we provide a detailed protocol for the CRISPR/Cas9 off-target identification and validation in $CD34^+$ hematopoietic stem and progenitor cells (HSPCs). Using this protocol, researchers can perform computational off-target site searches using a web-based tool *C*RISPR *O*ff-target *S*ites with *M*ismatches, *I*nsertions, and *D*eletions (COSMID) [5], and experimental identification of potential off-target sites using *G*enome-wide, *U*nbiased *I*dentification of *D*SBs *E*nabled by sequencing (GUIDE-Seq) [6]. Identified off-target sites can be validated by PCR amplification, followed by next-generation sequencing with a detection limit of 0.1%. This method allows for multiple sites to be assessed simultaneously with a high degree of sensitivity. Although the CRISPR/Cas9 system derived from *Streptococcus pyogenes* (SpCas9) is used in the protocol here, our method can be readily adopted for off-target analysis of other CRISPR/Cas systems.

2 Materials

2.1 Expansion and Culture of HSPCs

Reconstitute cytokines in sterile 1× PBS containing 0.1% BSA. Distribute to aliquots and store at −20 °C to −80 °C (for 3–12 months).

1. Ficoll-Hypaque.
2. CD34 Microbeads.
3. Serum-free medium (or other growth medium) for expansion of hematopoietic cells.
4. Stem cell factor (SCF).
5. Thrombopoietin (TPO).
6. Flt3 ligand.
7. Interleukin 3 (IL3).
8. $CD34^+$ HSPC expansion medium supplemented with 300 ng/mL SCF, 100 ng/mL TPO, 300 ng/mL Flt3 ligand, 60 ng/mL IL3, and 1× Penicillin-Streptomycin. Store at 4 °C.
9. Fresh or frozen $CD34^+$ HSPCs.
10. 12-well plate.
11. Laminar flow hood with UV light source.
12. Cell culture CO_2 incubator.

13. 15 mL tubes.
14. Centrifuge for 15 mL tubes.

2.2 HSPC Genome-Editing

1. SpCas9 protein, 10 μg/μL stock.
2. Chemically modified CRISPR sgRNAs. Rehydrate in nuclease-free 1× TE. 2.5 μg/μL in stock. Distribute to aliquots and store at −80 °C.
3. Sterile PCR strip tubes (0.2 mL).
4. 1× phosphate-buffered saline (PBS), pH 7.4, without calcium.
5. 15 mL tubes.
6. Centrifuge for 15 mL tubes.
7. 24-well plate.
8. P3 Primary Nucleofection Kit.
9. Lonza Nucleofector 4-D.

2.3 Genomic DNA Extraction and Quantification

1. Genomic DNA extraction kit.
2. 1× phosphate-buffered saline (PBS), pH 7.4, without calcium.
3. 56 °C heat block.
4. Molecular biology grade ethanol (200 Proof).
5. DNase-free 1.5 mL microcentrifuge tubes.
6. Table-top microcentrifuge.
7. Fluorometric DNA quantification assay kit.
8. Fluorometer for DNA quantification.

2.4 GUIDE-seq [6]

1. Oligonucleotides for dsDNA tag. Rehydrate in nuclease-free 1×TE to make a 200 μM stock. Store at −20 °C.

   Sense oligonucleotide:

   5′- /Phos/G*T* TTAATTGAGTTGTCATATGTTAA TAACGGT*A*T -3′

   Anti-sense oligonucleotide:

   5′-/Phos/A*T* ACCGTTATTAACATATGACAACTCAAT TAA*A*C -3′

   * Indicates a Phosphorothioate Bond Modification.

2. Thermocycler.
3. SpCas9 protein.
4. Chemically modified CRISPR dsODNs. Rehydrate in nuclease-free 1×TE and store at −80 °C.
5. P3 Nucleofection kit (Lonza). Store at 4 °C.
6. Lonza Nucleofector 4-D.
7. $CD34^+$ HSPCs.

8. $CD34^+$ expansion medium.
9. 24-well plate.
10. Sterile PCR strip tubes (0.2 mL).
11. PCR cleanup magnetic beads. Light sensitive, do not freeze, store at 4 °C.
12. Magnetic 96-well plate.
13. 80% Ethanol.
14. Balance.
15. Agarose.
16. 1× TAE buffer: Dilute 1:50 from a 50× TAE buffer stock with DI water.
17. Microwave.
18. Agarose electrophoresis apparatus.
19. Nucleic acid gel stain.
20. Gel loading dye.
21. DNA ladder.
22. UV gel box with gel imager.
23. Covaris S220 microTube.
24. M220 Covaris.
25. Illumina Y-adapters (*see* Table 1 for oligonucleotide sequences).
26. T4 DNA ligase. Store at −20 °C.
27. dNTP mix. Store at −20 °C.
28. Ligation buffer, 10×. Store at −20 °C.
29. End-repair mix (low).
30. 10× Taq buffer, Mg^{2+}-free.
31. Taq polymerase (non-hot start).
32. 10× Taq buffer, Mg2+-free.
33. 25 mM $MgCl^{2+}$.
34. High fidelity Taq polymerase.
35. GSP oligonucleotide primers. Rehydrate in 1× TE to make a 100 μM stock and dilute to a 10 μM working stock. Store at −20 °C.

    GSP1+: GGATCTCGACGCTCTCCCTATACCGTTATTAA CATATGACA.

    GSP1-: GGATCTCGACGCTCTCCCTGTTTAATTGAGTT GTCATATGTTAATAAC.

    GSP2+: CCTCTCTATGGGCAGTCGGTGATACATATGAC AACTCAATTAAAC.

Table 1
Y-adapter sequences [6]

Illumina Y-adapters	Sequence (5′ → 3′)
Miseq common adapter	[Phos]GATCGGAAGAGC*C*A
A01	AATGATACGGCGACCACCGAGATCTACACTAGATCGCNNWNNWNNACACTCTTTCCCTACACGACGCTCTTCCGATC*T
A02	AATGATACGGCGACCACCGAGATCTACACCTCTCTATNNWNNWNNACACTCTTTCCCTACACGACGCTCTTCCGATC*T
A03	AATGATACGGCGACCACCGAGATCTACACTATCCTCTNNWNNWNNACACTCTTTCCCTACACGACGCTCTTCCGATC*T
A04	AATGATACGGCGACCACCGAGATCTACACAGAGTAGANNWNNWNNACACTCTTTCCCTACACGACGCTCTTCCGATC*T
A05	AATGATACGGCGACCACCGAGATCTACACGTAAGGAGNNWNNWNNACACTCTTTCCCTACACGACGCTCTTCCGATC*T
A06	AATGATACGGCGACCACCGAGATCTACACACTGCATANNWNNWNNACACTCTTTCCCTACACGACGCTCTTCCGATC*T
A07	AATGATACGGCGACCACCGAGATCTACACAAGGAGTANNWNNWNNACACTCTTTCCCTACACGACGCTCTTCCGATC*T
A08	AATGATACGGCGACCACCGAGATCTACACCTAAGCCTNNWNNWNNACACTCTTTCCCTACACGACGCTCTTCCGATC*T
A09	AATGATACGGCGACCACCGAGATCTACACGACATTGTNNWNNWNNACACTCTTTCCCTACACGACGCTCTTCCGATC*T
A10	AATGATACGGCGACCACCGAGATCTACACACTGATGGNNWNNWNNACACTCTTTCCCTACACGACGCTCTTCCGATC*T
A11	AATGATACGGCGACCACCGAGATCTACACGTACCTAGNNWNNWNNACACTCTTTCCCTACACGACGCTCTTCCGATC*T
A12	AATGATACGGCGACCACCGAGATCTACACCAGAGCTANNWNNWNNACACTCTTTCCCTACACGACGCTCTTCCGATC*T
A13	AATGATACGGCGACCACCGAGATCTACACCATAGTGANNWNNWNNACACTCTTTCCCTACACGACGCTCTTCCGATC*T
A14	AATGATACGGCGACCACCGAGATCTACACTACCTAGTNNWNNWNNACACTCTTTCCCTACACGACGCTCTTCCGATC*T
A15	AATGATACGGCGACCACCGAGATCTACACCGCGATATNNWNNWNNACACTCTTTCCCTACACGACGCTCTTCCGATC*T
A16	AATGATACGGCGACCACCGAGATCTACACTGGATTGTNNWNNWNNACACTCTTTCCCTACACGACGCTCTTCCGATC*T

Rehydrate in 1×TE to make a 100 μM stock and store at −20 °C

Table 2
P7 primer sequences [6]

P7 adapters	Sequence (5′ → 3′)
P701	CAAGCAGAAGACGGCATACGAGATTCGCCTTAGTGACTGGAGTCCTCTCTATGGGCAGTCGGTGA
P702	CAAGCAGAAGACGGCATACGAGATCTAGTACGGTGACTGGAGTCCTCTCTATGGGCAGTCGGTGA
P703	CAAGCAGAAGACGGCATACGAGATTTCTGCCTGTGACTGGAGTCCTCTCTATGGGCAGTCGGTGA
P704	CAAGCAGAAGACGGCATACGAGATGCTCAGGAGTGACTGGAGTCCTCTCTATGGGCAGTCGGTGA
P705	CAAGCAGAAGACGGCATACGAGATAGGAGTCCGTGACTGGAGTCCTCTCTATGGGCAGTCGGTGA
P706	CAAGCAGAAGACGGCATACGAGATCATGCCTAGTGACTGGAGTCCTCTCTATGGGCAGTCGGTGA
P707	CAAGCAGAAGACGGCATACGAGATGTAGAGAGGTGACTGGAGTCCTCTCTATGGGCAGTCGGTGA
P708	CAAGCAGAAGACGGCATACGAGATCCTCTCTGGTGACTGGAGTCCTCTCTATGGGCAGTCGGTGA

Rehydrate in 1×TE to make a 100 μM stock and dilute to a 10 μM working stock. Store at −20 °C

GSP2-: CCTCTCTATGGGCAGTCGGTGATTTGAGTTGTCATATGTTAATAACGGTA.

36. P5 oligonucleotide primers. Rehydrate in 1× TE to make a 100 μM stock and dilute to a 10 μM working stock. Store at −20 °C.

    P5_1: AATGATACGGCGACCACCGAGATCTA.

    P5_2: AATGATACGGCGACCACCGAGATCTACAC.

37. P7 oligonucleotide primers (*see* Table 2 for primer sequences).
38. Tetramethylammonium chloride (TMAC).
39. Fluorometric DNA quantification assay kit.
40. Fluorometer for DNA quantification.
41. NGS library quantification kits.
42. MiSeq Reagent Kit v2 (300-cycles) (Illumina).
43. miSeq (Illumina).
44. GUIDE-seq read and index primers. Rehydrate in 1×TE to make a 100 μM stock and store at −20 °C.

    GUIDE-seq read2 primer: GTGACTGGAGTCCTCTCTATGGGCAGTCGGTGAT.

    GUIDE-seq index1 primer: ATCACCGACTGCCCATAGAGAGGACTCCAGTCAC.

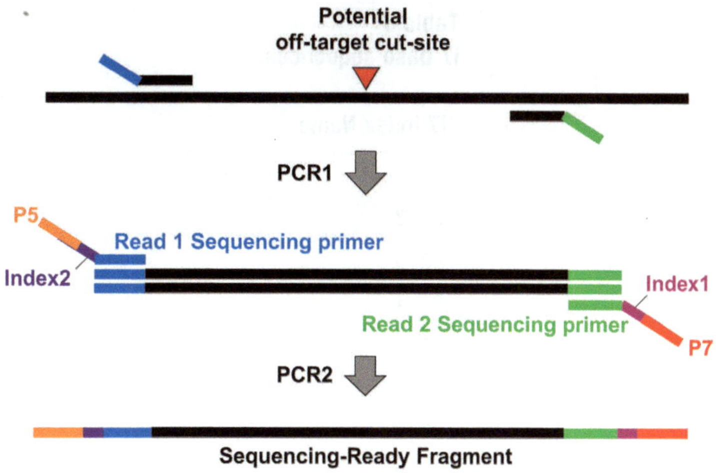

Fig. 1 Schematics of off-target library preparation for NGS. DNA flanking the gRNA off-target sites was amplified using locus-specific primers followed by a second PCR to introduce Illumina sequencing adaptors and sample barcodes

2.5 Off-Target NGS Library Preparation and Sequencing (See Fig. 1)

1. Thermocycler.
2. 5× OneTaq standard buffer. Store at −20 °C.
3. OneTaq Polymerase. Distribute to aliquots and store at −20 °C.
4. Oligonucleotide primers. Rehydrate in 1× TE to make a 100 μM stock and dilute to a 10 μM working stock. Store at −20 °C.

   PCR1:

   Forward primer: 5′- TCTACAGTCCGACGATCA $(N)_n$ - 3′ where $(N)_n$ denotes target specific primer sequence.

   Reverse primer: 5′ -GACGTGTGCTCTTCCGATC(N)n - 3′ where $(N)_n$ denotes target specific primer sequence.

   PCR2: (*see* Table 3 for i7 base sequences and *See* Table 4 for i5 bases sequences)

   Custom Index 1 (i7) adapters

   CAAGCAGAAGACGGCATACGAGA T[i7] ATGT GACTGGAGTTCAGACGTGTGCTCTTCCGATC.

   Custom Index 2 (i5) adapters

   AATGATACGGCGACCACCGAGATCTACAC [i5] TGTTCAGAGTTCTACAGTCCGACGATCA.
5. Magnetic 96-well plate.
6. Magnetic beads. Store at 4 °C.
7. 96-well plate.
8. Reagent reservoirs (for multichannel pipette).

Table 3
i7 base sequences

i7 Index Name	Bases in adapter
P7_1	CACTCACG
P7_2	ACACGATC
P7_3	TATCTGAC
P7_4	CACGTCGT
P7_5	TAGCGACG
P7_6	AGCTCTAG
P7_7	ACTAGAGC
P7_8	CGTACGCA
P7_9	CTACACTA
P7_10	TGCTGCTT
P7_11	TCACGCGT
P7_12	GTAGATCG
P7_13	GTGACGCA
P7_14	CATACTAG
P7_15	AGTGTAGA
P7_16	CGAGAGTT
P7_17	GACATAGT
P7_18	ACGCTACT
P7_19	ACTCACTG
P7_20	TGAGTACG
P7_21	CTGCGTAG
P7_22	TAGTCTCC
P7_23	ACTACGAC
P7_24	GTCTGCTA
P7_25	GTCTATGA
P7_26	CTCGACTT
P7_27	CGAAGTAT
P7_28	TAGCAGCT
P7_29	TCTCTATG
P7_30	GATCTACG
P7_31	GTAACGAG
P7_32	ATAGTACC
P7_33	GCGTATAC

(continued)

Table 3
(continued)

i7 Index Name	Bases in adapter
P7_34	TGCTCGTA
P7_35	AACGCTGA
P7_36	CGTAGCGA
P7_37	ATAGCGCT
P7_38	TCTAGACT
P7_39	TCCTCATG
P7_40	CGAGCTAG
P7_41	CTCTAGAG
P7_42	ATGAGCTC
P7_43	AGCATACC
P7_44	CGTCATAC
P7_45	TCAGTCTA
P7_46	CATCGTGA
P7_47	GAGCTCGA
P7_48	TACTAGGT
P7_49	ACGTACGT
P7_50	CGCGATAT
P7_51	CTATCGTG
P7_52	GCGATACG
P7_53	AGTCGCAG
P7_54	GTTACAGC
P7_55	TAACGTCC
P7_56	CTACGACC
P7_57	GAGACTTA
P7_58	ACTGTGTA
P7_59	TGCGTCAA

9. 80% Ethanol.
10. Balance.
11. Agarose.
12. 1× TAE buffer: Dilute 1:50 from a 50× TAE buffer stock with DI water.
13. Microwave.
14. Agarose electrophoresis apparatus.

Table 4
i5 base sequences

i5 index name	Bases in adapter
P5_0	GTTCAGAG
P5_1	CTGATCGT
P5_2	ACTCTCGA
P5_3	TGAGCTAG
P5_4	GAGACGAT
P5_5	CTTGTCGA
P5_6	TTCCAAGG
P5_7	CGCATGAT
P5_8	ACGGAACA
P5_9	CGGCTAAT
P5_10	ATCGATCG

15. Nucleic acid gel stain.
16. Gel loading dye.
17. DNA ladder.
18. UV gel box with gel imager.
19. Nanodrop.
20. Fluorometric DNA quantification assay kit.
21. Fluorometer for DNA quantification.
22. MiSeq Reagent Kit v2 (500-cycles) (Illumina).
23. MiSeq Instrument (Illumina).
24. Read and index primers. Rehydrate in 1×TE to make a 100 μM stock. Store at −20 °C.

    Custom read2 primer: 5′ - TGTGACTGGAGTTCA GACGTGTGCTCTTCCGATC - 3′.

    Custom index primer: 5′ - GATCGGAAGAGCACACGTCT GAACTCCAGTCACAT - 3′.

3 Methods

3.1 COSMID In Silico off-Target Site Prediction (See Fig. 2 [7])

1. Select the "Target Genome" for which off-target analysis should be performed (*see* **Note 1**).
2. Enter the gRNA sequence in the "Query Sequence" box.
3. Enter the PAM sequence of the Cas9 ortholog (NGG for SpCas9) and select the number of allowed indels and mismatches.

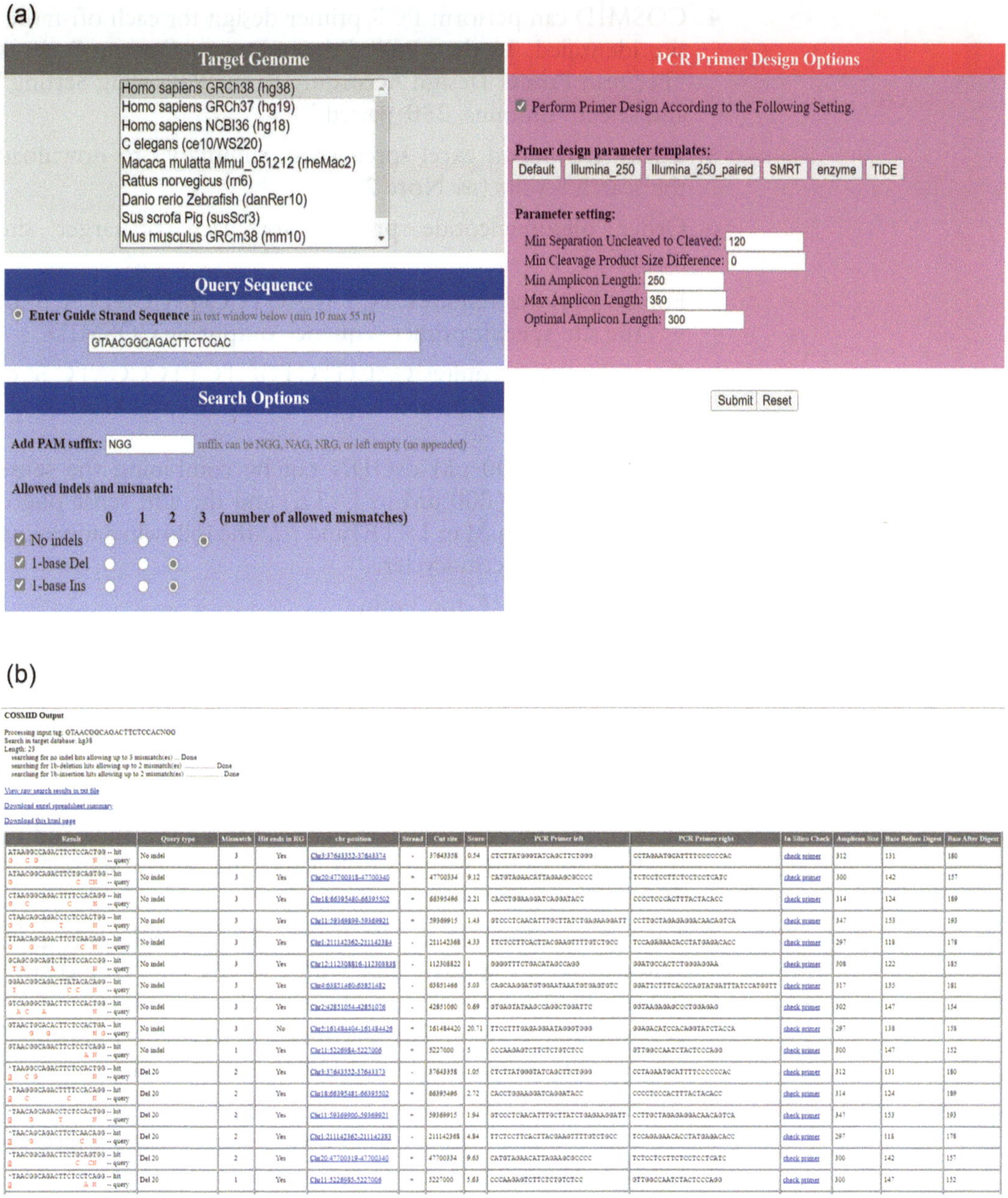

Fig. 2 COSMID in silico off-target screening. COSMID was performed for R-66 SCD gRNA that targets the *HBB* sequence with SCD mutation [7]. (**a**) Bioinformatic prediction of potential off-target sites for the gRNAs was carried out using the COSMID with up to 3 mismatches or with up to 2 mismatches and an insertion or deletion allowed in the 19 PAM proximal bases. The *Homo sapiens* genome assembly GRCh38/hg38 genome build was used as a reference. (**b**) Typical COSMID output. Only part of the output is shown

4. COSMID can perform PCR primer design for each off-target site identified. In the "PCR Primer Design Options," check "Perform Primer Design According to the Following Setting" and select "Illumina_250_paired."
5. Click "Download excel spreadsheet summary" to download the search results (*see* **Note 2**).
6. Order oligonucleotide primers for each off-target site identified.

   Forward primer requires TCTACAGTCCGACGATCA 5′ of the site-specific primer sequence output by COSMID.

   Reverse primer requires GACGTGTGCTCTTCCGATC 5′ of the site-specific primer sequence output by COSMID.

3.2 Experimental Identification of Off-Target Sites Using GUIDE-seq [6]

3.2.1 Genome Editing in CD34⁺ HSPCs for Integration of dsODN Tag at Cas9 Cut Sites (See Note 3 for GUIDE-seq in Other Cell Types)

1. Generate the 100 μM dsODN tag by combining the sense oligonucleotide (200 μM in 1×TE) and the anti-sense oligonucleotide (200 μM in 1×TE) and run the following annealing program on a thermocycler:
   (a) Heat to 95 °C for 2 min.
   (b) Ramp cool to 25 °C over 45 min.
   (c) Hold at 12 °C.
2. Aliquot and store annealed dsODN tag at −20 °C. Avoid freeze thaw cycles.
3. Harvest $CD34^+$ HSPCs and prepare 5×10^5 cells for each electroporation reaction. Transfer cells to 15 mL tube and fill the tube to 15 mL with 1×PBS. Pellet the cells by spinning at 300 × *g* for 5 min.
4. Meanwhile, assemble the RNP complex. Carefully mix 1 μL sgRNA, 0.5 μL SpCas9, and 1 μL dsODN tag (20 μM) in a 0.2 mL PCR tube per electroporation. Incubate at room temperature for 15 min. For the mock-treated control sample, prepare the same amount dsODN tag as treated samples without RNP.
5. Aspirate as much medium and PBS as possible without disturbing the cell pellet.
6. Add 22 μL P3 solution per 5×10^5 $CD34^+$ HSPCs for each electroporation and gently resuspend cells.
7. Transfer 20 μL of cells in P3 solution mixture to the 0.2 mL tube containing the RNP and dsODN tag.
8. Mix well by gentle flick. Transfer 20 μL of the mixture in one pipet into the center groove of Lonza 4D electroporation cuvette. Avoid bubbles.
9. Immediately perform electroporation using the CA137 program. Incubate cells for 10 min at room temperature.

10. Add 80 μL pre-warmed CD34⁺ HSPCs expansion medium into each cuvette. Transfer all cells from each cuvette to a 24-well plate containing 500 μL pre-warmed culture medium per well.
11. 48 h after electroporation, add 500 μL fresh pre-warmed medium to the wells.
12. After 4 days, harvest cells and extract genomic DNA.

3.2.2 Confirmation of dsODN Integration at the Target Site

1. Design primers that yield amplicon length between 400 and 500 bp.
2. Set up PCR using the genomic DNA of cells treated with RNP and from mock-treated cells for use as controls.
3. Send purified PCR products for Sanger sequencing using the forward or reverse primer used for PCR amplification.
4. Go to Inference of CRISPR Editing (ICE) [8] web-based tool that determines rates of CRISPR/Cas9 knockout and knockin editing: https://ice.synthego.com/#/
5. Enter the Sample Label, 17–23 nt guide RNA sequences without PAM and the Donor Sequence. The donor sequences should contain the 34-bp dsODN Tag sequence inserted at the cut site in forward or reverse orientation.
6. Upload a chromatogram file for both a Control File (mock-treated cells) and an Experiment File (RNP and dsODN tag treated cells).
7. Confirm targeted integration of dsODN Tag in forward or reverse orientation.
8. *See* Fig. 3 for typical ICE output.

3.2.3 Quantification and Shearing of Genomic DNA (See ***Note 4****)*

1. Determine the concentration of extracted genomic DNA by fluorometric DNA quantification assay.
2. For each genomic DNA sample, prepare two tubes (antisense/ sense or +/−) of 1 μg DNA in a total volume of 140 μL 1× TE buffer.
3. Place 130 μL of each tube in a Covaris S220 microTube and shear for 25 s on a M220 Covaris following manufacturer's instructions. Run settings should be:
   (a) Average incident power = 10 W.
   (b) Peak incident power = 50 W.
   (c) Duty factor = 20%.
   (d) Cycles/burst = 200 counts.
4. Move samples from microtube to a new plastic tube.
5. Run 5 μL of sheared DNA on a 1.5% agarose gel and verify that the bulk of the shearing is 500 bps.

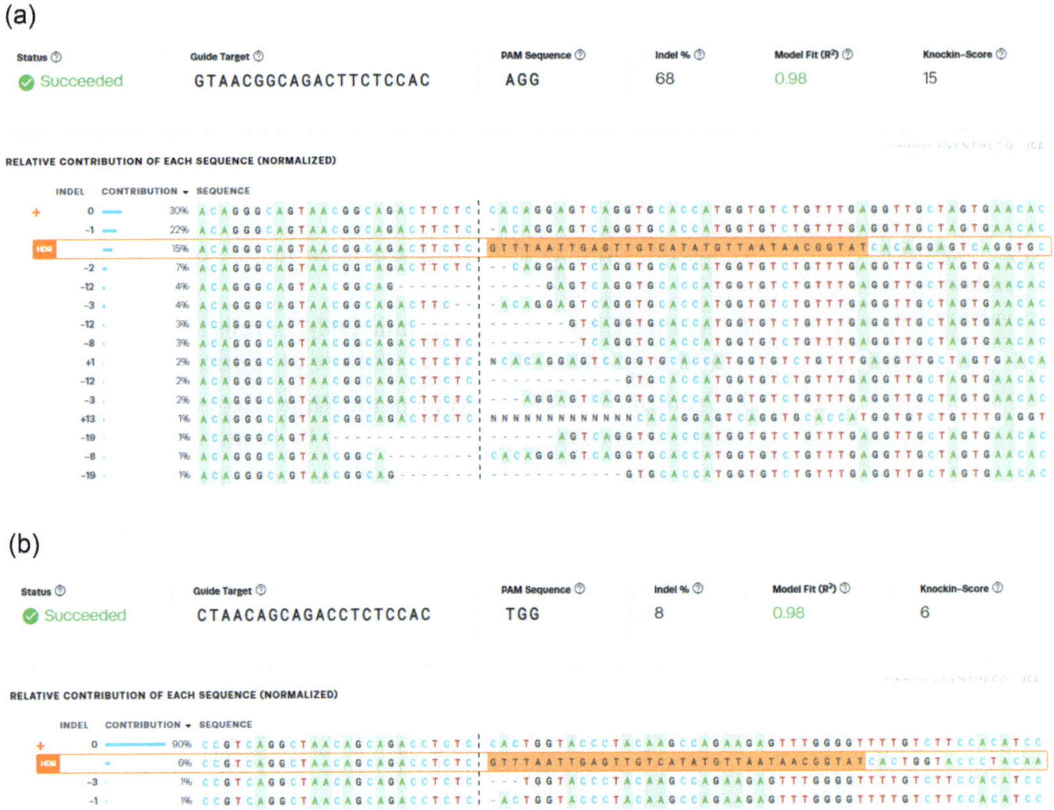

Fig. 3 dsODN tag knockin rate quantification at on and off-target sites in $CD34^+$ HSPCs by Inference of CRISPR Edits (ICE) [8]. GUIDE-seq was performed in sickle cell disease patient derived $CD34^+$ HSPCs for R-66 SCD gRNA that targets the HBB sequence with sickle mutation [7]. Relative contribution of alleles with dsODN tag knockin in forward orientation at (**a**) the on-target and (**b**) the active off-target CRISPR cleavage site

6. Clean up samples with 1× magnetic beads (120 μL) and elute with 15 μL water (*see* **Note 5** for detailed magnetic bead cleanup protocol). You may leave the samples with the beads on the magnet and move them when ready to add to tubes with end-repair master mix.

3.2.4 Y-Adapter Preparation

1. Prepare the following annealing reaction by mixing MiSeq Common Adapter with each of the A01–A16 oligos (*see* Table 1):

80 μL	1× TE
10 μL	100 μM MiSeq common adapter
10 μL	100 μM A## oligo

2. Run the following annealing program on a thermocycler:
   (a) Heat to 95 °C for 2 min.
   (b) Ramp cool to 25 °C over 45 min.
   (c) 4 °C hold.

3.2.5 End Repair

1. Make the following master mix at 1.2× number of samples and dispense 8.5 μL to each PCR tube or well:

0.5 μL	Nuclease-free water
1.0 μL	5 mM dNTP mix
2.5 μL	Ligation buffer, 10×
2.0 μL	End-repair mix (low)
2.0 μL	10× Taq buffer, Mg^{2+}-free
0.5 μL	Taq polymerase (non-hot start)

2. Add 14 μL of water eluted samples to tubes containing the master mix. You should have two PCR tubes for each sample, one sense and one anti-sense.
3. Mix with pipet set at 22 μL.
4. Run the following on a thermocycler with heated lid turned off:
   (a) 12 °C for 15 min.
   (b) 37 °C for 15 min.
   (c) 72 °C for 15 min.
   (d) 4 °C hold.

3.2.6 Adapter Ligation

1. To new PCR tubes, add 1.0 μL of one barcode of annealed A01–A16 oligos. Use the same barcode for the +/− strands of the same sample.
2. Add 2.0 μL T4 DNA ligase to each tube.
3. Add 22.5 μL of the end-repaired sample. Mix by pipetting, vortex, quick spin.
4. Run the following on a thermocycler with heated lid turned off:
   (a) 16 °C for 30 min.
   (b) 22 °C for 30 min.
   (c) 4 °C hold.
5. Clean up samples with 0.9× magnetic beads (22.95 μL). Elute with 22 μL water (*see* **Note 5**).

3.2.7 PCR1: Amplification of dsTag Integrated Sites

1. Make the following master mix for PCR1. Prepare two PCR1 reactions for each sample using GSP1+ or GSP1- primer. Add sufficient volume of all reagents for the appropriate number of PCR1 reactions. Only one GSP1 primer is used per PCR1 reaction. Dispense 20 μL master mix to each PCR tube:

10.7 μL	Nuclease-free water
0.6 μL	10 mM dNTP mix
3.0 μL	10× Taq buffer, Mg^{2+}-free
2.4 μL	25 mM $MgCl^{2+}$
0.3 μL	Platinum Taq polymerase
1.0 μL	10 μM GSP1 + or -
1.5 μL	0.5 M TMAC
0.5 μL	10 μM P5_1

2. Add 10 μL of adapter ligated sample prepared in 3.2.5 (f) to 20 μL PCR1 reaction. Make sure to set up two PCR1 reactions (GSP1+ or GSP1-) for each sample. Mix by pipetting, vortex, quick spin.
3. Run the following program on a thermocycler with heated lid on:
   (a) 95 °C for 5 min.
   (b) (95 °C for 30 s, 70 °C (−1 °C/cycle) for 2 min., 72 °C for 30 s) 15 cycles.
   (c) (95 °C for 30 s, 55 °C for 1 min, 72 °C for 30 s) 10 cycles.
   (d) 72 °C for 5 min.
   (e) 4 °C hold.
4. Clean up samples with 1.2× magnetic beads (36 μL) and elute with 15 μL water (*see* **Note 5**).

3.2.8 PCR2: Adaptor Labeling of PCR1 Amplicons

1. Make the following master mix for PCR2. Prepare two PCR2 reactions for each sample using GSP2+ or GSP2- primer. Add sufficient volume of all reagents for the appropriate number of PCR1 reactions. Only one GSP2 primer is used per PCR2 reaction:

4.2 μL	Nuclease-free water
0.6 μL	10 mM dNTP mix
3.0 μL	10× Taq buffer, Mg^{2+}-free
2.4 μL	25 mM $MgCl^{2+}$

(continued)

0.3 μL	Platinum Taq polymerase
1.0 μL	10 μM GSP2 + or -
1.5 μL	0.5 M TMAC
0.5 μL	10 μM P5_2

2. Dispense 13.5 μL to each PCR tube or well:
3. Add 1.5 μL of 10 μM P7_# to each tube to make a unique pair with A01–A16.
4. Add 15 μL of the PCR1 sample to the PCR2 master mix. Make sure to use GSP2+ primer for the GSP1+ PCR1 mix and GSP2- primer for the GSP1- PCR1 mix. Mix by pipetting, vortex, quick spin.
5. Run the following program on a thermocycler with heated lid on:
   (a) 95 °C for 5 min.
   (b) (95 °C for 30 s, 70 °C (−1 °C/cycle) for 2 min, 72 °C for 30 s) 15 cycles.
   (c) (95 °C for 30 s, 55 °C for 1 min, 72 °C for 30 s) 10 cycles.
   (d) 72 °C for 5 min.
   (e) 4 °C hold.
6. Clean up samples with 0.7× magnetic beads (36 μL) and elute with 30 μL TE (*see* **Note 5**).

3.2.9 Sample Quantification and Normalization

1. Quantify each sample using a Fluorometric DNA quantification assay fluorometer.
2. Use the following formula to convert from ng/μL to nM:

$$\frac{\left(\text{concentration in}\,\frac{\text{ng}}{\mu\text{L}}\right)}{660\,\frac{\text{g}}{\text{mol}} \times \text{PCR product size in bp}} \times 10^6 = \text{concentration in nM}$$

3. Choose as high a stock concentration as possible and normalize each sample to the same stock concentration.
4. Pool the normalized samples (*e.g.*, Pool 3 μL of each normalized sample into the one tube).
5. Quantify the pooled library using a fluorometric DNA quantification assay according to manufacturer instructions.
6. Dilute an aliquot of the pooled library to 4 nM in nuclease-free water.
7. Make a serial dilution of the pooled library and quantify using NGS library quantification kit as per manufacturer's instructions.

3.2.10 Illumina Sequencing

1. Thaw cartridge from MiSeq Reagent Kit v2 (300-cycles) >1.5 h prior to use in a room temperature water bath and thaw buffer HT1 at 4°C.
2. Dilute 5 μL of 4 nM pooled library with equal volume of freshly prepared 0.2N NaOH and vortex briefly. Centrifuge at 280 × *g* for 1 min and incubate at room temperature for 5 min.
3. Dilute denatured library to 20 pM by adding 990 μL of chilled buffer HT1.
4. Further dilute denatured library to 12.5 pM by mixing 225 μL prechilled HT1 and 375 μL of a 20 pM denatured library.
5. Invert to mix and store on ice until ready to load onto the MiSeq reagent cartridge.
6. Dilute each GUIDE-seq read2 and GUIDE-seq index primers to 0.5 μM with HT1.
7. Prepare MiSeq Sample Sheet according to the MiSeq Sample Sheet Quick Ref Guide (15028392 J).
8. Pierce the foil seal of position 17, 19, and 20 on the reagent cartridge with a clean 1 mL pipette.
9. Pipette 600 μL of the 12.5 pM denatured library to the position 17 designated as "Load Sample." Avoid buddle.
10. Pipette 600 μL of GUIDE-seq index1 primer to position 19 of the reagent cartridge.
11. Pipette 600 μL of GUIDE-seq read 2 primer to position 20 of the reagent cartridge.
12. Select "Sequence" on the software interface and follow the instruction on the screen.
13. Select the sample sheet containing the run specific information.
14. Rinse the flow cell with Milli-Q water. Dry with lint-free cleaning tissue.
15. Load the flow cell, cartridge, and PR2.
16. Sequence.

3.2.11 Bioinformatic Analysis of Sequencing Data

1. Install bioconda packages required to run GUIDE-seq script.
   (a) conda install -c bioconda bwa,
   (b) conda install -c bioconda bedtools,
   (c) conda install -c anaconda numpy,
   (d) conda install -c dranew bcl2fastq,
   (e) git clone --recursive https://github.com/aryeelab/guideseq.git
2. In the Illumina raw sequencing output folder, run bcl2fastq.

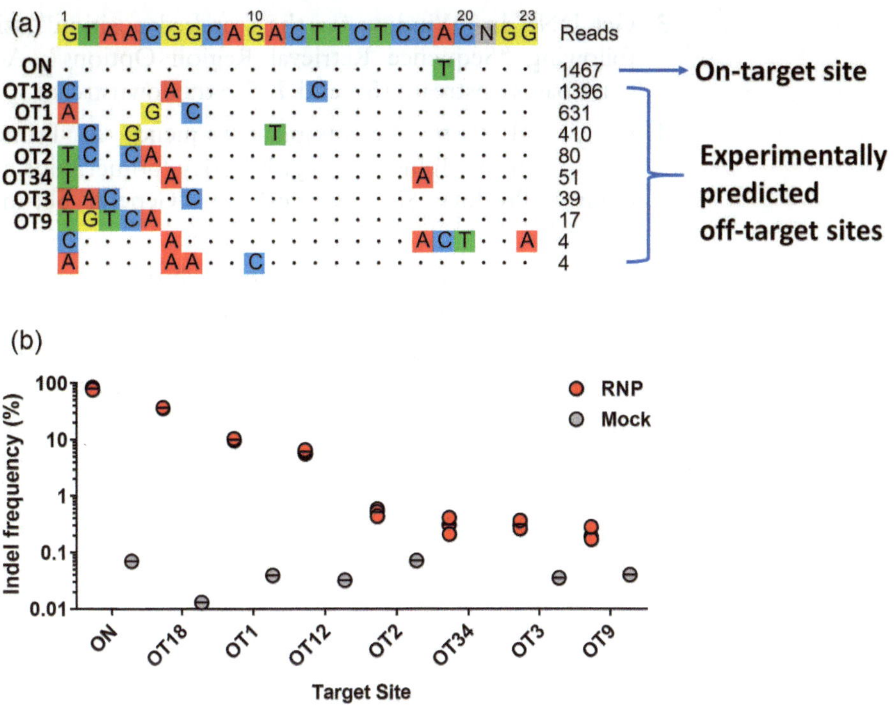

Fig. 4 Example of GUIDE-seq results and off-target activity quantification by NGS in CD34+ HSPCs. (**a**) Visualization of GUIDE-seq output. (**b**) The cleavage activities at the on-target site (ON) and 57 COSMID and GUIDE-seq identified off-target sites (OT) were analyzed by NGS. The dot-plot shows indel frequencies at the on-target site (HBB) and the 7 off-target sites with measurable cleavage activity [7]. GUIDE-seq read counts correlated highly with the indel frequency quantified by NGS, with on-target sites having the highest GUIDE-seq read counts

3. Copy and paste Read1, 2 Index1, 2 files to "GuideSeq_WorkingFolder."
4. Add yaml file to the same folder.
5. Run guide-seq.
   python guideseq/guideseq/guideseq.py all -m ~/GuideSeq_WorkingFolder/sample.yaml
6. *See* Fig. 4 for typical GUIDE-seq output [7].

3.2.12 Primer Design for GUIDE-seq Identified Off-Target Sites

1. In identified.txt GUIDE-seq output file, the chromosome position of the identified off-target site in the target genome can be found in column G (chromosome) and column H (position).
2. In UCSC genome browser (http://genome.ucsc.edu/cgi-bin/hgGateway), select the relevant human assembly (same as the reference genome used for GUIDE-seq alignment). Search the chromosome position.

3. Get DNA in "Window for the searched position" using the following "Sequence Retrieval Region Options": Add *175* extra bases upstream (5′) and *175* extra downstream (3′).
4. Copy and Paste the template sequence in Primer3Plus (https://primer3plus.com/cgi-bin/dev/primer3plus.cgi) and define "Product Size Ranges" in "General Settings" to *275–325*.
5. Order oligonucleotide primers for each off-target site identified with the following 5′ adaptor sequences:

   Forward primer requires TCTACAGTCCGACGATCA 5′ of the site-specific primer sequence output by Primer3Plus.

   Reverse primer requires GACGTGTGCTCTTCCGATC 5′ of the site-specific primer sequence output by Primer3Plus.

3.3 Identification of Off-Target Activity in CD34$^+$ HSPCs

3.3.1 CD34$^+$ Cell Culture [9]

1. Purify CD34$^+$ HSPCs from the peripheral blood. Separate the mononuclear fraction from the peripheral blood of patients with SCD by Ficoll-Hypaque density centrifugation. Extract CD34$^+$ cells from the mononuclear fraction by immunomagnetic separation using the CD34 Microbeads Kit. Assess the purity of CD34$^+$ by flow cytometric analysis.
2. Extract CD34+ HSPCs from the mononuclear fraction by immunomagnetic separation using the CD34 Microbeads Kit according to the manufacturers' instructions.
3. Culture the CD34$^+$ cells using HSPC expansion medium for 2–3 days before electroporation.

3.3.2 RNP Delivery Using Electroporation

1. Count the cells and prepare 2×10^5 CD34$^+$ cells for each electroporation reaction. Transfer cells to 15 mL tube and fill the tube to 15 mL with 1× PBS. Pellet the cells by spinning at 300 × *g* for 5 min.
2. Meanwhile, assemble the RNP complex by adding 5 μg (30.5 pmol) of SpCas9 protein and 2.5 μg (73 pmol) of chemically modified CRISPR sgRNA to a 0.2 mL PCR tube and incubate at room temperature for 10 min.
3. Aspirate as much medium and PBS without disturbing the cell pellet.
4. Add 22 μL P3 solution per 2×10^5 CD34$^+$ cells for each electroporation reaction.
5. Transfer 20 μL of cell suspension to each 0.2 mL tube containing RNP complexes. For mock-treated CD34$^+$ cells, electroporate of the same number of cells using the same program as the treated cells, but without RNP.

6. Mix well by gentle flick. Transfer 20 μL of the mixture in one pipet into the center groove of Lonza 4D electroporation cuvette. One transfection per well. Avoid bubbles.
7. Immediately perform electroporation using CA137 program. Wait 10 min at room temperature.
8. Add 80 μL pre-warmed $CD34^+$ expansion medium into each cuvette. Transfer all cells from each cuvette to a 24-well plate with 500 μL pre-warmed expansion medium per well.
9. Add fresh expansion medium every 2 days and culture at a density under 1×10^6 live cells/mL.
10. After 3–5 days, harvest cells and extract genomic DNA using genomic DNA extraction kit according to the manufacturer's protocol.

3.3.3 Off-Target Library Sample Preparation (See Fig. 1)

1. Prepare PCR1 reactions with 50 ng of gDNA for each off-target site as follows:

25 μL reaction	Component
5 μL	5× OneTaq standard buffer
0.5 μL	10 mM dNTPs
0.5 μL	10 μM forward primer
0.5 μL	10 μM reverse primer
2 μL	50 ng gDNA at 25 ng/μL
0.125 μL	OneTaq polymerase
16.375 μL	Nuclease-free water

   (a) 94 °C for 2 min.
   (b) (94 °C for 30s, 63 °C (−1 °C/cycle) for 30s, 68 °C for 30s) 7 cycles.
   (c) (94 °C for 30s, 57 °C for 30s, 68 °C for 25 s) 35 cycles.
   (d) 68 °C for 10 min.
   (e) 4 °C hold.
2. Use 2 μL to verify a single band at the proper size for each amplicon (usually 250–350 bps) on a 1.5% agarose gel (TAE running buffer).
3. Clean up PCR product with 0.8× magnetic beads. Add 18.4 μL beads to each PCR reaction and elute by adding 25 μL water. Transfer 22 μL to a clean tube or plate (*see* **Note 5**).
4. Prepare PCR2 reactions using 10 μL PCR 1 sample with adapters (barcoding) as follows. Use a unique combination of i7 and i5 index adapters for each amplicon.

25 μL reaction	Component
5 μL	5× OneTaq standard buffer
0.5 μL	10 mM dNTPs
1 μL	5 μM i5 adapter
1 μL	5 μM i7 adapter
10 μL	PCR I sample
0.125 μL	OneTaq polymerase
7.375 μL	Nuclease-free water

(a) 94 °C for 2 min.

(b) (94 °C for 30s, 60 °C for 30s, 68 °C for 35 s) 30 cycles.

(c) 68 °C for 5 min.

(d) 4 °C hold.

5. Use 2 μL to verify a single band at the proper size (usually 300 bp) on a 1.5% agarose gel (1× TAE running buffer).
6. Clean up PCR product with magnetic beads using 1.2× beads to PCR product ratio. Add 30 μL beads/ PCR rxn. Elute by adding 25 μL water. Move 22 μL to a clean tube or plate (*see* **Note 5**).

3.3.4 Quantification and Pooling of Off-Target Amplicons

1. Quantify purified amplicons by Nanodrop. Use the following formula to convert from ng/μL to nM:

$$\frac{\left(\text{concentration in}\,\frac{\text{ng}}{\mu\text{L}}\right)}{660\,\frac{\text{g}}{\text{mol}} \times \text{PCR product size in bp}} \times 10^6 = \text{concentration in nM}$$

2. Choose as high a stock concentration as possible and normalize all samples to the same stock concentration. Pool the normalized samples (*e.g.*, Pool 3 μL of each normalized sample into the one tube).
3. *See* **Note 6** for optional sample concentration and *see* **Note 7** for adapter-dimer removal.
4. Quantify the pool concentration using a Fluorometric DNA quantification assay® dsDNA HS Assay Kits according to manufacturer instructions.
5. Dilute an aliquot of the pooled library to 4 nM (*see* **Note 8**).

3.3.5 Illumina Sequencing

1. Thaw cartridge from MiSeq Reagent Kit v2 (500-cycles) ≥ 1.5 h prior to use in a room temperature water bath or overnight at 2–8 °C.
2. Ensure HT1 is thawed at 4 °C.

3. Use within 12 h freshly diluted 0.2 N NaOH.
4. Dilute 5 μL of 4 nM pooled library with equal volume of 0.2 N NaOH and vortex briefly.
5. Centrifuge at 280 × *g* for 1 min and incubate at room temperature for 5 min.
6. Add 990 μL chilled HT1. The result is 1000 μL of a 20 pM denatured library.
7. Dilute denatured library to 12.5 pM by mixing 225 μL prechilled HT1 and 375 μL of a 20 pM denatured library.
8. Invert to mix. The result is 600 μL of a 12.5 pM denatured library.
9. Set aside on ice until you are ready to load it onto the reagent cartridge.
10. Dilute each Custom read2 and Custom index primers to 0.5 μM with HT.
11. Prepare each Custom read2 and Custom index primers at 100 μM in nuclease-free water.
12. Add 3 μL primer to 597 μL chilled HT1.
13. Prepare MiSeq Sample Sheet according to the MiSeq Sample Sheet Quick Ref Guide (15028392 J).
14. Add 3 μL primer to 597 μL chilled HT1.
15. Pierce the foil seal of position 17, 19, and 20 on the reagent cartridge with a clean 1 mL pipette.
16. Pipette 600 μL of the 12.5 pM denatured library to the position 17 designated as "Load Sample." Avoid buddle.
17. Pipette 600 μL of custom index primer to position 19.
18. Pipette 600 μL of custom read 2 primer to position 20.
19. Select "Sequence" on the software interface and follow the instruction on the screen.
20. Select the sample sheet containing the run specific information.
21. Rinse the flow cell with Milli-Q water. Dry with lint-free cleaning tissue.
22. Load the flow cell, cartridge, and PR2.

3.3.6 Bioinformatic Analysis of Sequencing Data

1. CRISPR/Cas9 genome editing outcomes from deep sequencing data were analyzed using a custom pipeline [6]. Details of the custom pipeline can be found: https://github.com/piyuranjan/NucleaseIndelActivityScript
2. Install bioconda packages required to run nuclease activity script.

(a) -c bioconda perl-sort-naturally
(b) install -c bioconda perl-list-moreutils,
(c) conda install -c bioconda perl-List-Util,
(d) conda install -c bioconda cutadapt,
(e) conda install -c bioconda trim-galore,
(f) conda install -c bioconda prinseq,
(g) conda install -c bioconda bwa,
(h) conda install -c bioconda samtools openssl = 1.0.

3. In Illumina run folder, perform "bcl2fastq." Be sure to use the run specific sample sheet and that it is in this Illumina runs main folder. The Barcodes and samples in this sample sheet will demultiplex your reads into Read1 and Read2 .fastq files, output in "Data/Intensities/BaseCalls/."
4. These demultiplexed files exist as zipped .fastq files (fastq.gz). Unzip the files using the command: gunzip *fastq.gz.
5. In your working folder, be sure it contains: your unzipped R1 and R2 fastq's, your reference file (−ref.fa), your cutsites file (Cutsites.txt), the indelQuantificationFromFastqPaired2bp-1.0.1.pl script, and your parameters.config file.
6. parameter.config file contains information specific to each run. Change the sample name (First Column), the R1 and R2 names (Second and Third Column), the -ref.fa filename (the Eighth Column), and the Cutsite.txt filename (the Ninth Column) to reflect the files in your working folder.
7. Run the following command: perl indelQuantificationFromFastqPaired2bp-1.0.1.pl -c parameters.config -v. The output gives the level of activity at each off-target site analyzed.

4 Notes

1. Use the same Target Genome for COSMID and GUIDE-seq to easily identify overlapping sites.
2. COSMID identifies the same off-target site multiple times when there are multiple possible sequence alignment patterns. Users need to remove duplicated sites from the list manually.
3. Cell types other than $CD34^+$ HSPCs could be used for GUIDE-seq. To optimize genome editing conditions for GUIDE-seq in the cell type of interest:

   (a) Perform dsODN tag dose optimization using RNP and dsODN Tag. It is recommended to deliver 10 pmol to 100 pmol of dsODN tag.

(b) Determine the optimal dsODN tag dose that resulted in the highest rate of dsODN tag integration at the on-target cleavage site measured by ICE.

4. Alternatively, enzymatic fragmentation of genomic DNA, end-repair and adapter ligation can be performed using Lotus™ DNA Library Prep Kit (IDT) according to the manufacturer's protocol.
5. Magnetic beads cleanup protocol.
   (a) Determine the desired volumetric ratio of magnetic beads to sample for the cleanup.
   (b) Vigorously mix stock tube of beads until foamy before use.
   (c) Add beads to sample and mix by pipetting.
   (d) Incubate at room temperature for 5 min.
   (e) Place and leave on magnet through the remaining steps until indicated to be removed.
   (f) Room temperature for 10 min.
   (g) Remove liquid with a pipet.
   (h) Add 80% ethanol (200 μL for PCR reactions).
   (i) Room temperature for 1 min.
   (j) Remove liquid with a pipet.
   (k) Repeat **steps h–j**.
   (l) Use a pipette to remove any residual liquid from the bottom of the tubes.
   (m) Remove samples from the magnet.
   (n) Air dry for 2 min.
   (o) Mix sample by pipetting with desired volume and solution for elution. Make sure no air bubbles are present.
   (p) Return to the magnet.
   (q) Incubate for 5 min at room temperature.
   (r) Transfer eluate to new tube.
6. Clean up the pool with 2× magnetic beads and elute with 30 μL water (aimed at concentrating the normalized pool).
7. If adapter-dimers are present in the library, gel extract the pooled library from a 1.5% agarose gel. Use DNA gel extraction kit as per manufacturer's instructions.
8. If more than one library is being run at once (pool several libraries), determine the number of samples in each pool and pool the 4 nM libraries according to the number of samples. Keep the same ratio or percentage of samples/pool.

Acknowledgments

This work was supported by the National Institutes of Health (UG3HL151545, R01HL152314, and OT2HL154977 to G.B.) and the Cancer Prevention and Research Institute of Texas (RR140081 to G.B.).

References

1. Hsu PD, Scott DA, Weinstein JA et al (2013) DNA targeting specificity of RNA-guided Cas9 nucleases. Nat Biotechnol 31(9):827–832
2. Cradick TJ, Fine EJ, Antico CJ et al (2013) CRISPR/Cas9 systems targeting β-globin and CCR5 genes have substantial off-target activity. Nucleic Acids Res 41(20):9584–9592
3. Fu Y, Foden JA, Khayter C et al (2013) High-frequency off-target mutagenesis induced by CRISPR-Cas nucleases in human cells. Nat Biotechnol 31(9):822–826
4. Torres R, Martin MC, Garcia A et al (2014) Engineering human tumour-associated chromosomal translocations with the RNA-guided CRISPR-Cas9 system. Nat Commun 5:3964
5. Cradick TJ, Qiu P, Lee CM et al (2014) COSMID: a web-based tool for identifying and validating CRISPR/Cas off-target sites. Mol Ther Nucleic Acids 3(12):e214–e214
6. Tsai SQ, Zheng Z, Nguyen NT et al (2015) GUIDE-seq enables genome-wide profiling of off-target cleavage by CRISPR-Cas nucleases. Nat Biotechnol 33(2):187–197
7. Park SH, Lee CM, Dever DP et al (2019) Highly efficient editing of the beta-globin gene in patient-derived hematopoietic stem and progenitor cells to treat sickle cell disease. Nucleic Acids Res 47(15):7955–7972
8. Hsiau T, Conant D, Rossi N et al (2019) Inference of CRISPR edits from Sanger Trace Data. bioRxiv, p 251082
9. Ronzoni L, Sonzogni L, Fossati G et al (2014) Modulation of gamma globin genes expression by histone deacetylase inhibitors: an in vitro study. Br J Haematol 165(5):714–721

Chapter 20

Genome Engineering of Hematopoietic Stem Cells Using CRISPR/Cas9 System

Nivedhitha Devaraju, Vignesh Rajendiran, Nithin Sam Ravi, and Kumarasamypet M. Mohankumar

Abstract

Ex vivo genetic manipulation of autologous hematopoietic stem and progenitor cells (HSPCs) is a viable strategy for the treatment of hematologic and primary immune disorders. Targeted genome editing of HSPCs using the CRISPR-Cas9 system provides an effective platform to edit the desired genomic locus for therapeutic purposes with minimal off-target effects. In this chapter, we describe the detailed methodology for the CRISPR-Cas9 mediated gene knockout, deletion, addition, and correction in human HSPCs by viral and nonviral approaches. We also present a comprehensive protocol for the analysis of genome modified HSPCs toward the erythroid and megakaryocyte lineage in vitro and the long-term multilineage reconstitution capacity in the recently developed NBSGW mouse model that supports human erythropoiesis.

Key words HSPCs, Nucleofection, Transduction, Erythroid differentiation, Hemoglobinopathies, Megakaryopoiesis, NHEJ, HDR, Base editing, Clonal analysis, NBSGW mouse

1 Introduction

Hematopoietic stem and progenitor cell (HSPC) transplantation is the current primary treatment strategy for many genetic disorders of blood and immunologic origin [1]. The significant limitations of finding a matched-unrelated donor (MUD) for allogeneic transplantation have shifted the focus towards autogenic ex vivo gene therapy as an option for the treatment of monogenic disorders. HSPCs are readily isolated from mobilized peripheral blood from the donor after G-CSF (granulocyte-colony stimulating factor) stimulation and are able to tolerate gene modification in ex vivo conditions before reinfusion into the patients [2]. Most of the current gene therapy studies in HSPCs are performed using lentiviral vectors (LV), as these vectors are most efficient in transfecting the transgene into recipient cells. Gene therapy based on viral vectors can be effectively used to improve a spectrum of disorders

Nagarajan Kannan and Philip Beer (eds.), *Stem Cell Assays: Methods and Protocols*, Methods in Molecular Biology, vol. 2429, https://doi.org/10.1007/978-1-0716-1979-7_20,

[3]; however, insertional mutagenesis with viral vectors leading to oncogenesis is a major hurdle for translating this approach into the clinic.

Genome editing technologies based on programmable nucleases are emerging as a potential alternative to viral-mediated gene modification. Programmable nucleases such as meganucleases, zinc finger nucleases (ZFN), transcription activator-like effector nucleases (TALENS), and clustered regularly interspaced short palindromic repeat (CRISPR)-associated nuclease Cas9 have enabled the possibility of site-specific genome editing by correction of the disease-causing mutation, the addition of transgenes, or the disruption of regulatory regions. Double-strand DNA breaks caused by gene modification are typically processed by the endogenous DNA repair pathways like nonhomologous end-joining (NHEJ) and homology directed repair (HDR) [4]. NHEJ is the predominant repair pathway occurring throughout the cell cycle phase, whereas the HDR pathway is used in G2 and S phase preferentially.

The target specific endonuclease activity of the cas9 sgRNA (RNP) complex results in double stranded DNA breaks which activate NHEJ, resulting in gene disruption and deletion. The modular nature of the CRISPR system and the ease of designing RNA sequences over the protein sequences underline the popularity of CRISPR over other gene-editing tools. The major limiting factor of protospacer adjacent motif (PAM) availability is addressed by the newer variants of cas protein which recognizes DNA sequences with any specific PAM. The efficiency of the cas9 protein in creating double stranded DNA breaks is determined by the efficiency of the protein to localize into the nucleus. The function is enhanced by coupling a nuclear localizing signal (NLS) gene along with the cas9 gene. Fusion of 3×NLS to the cas9 protein is the commonly used to achieve higher efficiency. In this chapter, we describe the optimized protocol to obtain higher editing efficiency in the primary cells with the 1×NLS-Cas9 protein.

Gene correction and addition are achieved through the HDR pathway, which requires a donor DNA cassette containing homology arms to act as a template for the repair mechanism during the double-strand break. Donor template size, mode of delivery, and the dependence on the cell cycle phase are the major limiting factor for the efficiency of HDR. Introduction or correction of point mutation is mostly preferred by single-stranded oligo donor (ssODN) delivery (with asymmetric homology arms [5]) even though recent studies report higher efficiency of recombinant adeno-associated virus serotype 6 (rAAV6) [6]. Large gene delivery (up to 4 kb), with or without its promoter, in the intronic or promoter regions of the target gene is facilitated by the rAAV6 donor delivery system [7].

The fusion of cas9 nickase with specific deaminase enzyme is employed in base editing to create or correct certain transition mutations like C*G to T*A or A*T to G*C. Cytosine base editor (CBE) converts cytosine to thymine base, and the adenosine base editor (ABE) converts adenosine to guanine in the target region [8]. The major advantages of base editing over HDR are the independence of cell cycle phase and higher desired editing with lower off-target activity (due to single-strand nick instead of DSB). Base editing is limited by the transversions, bystander effect, editing window, and PAM availability which has been overcome by the recently developed base editor variants.

The delivery of the editing tools is achieved by viral (lentivirus and AAV), nonviral (electroporation), or by a hybrid system involving both (viral and electroporation) into HSPCs. Each delivery method has advantages (and disadvantages) which need to be considered cautiously based on the desired editing requirement to achieve high gene editing efficiency. Various delivery strategies and their applications are listed in Table 1, along with the optimal off-target analysis suitable for each method and its application.

This chapter provides different strategies for efficient genome engineering of human HSPCs for therapeutic purposes. The possible effects of targeted genome engineering in HSPCs are evaluated by multiple functional assays during in vitro differentiation (myeloid lineage-erythroid/megakaryocyte) in pooled cell population and at the single-cell level to understand the biology of the gene modification in the individual cells.

Further, we also elucidate the methodology to determine the long-term engraftment and multilineage repopulation efficiency of genome engineered HSPCs in NBSGW mouse model which exhibits higher human chimerism and supports better erythropoiesis of engrafted gene-modified cells than NSG mice [13].

2 Materials

2.1 Equipment

1. Laminar airflow.
2. Magnetic separator.
3. Cell strainer (40 μ).
4. FACS tubes.
5. FACS analyzer.
6. Lonza 4D nucleofector.
7. 16-well nucleofection strips.
8. Maxcyte electroporator.
9. OC25 × 3 MaxCyte cuvette and buffer.
10. Swinging bucket centrifuge.

Table 1
Different genome engineering strategies in HSPCs and their applications

Scheme	Approach	Delivery Strategy	Pathway	Application	Efficiency	Off-target analysis
Gene disruption	Cas9-sgRNA	Electroporation	NHEJ	Knockout of gene of interest with high efficiency and specificity. Usually preferred for exons, promoter regions, and enhancers	Very high efficiency	Guide seq [9]
Gene deletion	Cas9-sgRNA(s)	Electroporation	NHEJ	Deletion of gene or gene sequences	Efficiency depends on the deletion fragment size	Guide Seq Chromosome karyotyping
Gene correction/ addition	Cas9-sgRNA+ ssODN	Electroporation	HDR	• Point mutation insertion, correction up to 200bases • Usually preferred at coding/ regulatory regions	Low efficiency	Guide seq
	Cas9-sgRNA+ rAAV6	Electroporation and viral transduction	HDR	Point mutations insertions and also to insert any gene of interest up to 4 kb	Low efficiency	Guide seq
Base editing	ABE-nCas9+sgRNA	Lentiviral		Conversion of A*T to G*C in regulatory and CDS regions Usually preferred at the coding regions to introduce a desired mutation	High efficiency	Orthogonal R-loop assay [10], RNA-seq, CIRCLE-seq [11] LAM-PCR [12]
		mRNA			High efficiency	Orthogonal R-loop assay RNA-seq CIRCLE-seq

11. Cytospin centrifuge.
12. CO_2 incubator.
13. Inverted microscope.
14. Thermocycler.
15. Falcon tubes.
16. Serological pipettes.
17. Culture dishes.
18. Glass slides.

19. Dissection scissors.
20. Scalpel and forceps.
21. Beckman Coulter "Class S" Ultracentrifuge.

2.2 Reagents

1× PBS.

1× RBC lysis buffer.

Easy sep CD34 positive selection kit (Stem cell technologies, 17856).

Trypan Blue stain.

Giemsa stain.

Propidium iodide.

NucRed (Thermo Scientific, R37106).

1 × NLS Cas9 protein (TAKARA, 632640).

Chemically modified sgRNAs (Synthego), ssODN (Integrated DNA Technologies).

P3 Nucleofection solution (Lonza, V4XP-3024).

Methocult Optimum medium (Stem cell technologies, H4034).

Cyclosporine H (Sigma, SML1575).

Megakaryocyte expansion supplement (Stem cell technologies #02696).

Sodium Meta Bisulfide.

Transfection reagents—polyethylenimine (PEI) (Polysciences), TransIT-LT1(Mirus Bio).

0.45 μm filter (PVDF).

HEPES buffer.

Polybrene.

Opti-MEM (Gibco).

Benzonase, Tris–HCl.

NaCl.

HiScribe mRNA synthesis Kit, CleanCap AG (Trilink).

2.3 Media

1. HSC expansion medium [14]—Stemspan STEM II medium (Stem cell technologies,09655) supplemented with 240 ng/ml SCF (300-07), 40 ng/ml IL6 (200-06), 240 ng/ml FLT3 (300-19), and 80 ng/ml TPO (300-18) (All the cytokines are obtained from Pepro Tech).
2. Erythroid differentiation medium [15]—Step I medium—1× IMDM–GlutaMAX (Thermo Scientific, 31980030), 100 U/ml penicillin–streptomycin (Thermo Scientific 10378-016), 5% Human AB serum (MP Biomedicals, 2930649), 330 mg/ml human holotransferrin (BBI Solutions, T101-5), 20 mg/ml

human insulin (Sigma, SLCC5730), 2 U/ml heparin (Sigma, H3149), 1 mM hydrocortisone (MP Biomedicals, 101996), 3 U/ml recombinant human erythropoietin (Zyrop 4000), 100 ng/ml stem cell factor, and 5 ng/ml interleukin-3 (Peprotech, 200-03). Step II medium—Step I medium without IL-3 and hydrocortisone. Step III medium—Step II medium without SCF (*see* **Note 1**).

3. Megakaryocyte expansion medium: 100 μl of 100× megakaryocyte supplement in 10 ml of SFEM II medium.
4. Freezing Medium—7% IMDM (Hyclone SH30228.01), 2% Filtered FBS (Thermo Scientific, A4766801) and 1% DMSO (7:2:1) or Cryostor (Stemcell Technologies, 07941).
5. AAV lysis buffer: 50 mM Tris–HCl, 150 mM NaCl adjusted to pH 8.0.

2.4 Antibodies

1. hCD34 PE (550761), hCD90 BV421(562556), hCD133 APC (372806) for HSC characterization; hCD45 FITC (55548), mCD45 APC (559864) for engraftment analysis, multi lineage reconstitution analysis requires CD19 PE (340364) and hCD3 APC (340440) [lymphoid markers], hCD56 APC (341025) and hCD16 PE (347617) [NK cells markers], hCD33 APC (340474) and hCD3APC (340440) [myeloid markers], hCD71 BV421 (562995) and hCD235a PE (21812354) [erythroid markers], hCD41a APC (559777) [platelet marker]. Annexin V,7-AAD. All these antibodies are obtained from BD biosciences. Human fetal HB APC (Thermo Scientific, MHFH05) (*see* **Note 2**).

2.5 Animal Requirement

Female Nonirradiated NOD, B6. SCID $Il2r\gamma^{-/-}Kit^{W41/W41}$ (NBSGW) at 6–8 weeks of age weighing around 25 g

3 Methods

All the cell culture experiments are to be followed under aseptic conditions inside the laminar airflow chamber.

3.1 PBMNC Isolation

3.1.1 Normal Donors

Peripheral blood mononuclear cells (PBMNCs) comprises of lymphocytes, monocytes, and dendritic cells. These cells are fractionated from mobilized peripheral blood by density gradient centrifugation method.

1. Transfer the G-CSF mobilized peripheral blood from blood bag into a 50 ml falcon tube. Make a note of donor details for future reference.

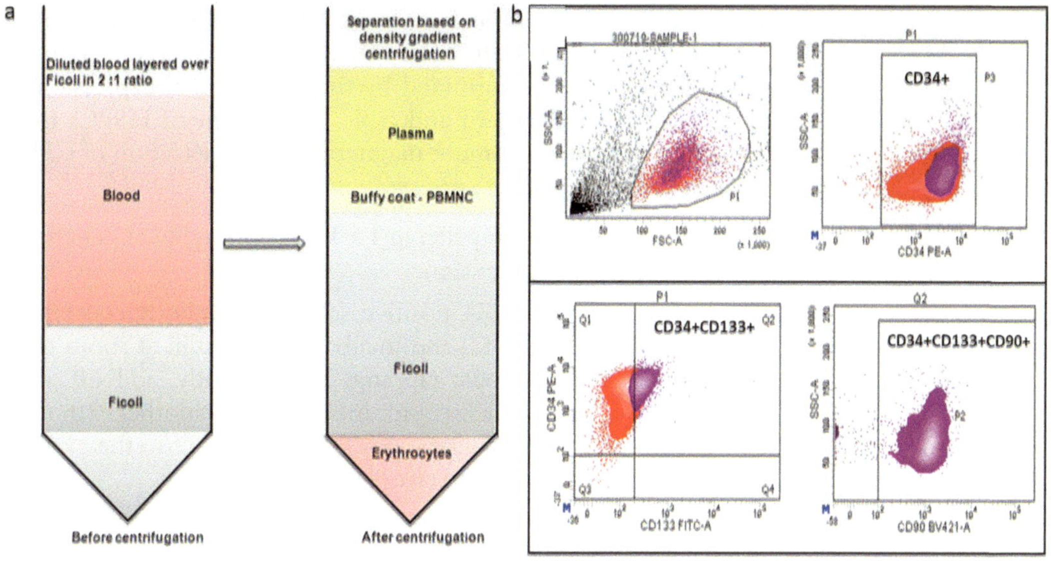

Fig. 1 (**a**) Isolation of PBMNC from mobilized peripheral blood by density gradient centrifugation using Ficoll-Paque. (**b**) Determination of long term and primitive HSC population by flow cytometry

2. Dilute the blood with an equal volume of 1× PBS (*see* **Note 3**). Gently layer the diluted blood above Ficoll-Paque in 2:1 ratio in a 50 ml falcon tube.
3. Spin the tubes at 400 × *g* for 30–40 min at room temperature with minimal acceleration and deacceleration (Fig. 1a).
4. Carefully discard the upper layer containing serum and aspirate the white buffy coat comprising PBMNCs into a new falcon tube (*see* **Note 4**).
5. In case of RBC contamination, incubate the cells with 1× RBC lysis buffer for 10 min on ice and spin at 200 × *g* for 10 min (*see* **Note 5**).
6. Wash the PBMNC cell pellet with 1× PBS and count the cells using trypan blue staining (*see* **Note 6**).
7. Cryopreserve the isolated PBMNCs in ice-cold freezing medium or proceed for HSPC isolation immediately.

3.1.2 Beta-Hemoglobinopathy Donors

1. Mobilized peripheral blood of β-hemoglobinopathy patient is processed directly by centrifuging at 600 × *g* for 10 min to separate out the buffy coat followed by RBC lysis for 5mins (twice) to remove the RBCs and obtain PBMNCs population.
2. This is because Ficoll-Plaque layering would result in thin PBMNC layer contaminated with defective erythrocytes.

3.2 HSPC Isolation

Hematopoietic stem and progenitor cells (HSPCs) constitute about 1–3% of PBMNCs derived from mobilized peripheral blood. These HSPCs are heterogeneous in population, and the state of each cell is determined by their differential expression of surface antigens. Standard and rapid enrichment of HSPCs from PBMNC is based on immunomagnetic positive selection of CD34 antigen expressing cells.

1. Resuspend PBMNC pellet in 1× PBS thoroughly to generate a homogeneous suspension.
2. Add 100 μl of CD34 positive selection antibody cocktail to 100 million PBMNCs and incubate for 30 min at room temperature with periodic tapping. Subsequently, add 60 μl of magnetic nanoparticles and incubate for another 20 min. These magnetic nanoparticles will bind to the CD34 antibody-bound target cells.
3. Transfer the labelled cells to an FACS tube and place inside the magnetic stand followed by 5 min incubation at room temperature. Invert the FACS tube along with the magnetic stand into a discard jar to remove the unbound cells after the incubation period.
4. Rinse the sides of the tube with 1× PBS and repeat **step 3** for a total of 4–5 washes with 1× PBS until the discard solution becomes transparent (*see* **Note** 7).
5. Remove the FACS tube from the magnetic stand and collect the CD34 positive cells bound to magnetic nanoparticles attached to the tube walls by flushing them with 1 × PBS.
6. Transfer the cells into a 15 ml centrifuge tube and pellet down the CD34 positive cells at 200 × *g* for 5 min, resuspend them in HSPC expansion medium and seed them in a culture dish at appropriate cell density, preferably 0.5 million/ml medium.

3.3 Long Term HSC Characterization

The isolated CD34 positive cells are further characterized for the presence of primitive and long-term repopulating HSCs based on the expression of other surface antigens such as CD90 (Thy-1) and CD133 (Prominin-1) as their role is elucidated in maintaining HSPC at quiescence state and enhancing engraftment capacity respectively [16–18].

1. Add 2 μl of CD34, CD90 and CD133 antibody to approximately 50,000 isolated cells after CD34 enrichment and incubate for 15–20 min at room temperature.
2. Wash the cells with 1×PBS to remove unbound antibodies. Resuspend the cells with 200 μl of 1×PBS and analyze the cells by flow cytometry (Fig. 1b)

3.4 Genome Engineering of HSPCs

Based on the desired type of target modification, the suitable strategy for genome editing is used to achieve maximum efficiency and for desired modification. Selecting the optimal procedure is the most vital step in gene editing.

3.4.1 Gene Disruption/Deletion

Gene disruption is performed by combining the endonuclease activity of cas9 with the endogenous NHEJ repair pathway. RNP complex electroporation is the most preferred method of cas9-sgRNA delivery, resulting in higher on-target and low off-target activity. Dual sgRNAs flanking the target region can be used to delete the desired loci.

1. Reconstitution of sgRNA—Spin the vial containing lyophilized sgRNA. Reconstitute the sgRNA with the appropriate volume of 1× TE buffer to achieve 100pmole per microliter concentration. Mix thoroughly and make aliquots of the sgRNA in smaller volumes to avoid multiple freeze-thawing (*see* **Note 8**). Nucleofection solution (NS) preparation—Add 16.4 μl P3 solution with 3.6 μl of supplement to prepare 20 μl of nucleofection solution per reaction (*see* **Note 9**).
2. Ribonucleoprotein (RNP) complex assembly—Mix 50 pmoles of Cas9 with 100 pmoles of reconstituted sgRNA and incubate at room temperature for 15 min (*see* **Note 10**).
3. Pellet 0.2 million of HSPCs and wash the cell pellet with 1× PBS to remove medium components completely. Gently resuspend the cell pellet in nucleofection solution and mix thoroughly with cas9: sgRNA complex. Carefully dispense the cells onto the sides of a well in a 16-well strip without generating air bubbles (*see* **Note 11**).
4. Cover the strip with a lid and insert into the X-unit of Lonza 4D nucleofector. Set up the Nucleofector program by choosing the appropriate wells and the buffer used. The cells are subjected to high-voltage electrical pulse using the defined pulse code DZ100 [19].
5. After the successful electroporation (*see* **Note 12**), add 50 μl of culture medium without antibiotic to each well and leave the cells undisturbed (this step will stabilize the edited cells from the shock created) for 10–15 min. Transfer the cells into a 12-well plate supplemented with complete expansion medium and incubate in a CO2 incubator at 37 °C.

3.4.2 Gene Correction/Addition

The donor DNA provides the template for the DNA repair mechanism to follow after the double-strand break by the cas9-sgRNA complex. The mode of delivery of the donor DNA is by either ssODN or rAAV6.

ssODN-Based Donor DNA Delivery

1. ssODN Design: DNA sequences containing the desired alteration is flanked by 111-27 bp homology arms (asymmetric) [5] from the antisense strand (against the sgRNA sequence) is designed along with a silent mutation in the sgRNA recognition domain or PAM motif to prevent recutting of the edited DNA sequence (*see* **Note 13**).
2. Reconstitution of ssODN—Dissolve the lyophilized ssODN contents from Integrated DNA Technologies (IDT) with the appropriate volume of 1× TE buffer to achieve 100 pmoles per microliter concentration. Mix thoroughly and make aliquots of the ssODN in smaller volumes to avoid multiple freeze-thawing (*see* **Note 8**).
3. Reconstitution of sgRNA—Spin the vial containing lyophilized sgRNA. Reconstitute the sgRNA with the appropriate volume of 1× TE buffer to achieve 100 pmoles per microliter concentration. Mix thoroughly and make aliquots of the sgRNA in smaller volumes to avoid multiple freeze-thawing.
4. Nucleofection solution (NS) preparation—Add 16.4 μl P3 solution with 3.6 μl of supplement to prepare 20 μl of nucleofection solution per reaction (*see* **Note 9**).
5. Ribonucleoprotein (RNP) complex assembly—Mix 50 pmoles of Cas9 with 100 pmoles of reconstituted sgRNA and incubate at room temperature for 15 min. 100 pmoles of ssODN (1 μl) is added along with the RNP complex right before mixing with the cells for nucleofection.
6. Pellet 0.2 million of HSPCs and wash the cell pellet with 1× PBS to remove medium components completely. Gently resuspend the cell pellet in nucleofection solution and mix thoroughly with cas9–sgRNA complex. Carefully dispense the cells onto the sides of a well in a 16-well strip without generating air bubbles (*see* **Note 11**).
7. Cover the strip with the lid and insert it into the X-unit of Lonza 4D nucleofector. Set up the Nucleofector program by choosing the appropriate wells and the buffer used. The cells are subjected to high-voltage electrical pulse using the defined pulse code CM149 [6].
8. After the successful electroporation (*see* **Note 12**), add 50 μl of culture medium without antibiotic to each well and leave the cells undisturbed (this step will stabilize the edited cells from the shock created) for 10–15 min. Transfer the cells into a 12-well plate supplemented with complete expansion medium and incubate in a CO_2 incubator at 37 °C.

rAAV-Based Donor DNA Delivery

1. Clone the donor DNA template in rAAV6 vector by Gibson assembly. More detailed donor DNA cloning in rAAV is available in [7].

2. Plate approximately 25–30 million HEK293T cells in a 15 cm dish on day 1. Sixteen hours later when the cells are at 80% confluency transfect them using 15 μg of serotype helper, rAAV6 vector, and adeno helper plasmids with 135 μl of PEI (transfection reagent) [6].
3. After 72 h, harvest the cells using a cell scraper and centrifuge for 5 min at 3000 × *g*.
4. Discard the supernatant and resuspend the cell pellet in 2 ml of AAV lysis buffer.
5. Perform three cycles of freeze–thaw using dry ice and 37 °C water bath. Add benzonase (10 U/ml of lysate) and incubate at 37 °C for 30 min.
6. Centrifuge at 10,000 × *g* for 10 min and perform an iodixanol density gradient ultracentrifugation with the supernatant to purify the virions with recombinant rAAV6 genomes.
7. Wash the AAV virus pellet with 1×PBS (supplemented with 0.001% Tween-20) thrice and concentrate to 200 μl and store at −80 °C.
8. After nucleofection of the cells with RNP complex, seed the edited cells in medium containing the AAV virus (at 3% of culture volume) [6] and incubate in a 5% CO_2 incubator.
9. After 24 h, change the medium to HSC expansion medium supplemented with the cytokines.

3.5 Base Editing Mediated Gene Modification

The base editor is involved in the conversion of nucleobase (C*G to T*A or A*T to G*C) within the editing window of the target region. sgRNAs for base editing is designed containing the target base within the editing window for ABE8e (4–8 bases from the opposite end of PAM) and CBE (2–9 bases from the opposite end of PAM). The base conversion protocol for the recently evolved ABE 8 through lentiviral and mRNA delivery is given below (*see* **Note 14**).

3.5.1 Lentiviral Approach

Plasmid Details: Cas9 coding sequence in the Addgene Plasmid (#57818) [20] is replaced by the ABE8e segment from the Addgene Plasmid (#138495) [21] by restriction digestion and subsequent cloning to achieve a lentiviral vector containing ABE8e and sgRNA scaffold along with the GFP reporter gene.

1. Plate approximately 4.5million 293 T cells in a 100 mm dish on day 1 (*see* **Note 15**).
2. Sixteen hours later, when the cells are at 80% confluency transfect them using second-generation lentiviral packaging plasmids and the plasmid of interest with 50 μl of TransIT LT1 (transfection reagent) mixed in 500 μl Opti-MEM medium.

3. Gently add the solution dropwise onto the plate and incubate at 37 °C.
4. After 24 h, replace the plate with new medium. Forty-eight hours after transfection, collect the supernatant and store it at 4 °C. In the same way, collect 60 h and 72 h supernatant.
5. Combine the supernatants in a 50 ml falcon tube and filter using a 0.45 μm filter (PVDF). Carefully add the filtered supernatant to the ultracentrifugation tube and close it using a cap.
6. Measure the weight of the ultracentrifuge tubes to ensure balancing and centrifuge at 40,000 × *g* using 70Ti rotor at 4 °C for 2 h.
7. After 2 h, remove the supernatant and carefully resuspend the virus pellet in IMDM–GlutaMAX medium (*see* **Note 16**).
8. Incubate the virus on ice for 1 h for homogeneous suspension and store at −80 °C.
9. Seed around 0.2million Day 1 HSPCs after characterization in one well of 12-well plate. Add 0.8 μl polybrene (6 mg/ml) and 10 μl of HEPES per ml of medium to aid transduction.
10. To cells add 40 μl of 300 × concentrated virus and plate centrifuge at 800 × *g* for 30 min at room temperature.
11. To enhance transduction efficiency, preincubate the isolated HSPCs in medium containing CsH (Concentration—2 μM (micromolar) of medium) for 16 h and perform transduction (*see* **Note 17**).
12. Change medium after 24 h of transduction and seed the cells in a 6-well plate at a seeding density of 0.2 million per ml.
13. Observe selection marker (GFP) in the transduced cells by FACS after 48 h and then set up differentiation using the required protocol.

3.5.2 Electroporation [8]

1. Amplify the template for mRNA synthesis from plasmid comprising base editor using forward and reverse primer with Q5 Hot Start 2× Master Mix (*see* **Note 18**).
2. Purify the PCR product using Zymo Research 25 μg DCC column.
3. With the obtained product as a template, synthesize mRNA using HiScribe High-Yield Kit with uridine modification and capping with CleanCap AG (Trilink) following the protocol given in the manual.
4. Pellet down 1.25 million Day 2 HSPCs cells and wash with 1 ml of MaxCyte buffer (HyClone) with 0.1% HAS.
5. Resuspend the cells in 1 ml ice-cold MaxCyte buffer and mix thoroughly. Split this cell suspension into multiple 20 μl aliquots.

6. Add 0.15 μM ABE8 mRNA and 4.05 μM sgRNA in a tube and make the final volume to 5 μl using MaxCyte buffer.
7. Load a set of 3 reactions containing 20 μl of cells and 5 μl of RNA composition are mixed uniformly by pipetting up and down without creating air bubbles into the individual chamber of an OC25 × 3 MaxCyte cuvette.
8. Select HSC-4 code to electroporate the RNA mixture into the cells.
9. Soon after electroporation transfer the cells into a 24well plate containing basal medium.
10. After 20 min of recovery, add complete medium with cytokines and incubate 37 °C (*see* **Note 19**).

The above protocol can be adapted for base conversion using cytosine base editors.

3.6 Ex Vivo Clonal Analysis

The pool of genome-modified HSPCs displays molecular and functional heterogeneity which creates a challenge in understanding the biology of individual cells. The development of methods to isolate and culture individual gene-modified cells by flow cytometry based single-cell index sorting is preferred for understanding heterogeneity. The homozygous and heterozygous edited clonal biology can be studied by analyzing the clonal editing percentage [22].

1. Add 150 μl 1×PBS to the side wells of 96-well Nunc flat bottom plate to maintain sufficient humidity within the plate.
2. Add 120 μl of MethoCult (semisolid) medium or differentiation medium to the other wells and incubate at 37 °C for 1 h before sorting.
3. Test sort using a test plate (96-well Nunc flat bottom) and verify proper placement of the cells (toward the center of a well). Place the cells to be sorted on ice before taking for sorting.
4. Gate only the viable cells by adding 2 μl of 7 AAD (viability dye) and index sort the cells using 100 μm diameter nozzle. Incubate the plate 37 °C with 5% CO2 (*see* **Note 20**).
5. Leave the plate undisturbed for 7 days and then visualize under a microscope for the appearance colonies. Score the colonies based on their morphology after 14 days of culture and genotype individual colonies.
6. Monitor the sorted gene-modified cells cultured in differentiation medium on alternative days. Based on the size and number of cells expanded, score the clones as small, medium or large. Top up the well with another 50 μl of medium if required (*see* **Note 21**).

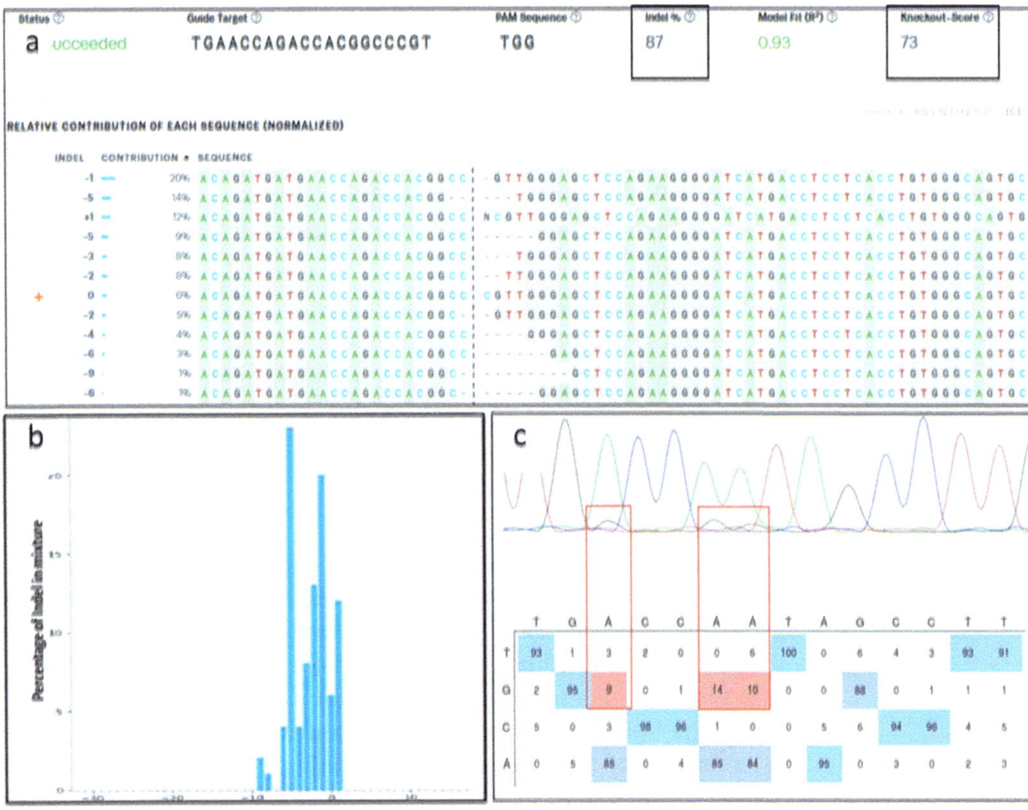

Fig. 2 (**a**) Cas9 RNP–mediated disruption of target region in HSPCs by Synthego ICE software. (**b**) Editing frequency and size of the Insertions/Deletions generated at the target region. (**c**) Adenosine base editor mediated A to G substitution in the target region in human HSPCs

7. Transfer the cells to a 48-well plate followed by 12-well as the cell number increases and perform genotyping and functional assays.

3.7 Analysis of Genome Engineered HSPCs

1. Isolate DNA from approximately 50,000 nucleofected HSPCs (from pooled/clonal populations obtained).
2. Perform Sanger sequencing of the target region using suitable primer and quantify the level of on-target gene editing efficiency using Synthego ICE software (for NHEJ and HDR) (Fig. 2 a, b) and EditR (for Base editing) (Fig. 2c) by aligning with the unedited reference sequence.
3. To analyze the in vitro colony-forming potential of gene-edited HSPCs by MethoCult colony-forming assay, premix the thawed MethoCult (CFU assay medium) with optimized cell number (500 cells/ml of medium) vigorously to obtain homogeneous cell suspension.
4. Keep the cell suspension undisturbed for approximately 30 min at room temperature for the air bubbles to settle.

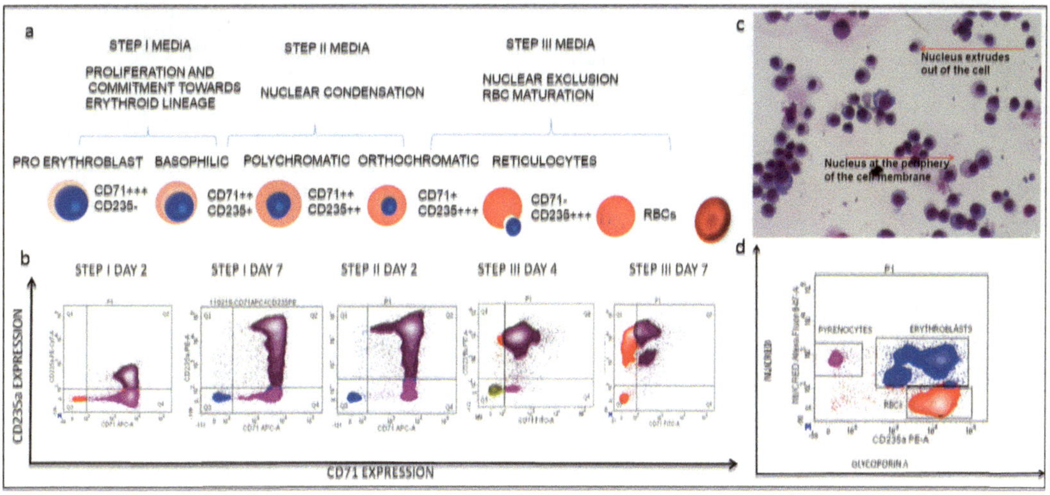

Fig. 3 (**a**) Image representing the sequential stages of HSC toward erythroid lineage differentiation. (**b**) Differential expression of erythroid markers during in vitro erythropoiesis. c) Morphology analysis of erythroid maturation using Giemsa stain. (**d**) Evaluation of percentage of nucleated/enucleated cells with erythroid differentiation marker CD235a

5. Layer the cell suspension evenly throughout the MethoCult plate (or 35 mm TC dish) without generating air bubbles. Incubate the plates at 37° C for 14–16 days.
6. Examine the plates under a light field microscope with 10× magnification and score the colonies formed based on their morphology, colour, and number (*see* **Note 22**) (Fig. 5a).
7. Genotype the individual colonies to verify the sustainability of the editing during in vitro multilineage differentiation.

3.8 Ex Vivo Differentiation of Genome Engineered HSPCs

The differentiation of edited HSPCs into different lineages is achieved by altering the culture conditions. The sequential maturation of the edited HSPCs in the myeloid lineage (Erythroid, Fig. 3a, and Megakaryocyte, Fig. 4a) is characterized by analyzing the differential expression of lineage-specific surface markers.

3.8.1 In Vitro Erythropoiesis

1. Gene modified HSPCs are seeded in varying culture conditions with different concentrations of cytokines that support erythrocyte maturation (*see* **Note 23**).
2. The gene-modified HSPCs proliferate and expand logarithmically in Step-1 erythroid differentiation medium for 7 days (48 h after nucleofection) with a regular medium change every 3–4 days.
3. On day 8, replace the culture medium into the Step 2 erythroid differentiation medium and culture for another 4–5 days. Subsequently, replate the cells in a Step-3 medium for 7–10 days or until enucleation is observed.

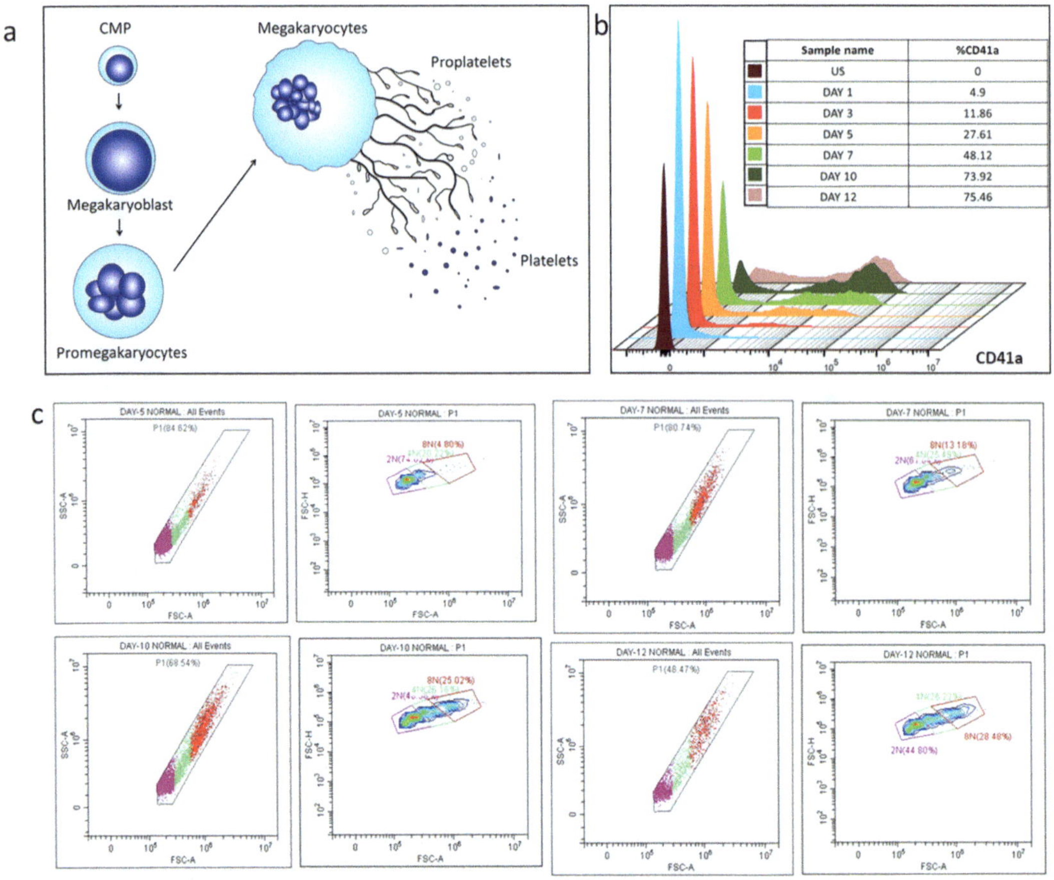

Fig. 4 (**a**) Schematic representation of various stages of megakaryocytic differentiation from human HSPCs. (**b**) Differential expression of megakaryocyte markers during in vitro megakaryocyte differentiation. (**c**) Assessment of DNA ploidy during in vitro megakaryocyte differentiation

4. Monitor the differential expression level of surface markers like transferrin (CD71) and Glycophorin A (CD235a) using labelled antibodies through FACS analysis at different stages of Erythroid maturation (Fig. 3b).
5. Assess the level of enucleation, by adding 10 μl of NucRed Live-cell stain along with 3 μl of CD235a antibody (Erythroid surface marker) (*see* **Note 24**).
6. Incubate the cells in the dark for 15 min and wash them with 1×PBS to remove unbound antibodies.
7. Analyze the percentage of enucleated cells in flow cytometry [14]. Cells positive for the Glycophorin A surface marker and negative for the nuclear stain are denoted as enucleated cells (CD235a + veNucRed-ve) (Fig. 3d).
8. Evaluate the growth profile of the gene-edited cells during erythroid expansion by counting the number of viable cells

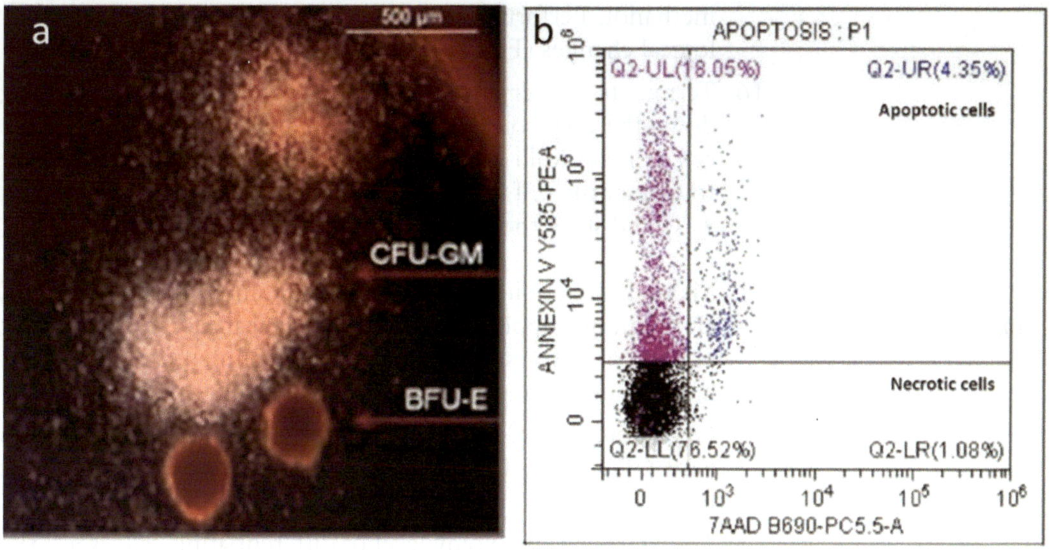

Fig. 5 (**a**) Representative images of colonies observed in Methocult media after 14 days. (**b**) Determination of cell viability using AnnexinV/7AAD

periodically using Trypan blue staining at every medium change.

9. Calculate fold expansion based on the ratio of total number cells at a particular time point to the initial number of cells seeded at the beginning of the experiment.
10. To determine the percentage of cell death during differentiation via apoptosis, add 2 µl of Annexin V and incubate for 15 min.
11. Acquire the sample immediately 5 min after the addition of 2 µl 7AAD in flow cytometry. Cells positive for Annexin V and 7 AAD undergo apoptosis, whereas cells positive for 7AAD and negative for Annexin V undergo necrosis (*see* **Note 25**) (Fig. 5b).
12. Isolate the RNA and protein from the gene-modified cells by using either commercially available kits or standard laboratory-based methods before the cell undergoes terminal differentiation.
13. Perform qRT-PCR and western blot analysis to validate the effect of gene modification on transcriptome and translatome level.
14. Morphological changes of cells during erythroid differentiation are observed by cytospin analysis.
15. Cytocentrifuge approximately 50,000 erythroid cells onto the slides at 200 × *g* for 5 min and fix them using ice-cold

methanol. Perform Giemsa staining to visualize the morphological changes (Fig. 3c).

16. Terminate the erythroid differentiation culture by pelleting down the erythroid cells at 200 × *g* for 5 min. Resuspend the cell pellet in Milli-Q water of appropriate volume and sonicate to extract the cell lysate. Remove the cell debris by spinning at 4 °C and store the supernatant at −80 °C until HPLC is run.

3.8.2 In Vitro Megakaryopoiesis

1. Maintain 0.5million HSPCs after gene modification in SFEM II medium supplemented with megakaryocyte supplement (100 μl of 100× supplement in 10 ml of medium). Carry out half medium change every alternative day up to Day 14.
2. On different days of differentiation stain around 50,000 cells with CD41a for 15 min and wash with 1 × PBS containing 0.5% BSA and 2 mM EDTA (Fig. 4b).
3. To determine the ploidy level of differentiating cells, wash the cells after CD41a staining with 1 × PBS containing 0.5% BSA and 2 mM EDTA (Fig. 4c).
4. Resuspend the pellet in 200 μl of buffer containing 1.25 mM sodium citrate, 2.5 mM sodium chloride, 3.5 mM dextrose along with 20 μg/ml propidium iodide and 0.05% Triton X-100 followed by 15–20 min incubation at 4 °C in the dark.
5. To this add RNase at 0.03 mM concentration and incubate in the dark for 20–30 min. Analyse the intensity of propidium iodide in CD41a expressing cells by flow cytometry.

3.9 Engraftment of Genome Engineered Cells in NBSGW Mice

Current mouse models used for transplantation studies show lower chimerism of human HSCs even after myeloablation. Further, the analysis of multilineage differentiation potential of engrafted genome modified cells is challenging as they do not support in vivo human erythropoiesis, and the mortality rate is also higher due to damage caused during whole-body irradiation. The NBSGW (NOD, B6. SCID Il2rg-/-Kit-W41/W41) mouse model is superior to NSG ((NOD SCID gamma mouse), exhibiting higher chimerism without irradiation along with enhanced human erythropoiesis. For long-term engraftment analysis, harvest the spleen, blood, and bone marrow of the transplanted mice after sacrificing 16 weeks of transplantation.

3.9.1 Xeno-transplantation

1. Select 6–8-week-old female NBSGW mice for transplantation studies.
2. Transplant around 0.5-1 million of gene-modified HSPCs through the tail vein or retro-orbital route of administration

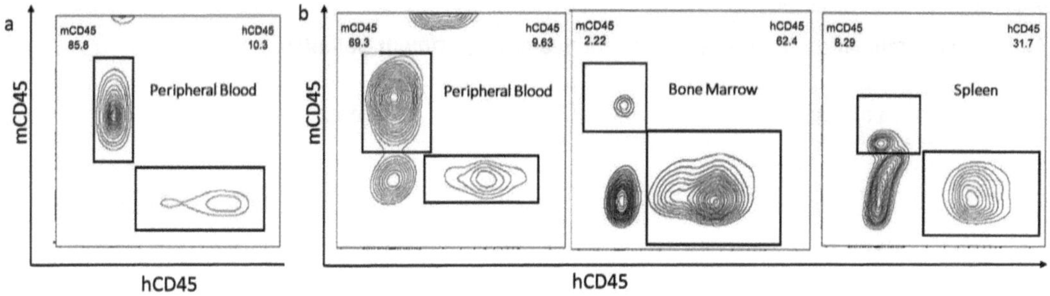

Fig. 6 (**a**) Short term engraftment potential of genome edited HSPCs in NBSGW mice after 8 weeks of transplantation. (**b**) Long term repopulation capacity of genome edited HSPCs in the Peripheral blood, bone marrow and spleen after 16 weeks of transplantation in NBSGW mice

using an insulin syringe with a 31 gauge X ¼ (0.25 mm × 6 mm) inch needle.

Short-Term Engraftment Analysis (Fig. 6a)

1. Collect peripheral blood via retro-orbital route 8 weeks after transplantation using a heparin-coated capillary tube and transferring blood into a tube containing heparin.
2. Wash the blood sample with 1×PBS.
3. To the red pellet obtained, add 1×RBC lysis buffer and place the tube on ice for 10 min followed by centrifugation at 110 × *g* for 5 min.
4. Repeat RBC lysis until the pellet becomes transparent.
5. Resuspend the pellet in 150 μl of 1×PBS; add 3 μl mCD45 APC antibody and 5 μl of hCD45 FITC antibody (*see* **Note 26**).
6. Incubate the tubes in the dark at room temperature with periodic tapping for 20 min.
7. Remove unbound antibodies by washing the cells with 1× PBS.
8. Re suspended the final cell pellet in 1× PBS and analyze using FACS.

Long-Term Bone Marrow Engraftment Analysis (Fig. 6b)

1. Collect both the tibia and femur bones of transplanted mice and remove the muscle along with other tissues using a sterile scalpel or blade.
2. Collect the bone marrow cells by syringe flushing or by centrifugation process (*see* **Note 27**) and seed the cells in a 65 mm dish with IMDM–GlutaMAX medium.
3. Pipette gently and make sure cells are mixed uniformly then incubate at 37 °C.

Table 2
Antibodies combination to analyse for multilineage engraftment potential

S.No	Lineage	Antibody combination
1	Lymphoid	hCD45 + hCD3 + hCD19
2	NK cells	hCD45 + hCD56 + hCD16
3	HSPCs	hCD45 + hCD133 + hCD34
4	Myeloid cells	hCD45 + hCD13 + hCD33
5	Megakaryocyte–erythroid progenitors	hCD45 + hCD41a(platelets) + hCD71
6	Erythroid–erythroid progenitors	hCD45 + hCD235a + hCD71

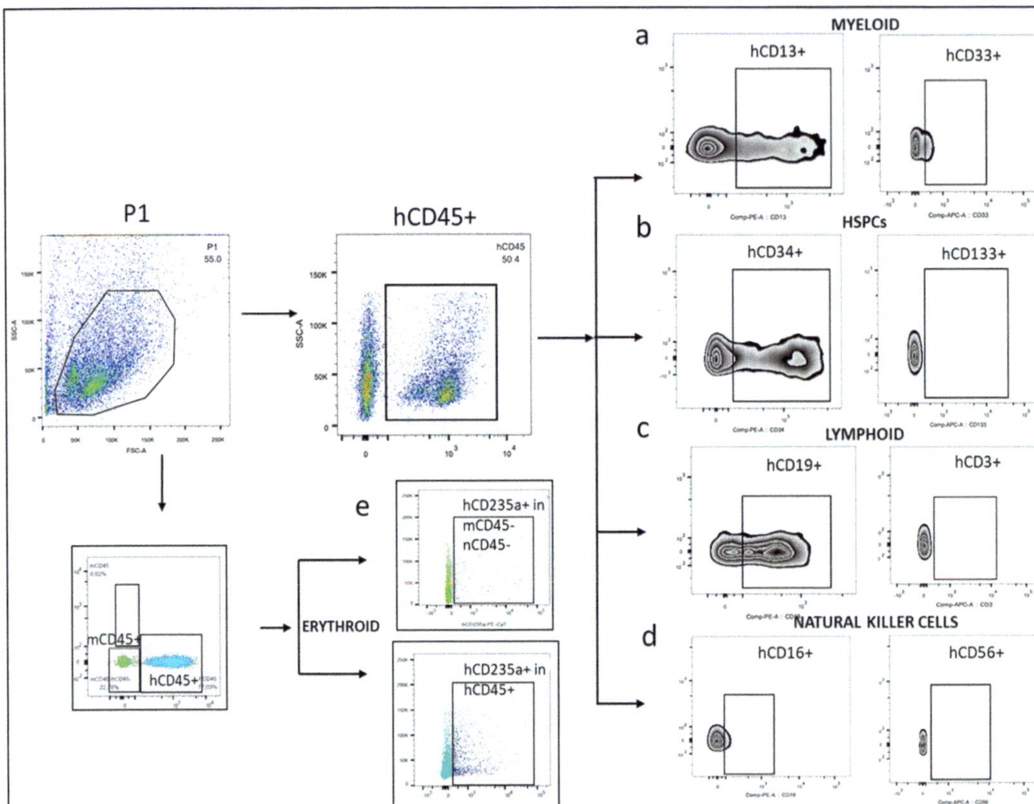

Fig. 7 Multilineage reconstitution potential of genome-edited HSPCs in bone marrow compartment of NBSGW mice after 16 weeks of transplantation depicting the frequency of (**a**) myeloid cells, (**b**) HSPC, (**c**) lymphoid, (**d**) natural killer cells, (**e**) erythroid

4. To 50,000 cells, add 5 μl of hCD45 FITC and 3 μl of mCD45 APC antibody. Wash the cells after 15 min of incubation and analyze by flow cytometry (*see* **Note 28**).
5. Incubate 50,000 bone marrow cells with appropriate antibodies as depicted in Table 2 for multilineage analysis and analysis in flow cytometry (Fig. 7) (*see* **Note 29**).
6. Stain 50,000 bone marrow cells with 2 μl of CD235a antibody and analysis through flow cytometry to estimate the percentage of human erythroid cells present in mouse bone marrow (NBSGW is reported to support human hematopoiesis due to hypomorphic mutation in the c-kit gene [23, 24]).

Long-Term Peripheral Blood Engraftment Analysis

1. Collect blood by cardiac puncture and process similarly as in short term engraftment analysis.

Long-Term Spleen Engraftment Analysis

1. Collect the spleen and place it in a well of a 12-well plate with 500 μl 1 XPBS.
2. Crush the spleen tissue with the end of the syringe plunger and wash with RBC lysis buffer.
3. Add human CD45 and mouse CD45 antibodies and analyze by flow cytometry.

4 Notes

1. Store all the supplements, including cytokines, at an appropriate temperature in multiple small aliquots to minimize repeated freeze–thaw cycles.
2. Fluorescence conjugated antibodies are highly sensitive to temperature and light; therefore, maintain them at 4 °C in the dark.
3. Increase the purity of PBMNCs by diluting the blood sample with a higher volume of 1×PBS.
4. Aspirate carefully the buffy coat containing the PBMNCs to minimize the contamination with other cells. The presence of Ficoll in the buffy coat hinders the cells from pelleting down efficiently during subsequent processes. Eliminate the Ficoll contamination by multiple washes with 1×PBS.
5. Avoid longer incubation of cells in 1× RBC lysis buffer to reduce cellular stress.
6. To prevent clump formation, perform DNAse-1 treatment for 10mins during the initial resuspension of PBMNCs in 1×PBS. Remove persisting clumps after DNAse-1 treatment by filtering the cell suspension through the 40 μm cell strainer to generate a homogeneous cell suspension.

7. Decreasing the number of washes with 1×PBS generates a higher cell counts. On the other hand, the purity of isolated cells dramatically improves by increasing the number of washes with 1× PBS.
8. For long term storage, store sgRNA aliquots in −80 °C immediately after reconstitution. Handle the sgRNA aliquots in an RNAse-free environment to prevent degradation.
9. Equilibrate nucleofection solution at room temperature for 15 min before the experiment. Adding cold nucleofection solution will affect the viability of the cells and decreases the editing efficiency.
10. Deletion depends on the individual sgRNA efficiency in causing double-strand break and also the total size of the deletion. RNP complex for each sgRNA is prepared separately and mixed before electroporation.
11. Wash the cell pellet with 1× PBS to remove any medium component like salts that would interfere with electroporation. Incubation of cells in nucleofection solution for an extended period is toxic to cells. Make sure that the final solution makes proper contact with the sides of the well, and the volume does not exceed 20–25 μl inside each well.
12. After the electroporation, a summary of the nucleofection process results will be displayed on the Lonza 4D nucleofector system. A colour code depicts the overall result: Green "+" symbol denotes successful nucleofection, whereas yellow "+" means nucleofection passed with reduced efficiency, and red "+" means unsuccessful nucleofection.
13. The total size of ssODN ranges from 50 to 400 bp inclusive of the homology arms but rAAV6 is approximately 4.2 kb with ~400 bp homology arm. Titrate the ssODN concentration to achieve good viability after nucleofection. Large gene insertions can be performed efficiently with the rAAV6 donor delivery system. Successive HR with two delivery vectors can be performed for the addition of genes larger than 4 kb. The efficiency of integration will be higher when the modification is proximal to the DNA cleavage site.
14. IDLV (integrase-deficient lentivirus) based delivery of ABE 8 is better suited in clinical applications over lentiviral delivery.
15. For efficient virus production, low passage 239 T cells are preferred, and transfection is carried out at optimal confluency (80–90%).
16. Make a circular mark in the ultracentrifugation tubes at one bottom corner and place the tube accordingly so that the virus pellet would be confined with a circle and is easy to obtain.

17. Adding Cyclosporin H to HSPCs before transduction at a higher concentration will affect the viability of cells.
18. Linearizing the plasmid using suitable restriction enzymes (giving blunt end or 5′ staggered cut) before amplification would be a better strategy to obtain template mRNA synthesis.
19. The rate of base conversion varies between different versions of base editors. ABE7.1 takes 5 days, whereas ABE 8e shows optimal conversion after 24 hrs of nucleofection.
20. Exclude preapoptotic cells while performing single-cell sorting by gating only the live population using a viability stain.
21. Monitor every alternative day for the appearance of colonies and check for the level of PBS in the side wells of the plate.
22. Methocult medium is highly susceptible to bacterial contamination it is enriched with nutrients and cytokines. Therefore, handling the medium in sterile conditions with utmost care is essential. Add water to outer space surrounding the wells in the Methocult plates or place additional wells containing water to prevent evaporation. Stemgrid (#27000) from Stemcell Technologies provides a grid background for easy counting of the colonies.
23. The cell pellet becomes red in colour during the initial erythroid expansion in Step 2 medium, and progressively the colour changes into dark brick red colour during the terminal differentiation stage. Cells are maintained at high culture density in subsequent stages of erythroid differentiation. Each donor exhibits differences in the duration of erythroid expansion and erythroid differentiation profile.
24. Hoechst staining can be used as an alternative to NucRed to stain the nucleus.
25. Cells positive for Annexin V and 7 AAD undergo apoptosis, whereas cells positive for 7AAD and negative for Annexin V undergo necrosis.
26. The total number of events in the flow-cytometry analysis increases if there are any residual RBCs present in the sample, which leads to the reduced live cell population. To overcome the issue, increase the number of PBS washes and acquire a higher number of total events.
27. Syringe flushing of bone marrow is performed by dissecting the ends of the bone and flushing the cells out using an insulin syringe with 1× PBS. Pellet down the bone marrow cells from the final soup. In the centrifugation method, the bones are cut down proximal to the knee-end and placed downward in a 0.5 ml centrifuge tube perforated at the bottom. Stack the perforated tube inside a 1.5 ml centrifuge tube and spin down for 30 s at maximum RPM. Culture the red pellet

containing bone marrow cells in the appropriate medium for further processing. [25].

28. The ratio of human CD45 to the total CD45 expressing cells (including mouse and human) determines the percentage of human engraftment [26].
29. Proceed with multilineage engraftment analysis only if there is a significant amount of human CD45 expression.

References

1. Morgan RA, Gray D, Lomova A, Kohn DB (2018) Hematopoietic stem cell gene therapy—Progress and lessons learned. Cell Stem Cell 21(5):574–590
2. Melve GK, Ersvaer E, Eide GE, Kristoffersen EK, Bruserud O (2018) Peripheral blood stem cell mobilization in healthy donors by granulocyte colony-stimulating factor causes preferential mobilization of lymphocyte subsets. Front Immunol 9:845
3. Cavazzana M, Bushman FD, Miccio A, André-Schmutz I, Six E (2019) Gene therapy targeting haematopoietic stem cells for inherited diseases: progress and challenges. Nat Rev Drug Discov 18(6):447–462
4. Gaj T, Staahl BT, Rodrigues GMC, Limsirichai P, Ekman FK, Doudna JA, Schaffer DV (2017) Targeted gene knock-in by homology-directed genome editing using Cas9 ribonucleoprotein and AAV donor delivery. Nucleic Acids Res 45(11):e98
5. Richardson CD, Ray GJ, DeWitt MA, Curie GL, Corn JE (2016) Enhancing homology-directed genome editing by catalytically active and inactive CRISPR-Cas9 using asymmetric donor DNA. Nat Biotechnol 34(3):339–344
6. Pattabhi S, Lotti SN, Berger MP, Singh S, Lux CT, Jacoby K, Lee C, Negre O, Scharenberg AM, Rawlings DJ (2019) In vivo outcome of homology-directed repair at the HBB gene in HSC using alternative donor template delivery methods. Mol Ther Nucleic Acids 17:277–288
7. Bak RO, Dever DP, Porteus MH (2018) CRISPR/Cas9 genome editing in human hematopoietic stem cells. Nat Protoc 13(2): 358–376
8. Gaudelli NM, Lam K, Rees HA, Solá-Esteves NM, Barrera LA, Born DA, Edwards A, Gehrke JM, Lee S-J, Liquori AJ, Murray R, Packer MS, Rinaldi C, Slaymaker IM, Yen J, Young LE, Ciaramella G (2020) Directed evolution of adenine base editors with increased activity and therapeutic application. Nat Biotechnol 38(7):892–900
9. Tsai SQ, Zheng Z, Nguyen NT, Liebers M, Topkar VV, Thapar V, Wyvekens N, Khayter C, John Iafrate A, Le LP, Aryee MJ, Joung JK (2014) GUIDE-seq enables genome-wide profiling of off-target cleavage by CRISPR-Cas nucleases. Nat Biotechnol 33(2):187–197
10. Doman JL, Raguram A, Newby GA, Liu DR (2020) Evaluation and minimization of Cas9-independent off-target DNA editing by cytosine base editors. Nat Biotechnol 38(5): 620–628
11. Tsai SQ, Nguyen NT, Malagon-Lopez J, Topkar VV, Aryee MJ, Keith Joung J (2017) CIRCLE-seq: a highly sensitive *in vitro* screen for genome-wide CRISPR-Cas9 nuclease off-targets. Nat Methods 14(6):607–614
12. Gabriel R, Kutschera I, Bartholomae CC, von Kalle C, Schmidt M (2014) Linear amplification mediated PCR – localization of genetic elements and characterization of unknown flanking DNA. J Vis Exp (88):e51543
13. McIntosh BE, Brown ME, Duffin BM, Maufort JP, Vereide DT, Slukvin II, Thomson JA (2015) Nonirradiated NOD,B6.SCID Il2rγ-/-KitW41/W41 (NBSGW) mice support multilineage engraftment of human hematopoietic cells. Stem Cell Reports 4(2):171–180
14. Tajer P, Pike-Overzet K, Arias S, Havenga M, Staal FJT (2019) Ex vivo expansion of hematopoietic stem cells for therapeutic purposes: lessons from development and the niche. Cell 8:169
15. Psatha N, Reik A, Phelps S, Zhou Y, Dalas D, Yannaki E, Levasseur DN, Urnov FD, Holmes MC, Papayannopoulou T (2018) Disruption of the BCL11A erythroid enhancer reactivates fetal hemoglobin in erythroid cells of patients with b-thalassemia major. Mol Ther Methods Clin Dev 10:313–326

16. Radtke S, Pande D, Cui M, Perez AM, Chan Y-Y, Enstrom M, Schmuck S, Berger A, Eunson T, Adair JE, Kiem H-P (2020) Purification of human CD34+ CD90+ HSCs reduces target cell population and improves lentiviral transduction for gene therapy. Mol Ther Methods Clin Dev 18:679–691
17. Drake AC, Khoury M, Leskov I, Iliopoulou BP, Fragoso M, Lodish H, Chen J (2011) Human $CD34^{+}CD133^{+}$hematopoietic stem cells cultured with growth factors including Angptl5 efficiently engraft adult NOD-SCID $Il2r\gamma^{-/-}$ (NSG) mice. PLoS One 6:e18382
18. Radtke S, Görgens A, Kordelas L, Schmidt M, Kimmig KR, Köninger A, Horn PA, Giebel B (2015) CD133 allows elaborated discrimination and quantification of hematopoietic progenitor subsets in human hematopoietic stem cell transplants. Br J Haematol 169:868–878
19. Gomez-Ospina N, Scharenberg SG, Mostrel N, Bak RO, Mantri S, Quadros RM, Gurumurthy CB, Lee C, Bao G, Suarez CJ, Khan S, Sawamoto K, Tomatsu S, Raj N, Attardi LD, Aurelian L, Porteus MH (2019) Human genome-edited hematopoietic stem cells phenotypically correct Mucopolysaccharidosis type I. Nat Commun 10(1):4045
20. Heckl D, Kowalczyk MS, Yudovich D, Belizaire R, Puram RV, McConkey ME, Thielke A, Aster JC, Regev A, Ebert BL (2014) Generation of mouse models of myeloid malignancy with combinatorial genetic lesions using CRISPR-Cas9 genome editing. Nat Biotechnol. 10(1):4045
21. Richter MF, Zhao KT, Eton E, Lapinaite A, Newby GA, Thuronyi BW, Wilson C, Koblan LW, Zeng J, Bauer DE, Doudna JA, Liu DR (2020) Phage-assisted evolution of an adenine base editor with improved Cas domain compatibility and activity. Nat Biotechnol 38(7): 883–891
22. Wagenblast E, Azkanaz M, Smith SA, Shakib L, McLeod JL, Krivdova G, Araújo J, Shultz LD, Gan OI, Dick JE, Lechman ER (2019) Functional profiling of single CRISPR/Cas9-edited human long-term hematopoietic stem cells. Nat Commun 10(1):4730
23. Hess NJ, Lindner PN, Vazquez J, Grindel S, Hudson AW, Stanic AK, Ikeda A, Hematti P, Gumperz JE (2020) Different human immune lineage compositions are generated in non-conditioned NBSGW mice depending on HSPC source. Front Immunol 11:573406
24. Fiorini C, Abdulhay NJ, McFarland SK, Munschauer M, Ulirsch JC, Chiarle R, Sankaran VG (2017) Developmentally-faithful and effective human erythropoiesis in immunodeficient and Kit mutant mice. Am J Hematol 92(9):E513–E519
25. Amend SR, Valkenburg KC, Pienta KJ (2016) Murine hind limb long bone dissection and bone marrow isolation. J Vis Exp (110):53936
26. Futrega K, Lott WB, Doran MR (2016) Direct bone marrow HSC transplantation enhances local engraftment at the expense of systemic engraftment in NSG mice. Sci Rep 6:23886

Chapter 21

Generation of Rat Neural Stem Cells to Produce Different Astrocyte Phenotypes

Rebecca Sherrard Smith, Susan C. Barnett, and Susan L. Lindsay

Abstract

Striatum-derived neural stem cells have been used to generate a variety of neural cell populations. They are composed of free-floating clusters of clonal neural stem cells, termed neurospheres, and can be expanded under growth factor stimulation in vitro. The multipotent nature of neurospheres means that under certain growth conditions they can differentiate into neuronal and glial progenitors of the central nervous system (CNS).

Here, we describe a method for creating a population of astrocytes derived from rat striatum neurospheres, which in turn can be used to generate astrocytes with different reactivity phenotypes. Several methods and techniques are already available for the generation of neurospheres, but the method detailed herein provides an accessible, reproducible protocol for large numbers of astrocyte cultures, that can then be manipulated in an experimental format for further investigation.

Key words Astrocytes, Neural stem cells, Neurospheres, Reactivity, CNS

1 Introduction

Astrocytes are key players in many aspects of CNS disease and injury. Their diverse roles extend from energy metabolism, neurotransmitter homeostasis, axonal outgrowth, support of myelination to the maintenance of the blood–brain barrier (BBB) and synaptogenesis [1–7]. Astrocytes, which normally exist in a resting, quiescent state in vivo, transform into hypertrophic, reactive cells (astrocytosis) that express many different proteins after injury or during disease [8]. This reactive state is typically characterized by the upregulation of the intermediate filament proteins glial fibrillary acidic protein (GFAP) and nestin, cellular hypertrophy, and proliferation [9–12]. It has been suggested that astrocytes form a continuum of phenotypes ranging from activated to highly reactive [13] which may reflect whether they hinder or support CNS recovery [13, 14]. The activated phenotype of astrocytes secrete a variety of enzymes, growth and trophic factors, and antioxidants [2] and

Nagarajan Kannan and Philip Beer (eds.), *Stem Cell Assays: Methods and Protocols*, Methods in Molecular Biology, vol. 2429, https://doi.org/10.1007/978-1-0716-1979-7_21,

are associated with improved tissue remodelling and recovery in vivo [2, 13, 15]. More recently, reactive astrocytes have been classified into phenotypes, termed A1 and A2, which were induced by neuroinflammation and ischemia, respectively; the A2 phenotype being more protective to neural cells than the A1 phenotype [9]. However, it is likely that these two types are still part of several phenotypes that form a continuum [9].

It is well accepted that there is a need for a greater understanding of astrocyte reactivity and its role in CNS repair and disease. Indeed, several in vitro models which manipulate the extracellular matrix astrocytes grow on, have been proposed to allow the investigation of astrocyte phenotype [16, 17]. In vitro, astrocytes are typically plated on the standard electrostatic attachment compound, poly-L-lysine (PLL-astrocytes) and are thought to adopt a more activated phenotype than that typically seen in vivo [17]. However, when astrocytes are plated on top of the extracellular matrix glycoprotein, Tenascin C (TnC), it has been suggested that they acquire a quiescent phenotype, more akin to their resting in vivo state [16–18]. Interestingly, it was shown that when mixed embryonic spinal cord cells were plated on top of TnC-astrocytes myelination was significantly lower when compared to when plated on PLL-astrocytes [17]. In addition, other reactivity phenotypes can be mimicked by treating astrocyte cultures with cytokines, such as ciliary neurotrophic factor (CNTF), a cytokine known to induce an activated phenotype in vivo [2]. Moreover, CNTF treated astrocytes were found to better promote in vitro myelination compared to PLL-astrocytes, suggesting again the importance of astrocyte phenotype in CNS repair [17]. Therefore, here we will describe a method to produce consistent astrocyte cultures derived from rat striatum neurospheres, which can then be used to further investigate a range of functional astrocyte phenotypes already reported to influence in vitro myelination.

2 Materials

2.1 Cell Culture

1. Neurosphere culture medium (NSM): Dulbecco's Modified Eagle's Medium/Nutrient Mixture F-12 (DMEM/F12 1:1, containing 4500 mg/L glucose), supplemented with 0.105% $NaHCO_3$, 2 mM glutamine, 5000 IU/mL penicillin, 5 μg/mL streptomycin, 5.0 mM HEPES, 25 μg/mL human insulin solution, 100 μg/mL apotransferrin, 60 μM putrescine, 20 nM progesterone, and 30 nM sodium selenite (*see* **Note 1**).
2. Recombinant murine epidermal growth factor (EGF): final concentration 20 ng/mL.

3. Astrocyte medium: DMEM containing 1 g/mL glucose supplemented with 10% heat-inactivated fetal bovine serum (FBS) and 2 nM L-glutamine.
4. Leibovitz's L-15 Medium.
5. Gentamycin solution (50 mg/ml): final concentration 50 μg/mL.

2.2 Immunocytochemistry

1. Phosphate-buffered saline (PBS): 137 mM NaCl, 2.7 mM KCl, 10 mM Na_2HPO_4, 1.8 mM KH_2PO_4, pH 7.4.
2. Fixative: 4% paraformaldehyde in phosphate-buffered saline (PFA).
3. Blocking buffer: 0.1% Triton X-100 and 0.2% gelatin from porcine skin in PBS.
4. Primary antibodies: rabbit anti-glial fibrillary acidic protein (GFAP; 1:500, Z0334, DAKO, UK, Intracellular); mouse anti-nestin, clone rat-401 (1:200, MAB353, Merck-Millipore, UK, Intracellular); mouse anti-O4 premyelinating oligodendrocytes (O4; 1:1, hybridoma, Cell Surface).
5. Secondary antibodies: goat anti-rabbit IgG (H + L) highly cross-adsorbed secondary antibody, Alexa Fluor 488); goat anti-mouse IgG1 cross-adsorbed secondary antibody, Alexa Fluor 568 (1:500) and Alexa Fluor 488); goat anti-mouse igm (heavy chain) cross-adsorbed secondary antibody, Alexa Fluor 488 (1:500) and Alexa Fluor 568 (1:500).
6. Antifade mounting medium: Vectashield containing DAPI.

2.3 Tissue Dissection

1. Dissection stereo microscope with cold light source.
2. Sterile surgical scalpel blades (No. 22).
3. Dumont superfine straight tweezers (N°5), Dumont tweezers curved (N°5), and Knapp curved scissors.
4. 1 × 90 mm tissue culture dish containing 20 mL L-15
5. 1 × Ice box
6. 1 × 7 mL bijou containing 1 mL L-15.

2.4 Cell Culture

1. Water bath set to 37 °C.
2. 1 × T-75 vented tissue culture flask, canted neck, surface area 75 cm^2, capacity 250 mL
3. P10 and P1000 pipettes.
4. Serological pipettes.
5. Electronic pipette controller.
6. Benchtop centrifuge with the capacity to spin at 1200 rpm.
7. 5 and 10 mL syringes

8. 21G needles or glass Pasteur pipettes
9. 15 mL and 50 mL centrifuge tubes
10. 13 mm round, size 0 cover glasses
11. 24-well clear tissue culture-treated plates.

2.5 Immunocytochemistry

1. Humidified staining tray (*see* **Note 2**).
2. Dumont superfine curved tweezers (N°5).
3. Microscope slides: Frosted tips (Ground 90°, 20 mm).
4. Fluorescence microscope (e.g., Olympus BX51 [Olympus, UK]).

3 Methods

3.1 Animal Preparation

1. Neurospheres are generated from the striata of postnatal day 1 (P1) Sprague-Dawley (SD) rats. The SD rats are housed under a 12 h light–dark cycle. Animals are euthanized using an overdose of pentobarbitone solution (200 mg/mL) administered via 50 mL (100 mg/kg) intraperitoneal injection, followed by cutting of the femoral artery to confirm euthanasia (*see* **Note 3**).
2. Once animals are euthanized, perform decapitation with a single incision using Knapp curved scissors at the base of the brainstem. Heads of animals should be kept immersed in fresh L-15 medium containing gentamycin in 50 mL tubes and kept on ice.

3.2 Dissection

1. Prepare a bijou with 1 mL and a 90 mm tissue culture dish with 20 mL sterile L-15 containing gentamycin (*see* Fig. 1).
2. Sterilize the dissection tools by soaking in 70% methylated spirits. Make sure the dissecting instruments are free of methylated spirits before commencing the dissection by air-drying on a sterile surface.
3. Secure the head on a cork board using a 21G needle through the nasal cavity or hold securely between your thumb and first forefinger (use the skin to help maintain a tight grip).
4. Remove the outer skin from the head by making a midline longitudinal cut from the base of the head to the top using the curved scissors. Peel skin away to reveal the skull. The loose skin created can be further used to retain good grip.

   Next, remove the skull by making a clean incision at the base of the skull and cutting horizontally around the side toward the eyes longitudinally to just above the olfactory bulb.

   Do this in both directions so that the cuts meet each other. Avoid damaging the brain by keeping the scissors low to the

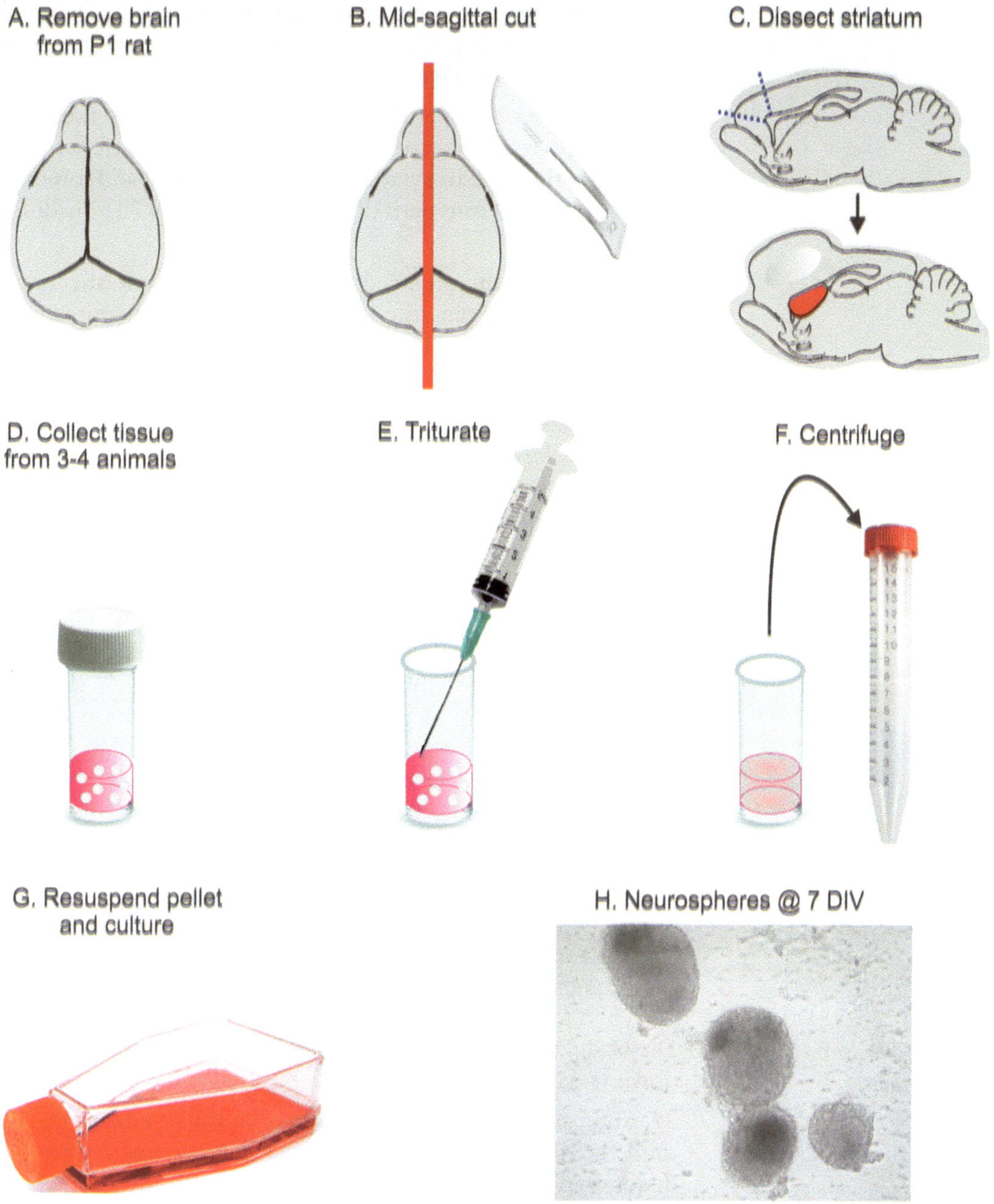

Fig. 1 Schematic depicting the generation of neurospheres from P1 rat brains. (**a**) Carefully dissect the brain from 1-day postnatal (P1) rat pups and place in a 90 mm petri dish containing 20 mL L-15 and gentamycin (50 μg/mL). (**b**) Using a number 22 sterile scalpel blade, half the brain along the mid sagittal plane, separating the two brain hemispheres. (**c**) Orientate each half so the inner sagittal surface is facing upward. Using fine forceps gently make two cuts in the outer cortex (purple dotted lines) and pull open the cortex tissue flap created to help expose the striatum region, which is bean shaped and located near the corpus callosum (red circle, 3). (**d**) Remove the striatum carefully using fine forceps and place tissue collected from 3 to 4 animals into a bijou containing 1 mL L-15 containing gentamycin (50 μg/mL). (**e**) Dissociate the tissue pieces by triturating through a 23G needle and syringe (alternatively use a glass Pasteur pipette and rubber teat). Be careful not to cause excessive bubbles. (**f**) When the medium appears cloudy transfer the dissociated tissue

skull outer edge and by pointing them outward away from the brain during cutting. Gently peel away and discard the skull using the curved forceps, exposing the brain.

5. Using either a small spatula, or curved forceps, scoop the brain out (being careful to keep your spatula or forceps as close to the bottom of the skull to avoid damaging the brain) and carefully transfer into a 90 mm petri dish containing fresh L15 medium (*see* Fig. 1a).
6. Position the brain on its dorsal surface and cut the brain mid-sagittal between the two hemispheres (hemisection) using a sterile scalpel blade (*see* Fig. 1b).
7. Under a stereo microscope at ×10 magnification using the superfine straight forceps, gently position each hemisphere onto its outer side with the inner sagittal surface facing up (medial view, *see* Fig. 1c).
8. Locate the corpus striatum brain region which sits just under the corpus collosum and looks a little darker than the surrounding tissue. The striatum can be better visualized by making two cuts in the frontal cortex using the superfine forceps (*see* Fig. 1c, dashed purple lines) and open and fold outward the cortex tissue flap created (*see* Fig. 1c).
9. Remove the striatum by gently teasing around the bean shaped area (*see* Fig. 1c, red region) using the superfine straight forceps and remove the tissue and place it into a 5 mL bijou containing 1 mL L-15 medium. Remove the striatum from the other hemisphere following the same method. To the bijou add the striata of 3–4 animals. Keep the bijou on ice as you collect sufficient tissue (*see* Fig. 1d).

The next section of this method should be performed in a class 2 safety cabinet under aseptic conditions.

3.3 Tissue Dissociation

1. Dissociate the tissue using a P1000 pipette by pipetting up and down (*see* **Note 4**).
2. Triturate gently using a 21G needle (or a glass Pasteur pipette) until the medium goes cloudy, a sign that the tissue has been dissociated to a single cell suspension (*see* Fig. 1e).
3. Transfer the suspension to a 15 mL tube and centrifuge for 5 min at 300 × *g* (*see* Fig. 1f).

Fig. 1 (continued) into a 15 mL tube and centrifuge for 5 min at 1200 rpm. (**g**) Discard the supernatant and resuspend the tissue pellet in 20 mL neurosphere medium (NSM) containing epidermal growth factor (EGF, 20 ng/mL). Feed every other day by adding 5 mL NSM medium and EGF (20 ng/mL). Do not remove any liquid during feeding. (**h**) After 7 days in vitro (DIV) free floating neurospheres will form which can be used to generate astrocyte monolayers (*see* Fig. 2)

4. Remove the supernatant and discard. Resuspend the pellet in 2 mL NSM culture medium.
5. Add 18 mL NSM to a T-75 cm^2 vented flask along with 4 μL EGF. Add the cell suspension to the flask and incubate at 37 °C in a humidified incubator under 7% CO^2 for 7–10 days until free floating neurospheres form (*see* Fig. 1g).
6. Cultures are maintained by feeding every 2 days by adding 5 mL NSM and 4 μL EGF. Do not remove any medium when feeding (*see* Fig. 1h and **Note 5**).

3.4 Differentiation of Neurospheres into Astrocytes

1. After 7–10 days in vitro, transfer all the medium containing the free floating neurospheres from the flask to a 50 mL tube using a serological pipette (*see* Fig. 2c and **Note 6**).
2. Using a benchtop centrifuge spin down for 5 min at 300 × *g*.
3. Carefully remove the supernatant and discard.
4. Resuspend the pellet in 2 mL low glucose DMEM supplemented with 10% fetal bovine serum and 2 mM L-glutamine (astrocyte medium).

   Using a 5 mL syringe and 21G needle, carefully triturate the cells until cloudy (*see* Fig. 2d). This should take between 5 and 7 triturations. Repeat if larger clumps are still visible in the cell suspension. After trituration, add an additional 46 mL astrocyte medium to make a final volume of 48 mL (*see* Fig. 2e).
5. The cell suspension from one flask of neurospheres can be used to set up 4 × 24-well plates that contain precoated poly-l-lysine (PLL) (*see* Subheading 3.5) or Tenascin C coated coverslips (*see* Subheading 3.5). Plate each coverslip with 500 μL of the cell suspension (*see* Fig. 2f).
6. Maintain plates in an atmosphere of 7% CO^2 at 37 °C and feed by removing 400 μL of medium and adding 500 μL astrocyte medium every 3 days and incubate for a further 7–10 days in vitro (DIV) until confluent (*see* Fig. 2g).

3.5 Astrocyte Phenotypes

Various astrocyte phenotypes, which have been previously defined on their ability to enhance or inhibit myelination [17], can be induced at this stage using the below methods/treatments. *Carry out each procedure in a laminar flow cabinet under aseptic conditions.*

1. *Control astrocytes:* Neurospheres should be plated onto 13 mm round coverslips, size 0 which have been precoated with PLL (13.3 μg/mL, in distilled water). To coat coverslips with PLL, place glass coverslips in a petri dish containing PLL (13.3 μg/mL) in 20 mL distilled autoclaved water submerging all the coverslips. Using the tip of a pipette, spread the coverslips

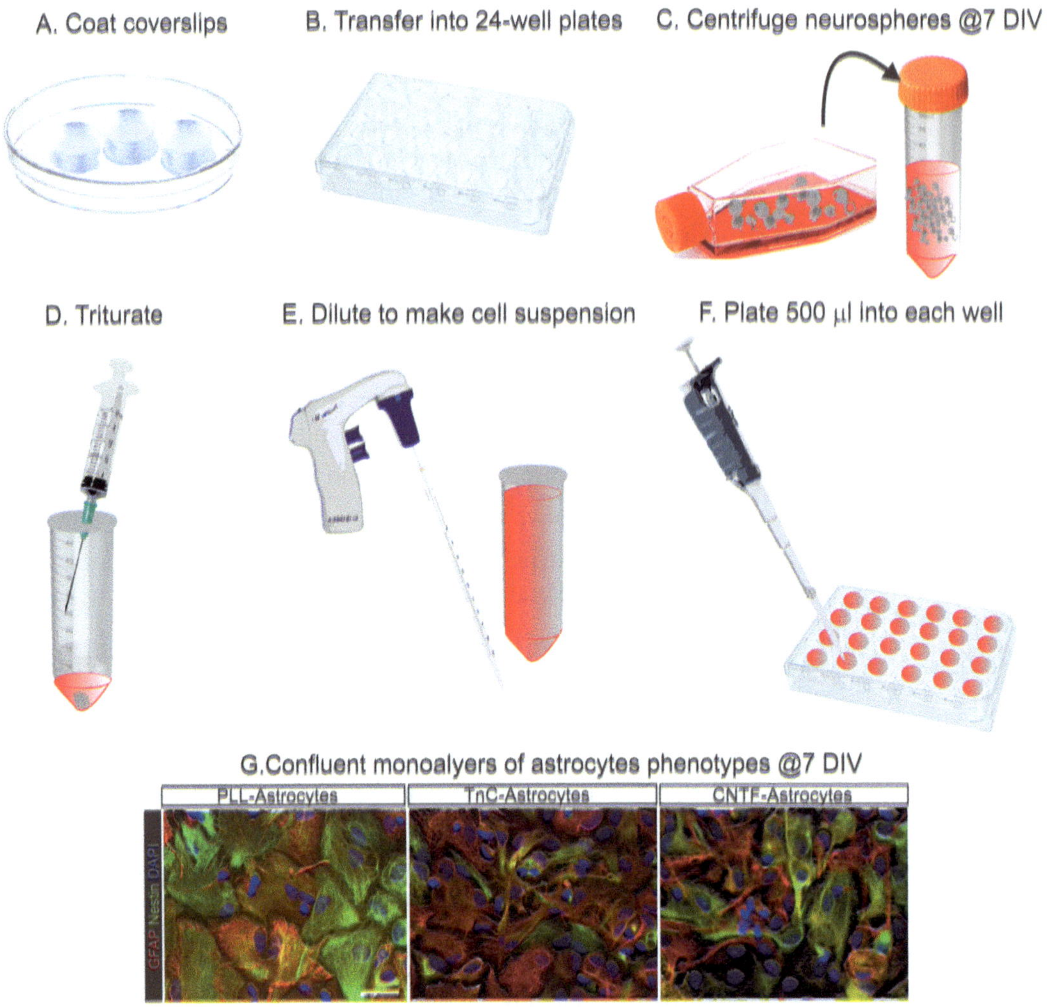

Fig. 2 Schematic depicting the generation of astrocyte monolayers grown on coated glass coverslips derived from neurospheres. (**a**) Prepare poly-L-lysine (PLL) coated glass coverslips by placing them in a 90 mm petri dish containing PLL (13 mg/mL) in 20 mL water. Incubate at 37 °C for a minimum of 30 min. Alternatively coat in Tenascin C (TnC) following the method detailed in Subheading 3.5. (**b**) Using a set of curved forceps place one coverslip into each well of a 24-well plate. One batch of coverslips will fill 4 × 24-well plates. Leave to air-dry in a tissue culture hood. Plates can be stored wrapped in Parafilm at 4 °C, until required. (**c**) Harvest neurospheres at 7 days in vitro (DIV) (as detailed in Fig. 1) and transfer growth medium into a 50 mL tube and centrifuge for 5 min at 300 × *g*. (**d**) Discard the supernatant and resuspend pellet in 2 mL astrocyte medium (DMEM containing fetal bovine serum (FBS) and gentamycin (50 mg/mL)). (**e**) Triturate spheres using a 23 G needle and syringe until dissociated, and the medium is cloudy. (**f**) Make volume to 48 mL using astrocyte medium and seed 500 μL onto each coverslip. (**g**) After 7 DIV confluent astrocyte monolayers can be stained for GFAP (shown in green) and nestin (shown in red). Astrocytes grown on PLL (PLL-astrocytes) are control astrocytes; when plated on TnC (TnC-astrocytes) are a quiescent phenotype. When PLL-astrocytes are treated with CNTF (2 ng/mL; CNTF-astrocytes) they are akin to an activated phenotype. Scale bar represents 25 μm

evenly to ensure all coverslips are well coated. Incubate for up to 2 h at 37 °C (minimum of 30 min). Remove from incubator and wash three times with autoclaved water using an electronic pipette controller and a 25 mL serological pipette. Using sterile curved forceps, place a single coverslip into each well of a 24-well plate. Leave the plate in the hood to dry *(leaving lid off)*. After drying, plates can be stored wrapped in Parafilm and stored at 4 °C, until required. (*see* Fig. 2a).

2. *Quiescent astrocytes*: Neurospheres should be plated onto 13 mm round coverslips, size 0 which have been precoated in TnC (5 μg/mL, in distilled water). To coat coverslips with Tenascin C (TnC), use sterile forceps to place a single glass coverslip into each well of a 24-well plate. In a 50 mL centrifuge tube make up 5 μg/mL of TnC with autoclaved sterile water. Pipet 300 μL of TnC matrix onto the centre of each coverslip. Replace lid. Incubate plate for 20 min at 37 °C. Return plate to laminar flow hood and carefully remove the TnC matrix using either a Gilson pipette or an electronic pipette controller with a serological pipette. Leave the plate in the hood to dry *(leaving lid off)*. Plate the astrocytes by transferring 500 μL of the single cell suspension (*described in methods*) into the center of the coverslip.
3. *Activated astrocytes*: to induce activated astrocytes; neurospheres plated on PLL coated astrocytes after 7 DIV can be treated with recombinant rat CNTF (2 ng/mL) for 24 h. Carefully remove all medium from each well of the 24-well plate containing astrocyte monolayers using either a Gilson pipette or an electronic pipette controller with a serological pipette. Make up CNTF treatment using the astrocyte medium at a concentration of (2 ng/mL). Carefully add 500 μL of treatment to each well and leave for 24 h. Maintain plates in an atmosphere of 7% CO^2 at 37 °C.

3.6 Immunocytochemistry

Purity and morphology of astrocyte monolayers can be assessed by staining using the following protocol. Using curved forceps, carefully remove each coverslip from the 24-well plate and remove excess medium by blotting the edge of the coverslip against tissue paper.

1. Wash the coverslip 3 times by gentle immersion in three individual containers with a volume of at least 50 mL PBS, sequentially. Place each coverslip in a humified staining tray (*see* **Note 2**).
2. For cell surface markers, antibodies should be diluted in astrocyte medium and added to live cells for 20 min prior to fixation. Note: typically, a volume of 50 mL is sufficient to cover the entire surface.

3. For all intracellular antibodies (or after live cell surface staining), coverslips should be fixed using 4% PFA for 20 min at room temperature (RT). *Note:* if only staining live cell surface markers move to **step 9** after fixation.
4. Coverslips should then be washed 3 times in PBS using the method described above.
5. Following this, cells should be permeabilized with 0.2% Triton X-100 at RT for 15 min.
6. After washing 3 times in PBS, cells should be blocked for 20 min in blocking buffer (BB).
7. Drain coverslips on tissue and add intracellular marker antibodies diluted in BB for 1 h.
8. Coverslips should then be washed 3 times in PBS using the method described above.
9. Appropriate secondary fluorochrome class specific antibodies are then incubated for 30–40 min diluted in BB.
10. Coverslips should be washed 3 times in PBS and mounted inverted onto a glass slide using Vectashield mounting medium.
11. Coverslips should then be sealed using nail varnish and stored at 4 °C until image acquisition.
12. In general, cells derived from neurospheres are approximately 80% positive for GFAP when plated directly onto glass coverslips. There are other small populations of contaminating glial cells types, such as O4 positive oligodendrocytes and microglia. Should purified astrocytes be required, neurospheres can instead be plated directly into 4 × T75 cm^2 flasks after trituration, rather than directly on coverslips (at the point shown in Fig. 2f). Upon reaching confluency, contaminating cells that lie on the astrocyte surface can easily be shaken off by tapping the flask firmly on the side 3 to 4 times or by using an orbital shaker for a few hours at a low speed. Medium should be removed, and cells washed using astrocyte medium. Purified astrocytes can then be passaged following standard methods and seeded onto coverslips following the protocol described in Fig. 2.

4 Notes

1. If making neurospheres regularly it is useful to make a 10× stock of the mixture of hormones that are used in the medium (hormone mix). This can be stored frozen at −20 °C and thawed when required. To make 250 mL 10× hormone mix, combine 215 mL DMEM/F12 (4500 mg/mL glucose) supplemented with 2 mM L-glutamine and penicillin–

streptomycin (100 units/mL). Add 5 mL 30% glucose, 3.75 mL 7.5% $NaHO_3$, 1.2 mL 1 M HEPES, 250 mg apo-transferrin. Dilute 6.25 mL human insulin solution (concentration 10 mg/mL) in 19.75 sterile double distilled water (ddH_20), add the 25 mL to mix. Add 25 mL putrescine solution (600 mM stocks are made by adding 96.6 mg putrescine to 100 mL ddH2O, store in 25 mL aliquots at −20 °C until required). Add 25 μL of selenium (3 mM stocks are made by adding 1 mg sodium selenite to 1.93 mL sterile water, store at −20 °C in 25 mL aliquots), and finally add 25 μL of progesterone (2 mM stocks are made by adding 1 mg progesterone to 1.59 mL 95% EtOH, store at −20 °C in 25 mL aliquots). Filter through a sterile 500 mL capacity 0.22 μm filter unit and aliquot the 10× hormone mix in 25 mL. Label and store at −20 °C. Stocks should be kept for no longer than 1 year. When preparing fresh neurosphere culture medium thaw the 25 mL aliquot of 10× hormone stock and combine with an additional 5 mL 30% glucose, 3.75 mL 7.5% $NaHCO_3$, 1.25 mL 1 M HEPES, and 215 mL DMEM/F12 (4500 mg/mL glucose) supplemented with 2 mM L-glutamine and penicillin-streptomycin (100 units/mL).

2. To minimize the amount of antibody used in the immunohistochemistry protocol avoid staining coverslips directly in the 24-well plates. Instead, coverslips can be inverted onto a small 50 μL drop of antibody solution on Parafilm strips placed within a coverable box. Ensure the box can either be placed in a dark area or wrapped with foil. Alternatively, you can create a staining tray by gluing 24 eppendorf lids in rows onto the lid of a polystyrene lab box (120 L × 75 H mm). The lids allow the coverslips to sit proud and receive small 50 μL volumes of antibody on the cell surface. The tray can be kept humidified by placing wet tissue paper along the box edges. The box can be wrapped in foil during the staining process.
3. In general, a single flask of neurospheres requires the tissue from 3–4 animals, so arrange the required number of animals ahead of dissection to reduce waste.
4. Do not pipet too aggressively as excessive mechanical dissociation can lead to increased cell death. Aggressive pipetting can also lead to the production of bubbles which can destroy cells during centrifugation. Slow pipetting action should reduce this, although a few bubbles can still occur.
5. Neurospheres remain free floating while being maintained and therefore it is important to never remove any of the medium, as doing so would remove cells from the culture. Instead, as described, only add fresh prewarmed medium to the flask.

6. After 7 days some cells can begin to stick to the base side of the flask. Therefore, when transferring the medium to a fresh centrifuge tube, gently tap the flask and rinse the cells off the base using a 25 mL pipette and the medium in the flask.

References

1. Ullian EM, Sapperstein SK, Christopherson KS et al (2001) Control of synapse number by glia. Science 291:657–661
2. Liberto CM, Albrecht PJ, Herx LM et al (2004) Pro-regenerative properties of cytokine-activated astrocytes. J Neurochem 89:1092–1100
3. Silver J, Miller JH (2004) Regeneration beyond the glial scar. Nat Rev Neurosci 5: 146–156
4. Pellerin L (2005) How astrocytes feed hungry neurons. Mol Neurobiol 32:59–72
5. Nair A, Frederick TJ, Miller SD (2008) Astrocytes in multiple sclerosis: a product of their environment. Cell Mol Life Sci 65:2702–2720
6. Sorensen A, Moffat K, Thomson C et al (2008) Astrocytes, but not olfactory ensheathing cells or Schwann cells, promote myelination of CNS axons in vitro. Glia 56:750–763
7. Watkins TA, Emery B, Mulinyawe S et al (2008) Distinct stages of myelination regulated by gamma-secretase and astrocytes in a rapidly myelinating CNS coculture system. Neuron 60:555–569
8. Zamanian JL, Xu L, Foo LC et al (2012) Genomic analysis of reactive astrogliosis. J Neurosci 32:6391–6410
9. Liddelow SA, Barres BA (2017) Reactive astrocytes: production, function, and therapeutic potential. Immunity 46:957–967
10. Oberheim NA, Goldman SA, Nedergaard M (2012) Heterogeneity of astrocytic form and function. Methods Mol Biol 814:23–45
11. Oberheim NA, Takano T, Han X et al (2009) Uniquely hominid features of adult human astrocytes. J Neurosci 29:3276–3287
12. Pekny M, Nilsson M (2005) Astrocyte activation and reactive gliosis. Glia 50:427–434
13. Sofroniew MV, Vinters HV (2010) Astrocytes: biology and pathology. Acta Neuropathol 119: 7–35
14. Anderson MA, Burda JE, Ren Y et al (2016) Astrocyte scar formation aids central nervous system axon regeneration. Nature 532: 195–200
15. Faulkner JR, Herrmann JE, Woo MJ et al (2004) Reactive astrocytes protect tissue and preserve function after spinal cord injury. J Neurosci 24:2143–2155
16. Holley JE, Gveric D, Whatmore JL et al (2005) Tenascin C induces a quiescent phenotype in cultured adult human astrocytes. Glia 52: 53–58
17. Nash B, Thomson CE, Linington C et al (2011) Functional duality of astrocytes in myelination. J Neurosci Off J Soc Neurosci 31: 13028–13038
18. Nash B, Ioannidou K, Barnett SC (2011) Astrocyte phenotypes and their relationship to myelination. J Anat 219:44–52

Chapter 22

In Situ Quantification and Isolation of Müller Glial Cells by Fluorescence-Activated Cell Sorting from the Regenerating Larval Zebrafish Retina

Jeffrey Stulberg and Vincent Tropepe

Abstract

Müller glia (MG) are a relatively quiescent radial glial cell population capable of dedifferentiating to regenerate cells in the zebrafish retina that are lost due to damage. Here, we provide a protocol to both quantify MG cell dedifferentiation behavior during a regenerative response and isolate MG cells by fluorescence activated cell sorting (FACS). First, the retina is exposed to high-intensity light to induce retinal damage and either processed for immunohistochemistry or live MG cells are isolated by FACS that can be used for subsequent genomic or transcriptomic analyses. This method allows us to correlate MG cell behavior observed in situ with their transcriptomic profile at different stages during the regenerative response.

Key words Neural stem cells, Retina, Müller glia, Zebrafish, Regeneration, FACS, Cell quantification

1 Introduction

Some types of mature vertebrate cells can undergo a genomic reprogramming event (often referred to as dedifferentiation) by reverting to a stem cell-like state [1]. Müller glial (MG) cells of the zebrafish retina have this unique property [2–5]. Following retinal injury, MG cells will acquire stem cell characteristics by dedifferentiation followed by cell cycle reentry to generate multipotent progenitor cells that produce new tissue to repair the retina [3, 5].

MG cell behavior is highly dynamic during regeneration [3, 6, 7]. This requires having methods to interrogate both the cellular and molecular phenotype at specific stages during the regenerative response. In addition, the retina is a highly heterogeneous tissue which limits the ability to quantify changes in gene expression from total retinal RNA extracts. Here, we demonstrate a protocol to both quantify MG cells in histological sections using 3D image

Nagarajan Kannan and Philip Beer (eds.), *Stem Cell Assays: Methods and Protocols*, Methods in Molecular Biology, vol. 2429, https://doi.org/10.1007/978-1-0716-1979-7_22,

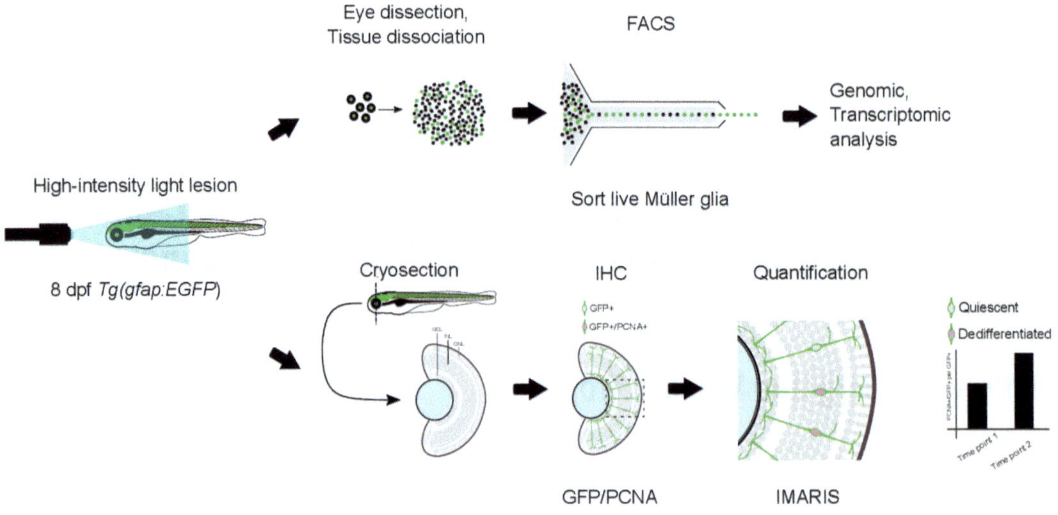

Fig. 1 Flow diagram illustrating the steps involved in isolating live MG cells (top) and processing lesioned retinal tissue for quantification (bottom)

analysis software (IMARIS), and isolate MG cells from the retina using an MG-specific reporter by fluorescence activated cell sorting (FACS).

We adopted a high-intensity light lesion paradigm that specifically ablates photoreceptors to experimentally induce MG cell dedifferentiation and cell cycle re-entry [3, 8]. Zebrafish larvae harboring the MG-specific (in the retina) transgenic, *Tg(gfap:EGFP)* [4] are processed postlesion to either perform immunohistochemistry (IHC) using antibodies that label proliferating MG cells (PCNA), or larval eyes are dissected and dissociated to isolate MG cells via FACS. Viable MG cells are sorted based DAPI exclusion ($GFP^{+}/DAPI^{-}$) (Fig. 1). The resulting MG cell suspension can be used for downstream applications such as RT-qPCR or RNA-sequencing.

This method allows us to correlate MG cell behavior observed in situ with their transcriptomic profile obtained from isolating MG cells by FACS at the sampled stages during the regenerative response.

2 Materials

2.1 High-Intensity Light Lesion

1. 50 mL and 600 mL glass beakers.
2. 2 mL Pasteur pipette.
3. UV goggles.
4. GEMMA Micro 75 food (Skretting).
5. LED light source (e.g., X-Cite 120LED Boost; Excelitas Technologies).

2.2 Eye Dissection

1. 100 mm × 15 mm polystyrene petri dish.
2. Fine forceps.
3. 0.016% Tricaine.
4. Calcium-free Ringer's solution: 116 mM, NaCl, 2.9 mM KCl, 5.0 mM HEPES. Add 500 mL of distilled water to a beaker. Add 6.78 g NaCl, 0.22 g KCl, and 50 mL 100 mM HEPES (pH 7.2) to the beaker. Add water to bring the final volume to 1 L. Adjust the pH to 7.4 and then pour the solution through a 0.22-μM filter. Autoclave prior to use.

2.3 Tissue Dissociation

1. Leibovitz's L-15 Medium (1×), with L-glutamine but no phenol red.
2. Stem cell grade fetal bovine serum (FBS) (e.g., EmbryoMAX ES cell qualified; Sigma-Aldrich).
3. Papain Dissociation System (Worthington Biochemical Corporation LK003150) (*see* **Note 1**).
   (a) DNase (2000 U/mL): Resuspend lyophilized vial in 0.5 mL Leibovitz's L-15.
   (b) Papain (1 mM L-cysteine, 0.5 mM EDTA, 0.005% DNase): Resuspend in 5 mL Leibovitz's L-15 and place in 37 °C incubator for 10 min or until the papain is fully dissolved. Add 250 μL DNase (above) and pH to 7.4. Prepare fresh.
   (c) Ovomucoid albumin inhibitor (10 mg/mL): Resuspend lyophilized vial in 32 mL Leibovitz's L-15.
4. Resuspension buffer: 86% Leibovitz's L-15, 9.5% albumin ovomucoid inhibitor, 4.7% DNase. Add 540 μL Leibovitz's L-15 with 60 μL albumin ovomucoid inhibitor and 30 μL DNase per tube.
5. Cell sort buffer: 10% FBS, 90% Leibovitz's L-15. Mix 100 μL FBS with 900 μL Leibovitz's L-15. Prepare fresh.
6. 5 ¾″ glass Pasteur pipette.
7. 5 mL polypropylene round-bottom tubes with 35 μM strainer cap.

2.4 Cryosectioning and Immunohistochemistry

1. Paraformaldehyde: 4% paraformaldehyde in PBS. Add 2 g paraformaldehyde to 30 mL PBS. Top up to 50 mL. Store at 4 °C for up to 1 month.
2. Adhesive microscope slides (75 × 25 mm).
3. Optimal cutting temperature compound (OCT).
4. Sucrose: 30% sucrose in PBS. Mix 15 g sucrose in 15 mL PBS. Top up to 50 mL.

5. 1× phosphate-buffered saline (PBS): Make a 10× solution by adding 800 mL purified water to a 1 L beaker. Weigh and dissolve 80 g NaCl, 2 g KCl, 14.4 g Na_2HPO_4 and 2.4 g KH_2PO_4. pH to 7.4 and autoclave. Mix 5 mL 10× PBS with 45 mL purified water.
6. Normal goat serum.
7. PAP Pen.
8. Sodium citrate buffer: 10 mM sodium citrate, 0.05% Tween 20, pH 6.0. Add 950 mL autoclaved ddH2O to a 1 L storage beaker. Weigh 2.94 g trisodium citrate (dihydrate) and dissolve in beaker. Add 0.5 mL Tween 20. Adjust to pH 6 with 1 M HCl then adjust water to 1 L. Can be stored for 3 months at room temperature.
9. PBS, Tween 20 (PBT) (1× PBS, 0.1% Tween 20): Add 5 mL 10× PBS, 50 μL Tween 20 and 45 mL autoclaved ddH_2O to a 50 mL Falcon tube.
10. Blocking solution: 1× PBS, 0.2% Triton X/Tween 20, 2% normal goat serum. Add 977 μL 1× PBS, 2 μL Triton X-100 or Tween 20 and 20 μL normal goat serum. Use Triton X-100 and Tween 20 in blocking solution on day 1 and day 2, respectively.
11. Hoechst 33342, 10 mg/mL.
12. Anti-GFP rabbit antibody (e.g., G10362; Life Technologies).
13. Anti-PCNA mouse antibody (P8825, Sigma-Aldrich).
14. Anti-Cy3 rabbit antibody.
15. Anti-Cy5 mouse antibody.

2.5 Zebrafish Lines and Larval Care

1. All experiments are described using the *Tg(gfap:EGFP)*(*see* **Note 2**) [4] transgenic line.
2. Male and female adult transgenic fish are bred according to the zebrafish book [9].
3. Embryos are collected from breeding tanks after 1 h of mating.
4. Embryos are manually dechorionated in embryo medium (or system water) using fine forceps at 1 day postfertilization (dpf).
5. Larvae are fed 2× per day using Gemma Micro 75 food from 6 dpf onward and maintained in a 28 °C incubator.
6. No more than 50 larvae are kept in a single 100 × 20 mm glass petri dish.
7. Larvae are kept incubated in dark conditions until the lesion. Exceptions include feeding and cleaning the petri dish to remove debris or to euthanize sick larvae.

2.6 Fluorescence Activated Cell Sorting

1. Recommended machines: BD FACS Aria IIIu (100 μM nozzle, 20 psi) or BD Influx sorter (100 μM nozzle, 17 psi).
2. Laser requirements: 405 nm (Hoechst) and 488 nm (GFP).

2.7 Confocal Microscopy

1. Recommended instrument: Leica SP8 confocal microscope with a 40× objective.
2. Laser requirements: 405 nm (Hoechst) and 638 nm (Cy5) with HyD detectors; 552 nm (Cy3) laser with a PMT detector.

3 Methods

In addition to the light-lesioned treatment, a nonlesioned control should be processed in parallel.

3.1 Photoreceptor-Specific Lesion Using High-Intensity Light on Zebrafish Larvae

1. Grow *Tg(gfap:EGFP)* larvae to 8 dpf following the guidelines in Subheading 2.5.
2. Place 10–100 larvae in a 50 mL beaker with 20 mL of system water.
3. Place the 50 mL beaker in a larger 600 mL beaker containing 150 mL of room temperature water (*see* **Note 3**).
4. Align the beaker with the LED light source such that the light source is level with the water and is 3 cm away from the center of the beaker.
5. While wearing UV safety goggles turn on the light source and set the intensity to 120,000 lux (or 40% on the dial) (*see* **Note 4**). Illuminate larvae for 30 min (*see* **Note 5**) (Fig. 2).
6. Turn off light source and return larvae to petri dish with fresh system water.
7. During the course of the experiment remove any dead larvae or euthanize those that develop edema.
8. If performing IHC, place larvae in a glass vial and fix in 1 mL 4% paraformaldehyde overnight at 4 °C at desired time postlesion and proceed to Subheading 3.5. If isolating MG cell suspension for gene expression analyses (RNA-seq, RT-qPCR, etc.) proceed to Subheading 3.2.

3.2 Eye Dissection of Zebrafish Larvae

Non–GFP-expressing larvae should be dissected and dissociated in parallel to be used as a negative control during FACS (*see* **Note 6**).

1. Place larvae in 28 °C Calcium-free Ringer's Solution in plastic petri dish and anesthetize with 0.016% tricaine. Begin dissection once larvae no longer exhibit a touch response.
2. With clean forceps, grab the base of the head and anchor the larva to the bottom of the dish. Then, gently pull off each eye. Be careful not to damage eyes.

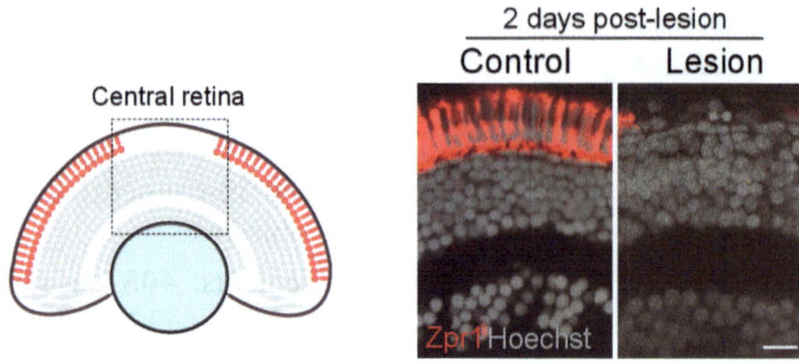

Fig. 2 Light damage induces loss of photoreceptors in the central retina. (Left) schematic of the zebrafish retina with a dashed box highlighting the central region. (Right) Loss of Zpr1 (double-cone marker) and Hoechst immunoreactivity in the central retina by 2 dpl (days postlesion). Scale bar = 10 μM

3. Using a glass Pasteur pipette, transfer eyes into 1.5 mL Eppendorf tube on ice.
4. Repeat this process until all eyes have been dissected and collected (*see* **Note 7**).
5. Immediately proceed to Subheading 3.3.

3.3 Dissociation of Cells from the Zebrafish Eye

1. Remove Ringer's solution from Eppendorf tube containing eyes and add 1 mL of room temperature papain. Incubate at 37 °C for ~30 min. For the first 10 min of incubation, agitate the solution by gentle inversion every 3 min (this prevents the tissue from clumping). For the remaining 20 min, using a glass Pasteur pipette gently pipette solution up and down 10 times every 3 min (*see* **Note 8**). Tissue should be incubated until completely dissociated (*see* **Note 9**).
2. Remove samples from incubator and spin down in a centrifuge at 300 × *g* for 5 min at 4 °C.
3. Remove supernatant and resuspend pellet in 630 μL resuspension buffer. The resuspension buffer stops the activity of papain.
4. In a 2 mL Eppendorf tube, add 1 mL of albumin ovomucoid inhibitor and then add the cell suspension by pipetting slowly onto the side of the tube.
5. Spin down at 100 × *g* for 6 min at 4 °C.
6. Carefully remove supernatant (pellet can be easily dislodged) and resuspend pellet in 0.2 mL cell sort buffer, then pipette solution through a 35 μ*M* strainer cap into a 5 mL polypropylene round-bottom tube.
7. Add DAPI to a final concentration of 0.2 ng/nL 5–10 min before performing the sort (*see* **Note 10**).

8. Prepare the collection tubes by adding 100 μL FBS and 100 μL Leibovitz's L-15 into a 1.5 mL Eppendorf tube (*see* **Note 11**).
9. Promptly sort cells (*see* Subheading 3.4).

3.4 Isolation of Live MG Cells by Fluorescence Activated Cell Sorting

1. Vortex samples immediately before placing into the FACS instrument. Cells are gated based on forward scatter (FSC) and side scatter (SCC) to remove cell debris, and pulse width to remove cell doublets and aggregates (Fig. 3a, b).
2. The nonfluorescent sample is used to setup the gating parameters for GFP and DAPI (Fig. 3b, c).
3. Live MG cells are sorted based on GFP inclusion and DAPI exclusion ($GFP^+/DAPI^-$) (Fig. 3d). The cells are sorted into 1.5 mL collection tubes containing 200 μL 1:1 FBS–Leibovitz's L-15 (*see* **Notes 12** and **13**).
4. The cells are immediately used for genomic or transcriptomic analysis by lysing the cells and isolating DNA or RNA.

3.5 Cryosectioning and Immunohistochemistry of Larval Zebrafish

1. Remove paraformaldehyde solution from vial (*see* Subheading 3.1, **step 8**) and wash larvae 3× with 1× PBS for 5 min.
2. Cryoprotect samples by performing 5× washes (30 min each) using increasing concentrations of 5% to 30% sucrose in 1× PBS. Keep samples in 30% sucrose at 4 °C overnight (*see* **Note 14**).
3. The following morning, remove 30% sucrose and add a 2:1 mixture of 30% sucrose and OCT; incubate for 2 h at room temperature. After incubation, samples can be kept at −20 °C until sectioning.
4. Section larval eyes using a freezing cryostat (20 μM sections) and mount on Superfrost Plus slides (Sigma-Aldrich). Allow sections to dry for 2 h at room temperature.
5. If using anti-PCNA for IHC proceed to **step 6**, otherwise rehydrate samples in 1× PBS for 30 min and proceed to **step 11** (*see* **Note 15**).
6. Preheat sodium citrate in Coplin jar by microwaving ~1 min (or until it begins to boil).
7. Place Coplin jar in hybridization oven set to 95 °C and allow it to equilibrate for 10 min.
8. Place slides containing histological sections in Coplin jar and incubate at 95 °C for 15 min (*see* **Note 16**), then immediately place slides in new Coplin jar filled with room temperature PBT and let cool 10 min.
9. Working quickly, take slides out of Coplin jar, dry perimeter of the slide with a Kimwipes™ and draw boundary with PAP pen. Do not allow slides to dry out.

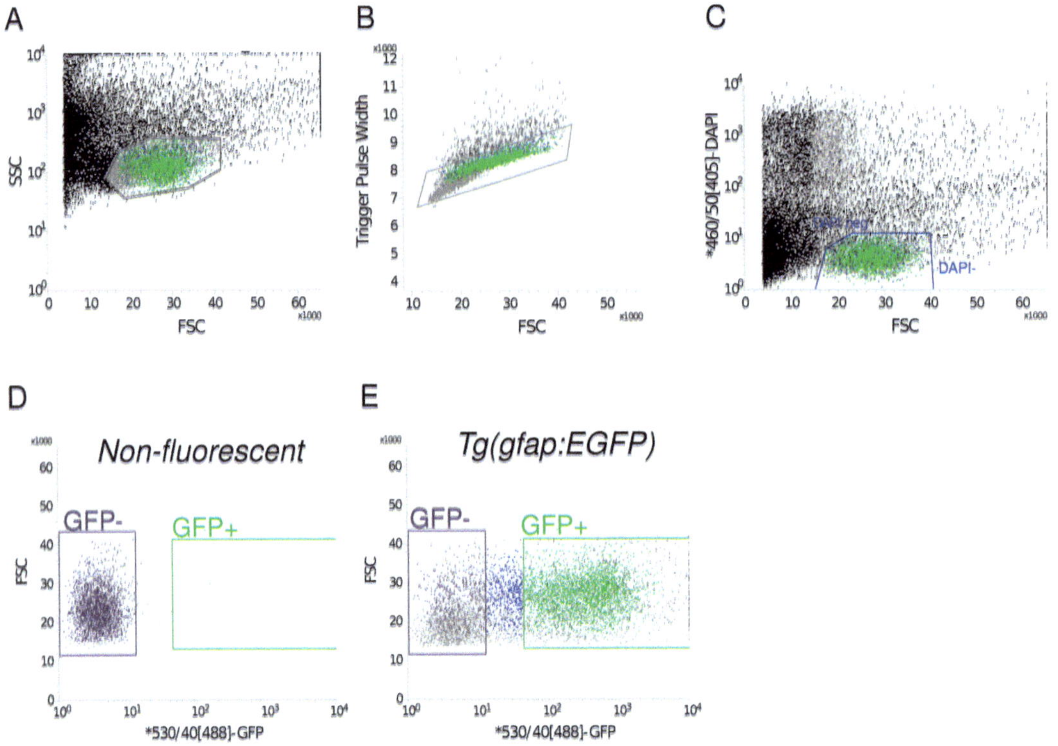

Fig. 3 FACS plots for gate setting to sort live MG cells. (**a**) Cells are identified based on FSC and SSC. (**b**) Cellular aggregates are excluded based on trigger pulse width and FSC. (**c**) Live cells are identified by DAPI-negative staining. (**d**) Nonfluorescent fish are used to set gates for GFP-positive and GFP-negative bins. (**e**) GFP + cells from gfap:EGFP fish are isolated based on the gating parameters established

10. Wash 2× with 300 μL PBT for 5 min at room temperature.
11. Add 300 μL blocking solution and incubate for 1 h at room temperature.
12. Add primary antibodies to blocking solution (anti-PCNA 1: 1000, anti-GFP 1:1000) and incubate samples with 300 μL overnight at 4 °C (*see* **Note 17**).
13. The following morning remove the primary antibody solution and wash 4× with 300 μL PBT for 10 min at room temperature.
14. Add secondary antibody to blocking solution (Cy3 1:500, Cy5 1:200) and incubate samples with 300 μL for 2–4 h at room temperature or 1–2 h at 37 °C in the dark.
15. Remove secondary antibody and wash 4× with 300 μL PBT for 10 min at room temperature.
16. Add 300 μL Hoechst 33342 (10 μg/mL) diluted in ddH_2O and incubate for 15 min at room temperature.
17. Remove Hoechst 33342 and wash 2× with 300 μL PBS for 5 min at room temperature.

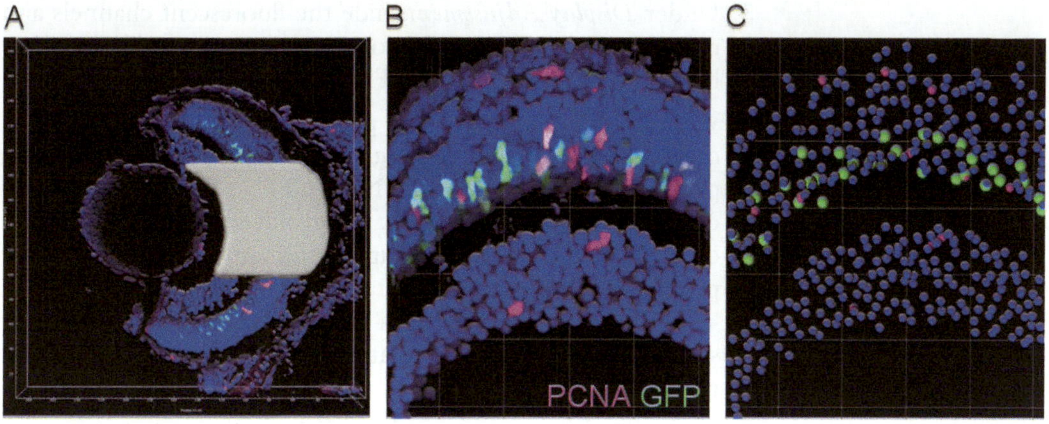

Fig. 4 Quantification of cells using IMARIS. (**a**, **b**) The central retina is cropped. (**c**) Hoechst+, GFP+, PCNA+ cells are quantified using the *spots* function

18. Remove solution and apply 40 μL 80% glycerol. Place coverslip on and seal edges with clear nail polish in a fume hood.
19. Slides can be kept in the dark at 4 °C for several months.

3.6 Confocal Imaging of Larval Zebrafish Retinal Sections

1. The three central retinal sections of each retina are imaged with a Leica SP8 confocal microscope using a 40× objective to obtain z-stack images (1 μM) through the entire thickness of the sections.
2. The 405 nm (Hoechst), 638 nm (Cy5), and 552 nm (Cy3) lasers are used sequentially with a 600 Hz bidirectional scan at 1024 × 1024 resolution.

3.7 Quantification of Proliferating MG Cells

1. Open Imaris 7.7.1.
2. Using the *surfaces* feature, crop each channel (Hoechst, PCNA, GFP) to include only the central region of the retina (*see* **Note 18**). This is done by using *Draw* (*Surpass* > *Surfaces* > *Skip automatic creation, edit manually* > *Contour* tab > *Draw*) to define the boundaries for cropping the central retina. Next, select *Copy* and set the slice position to 1, then *Paste* the boundary and select *create surface*—this will create the region to crop. Finally, select *edit* > *mask all* and the select channel to crop (this must be performed for each channel) (Fig. 4a, b).
3. In the *Statistics* tab, the area of the cropped region can be recorded to normalize cell counts (*see* **Note 19**).
4. Using the *Spots* feature, set the diameter for spot detection (Hoechst = 2.84 μM, PCNA = 3 μM, gfap:EGFP = 4 μM) (*see* **Note 20**) and automatically quantify cells for each channel. The cell counts are recorded in the *Statistics* tab.

5. Under *Display Adjustment*, hide the fluorescent channels and manually count the double-positive spots to quantify the number of PCNA+/GFP+ cells (Fig. 4c).
6. We will often quantify the percentage of proliferating and quiescent MG cells within a given retina: PCNA+/GFP+ cells per total GFP+ cells and PCNA-/GFP+ cells per total GFP+ cells, respectively. To compare the proliferative response of MG cells between treatment groups we quantify the percentage of proliferating MG cells per total Hoechst+ cells or area: PCNA +/GFP+ per total Hoechst+ cells, or per defined area (per section averaged over 3 central sections for each biological replicate).

4 Notes

1. We use Leibovitz's L-15 opposed to Earl's Balanced Salt Solution (EBSS) with the papain kit because it improves the viability of MG cells during the tissue dissociation and FACS, and does not need to be equilibrated with CO_2.
2. This line can be purchased from the Zebrafish International Resource Center (ZIRC): http://zebrafish.org/fish/lineAll.php?OID=ZDB-GENO-080606-263.
3. The 150 mL water helps mitigate large changes in temperature from the light source during the lesion.
4. Higher intensity light or longer light exposure results in more extensive retinal damage; however, larval survival declines significantly.
5. Nonlesion control larvae are exposed to identical conditions except without light exposure.
6. A negative (nonfluorescent) control is needed to set the gating parameters for fluorescence intensity during FACS.
7. Eye dissection should be done in <1 h.
8. The tissue is less likely to stick to glass compared to a plastic Pasteur pipette. Avoid bubbling the sample when pipetting.
9. Once tissue is dissociated keep the cells on ice throughout the procedure.
10. Do not add DAPI to the negative (nonfluorescent) control sample.
11. If RNA is to be extracted following FACS, TRIzol can be added to the collection tube instead of FBS. This will immediately lyse cells and stabilize the RNA. If TRIzol is not permitted to be used because the sorter is not housed in an appropriate fume hood then the cells can be sorted into FBS, spun down at

500 × *g* for 5 min and resuspended in an appropriate volume for TRIzol.

12. From 100 eyes, or 50 larvae (8 dpf), as starter tissue we often get 100,000–200,000 $GFP^+/DAPI^-$ cells from the sort. We have found that ~80% of cells remain viable in FBS/L-15 based on DAPI exclusion for up to 30 min postsort. Longer times have not been empirically tested.
13. The $GFP^+/DAPI^-$ population ranges from 40–80% of the parent population—a much higher proportion than our in situ estimates of 5–10% in the retina. Additional sorting experiments have indicated that this is due to neuronal cell death during the dissociation and sorting procedure.
14. We perform successive washes with 5%, 2:1 5%:30%, 1:1 5%:30%, 1:2 5%:30%, and 30% sucrose in 1× PBS.
15. The anti-PCNA antibody requires a heat-induced epitope retrieval (HIER) to expose the PCNA antigen site prior to IHC.
16. We have found that incubating the sample for 5–15 min on sectioned tissue is sufficient to expose the PCNA antigen. However, prolonged heat exposure may destroy antigen sites of other antibodies. If immunolabeling using multiple antibodies is desired, it must be empirically tested to determine if each antibody works after heat treatment. Antibody staining using GFP (as in this protocol) works well after 15 min of heat treatment, but the intensity of the label is often weaker compared to no heat treatment. Generally, a shorter heat treatment can improve labeling.
17. Keep samples in a humid chamber to prevent evaporation. We place our samples in Tupperware containers filled with distilled water with the lid closed.
18. The central region of the retina is the where retinal damage is most prominent. We define the central retina by the width of the lens in the corresponding section. However, any region of interest can be used.
19. We normalize cell counts either to the total number of Hoechst + cells, GFP+ cells, or area measured.
20. We have found that setting the diameter to a value 0.15–1 μM less than the average cell diameter tends to produce more accurate results. For example, Hoechst nuclei are approximately 3 μM; however, we have found that spot detection performs much better when underestimating the size of the nucleus to be 2.84 μM.

Acknowledgments

We would like to thank the Faculty of Medicine Flow Cytometry Facility and the Department of Cell & Systems Biology Imaging Facility at the University of Toronto.

Funding: This work was supported by a Vision Science Research Program Scholarship (J.S.), and a OIRM/CFREF-Medicine by Design New Ideas Grant (V.T.) from the University of Toronto.

References

1. Takahashi K, Yamanaka S (2006) Induction of pluripotent stem cells from mouse embryonic and adult fibroblast cultures by defined factors. Cell 126:663–676
2. Fausett BV, Goldman D (2006) A role for α1 tubulin-expressing Müller glia in regeneration of the injured zebrafish retina. J Neurosci 26: 6303–6313
3. Bernardos RL, Barthel LK, Meyers JR, Raymond PA (2007) Late-stage neuronal progenitors in the retina are radial Müller glia that function as retinal stem cells. J Neurosci 27: 7028–7040
4. Bernardos RL, Raymond PA (2006) GFAP transgenic zebrafish. Gene Expr Patterns 6: 1007–1013
5. Ramachandran R, Fausett BV, Goldman D (2010) Ascl1a regulates Müller glia dedifferentiation and retinal regeneration through a Lin-28-dependent, let-7 microRNA signalling pathway. Nat Cell Biol 12:1101–1107
6. Nagashima M, Barthel LK, Raymond PA (2013) A self-renewing division of zebrafish Müller glial cells generates neuronal progenitors that require N-cadherin to regenerate retinal neurons. Development 140:4510–4521
7. Lahne M, Li J, Marton RM, Hyde DR (2015) Actin-cytoskeleton- and rock-mediated INM are required for photoreceptor regeneration in the adult zebrafish retina. J Neurosci 35: 15612–15634
8. Meyers JR, Hu L, Moses A et al (2012) β-catenin/Wnt signaling controls progenitor fate in the developing and regenerating zebrafish retina. Neural Dev 7:30
9. Westerfield M (2000) The zebrafish book. A guide for the laboratory use of zebrafish (Danio rerio), 4th edn. University of Oregon Press, Eugene

Chapter 23

Quantification and Clonal Culture of Neural Stem Cells from the Hippocampus of Adult Mouse

Yoko Amagase, Hiroko Izumi-Nakaseko, Atsushi Sugiyama, and Yoshinori Takei

Abstract

Visualizing markers for neural stem cells (NSCs) and morphological analysis are frequently used for identification of NSCs in tissues. However, NSCs are defined as cells with the ability to both self-renew and produce descendants that can differentiate into neurons, astrocytes and oligodendrocytes. The neural colony forming cell (NCFC) assay is a single-step semisolid based assay for the identification of NSCs. In this assay, NSCs generate clonally derived colonies due to their high proliferative potential. The relative comparison of NSC populations between tissues is possible by counting the colonies obtained from the NCSC assay. Furthermore, the colonies can be isolated to establish monolayer cultures of clonal NSCs. Using clonal cultures of NSCs, it is possible to assess differentiation stage and differentiation potential of each NSC. Here, we describe a semi quantitative method for the enumeration of NSCs using the NCFC assay, with slight modification from the original protocol (Louis et al., Stem Cells 26:988–996, 2008). A method to establish monolayer culture of NSCs from a colony derived from NCFC assay is also described.

Key words Adult neurogenesis, Hippocampus, Neural stem cells, Accumax, NCFC, Colony

1 Introduction

Adult neurogenesis in the brain was described in postnatal rats about a half century ago [1]. It took more than a decade to confirm the original report [2, 3] and to describe neurogenesis in adult mice [4–6]. In the adult mammalian hippocampus, neurogenesis is observed in the subgranular zone (SGZ) of the dentate gyrus, which is a narrow region between the granular cell layer and the hilus. In the neurogenic niche, neural stem cells (NSCs) called Type-1 cells produce both neurons and glia. Type-1 cells can be subdivided into quiescent and active populations [7–9]. Quiescent Type-1 cells typically exhibit radial morphology [9, 10]. A glial cell marker, glial fibrillary acidic protein (GFAP), and a neural progenitor marker, nestin, are both expressed in these cells [7, 9, 11]. Active

Nagarajan Kannan and Philip Beer (eds.), *Stem Cell Assays: Methods and Protocols*, Methods in Molecular Biology, vol. 2429, https://doi.org/10.1007/978-1-0716-1979-7_23,

Type-1 cells express markers similar to those expressed by quiescent Type-1 cells, but typically proliferate at greater frequency [7, 9, 11].

Type-1 cells pass through transit amplifying cells, referred to as Type-2a, Type-2b and Type-3 NSCs, before differentiating into postmitotic neurons [10, 12]. Nestin is expressed in Type-2a and Type-2b as well as Type-1 cells. However, doublecortin, of which expression generally correlates with commitment of cells to the neuronal lineage, is expressed only in Type-2b cells [10, 13]. Type-2a and Type-2b cells are morphologically small and display very short tangential processes, and are often found in clusters in the SGZ [10]. Although Type-3 cells do not express nestin, they are mitotically active and express more markers of neuronal differentiation than Type-2 cells [10].

As described above, NSCs in the hippocampus are heterogeneous and their differentiation stages are identifiable by expressions of stem cell markers and stage-specific markers. However, when those cells are needed to be cultured in vitro, such detection methods are not applicable. Neural colony forming cell (NCFC) assay followed by clonal culture of obtained NSC colonies is the only option for their identification. The NCFC assay is a single-step semisolid based assay. In the assay, clonally derived colonies in different sizes are generated in a semisolid medium. Colony size is used to identify stem cells; due to high proliferative potential of NSCs, they generate bigger colonies than those of progenitors and mature cells.

2 Materials

2.1 Mouse Dissection

1. Apparatus for inhalation anesthesia.
2. Isoflurane.
3. Pentobarbital.
4. 1 mL syringe for i.p. application of pentobarbital
5. 26G needle for i.p. application of pentobarbital
6. Dissection tools.
7. Phosphate-buffered saline, Ca^{2+} and Mg^{2+}-free (PBS).
8. 10 cm culture dish
9. Ice-bucket.

2.2 Preparation of Single Cell Suspension

1. Microscissors.
2. 5 mL snap-cap centrifuge tube
3. Accumax™ (Innovative Cell Technologies, Inc., San. Diego, USA).
4. Centrifuge for 5 mL snap-cap centrifuge tube.

5. Blue tips and 1.0 mL micropipette.
6. Water bath incubator shaker.
7. 100 μm cell strainer.

2.3 NCFC Assay

We recommend using NeuroCult™ NCFC Assay Kit (#05740 Stem Cell Technologies Inc., Vancouver, Canada).

1. NeuroCult™ NSC Basal Medium: DMEM/Ham's F-12 medium containing 0.6% (w/v) glucose, 0.1% (w/v) sodium bicarbonate, and 5 mM HEPES.
2. NeuroCult™ NSC proliferation supplement (10x): NeuroCult™ NSC Basal Medium containing 20 mM glutamate 230 μg/mL insulin, 1 mg/mL transferrin, 200 nM progesterone, 90 μg/mL putrescine, and 300 nM sodium selenite.
3. Fundamental proliferation medium: 9:1 mix of NeuroCult™ NSC Basal Medium and NeuroCult™ NSC proliferation supplement.
4. Trypan blue.
5. 10 μg/mL epidermal growth factor in PBS
6. 10 μg/mL fibroblast growth factor 2 (FGF2) in PBS
7. 2 mg/mL heparin in PBS
8. Complete NeuroCult™ Proliferation Medium (Adult Mouse): Fundamental Proliferation Medium containing 20 ng/mL EGF, 10 ng/mL FGF2 and 2 μg/mL heparin.
9. Semisolid basal medium: The appropriate volume of the following components is mixed in order, depending on the number of replicates. Here we mix the amount needed for two replicates or duplicate dishes.
   - 1.7 mL of NeuroCult™ NCFC serum-free medium without cytokines
   - 330 μL of NeuroCult™ NSC proliferation supplements
   - 6.6 μL of 10 μg/mL epidermal growth factor in PBS
   - 32 μL of penicillin–streptomycin (100x)
   - 3.3 μL of 10 μg/mL FGF2 in PBS
   - 3.3 μL of 2 mg/mL heparin in PBS.
10. NeuroCult™ Collagen solution: 3% (w/v) collagen.
11. Complete NeuroCult™ Replenishment Medium (Adult Mouse): Fundamental Proliferation Medium containing 0.5 μg/mL EGF, 0.25 μg/mL FGF2, and 0.1 mg/mL heparin.
12. 35 mm and 100 mm culture dish
13. CO_2 incubator.

2.4 Quantification of NCFC Assay Derived Colonies

1. Microscope with low-power objective lens (2×–5×).
2. Ocular micrometer.

2.5 Colony Isolation and Monolayer Culture

1. N2B27 medium containing EGF and FGF and heparin: DMEM/Ham's F-12 medium containing 0.5× N-2 supplement, 0.5× B27 supplement, 20 ng/mL EGF, 10 ng/mL bFGF, 2 μg/mL heparin, and 1× penicillin–streptomycin.
2. Complete NeuroCult™ collagenase solution.
3. 1.5 mL microtube
4. Centrifuge for 1.5 mL microtube.
5. PBS.
6. 24-well and/or 96-well culture plate

   Wells for NSC culture should be coated with poly-L-ornithine and laminin. A protocol of coating wells is as follows. All the steps in the protocol should be performed in sterile conditions.

   (a) Dilute poly-L-ornithine on ice using tissue culture-grade water (2 mg/mL). Aliquot the diluted poly-L-ornithine and store the aliquots at −80 °C.

   (b) Aliquot laminin on ice and store the aliquots at −80 °C. Thaw the laminin slowly at 2–8 °C. If the solution has been warmed up to room temperature and it gels, it cannot be used.

   (c) Coat wells of 24- or 96-well culture plate with poly-L-ornithine (20 μg/mL in tissue culture-grade water) and incubate at 37 °C overnight (or, at least for more than 2 h).

   (d) Aspirate poly-L-ornithine and rinse the well with sterile PBS. Do not let the well dry.

   (e) Add laminin (5 μg/mL in tissue culture-grade water) in the well and incubate at 37 °C for 2 h or longer. The coated wells can be stored at 4 °C until use.

   (f) Aspirate the laminin before plating cells.
7. Poly-L-ornithine.
8. Laminin.
9. Tissue culture-grade water.
10. Yellow tip and 200 μL micropipette.
11. Felt tip pen.
12. CO_2 incubator.

3 Methods

3.1 Dissection of Mouse

1. Decapitate adult mice under deep anesthesia. Isolate the brain and wash with ice-cold PBS.
2. Divide the brain into two, along the longitudinal fissure of the cerebrum. Remove the cerebellum and the brain stem to access the hippocampus. Then, separate the hippocampus from the cerebral cortex with a scalpel (*see* **Note 1**).
3. Transfer the hippocampus to a 100 mm culture dish containing fresh, sterile ice-cold PBS, and rinse.

3.2 Preparation of Single-Cell Suspension

1. Transfer the hippocampus to a sterilised 5 mL snap-cap centrifuge tube.
2. Using microscissors, cut the tissue into small pieces approximately 1 mm cube in size (*see* **Note 2**).
3. Add 3 volumes of Accumax™ to the snap-cap centrifuge tube (3 times volume of the tissue) and gently agitate the sample by pipetting (*see* **Note 3**).
4. Incubate the samples in a water bath incubator shaker at 30 °C for 60 min. To release more cells, gently agitate the sample by pipetting several times (*see* **Notes 3** and **4**).
5. Filter the sample through a 100 μm cell strainer to remove remaining tissues. Wash the filter once with PBS.
6. Centrifuge at 300 × *g* for 3 min to pellet the cells.
7. Carefully remove the supernatant (*see* **Note 5**).

3.3 NCFC Assay

1. Resuspend the cell pellet in fundamental proliferation medium.
2. Count viable cells using trypan blue on a hemocytometer.
3. Dilute cell suspension to a concentration of 1–2 × 10^7 cells/mL in Complete NeuroCult™ Proliferation Medium (Adult Mouse) (*see* **Note 6**).
4. For a 35 mm culture dish, mix 10 μL of the cell suspension with 1 mL of semisolid basal medium.
5. Suspend the cells gently in the medium, add 0.7 mL of cold NeuroCult™ Collagen solution to the cell suspension and then mix well by gentle pipetting to avoid introducing any bubbles.
6. Dispense the cell suspension to the center of a 35 mm culture dish and gently swirl the dish to spread the cells evenly. (*see* **Note** 7).
7. Place the cell-seeded 35 mm culture dish with a lid, and the water-filled 35 mm culture dish without the lid, into a 100 mm culture dish. Cover the 100 mm culture dish with a lid to maintain the humidity.

8. Transfer the 100 mm culture dish to an incubator set at 37 °C, 5% CO_2 and 100% humidity. The collagen congeals with increasing temperature (*see* **Note 8**).
9. Incubate the cell culture for 21 days (*see* **Note 9**).
10. As cultures are incubated for an extended period of time, replenish the cultures with the appropriate Complete NeuroCult™ Replenishment Medium (Adult Mouse). Gently (so as not to disrupt the gel) add 60 μL of the Complete NeuroCult™ Replenishment Medium (Adult Mouse) into the centre of each NCFC Assay dish once every 7 days during the entire NCFC Assay culture incubation.

3.4 Quantification of NCFC Assay Derived Colonies

Using low-power (2×–5×) objective lens and ocular micrometer, scan each dish and score the colonies based on their sizes (*see* **Notes 10** and **11**).

3.5 Colony Isolation and Monolayer Culture

1. Take a yellow-tip with clean tweezers and cut the tip to widen the diameter enough to capture a colony.
2. Mark the colony that is to be isolated with a felt tip pen on the bottom of the dish (*see* **Note 12**).
3. Aliquot 100 μL of NeuroCult™ Collagenase Solution into individual sterile 1.5 mL microtubes (one tube for each colony being isolated).
4. Pick the colony up by sticking the wide-hole yellow tip (prepared in **step 1**), transfer to 1.5 mL microtube containing collagenase (prepared in **step 3**) and at 37 °C for 10 min (*see* **Note 13**). Gently agitate the sample by pipetting several times.
5. Centrifuge at 300 × g for 3 min to pellet the cells.
6. Wash the cell pellet twice with PBS.
7. Resuspend the cell pellet in 100 μL of N2B27 medium containing EGF, FGF2 and heparin.
8. Aliquot an appropriate volume of N2B27 medium containing EGF, FGF2 and heparin into each well of a 24- or 96-well culture plate coated with poly-L-ornithine and laminin. Generally, cells isolated from colonies >2 mm in diameter should be plated into a 24-well plate, cells isolated from colonies <2 mm in diameter should be plated into a 96-well plate.
9. Plate all of the cells isolated from a single colony into an individual well of a 24- or 96-well culture plate prepared in **step 8**.
10. Incubate at 37 °C in 5% CO_2 with 100% humidity (*see* **Notes 14** and **15**).

4 Notes

1. Stereoscopic microscopes may help for the dissection.
2. The tube should be on ice to avoid temperature rise.
3. To avoid damages to the cells, we use 1 mL micropipette with blue-tip of which the tip is cut to widen the diameter. After cutting the tip, the tip should be exposed to a naked flame to make the edge smooth.
4. Accumax™ can be replaced by collagenase, with following modification. Incubation temperature is 37 °C and the cells should be washed at least twice with PBS after collagenase treatment. Residual collagenase can be harmful for NSCs.
5. In our hands, tissue digestion by Accumax™ can be also applied to the preparation of single-cell suspensions from the cortex, the spinal cord and the adipose tissue of adult mouse.
6. The cell density is set higher than the original NCFC assay protocol [14], because the number of NSCs in the hippocampus gradually decreases in an age-dependent manner [9, 12]. The final cell number plated can be approximately 1–2 × 10^5 cells for a 35 mm culture dish for the hippocampus derived from a 35–40-week-old male C57BL/6 mouse. When older mice are used, the number of cells for a 35 mm culture dish should be increased. In some cases, the final cell plating density needs to be adjusted by performing a cell titration experiment. For statistical analyses, the total number of colonies detected after 21 days of culture should be within the range of at least 50–150 colonies. Typical results of colony counting are described in Table 1.
7. We recommend placing the 35 mm dishes on ice to avoid unwanted gel formation.
8. We recommend leaving the cultures overnight without any disturbance.
9. Cultures should be visually assessed regularly for overall colony growth and morphology using an inverted microscope. Do not leave cultures at room temperature for an extended period of time as the collagen gel will begin to liquefy and the colonies will not stay intact.
10. This method allows calculation of the frequency of NSCs, but not the actual number in the tissue. To identify differentiation stage of NSCs, immunofluorescence with antibodies against markers specific for each differentiation stages is required.
11. Instead of an ocular micrometer, a gridded scoring dish, which is included in NeuroCult™ NCFC Assay Kit, can be used for estimation of colony size. The scoring dish has 5 mm grid at

Table 1
Typical results of colony counting. 2×10^5 cells derived from 40-week-old male C57BL/6 mice were seeded in a 35 mm cell culture dish, as described in the main text

Colony diameter	Mouse-1	Mouse-2	Mouse-3
>2 mm	1	0	2
1–2 mm	12	3	8
0.5–1 mm	15	23	18
<0.5 mm	21	28	31
(total colony number)	49	54	59

the bottom, and is not recommended in our experience for estimation of colonies of 1–2 mm diameter. Typical results of colony counting are described in Table 1.

12. When isolating colonies, select the colonies that are not close to other colonies in order to avoid contamination with other colonies. Therefore, the colony number in a 35 mm culture dish should be around 20–50 for isolation of colonies.
13. When the colony is not digested enough, dissociate the colony by pipetting and incubate for a further 5 min to get single-cell suspension. Longer incubation time may cause damage to the cells.
14. We recommend not to disturb the culture for the first 3 days, to allow adhesion of NSCs. After cell adhesion is observed, medium should be changed every day. If neurospheres are observed, harvest neurospheres and repeat the **steps 3** to **10** of Subheading 3.5.
15. For immunofluorescence of cells derived from each colony to identify their differentiation stages, we recommend seeding the cells in a well of μ-Slide 8 well™ (#80826, iBidi GmbH, Gräfelfing, Germany). This is a chambered polymer coverslip with eight wells. It allows better coating efficiencies with poly-L-ornithine and laminin than a glass coverslip, which results in better cell adhesion. These features are helpful for observing cell morphology and to avoid loss of cells during immunofluorescence experiments. Typically, cells derived from colonies of more than 1 mm diameter express Type-1 markers, such as SRY-box 2 (Sox2), nestin, and GFAP. In our hands, expression of Type-2 markers, such as neurogenin 2 and doublecortin, is observed in roughly 10–15% colonies which are more than 1 mm in diameter.

Acknowledgments

This work was supported by JSPS Grant-in-Aid for Exploratory Research (Exploratory), Grant Number 18K18454.

References

1. Altman J, Das GD (1965) Autoradiographic and histological evidence of postnatal hippocampal neurogenesis in rats. J Comp Neurol 124:319–335
2. Kaplan MS, Hinds JW (1977) Neurogenesis in the adult rat: electron microscopic analysis of light radioautographs. Science 197:1092–1094
3. Bayer SA, Yackel JW, Puri PS (1982) Neurons in the rat dentate gyrus granular layer substantially increase during juvenile and adult life. Science 216:890–892
4. Reynolds BA, Weiss S (1992) Generation of neurons and astrocytes from isolated cells of the adult mammalian central nervous system. Science 255:1707–1710
5. Lois C, Alvarez-Buylla A (1993) Proliferating subventricular zone cells in the adult mammalian forebrain can differentiate into neurons and glia. Proc Natl Acad Sci U S A 90:2074–2077
6. Lois C, Alvarez-Buylla A (1994) Long-distance neuronal migration in the adult mammalian brain. Science 264:1145–1148
7. Suh H, Consiglio A, Ray J et al (2007) In vivo fate analysis reveals the multipotent and self-renewal capacities of Sox2+ neural stem cells in the adult hippocampus. Cell Stem Cell 1:515–528
8. Ehninger D, Kempermann G (2008) Neurogenesis in the adult hippocampus. Cell Tissue Res 331:243–250
9. Lugert S, Basak O, Knuckles P et al (2010) Quiescent and active hippocampal neural stem cells with distinct morphologies respond selectively to physiological and pathological stimuli and aging. Cell Stem Cell 6:445–456
10. Kempermann G, Jessberger S, Steiner B et al (2004) Milestones of neuronal development in the adult hippocampus. Trends Neurosci 27:447–452
11. Encinas JM, Vaahtokari A, Enikolopov G (2006) Fluoxetine targets early progenitor cells in the adult brain. Proc Natl Acad Sci U S A 103:8233–8238
12. Takei Y (2019) Age-dependent decline in neurogenesis of the hippocampus and extracellular nucleotides. Hum Cell 32:88–94
13. Uda Y, Xu S, Matsumura T et al (2016) P2Y4 nucleotide receptor in neuronal precursors induces glutamatergic subtype markers in their descendant neurons. Stem Cell Rep 6:474–482
14. Louis SA, Rietze RL, Deleyrolle L et al (2008) Enumeration of neural stem and progenitor cells in the neural colony-forming cell assay. Stem Cells 26:988–996

Chapter 24

Reprogramming Mouse Oviduct Epithelial Cells Using In Vivo Electroporation and CRISPR/Cas9-Mediated Genetic Manipulation

Matthew J. Ford and Yojiro Yamanaka

Abstract

Advances in gene editing tools such as CRISPR/Cas9 have made precise in vivo gene editing possible, opening up avenues of research into somatic cell reprograming to study adult stem cells, homeostasis, and malignant transformation. Here we describe a method for CRISPR/Cas9 mediated in vivo gene editing, in combination with Cre-based lineage tracing via electroporation in the mouse oviduct. This method facilitates the delivery of multiple plasmids into oviduct epithelial cells, sufficient for studying homeostasis and generation of high-grade serous ovarian cancer (HGSOC) models.

Key words Oviduct, Fallopian tube, CRISPR, Homeostasis, Ovarian cancer, HGSOC, In vivo gene editing

1 Introduction

Homeostasis is the process in which a cell, tissue, or organism maintains itself in a state of equilibrium. In the case of epithelia, which line the majority of organs, homeostasis consists of a balance between stem/progenitor cell maintenance, proliferation, and differentiation. Disruption of these processes has the potential to severely impede the function of organs and can lead to tumor formation [1]. Traditional methods to study the molecular control of homeostasis and malignant transformation in vivo required the generation of genetically engineered mouse models (GEMMs) [2, 3]. These models have provided valuable insight into gene function in vivo but require the propagation and mating of mouse lines which is expensive and time-consuming, making it difficult to screen the function of multiple genes. In the case of GEMMs of cancer, these models also suffer from non–tissue-specific Cre-lines which form tumors outside of the tissue of interest and do not accurately reflect the sporadic nature of malignant transformation

Nagarajan Kannan and Philip Beer (eds.), *Stem Cell Assays: Methods and Protocols*, Methods in Molecular Biology, vol. 2429,
https://doi.org/10.1007/978-1-0716-1979-7_24,

seen in human patients [4]. The recent advances in gene editing technologies such as CRISPR/Cas9 has opened new avenues for interrogating the function of multiple genes in vivo, which will prove particularly useful in the study of cancers that are genomically heterogenous such as high-grade serous ovarian carcinoma (HGSOC) [5, 6].

The oviduct, or fallopian tube as it is known in humans, is the most distal portion of the female reproductive tract connecting the ovary to the uterus. Epithelial cells lining the luminal surface of the oviduct provide an environment for successful fertilization and promote preimplantation embryonic development [7–9]. It is now recognized that a large portion of HGSOCs originate from the distal portion of the fallopian tube, igniting interest into understanding oviduct epithelial homeostasis and the mechanisms involved during malignant transformation [10–14].

Here we describe a highly flexible method for genome manipulation and lineage tracing of distal oviduct epithelial cells by a combination of in vivo electroporation, Cre-based lineage tracing and CRISPR-mediated gene editing. Adult mouse oviducts are surgically exposed and a solution of plasmids carrying Cas9, sgRNAs, and/or Cre recombinase is injected into the distal portion of the oviduct lumen, followed by electroporation with tweezer-type electrodes. Using this method, luminal epithelial cells in the distal region of the oviduct are electroporated without targeting stromal cells. The electroporation-based plasmid delivery method is sufficient to target many genes in a single cell and generate mouse models of pancreatic cancer and HGSOC [15, 16]. This method permits formation of sporadically mutated cells surrounded with unedited healthy normal cells, which is similar to the human tumorigenesis condition. Subsequent clonal selection of the mutated cells with a survival or proliferative advantage occurs with further genomic rearrangements leading to tumor formation.

The method presented is highly flexible for screening multiple gene combinations and for targeting different regions of the female reproductive tract. By combining with Cre-based lineage tracing and CRISPR-scar clonal tracing, it is possible to track the fate of targeted cells with high resolution [17]. This protocol will provide a versatile tool for the study of epithelial cell homeostasis and malignancies derived from the female reproductive tract.

2 Materials

2.1 Injection Solution

1. Filter-sterilized trypan blue.
2. Endotoxin free pX330-U6_BB-CBh-hSpCas9: encoding CAG promoter-driven codon-optimized Cas9 endonuclease and U6 promoter-driven sgRNA cassette (Addgene CAT# 42230).

Multiplex CRISPR-cas9 plasmids have also been developed, allowing for the expression of multiple gRNAs from a single plasmid (Addgene CAT#1000000055).

3. Endotoxin free pCS2 Cre-NLS: encoding CMV promoter-driven expression of Cre recombinase fused to a nuclear localization sequence [18].

2.2 Surgical Equipment

1. Sterile PBS.
2. Sterile saline.
3. Thin absorbent paper, such as Kimwipes (Kimberly-Clark).
4. Dumont forceps.
5. Dumont AA—Epoxy-coated forceps.
6. Adson forceps.
7. Curved iris serrated forceps.
8. Hartman hemostats.
9. Dissection scissors.
10. Sharp-blunt dissection scissors.
11. Glass capillary needles (1B100–3 (wpiinc) or B100–50-10 (Sutter) or equivalent).
12. Sterile petri dish.
13. Heat pad.
14. Micro bulldog clamp (3 cm).
15. Nonabsorbable silk braided sutures 5–0 (such as Stoelting #50487 or equivalent).
16. 1 mL syringe.
17. 27G needle.

2.3 Anesthetic, Antiseptic, and Pain Relief Solutions

1. Animal disinfectant solution (such as premise disinfectant).
2. Antiseptic solution (such as Baxedin (2% chlorhexidine in 70% isopropyl alcohol)).
3. Avertin (2,2,2-tribromoethanol). Stock solution is made at 1.6 g/mL in tert-amyl alcohol. On the day of surgery dilute 1.25% v/v in isotonic saline (20 mg/mL), mix vigorously and filter-sterilized through a 0.2-μm filter. Both stock solution and working dilutions are stored in the dark.
4. Carprofen (5 mg/kg dose) stored in the dark.
5. Eye lubricant.
6. Shaver or hair removal cream.

2.4 Mouse Lines

1. Cre-inducible fluorescent mouse lines such as Tdtomato, R26Fucci2a/R26Arl13bFucci2a (Cell cycle and cilia reporters) or Confetti (multicolor for clonal analysis) [19–22].

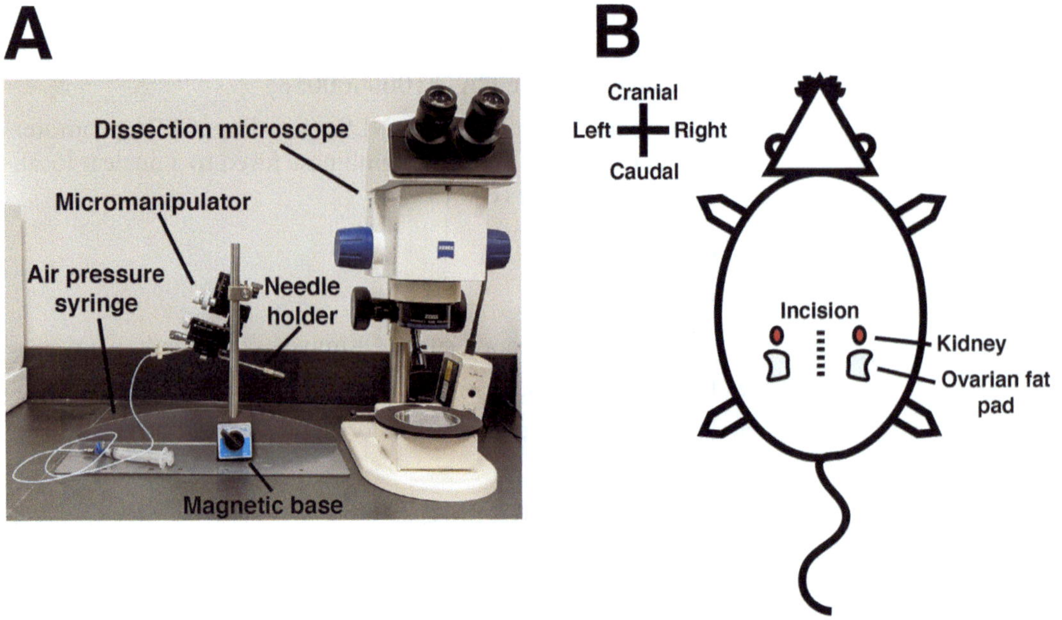

Fig. 1 Injection equipment and locating the ovarian fat pad. (**a**) Injection equipment used for in vivo oviduct electroporation, showing the micromanipulator, microscope, and air pressure syringe. (**b**) Schematic of a mouse showing the 1 cm dorsal midline incision and the location of the ovarian fat pads caudal to the kidneys

2.5 Other Equipment

1. Pulse generator—such as CUY21EDIT II electroporator (BEX) or equivalent.
2. 3 mm tweezer electrodes (BEX LF650P3 or equivalent).
3. Clamp mount micromanipulator with magnetic mount (such as (CAT# MM33R) with (CAT# MB-B) (AD Instruments)) and a 10 mL air pressure syringe with air valves (*see* Fig. 1a).
4. Dissection microscope.
5. Micropipette puller (such as flaming/brown P-97 (Sutter Instrument Co.) or equivalent).

3 Methods

3.1 Design and Cloning sgRNA Guides

1. Information on gene structure and genomic location can be found on the MGI database (http://www.informatics.jax.org).
2. Identify potential targets with the help of web-based sgRNA design platforms (*see* **Note 1**).
3. Design 2–3 potential targets based on predicted efficiency and specificity scores (*see* **Note 2**).
4. Order forward CACCG-sgRNA and reverse C-sgRNA-CAAA oligos and clone into pX330-U6_BB-CBh-hSpCas9 by BbsI

restriction cloning protocol using an endotoxin free DNA preparation kit for the DNA final isolation (https://www.addgene.org/crispr/zhang/).

3.2 Preparation of the Microinjection Needle and Injection Solution

1. Glass capillaries are pulled using a model P-97 micropipette puller or equivalent to create a sharp point for injection. Conditions for needle pulling should be optimized for each machine and capillary size following manufactures instructions. As a guide we used the following conditions: P = 500, Heat = 576, PULL = 50, VEL = 80, DEL = 70.
2. The injection solution is prepared on the day of surgery in a sterile 0.5 mL PCR tube. pX330-U6_BB-CBh-hSpCas9 and pCS2 Cre-NLS plasmids are combined with 0.05% trypan blue to a minimal final concentration of 200 ng/μL per plasmid.
3. The injection solution is stored at room temperature before injection to prevent precipitation of the trypan blue solution.

3.3 Preparation of Surgical Area

1. Surgeries should be performed in a sterile environment ideally in a biological safety cabinet. Disinfect all surfaces and sterilize surgical tools prior to starting the procedure. Maintain a clean and dirty side to your working area and clean surfaces and surgical tools between mice.
2. Roughly cut 1–2 cm^2 pieces of absorbent paper using sterile scissors and submerged in sterile PBS.
3. The injection needle is loaded onto an air pressured syringe attached to a micromanipulator and the tip broken using sterile forceps to allow pulling of the injection solution (*see* **Note 3**).
4. Pipet a 2 μL drop of well mixed injection solution onto a sterile petri dish and pull 1–2 μL into the injection needle using the air pressured syringe (*see* **Note 4**).

3.4 In Vivo Electroporation

1. Anaesthetize a 6–8-week-old mouse either by intraperitoneal injection of avertin solution (0.25–0.5 mg/g) or isoflurane gas followed by a subcutaneous injection of Carprofen (5 mg/kg) for preemptive analgesia.
2. Place the anaesthetized mouse onto a sterilized heated pad covered with a sheet absorbent paper (*see* **Note 5**).
3. Apply a drop of eye lubricant to each eye to prevent drying during the surgery and remove the hair around the incision site using either a shaver or hair removal cream.
4. Clean the back of the mouse with a piece of absorbent paper dipped in sterile saline solution and make a 1 cm cranial–caudal incision along the midline, 1–2 cm caudal to center by pinching the skin with forceps and cutting with sharp blunt scissors (*see*

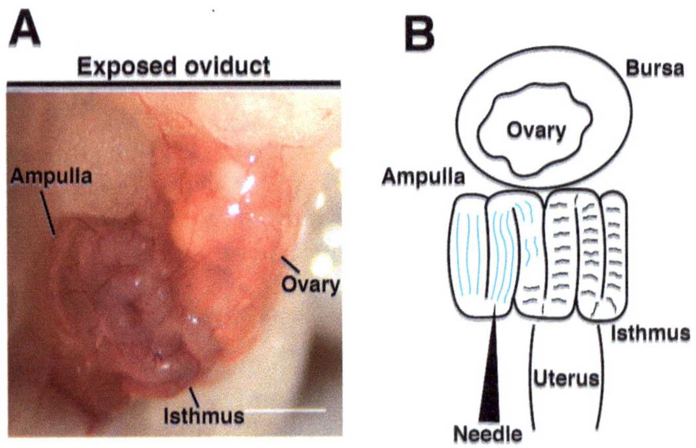

Fig. 2 Mouse oviduct injection. (**a**) An image of an exposed oviduct and ovary. (**b**) The injection solution (shown in blue) is injected into the ampulla with the ampulla orientated parallel to the injection needle. Scale bar in $A = 1000$ μm. Image **a** was kindly provided by Keerthana Harwalkar (McGill University)

Fig. 1b). Clean the cut site with a piece of absorbent paper dipped in disinfectant solution.

5. Using a pair of Dumont AA and a pair of curved iris serrated forceps separate the skin from the body wall and locate the left kidney by exposing the body wall 1–2 cm left of the dorsal incision. The kidney can be identified by its dark red color (*see* Fig. 1b).
6. Locate the fat pad caudal to the kidney and cut a small hole with sharp-blunt dissection scissors in body wall, taking care to avoid damaging any blood vessels running over the fat pad.
7. Insert and open blunt forceps to widen the hole in the body wall and grab the fat pad with the curved forceps to expose the ovary, oviduct and part of the uterus.
8. Clamp the fat pad with a bulldog clamp and rest on the mouse to keep the oviduct exposed and transfer to the dissecting microscope (*see* Fig. 2a) (*see* **Note 6**).
9. Use a pair of sharp forceps and the bulldog clamp to orientate the oviduct so that the ampulla is parallel to the injection needle.
10. Using the micromanipulator and a pair of sharp forceps insert the needle into the ampulla by picking up the ampulla with the forceps and simultaneously pushing the needle (*see* **Note 7**).
11. Slowly inject 1 μL of injection solution into the ampulla noticing the build-up of blue fluid and expansion of the ampulla (*see* Fig. 2b) (*see* **Note 8**).
12. Remove the injection needle and cover the ampulla with a $1 cm^2$ piece of absorbent paper presoaked in sterile PBS.

13. Grasp the ampulla with a pair of 3 mm electrode tweezers, pull slightly to minimalize electroporation of the ovary and electroplate using the following settings: 30 V, 3 pulses, 1 s interval, P. length = 50 ms, unipolar (*see* **Note 9**).
14. Remove the absorbent paper and place the mouse back onto the heat pad. Replace the oviduct and place a 1cm^2 piece of absorbent paper presoaked in sterile PBS over the incision while you prepare/replace the needle for the second injection.
15. Repeat the same protocol for the right oviduct and suture the wound. Apply sterile saline when required to prevent drying of the exposed tissue.
16. Transfer the mouse to a clean cage on top of a heat pad and allow to recover, monitoring the mouse frequently until she has awoken. The mouse should then be monitored over the next few days with administration of appropriate pain management medication.

3.5 Confirming Successful Electroporation and Gene Editing

1. Electroporation efficiency can be estimated by dissecting oviducts 5–7 days postelectroporation followed by visual inspection using a fluorescent microscope (*see* Fig. 3a, b) (*see* **Note 10**).
2. In order to determine successful gene editing of electroporated cells, fluorescent cells can be isolated by fluorescence-activated cell sorting (FACS) followed by DNA isolation (*see* **Note 11**).
3. The targeted region (300 bp) is then PCR amplified and sequenced using MiSeq targeted sequencing (Illumina). The resulting reads can be analysed to quantify the proportion of targeted alleles using custom designed tools or published online resources such as Cas-Analyzer (http://www.rgenome.net/cas-analyzer/#!) or CRISPResso2 (http://crispresso.pinellolab.partners.org). For quantification purposes, a minimum of 100 reads per sample is recommended using a 1% cutoff to filter sequencing errors. Allelic tracking can be performed by tracing the presence of unique indel mutation or CRISPR-scars between samples.

4 Notes

1. There are many available web-based tools for sgRNA design. The target area is usually between 200–300 bp containing the target sequence, appropriate PAM sequence and primer sites. The web-based tool CHOPCHOP (http://chopchop.cbu.uib.no) is a useful tool to search for target regions by genome and gene name. Another web-based tool, CRISPOR (http://crispor.tefor.net), reports potential target regions with

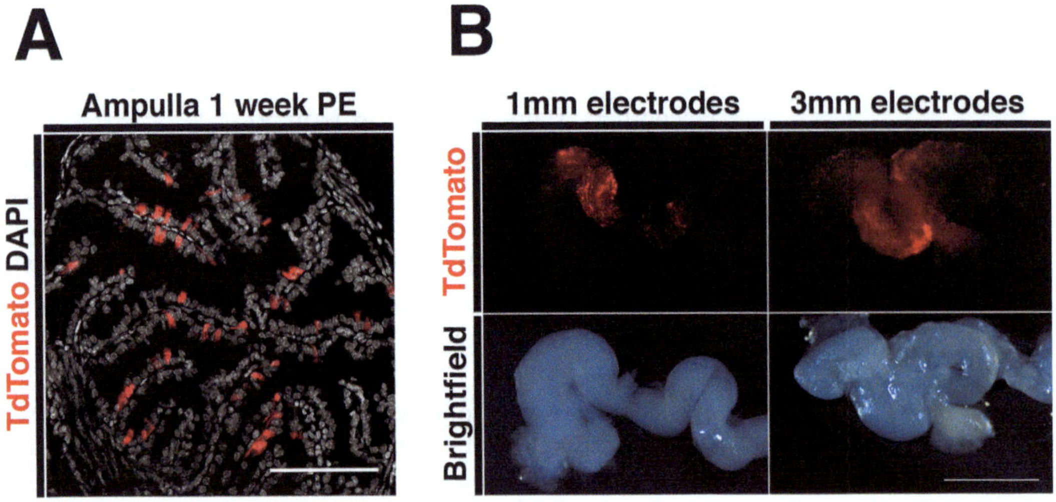

Fig. 3 Validation of successful oviduct epithelial electroporation. (**a**) A transverse section of a *Rosa-LSLtdTomato* oviduct 5 days postelectroporation with a pCS2 Cre-NLS plasmid using 3 mm tweezer electrodes. (**b**) Whole mount images of a *Rosa-LSLtdTomato* oviduct 5 days postelectroporation with a pCS2 Cre-NLS plasmid comparing target area between using 1 and 3 mm tweezer electrodes (scale bar $A = 100$ μm, $B = 1000$ μm)

efficiency and specificity scores from a manually input sequence limited to 2000 bp. Target sequences should be chosen based on the requirements of the experiment. In general, targeting an early exon to insert a frame-shift mutation is preferable for generating a knockout allele. However, the location of key functional domains and the presence of alternative start sites or splice variants should also be taken into consideration.

2. In vitro testing of sgRNA guide efficiency is recommended before cloning into pX330-U6_BB-CBh-hSpCas9 using the Takara Bio method (https://www.takarabio.com/assets/documents/User%20Manual/Guide-it%20sgRNA%20In%20Vitro%20Transcription%20and%20Screening%20Systems%20User%20Manual_040618.pdf) . In brief, sgRNA guides are either generated by in vitro transcription or ordered directly and combined at 37 °C with recombinant cas9 nuclease at a 1: 10 ratio to form a guide protein complex (sgRNA 50 ng/μL: cas9 500 ng/μL). 1.5 μl of this complex is then incubated with a 600–800 bp PCR amplified target region (100–250 ng) containing 1× BSA and 1× Cas9 reaction buffer for 1 h at 37 °C followed by inactivation at 80 °C for 5 min. Cutting efficiency can then be visualized by DNA separation using DNA gel electrophoresis.

3. Breaking the tip of the glass injection needle is essential to allow up uptake of the injection fluid. It is important to optimize the position at which the needle is broken to ensure the injection

fluid can move freely through the needle but not too large as to cause significant damage to the oviduct. The conditions used to make the needle can also be optimized to manipulate the size and shape of the tip.

4. Ensure when collecting the injection solution to avoid taking up any bubbles. Occasionally the needle can get blocked by precipitation from the trypan blue. If this occurs gently push some liquid back out of the needle to dislodge the blockage and repeat pulling of injection fluid.
5. The heated pad should be warm to touch but not too hot and extra attention should be paid to ensure the pad does not cool down too much during and between procedures as immobile mice can lose heat rapidly.
6. The oviduct can dry out quickly when exposed for extended periods. If required sterile PBS can be pipetted onto the oviduct or a PBS soaked piece of absorbent paper draped over exposed tissue. Some minor bleeding may occur if a vessel is damaged. If this occurs use a sterilized cotton swab to collect any blood. If extensive bleeding continues terminate the surgery.
7. It is preferable to rest wrists on a small box if possible, to stabilize the oviduct. The mouse can also be placed on a sterile petri dish to help attain a good angle for injection.
8. Take care to ensure the needle has penetrated the ampulla but has not come through the other side. Watch for any leakage of the injection fluid outside of the oviduct lumen and avoid injection of air bubbles.
9. The distance between the electrodes can be changed and should be set to roughly 1 mm apart when the tweezers are closed. This distance is important to optimize as too far apart will reduce electroporation efficiency and too close can squash the oviduct, damaging the tissue and pushing out the injected fluid. The electroporated location along the female reproductive tract can also be adapted by changing the injection and electroporation location. In addition, the size of the electroporated area and density of cells electroporated can be optimized by using different sized electrode tweezers and electroporation conditions. Using the quoted electroporation conditions with 3 mm electrodes we found a small number of cells located on the ovarian surface epithelium and/or hilum region were electroporated. If exclusive targeting of oviduct epithelial cells is required, we recommend using 1 mm electrodes (BEX LF650P1) and optimizing electroporation conditions (see Fig. 3b for comparison between 1 and 3 mm electrodes using the same electroporation conditions).

10. Oviducts are dissected by making an incision into the ventral wall of the peritoneal cavity and moving the gastrointestinal track cranially to expose the female reproductive track. Hold the uterus with a pair of forceps and slowly cut away the connective tissue moving toward the ovary. The ovary will be identifiable by its white and red color surrounded by fat tissue. Cut away the ovary from the fat and connective tissue and continue dissection of the oviduct in sterile PBS under a dissection microscope.
11. Isolation of oviduct cells for FACS has been previous published by Xie and colleagues [23].

References

1. Varga J, Greten FR (2017) Cell plasticity in epithelial homeostasis and tumorigenesis. Nat Cell Biol 19(10):1133–1141
2. Doyle A, Mcgarry MP, Lee NA, Lee JJ (2013) The construction of transgenic and gene knockout/Knockin mouse models of human disease. Transgenic Res 21(2):327–349
3. Kersten K, De Visser KE, Van Miltenburg MH (2017) Genetically engineered mouse models in oncology research and cancer medicine. EMBO Mol Med 9(2):137–153
4. Heyer J, Kwong LN, Lowe SW, Chin L (2010) Non-germline genetically engineered mouse models for translational cancer research. Nat Rev Cancer 10(7):470–480
5. Kroeger PT, Drapkin R (2017) Pathogenesis and heterogeneity of ovarian cancer. Curr Opin Obstet Gynecol 29(1):26–34
6. The Cancer Genome Atlas Research Network (2012) Integrated genomic analyses of ovarian carcinoma. Nature:0–7
7. Almiñana C, Bauersachs S (2019) Extracellular vesicles in the oviduct: progress, challenges and implications for the reproductive success. Bioengineering 6(32):1–25
8. Li S, Winuthayanon W (2017) Oviduct: roles in fertilization and early embryo development. J Endocrinol 232:R1–R26
9. Maillo V, Sánchez-calabuig MJ, Lopera-vasquez R, Hamdi M, Gutierrez-adan A, Lonergan P et al (2016) Oviductal response to gametes and early embryos in mammals. Reproduction 152(1):R127–R141
10. Soong TR, Howitt BE, Miron A, Horowitz NS, Campbell F, Feltmate CM et al (2018) Evidence for lineage continuity between early serous proliferations (ESPs) in the fallopian tube and disseminated high-grade serous carcinomas. J Pathol 246(September):344–351
11. Medeiros F, Muto MG, Lee Y, Elvin JA, Callahan MJ, Feltmate C et al (2006) The tubal fimbria is a preferred site for early adenocarcinoma in women with familial ovarian Cancer syndrome. Am J Surg Pathol 30(2):230–236
12. Shaw PA, Rouzbahman M, Pizer ES, Pintilie M (2009) Candidate serous cancer precursors in fallopian tube epithelium of BRCA1/2 mutation carriers. Mod Pathol:1133–1138
13. Lee Y, Miron A, Drapkin R, Medeiros F, Saleemuddin A, Garber J et al (2007) A candidate precursor to serous carcinoma that originates in the distal fallopian tube. J Pathol 211 (November 2006):26–35
14. Labidi-Galy SI, Papp E, Hallberg D, Niknafs N, Adleff V, Noe M et al (2017) High grade serous ovarian carcinomas originate in the fallopian tube. Nat Commun 8(1): 1093. http://www.nature.com/articles/s41467-017-00962-1
15. Friedrich M, Heid I, Maresch R, Mueller S, Veltkamp C, Rupert O et al (2016) Multiplexed pancreatic genome engineering and cancer induction by transfection-based CRISPR / Cas9 delivery in mice. Nat Commun 7(10770):1–13
16. Teng K, Ford MJ, Harwalkar K, Li Y, Pacis AS, Farnell D, Yamanaka N, Wang YC, Badescu D, Ton Nu TN, Ragoussis J, Huntsman DG, Arseneau J, Yamanaka Y (2021) Modeling High-Grade Serous Ovarian Carcinoma Using a Combination of In Vivo Fallopian Tube Electroporation and CRISPR-Cas9-Mediated Genome Editing. Cancer Res 81 (20):514–5160. https://doi.org/10.1158/0008-5472.CAN-20-1518. Epub 2021 Jul 23. PMID: 34301761

17. Wu SS, Lee J, Koo B (2019) Lineage tracing: computational reconstruction goes beyond the limit of imaging. Mol Cells 42(2):104–112
18. Chazaud C, Yamanaka Y, Pawson T, Rossant J (2006) Early lineage segregation between epiblast and primitive endoderm in mouse blastocysts through the Grb2-MAPK pathway. Dev Cell 10:615–624
19. Madisen L, Zwingman TA, Sunkin SM, Oh SW, Zariwala HA, Gu H, Ng LL, Palmiter RD, Hawrylycz MJ, Jones AR, Lein ES, Zeng H (2010) A robust and high-throughput Cre reporting and characterization system for the whole mouse brain. Neurosci 13(1):133–140. https://doi.org/10.1038/nn.2467. Epub 2009 Dec 20. PMID: 20023653; PMCID: PMC2840225.
20. Ford MJ, Yeyati PL, Jackson IJ, Mill P, Mort RL, Ford MJ et al (2018) A cell/cilia cycle biosensor for single-cell kinetics reveals persistence of cilia after G1 / S transition is a general property in cells and mice technology a cell / cilia cycle biosensor for single-cell kinetics reveals persistence of cilia after G1. Dev Cell 47(4):509–523.e5. https://doi.org/10.1016/j.devcel.2018.10.027
21. Mort RL, Ford MJ, Sakaue-sawano A, Lindstrom O, Casadio A, Douglas AT et al (2015) Fucci2a: a bicistronic cell cycle reporter that allows Cre mediated tissue specific expression in mice. Cell Cycle 4101
22. Livet J, Weissman TA, Kang H, Draft RW, Lu J, Bennis RA et al (2007) Transgenic strategies for combinatorial expression of fluorescent proteins in the nervous system. Nature 450(7166):56–62. http://www.ncbi.nlm.nih.gov/pubmed/17972876
23. Xie Y, Park E, Xiang D, Li Z (2018) Long-term organoid culture reveals enrichment of organoid-forming epithelial cells in the fimbrial portion of mouse fallopian tube. Stem Cell Res 32(February):51–60. https://doi.org/10.1016/j.scr.2018.08.021

Chapter 25

Generation of Human Liver Chimeric Mice and Harvesting of Human Hepatocytes from Mouse Livers

Rui Wei, Chi-Wa Cheng, Wai-In Ho, Kwong-Man Ng, Miguel A. Esteban, and Hung-Fat Tse

Abstract

Primary human hepatocytes (PHHs) are widely used as an in vitro model to evaluate various aspects of human hepatic physiology and pathology. However, PHHs isolated from the human liver have very limited ability for ex vivo expansion in culture. *Fah*$^{-/-}$/*Rag2*$^{-/-}$/*Il2rg*$^{-/-}$ (FRG) mice are proven to be an ideal bioincubator for repopulation of PHHs. The human liver chimeric FRG mouse is not only a humanized animal model for disease study and drug screening in vivo, but also a potential source of PHHs for cellular therapy. This chapter describes experimental protocols to generate chimeric FRG mice with humanized liver and to isolate PHHs from human liver chimeric FRG mice. Using these methods, PHHs can be expanded to more than 100-fold for harvesting.

Key words Hepatocyte, Liver regeneration, Humanized model, Hepatic function, Chimeras

1 Introduction

The human liver has remarkable regeneration capacity, as hepatocytes are highly proliferative under certain pathological stresses that lead to reduction of liver mass, such as after surgical excision or extensive liver injuries [1]. Therefore, it has been proposed that liver regeneration can be achieved by hepatocytes transplantation. Indeed, previous pilot studies have demonstrated that allogeneic hepatocyte transplantation appear to be safe and has modest clinical efficacy [2–4]. The major hurdles of allogeneic hepatocyte transplantation include the scarcity of donor organs, low quality of primary human hepatocytes (PHHs) isolated from cadaveric organs, and immune rejection of allogeneic grafts [1, 5, 6]. As a result, various methods have been attempted to maintain and expand PHHs in vitro [7, 8]. Alternatively, human liver chimeric mice using the *Fah*$^{-/-}$/*Rag2*$^{-/-}$/*Il2rg*$^{-/-}$ (FRG) strain have been developed as a humanized animal model for the study of disease and drug screening in vivo, as well as a potential source of PHHs for

Nagarajan Kannan and Philip Beer (eds.), *Stem Cell Assays: Methods and Protocols*, Methods in Molecular Biology, vol. 2429, https://doi.org/10.1007/978-1-0716-1979-7_25,

cellular therapy. This chapter describes experimental protocols to generate and isolate PHHs from human liver chimeric FRG mice.

The Fah-deficient (*Fah*$^{-/-}$) mouse model was firstly established by Groump et al. in 1993 [9]. In this animal model, the mouse undergoes progressive liver failure unless they are treated with 2-(2-nitro-4-trifluoromethylbenzoyl)-1, 3-cyclohexanedione (NTBC) to prevent the formation of hepatotoxic metabolites (*see* **Note 1**). Subsequently, Groump's lab generated an immunodeficient Fah-deficient (*Fah*$^{-/-}$ *Rag2*$^{-/-}$/*Il2rg*$^{-/-}$, FRG) mice to facilitate xenograft transplantation [10]. When NTBC is withdrawn, FRG mice develop progressive liver damage to facilitate the engraftment of the transplanted cells such as PHHs which are Fah-competent (*Fah*$^{+/+}$), which protects them from being injured. Therefore, NTBC withdrawal becomes a selective stress that favors the growth of transplanted cells. By adjusting the NTBC treatment cycle, Groump's lab managed to generate chimeric mice with over 90% PHHs engraftment rate [10].

By using the FRG mice, we are able to "culture" the PHHs in vivo using the mouse liver as a bioincubator. The transplanted PHHs expand by 100-fold in vivo, and maintain hepatic function in FRG mice. More importantly, these humanized FRG mice display typical human lipoprotein [11] and drug metabolism profiles [10], demonstrating that they can also serve as an ideal humanized model for drug screening or study of liver-related metabolic disorders.

Here, we describe the methods we set up in our laboratory relating to FRG chimeric mice generation and PHHs isolation. Using this protocol, we will be able to generate FRG chimeric mice with high engraftment efficiency (up to 90%), isolate PHHs from chimeric mouse livers with high cell viability (up to 98%), and cryopreserve the isolated PHHs for later use.

2 Materials

2.1 Materials for PHHs Preparation Before Transplantation

1. Ad:uPA Liver Tx Enhancer (Yecuris): referred as Ad:uPA in short hereinafter. The Ad:uPA stock solution is diluted with sterile saline into 1.25×10^{10} pfu/mL before injection (*see* Subheading 3.1).
2. Cryopreserved PHHs (LONZA).
3. Hepatocyte Wash Medium (Gibco). Hepatocyte Wash Medium is an enriched Williams's Medium E for the reduction of cell membrane leakage, ease of centrifugation, and high viability.
4. Matrigel.
5. Insulin syringe.

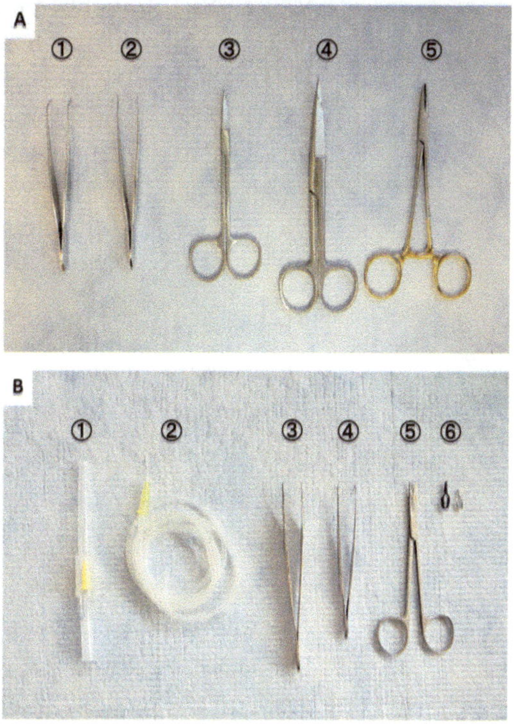

Fig. 1 Essential tools for intrasplenic injection and liver perfusion. (**a**) Surgical tools for intrasplenic injection: 1 tissue forceps (curved), 2 tissue forceps (straight), 3 iris scissors, 4 suture scissors, 5 needle holder. (**b**) Tools for liver perfusion: 1 I.V. Catheter, 2 perfusion tubes, 3 tissue forceps (straight), 4 tissue forceps (curved), 5 iris scissors, 6 surgical vein clips

6. Trypan Blue solution 0.4%.
7. 1× phosphate-buffered saline (PBS).

2.2 Surgical Materials for PHHs Transplantation by Intrasplenic Injection

1. Anesthetics (K + X): dilute 1 mL of 100 mg/mL Ketamine and 1 mL of 20 mg/mL Xylazine solution with 8 mL of double distilled water.
2. Tissue forceps, needle holder (*see* Fig. 1a).
3. Iris scissors, suture scissors (*see* Fig. 1a).
4. Hair removal cream.
5. Sterile gauze, cotton swabs.

2.3 Food and Medications for FRG Mice Husbandry

1. Mouse Diet 5058 (PicoLab): FRG mice are fed with this diet when breeding and daily maintenance.
2. Mouse Diet 5LJ5 (PicoLab): The chimeric mice are switched to 5LJ5 diet after PHHs transplantation. This chow is formulated to have high fat content and low protein content to avoid excessive levels of tyrosine that can lead to liver damage.

3. NTBC (Yecuris), add 16 mg/L NTBC in drinking water when breeding and daily maintenance. Switch to NTBC cycle after PHHs transplantation (*see* Subheading 3.3).
4. SMX/TMP (Yecuris): prepare 640 μg/mL SMX and 128 μg/mL TMP in drinking water every other week (*see* **Note 2**).
5. Human albumin ELISA kit (Bethyl): by measuring human albumin levels in the chimeric mice plasma, you can roughly evaluate PHHs repopulation efficiency.

2.4 Materials for Liver Perfusion, Hepatocyte Isolation, and Cryopreservation

1. Perfusion pump.
2. Mini water bath.
3. I.V. Catheter (24GA, 0.75IN) (*see* Fig. 1b).
4. Perfusion tubes (*see* Fig. 1b).
5. Isoflurane.
6. Tissue forceps, needle holder (*see* Fig. 1b).
7. Iris scissors (*see* Fig. 1b).
8. Surgical vein clips (*see* Fig. 1b).
9. Sterile gauze, cotton swabs.
10. HBSS-CMF (Gibco), Ca^{2+}/Mg^{2+} free HBSS.
11. HBSS-CM (Gibco), HBSS with Ca^{2+} and Mg^{2+} (Ca^{2+} 1.64 mM).
12. Trypan Blue solution 0.4%.
13. 0.5 M EDTA.
14. 5× collagenase (Yecuris), contains collagenase type 2 (5 mg/mL).
15. $CaCl_2$ solution: Prepare 2 M $CaCl_2$ stock solution, add to HBSS-CM when preparing perfusion solution B.
16. DNase I: prepare stock solution of 20 mg/mL.
17. Perfusion solution A: Prepare 50 mL of solution A for perfusion of each mouse: add 50 μL 0.5 M EDTA into 50 mL HBSS-CMF. (*see* **Note 3**).
18. Perfusion solution B: Prepare 40 mL solution B for each mouse: add 8 mL of 5× collagenase, 500 μL DNase I stock solution and 67 μL 2 M $CaCl_2$ into 32 mL HBSS-CM. (*see* **Note 4**).
19. Cell strainer: prepare both 100 μm and 70 μm mesh width cell strainers.
20. Isotonic Percoll: mix 1 part of sterile 10× PBS with 9 parts of Percoll.
21. Hepatocyte freezing medium: CryoStor CS 10 (StemCells) + 100 mM Fructose. (*see* **Note 5**).

22. Sterile cryogenic tubes.
23. Freezing container.

3 Methods

3.1 Preparations Before PHH Transplantation

3.1.1 Animal Preparation

FRG mice aged 6–8 weeks old are used as recipients. One day prior to PHHs transplantation, mice are injected with Ad:uPA to improve engraftment efficiency. The recommended dosage of Ad: uPA is 1.25×10^9 pfu/25 g body weight in 100 μL sterile saline via tail vein injection (*see* **Note 6**). The Ad:uPA injection should be performed inside a biosafety level II cabinet.

3.1.2 PHHs Preparation for Injection

1. Thaw Matrigel (MG) on ice 1 day prior experiment.
2. Prepare ice-cold PBS. Put on ice.
3. Prepare ice-cold hepatocyte wash medium, put on ice.
4. Prepare insulin syringes (1 syringe per mouse), place on ice.
5. Cryopreserved PHHs are retrieved from liquid nitrogen, quickly thaw at 37 °C. Repeatedly shake and observe the cell suspension in the vial, at the point of just being thawed, transfer to 15 mL falcon tube containing 10 mL of cold Hepatocyte Wash Medium.
6. Centrifuge at $140 \times g$ for 5 min at 4 °C, discard the supernatant and resuspend in 5 mL ice-cold PBS (*see* **Note 7**). Remove 10 μL cell suspension, mix with 10 μL of 0.4% Trypan blue and viable cells are counted with a hemocytometer.
7. For one FRG mouse, we usually inject 1×10^6 PHHs suspended in 55 μL MG-PBS [1:1 (v/v)]. Calculate the number of cells required according to animal number and aliquot the required amount of cell suspension (*see* **Note 8**). Spin down at $140 \times g$ for 5 min at 4 °C, discard the supernatant and resuspend with MG-PBS (*see* **Note 9**).
8. Take out an insulin syringe from the ice box, pull out the plunger, moisten the plunger head by simply dipping it in the PBS and place it aside. Cool down the pipette tip in ice-cold PBS before placing it into cell suspension. Carefully transfer 55 μL cell suspension into syringe barrel with a P200 pipette. Place the syringe plunger back and carefully drive away the bubbles (*see* **Note 10**). Place the syringe back on ice after cell package.

3.2 Transplantation of PHHs Via Intrasplenic Injection

1. Anesthetize FRG mouse by intraperitoneal injection of Anesthetics (K + X) at the dosage of 10 μL per 10 g body weight.
2. Place the mouse in left lateral decubitus position (*see* Fig. 2a). Remove fur of the surgical site on left flank by hair removal

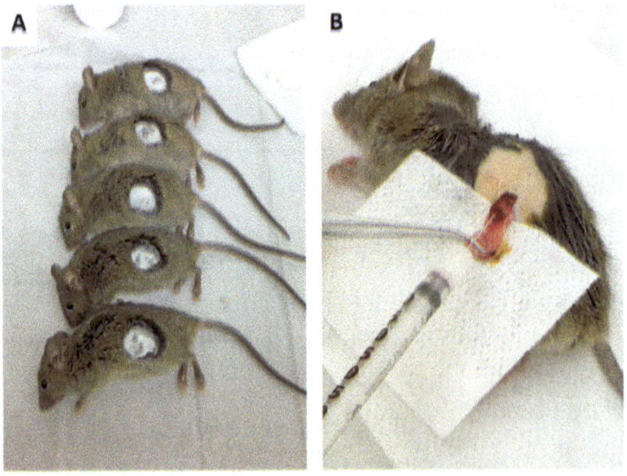

Fig. 2 Illustration of mouse position and surgical site of intrasplenic injection. (**a**) Position of mice before intrasplenic injection. Hair removal cream has been placed at the surgical site. (**b**) injection of PHHs into mouse spleen. Part of the spleen "blanched" after injection

cream. The spleen is a dark red organ and can be easily located through the skin after hair removal.

3. Disinfect the surgical site by scrubbing the skin with Betadine and 70% ethanol alternatively for 3 times, followed by a final soaking with Betadine.
4. Make a 0.5–1 cm incision through the skin and muscle layer with scissors. Clamp the adipose tissue underneath the spleen with forceps. Gently drag the adipose tissue out of the incision, and the spleen will come out along with it.
5. Insert the syringe needle through one end of the spleen 3–4 mm deep. Slowly inject the cell suspension into spleen parenchyma, wait for around 10 s before quickly retracting the needle. Place a Q-tip swab on the injection site and press for 1 min to stop bleeding. (*see* Fig. 2b).
6. Return the spleen back to abdominal cavity. Close up the incision by suturing.
7. After injection, place the cage on 34 °C warm chamber until they recover from anesthesia.

3.3 Food and Medications for Transplanted FRG Mice

1. FRG mice are kept on 16 mg/L NTBC and mouse diet 5058 before transplantation.
2. After cell transplantation, six NTBC short cycles starts (5 days NTBC-off and 3 days NTBC 8 mg/L), followed by long cycle (21 days NTBC-off and 3 days NTBC 8 mg/L).
3. Switch to high energy mouse diet (5LJ5) after PHHs transplantation. Monitor the body weight every twice a week in the first 2 months, then at least once a week. As stable or increasing body weight reflects balanced hydration statues, it is an indirect indicator of successful engraftment.

4. Collect blood sample every month and measure the plasma hALB level by ELISA to monitor PHHs repopulation. Normally, the hALB level should reach 1 μg/mL at first month posttransplantation, above 1 mg/mL after 3 months posttransplantation.

3.4 Liver Perfusion and Hepatocyte Isolation

3.4.1 Preparation

1. Calibrate the flow speed of an perfusion pump to around 3 mL/min (usually speed no. 10).
2. Set up all perfusion equipment (*see* Fig. 3a), and UV for disinfection.
3. Heat up the water bath to 40 °C.
4. Start pump and fill the tubes with perfusion solution A. Rinse the tube for a while and then stop the pump.

3.4.2 Perfusion and Collagenase Digestion

1. Euthanize the mouse with an overdose of isoflurane (*see* **Note 11**) or according to approved local procedures, place it on the perfusion pad (made with a 1-mL-tip box) and fix with elastic bands.
2. Use tissue scissors to open the abdominal cavity by cutting from the lower abdomen all the way up to the lower margin of ribs (*see* **Note 12**). Be careful not to injure the thoracic cavity.
3. Use a wet Q-tip cotton swab to move guts to the right side and expose the inferior vena cava (IVC), and then move liver lobes up to expose the portal vein (PV). Be careful not to injure the liver.
4. Cannulate PV (this should produce a fair amount of back flow into the needle when done properly), remove the needle and place a surgical vein clip near the end of the catheter to secure it (*see* Fig. 3b). (*see* **Note 13**).
5. Connect to the pump tubing, and start perfusion. The liver will immediately blanch if the cannulation is successful. Cut the IVC for drainage (*see* **Note 14**).
6. Perfuse with 20 mL solution A, 20 mL HBSS-CMF, 35 mL solution B accordingly (*see* **Note 15**). Avoid air bubble in the system at all cost during the whole process. (*see* **Note 16**).
7. After perfusion, remove gall bladder in situ (*see* **Note 17**). Clamp the portal area of the liver, carefully dissect it from the connective tissue without injuring the liver capsule.
8. Put the liver into petri dish containing 5 mL of perfusion solution B. Place the dish on ice and transfer to another hood for hepatocyte processing.

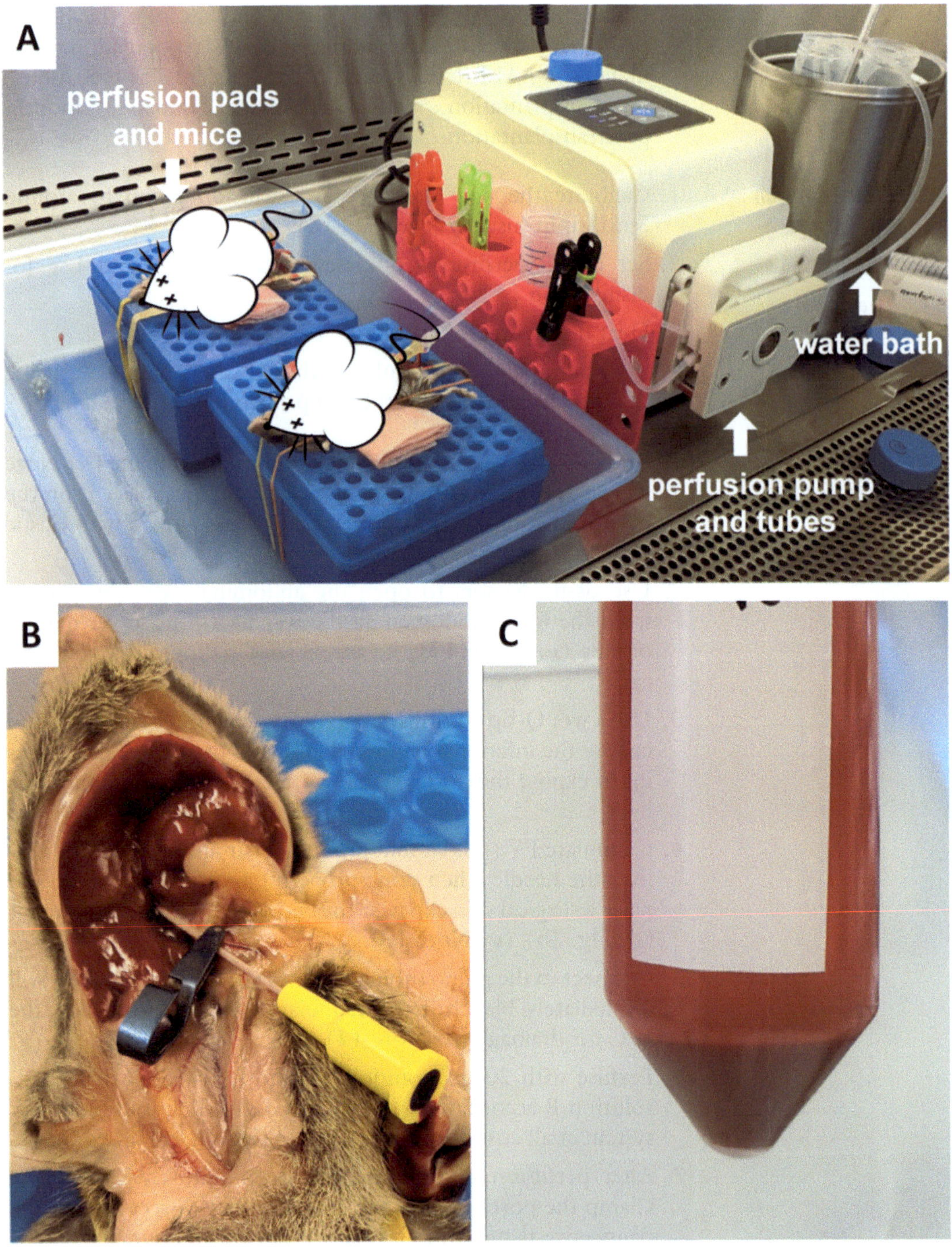

Fig. 3 Liver perfusion and hepatocytes isolation. (**a**) Set-up of the perfusion system. (**b**) Portal vein (PV) cannulation. Backflow of blood through the catheter indicates a good PV canulation. (**c**) Hepatocytes yield from one chimeric mouse (after first washing)

3.4.3 Hepatocyte Processing (Perform in Biosafety Level II Cabinet)

1. Prepare two 50 mL conical tubes and remove the lids. Place a 100 μm cell strainer and a 70 μm cell strainer on each tube respectively.
2. Disrupt liver capsule with sterile forceps and scissors and release the cells. (*see* **Note 18**).
3. Add ~10 mL ice-cold hepatocyte wash medium into the petri dish. Transfer the entire content of petri dish to the 50 mL conical tube through 100 μm cell strainer.
4. Repeat **step 3** and collect the remaining hepatocytes in petri dish.
5. Discard the 100 μm strainer and bring the filtered solution up to 45 mL with hepatocyte wash medium. Tightly cap the tube and invert several times to mix.
6. Filter the cells again by transfer the liver slurry to another 50 mL tube through the 70 μm strainer.
7. Centrifuge at 140 × *g* for 5 min at 4 °C. After centrifuging, over 2 mL of cells pellet should be achieved from one liver following optimal perfusion and digestion (*see* Fig. 3c). Gently aspirate supernatant, ensuring to not disturb the pellet.
8. Resuspend pellet in 21 mL hepatocyte wash medium plus 9 mL isotonic Percoll. Gently invert to mix. Spin down at 800 × *g* for 10 min at 4 °C. (*see* **Note 19**).
9. Discard supernatant and resuspend cells with 30 mL wash medium. (*see* **Note 20**).
10. Remove 10 μL homogenous cell suspension and mix with 10 μL of 0.4% trypan blue to distinguish live and dead cells. Count cell number with hemocytometer and calculate viability (*see* **Note 21**).
11. The isolated hepatocytes can be re-engrafted to new FRG mice for generation of new batches of chimeric mice (repeat Subheading 3.1) or cryopreserved in −80 °C for future use.

3.4.4 Cryopreservation of Isolated Hepatocytes

1. After cell counting, spin down the cells at 140 × *g* for 5 min at 4 °C.
2. Carefully discard supernatant and avoid disturbing the cell pellet.
3. Add hepatocyte freezing medium to the pellet at 1 mL/million cells. Mix gently by mildly pipette up and down several times.
4. Aliquot into cryogenic tubes and place the tubes into a freezing container.
5. Store the cells at −80 °C overnight and then transfer to liquid nitrogen tank for long-term storage.

4 Notes

1. Fumarylacetoacetate hydrolase (Fah) is a metabolic enzyme that catalyzes the last step of tyrosine metabolism. Deficiency of this enzyme leads to accumulation of toxic tyrosine metabolites, mainly maleylacetoacetate (MAA) and fumarylacetoacetate (FAA), which results in local tissue damage and eventually liver failure. The production of toxic metabolites can be blocked by administrating a liver-protective drug 2-(2-nitro-4-trifluoromethylbenzoyl)-1,3-cyclohexanedione (NTBC), which inhibits an upstream enzyme, 4-hydroxyphenylpyruvic acid dioxygenase (HPPD), and prevents the formation of MAA and FAA. Therefore, Fah-deficient animals can stay healthy when continuously given NTBC treatment.
2. This antibiotic product contains Sulfamethoxazole and Trimethoprim. First prepare the SMX/TMP stock solution (80 mg/mL SMX and 16 mg/mL TMP), then add 8 mL stock solution into every litter of drinking water to achieve final concentrations of 640 μg/mL SMX and 128 μg/mL TMP. The drinking water with antibiotics is given every other week before hepatocyte transplantation.
3. Perfusion solution A is used to flush out red blood cells, for the dissolution of intercellular junctions of hepatocytes, and to prevent blood clotting.
4. Perfusion solution B contains collagenase which is used for dissociation of hepatocytes. The activity of collagenase needs calcium ion (Ca^{2+}), and the recommended Ca^{2+} concentration is 5 mM. HBSS-CM contain 1.64 mM Ca^{2+} originally, so by adding 67 μL 2 M $CaCl_2$ into every 40 mL of the solution, the final Ca^{2+} concentration will be 5 mM.
5. The molecular weight of fructose is 180 Da. For preparation of 10 mL hepatocyte freezing medium, weight 180 mg fructose powder (irradiate to sterilize if necessary) and add into a 10 mL Cryostor CS10 solution.
6. Firstly, dilute Ad:uPA to the working concentration of 1.25×10^7 pfu/μL with sterile saline. Inject 100 μL per mouse of 25 g weight. Adjust the injection volume with different body weights.
7. According to provided cell number, adjust the cell suspension to one to two million cells/mL. Usually Lonza provided five to ten million cells/vial.
8. Aliquot 10% more cells than the required amount since a small number of cells will be lost when packaging.
9. Leave roughly half of the required amount of PBS after discarding the supernatant. Resuspend cells and use a pipette tip

to measure the volume of cell suspension at the same time. Replenish with cold PBS to half volume if needed, then slowly add the remaining half volume of MG into the vial and mix gently to avoid making any bubbles.

10. As injecting bubbles into the blood flow may cause air embolism, it is important to avoid air bubbles when drawing the cells into the syringe. Push the plunger up and down several times, as this can drive away the bubbles between plunger and cell suspension. Then slowly push the plunger up until a small drop of cell suspension coming up from the needle tip; this drives away the air on top.
11. Soak a sterile gauze with IsoFlo and put it into a 50 mL Falcon tube. Place the head of mouse into the tube for 1 min.
12. Wet the abdominal with 70% EtOH to avoid "flying around" of the fur when making incisions.
13. Once needle is removed, blood should flow into the catheter until it is completely full. If the back flow is too slow, use a syringe to fill the remaining space in the catheter with HBSS. Avoid creating bubbles.
14. Once liver blanching is observed, quickly cut the IVC to avoid creating high flow pressure damage the liver capillary vessels. You can see blood and fluid coming out of IVC.
15. Always stop the pump before changing solutions, which can avoid air entering the tubing and create bubbles.
16. During perfusion, you can gently massage the liver with a sterile swab to encourage flow of liquid to the extremities of lobes.
17. If subsequent experiments require "clean" PHHs, such as in vitro cell culture or retransplantation into new batch of FRG mouse livers, try not to injure the bladder capsule when removing bladder as it may contain bacteria.
18. After perfusion and collagenase digestion, the liver will become very fragile. Hold the portal area with forceps and rip the liver lobes with scissor tips repeatedly to dissociate the capsule. Cells will burst out and "flow" into the solution, and over 60 million cells can be obtained following optimal digestion.
19. Percoll "cushions" the cells, so high speed spinning is required to separate the live and dead cells. Dead hepatocytes float to the top of the Percoll.
20. This volume of medium used is for hepatocyte yields of about 60 million. If the estimated cell yield is less (estimate the cell yield by observing the cell pellet volume after centrifuge), resuspend with medium at one to two million cells/mL.
21. Hepatocytes are heavy and settle easily. Make sure to mix the cell suspension thoroughly before cells counting.

Acknowledgments

This project is supported by Innovation and Technology Support Programme (GHP/046/17GD ITSP TCFS project), the National Natural Science Foundation of China (81873521), Innovation and Technology Support Programme [GHP/130/18SZ], and the Innovative Team Program of the Bioland Laboratory (Guangzhou Regenerative Medicine and Health Guangdong Laboratory) (2018GZR110103001 to M.A.E.).

References

1. Liver repopulation by cell transplantation and the role of stem cells in liver biology (2020). In: The liver. pp 550–565. https://doi.org/10.1002/9781119436812.ch44
2. Fox IJ, Chowdhury JR, Kaufman SS, Goertzen TC, Chowdhury NR, Warkentin PI, Dorko K, Sauter BV, Strom SC (1998) Treatment of the Crigler-Najjar syndrome type I with hepatocyte transplantation. N Engl J Med 338(20): 1422–1426. https://doi.org/10.1056/NEJM199805143382004
3. Bilir BM, Guinette D, Karrer F, Kumpe DA, Krysl J, Stephens J, McGavran L, Ostrowska A, Durham J (2000) Hepatocyte transplantation in acute liver failure. Liver Transpl 6(1):32–40. https://doi.org/10.1002/lt.500060113
4. Schneider A, Attaran M, Meier PN, Strassburg C, Manns MP, Ott M, Barthold M, Arseniev L, Becker T, Panning B (2006) Hepatocyte transplantation in an acute liver failure due to mushroom poisoning. Transplantation 82(8):1115–1116. https://doi.org/10.1097/01.tp.0000232451.93703.ab
5. Nguyen MP, Jain V, Iansante V, Mitry RR, Filippi C, Dhawan A (2020) Clinical application of hepatocyte transplantation: current status, applicability, limitations, and future outlook. Expert Rev Gastroenterol Hepatol 14(3):185–196. https://doi.org/10.1080/17474124.2020.1733975
6. Strom SC, Chowdhury JR, Fox IJ (1999) Hepatocyte transplantation for the treatment of human disease. Semin Liver Dis 19(1): 39–48. https://doi.org/10.1055/s-2007-1007096
7. Zhang K, Zhang L, Liu W, Ma X, Cen J, Sun Z, Wang C, Feng S, Zhang Z, Yue L, Sun L, Zhu Z, Chen X, Feng A, Wu J, Jiang Z, Li P, Cheng X, Gao D, Peng L, Hui L (2018) In vitro expansion of primary human hepatocytes with efficient liver repopulation capacity. Cell Stem Cell 23:806–819. https://doi.org/10.1016/j.stem.2018.10.018
8. Xiang C, Du Y, Meng G, Soon Yi L, Sun S, Song N, Zhang X, Xiao Y, Wang J, Yi Z, Liu Y, Xie B, Wu M, Shu J, Sun D, Jia J, Liang Z, Sun D, Huang Y, Shi Y, Xu J, Lu F, Li C, Xiang K, Yuan Z, Lu S, Deng H (2019) Long-term functional maintenance of primary human hepatocytes in vitro. Science 364(6438):399–402. https://doi.org/10.1126/science.aau7307
9. Grompe M, Al-Dhalimy M, Finegold M, Ou CN, Burlingame T, Kennaway NG, Soriano P (1993) Loss of fumarylacetoacetate hydrolase is responsible for the neonatal hepatic dysfunction phenotype of lethal albino mice. Genes Dev 7(12a):2298–2307. https://doi.org/10.1101/gad.7.12a.2298
10. Azuma H, Paulk N, Ranade A, Dorrell C, Al-Dhalimy M, Ellis E, Strom S, Kay MA, Finegold M, Grompe M (2007) Robust expansion of human hepatocytes in fah−/−/Rag2−/−/Il2rg−/− mice. Nat Biotechnol 25(8):903–910. https://doi.org/10.1038/nbt1326
11. Ellis EC, Naugler WE, Parini P, Mork LM, Jorns C, Zemack H, Sandblom AL, Bjorkhem I, Ericzon BG, Wilson EM, Strom SC, Grompe M (2013) Mice with chimeric livers are an improved model for human lipoprotein metabolism. PLoS One 8(11):e78550. https://doi.org/10.1371/journal.pone.0078550

Chapter 26

Application of 3D Culture Assays to Study Breast Morphogenesis, Epithelial Plasticity, and Cellular Interactions in an Epithelial Progenitor Cell Line

Anna Karen Sigurdardottir, Bylgja Hilmarsdottir, Thorarinn Gudjonsson, and Gunnhildur Asta Traustadottir

Abstract

Capturing breast morphogenesis and cancer progression in 3D culture using cell lines with stem cell properties can greatly increase understanding of the underlying mechanisms involved in these processes, highlighting the importance of the culture method. D492 is a breast epithelial progenitor cell line that provides a model for branching morphogenesis when cultured in 3D reconstituted basement membrane matrix (rBM). Along with its derivate cell lines D492M and D492HER2, D492 also serves as a robust model for epithelial to mesenchymal transition (EMT) and tumorigenicity, respectively. Here, we describe the routine maintenance and application of the D492 cell lines in 3D culture for the study of branching morphogenesis, EMT and epithelial-endothelial interaction.

Key words D492 cell lines, 3D culture, Branching morphogenesis, Breast endothelial cells (BRENCs), Matrigel®

1 Introduction

During the reproductive period, breast epithelial stem cells maintain cellular remodeling in the female breast gland by inducing branching morphogenesis in each menstruation cycle. However, the human female breast gland does not reach its full developmental potential until pregnancy [1]. Breast epithelial stem cells, or their immediate progenitor cells, are believed to be the cells of origin for breast cancer [2], but signals from the microenvironment and stroma are increasingly being recognized as an important factor in both normal breast morphogenesis and malignant growth [3–5]. Three-dimensional (3D) cell culture systems have long been acknowledged as an integral part of mammary gland and breast cancer research (reviewed in [6]). Nevertheless, 3D culture is gaining increased recognition as an essential tool in relation to drug

Nagarajan Kannan and Philip Beer (eds.), *Stem Cell Assays: Methods and Protocols*, Methods in Molecular Biology, vol. 2429,
https://doi.org/10.1007/978-1-0716-1979-7_26,

discovery, drug testing, and tissue engineering by more closely mimicking the native environment of the cells within the body compared to conventional 2D culture systems.

D492 is an epithelial progenitor cell line established from tissue obtained from reduction mammoplasty [7]. A suprabasal population of MUC1^{-}/EpCAM^{+} cells was isolated from the primary breast tissue, expanded in culture and subsequently immortalized using lentiviral transduction of E6 and E7 oncogenes from human papilloma virus type 16, targeting and silencing the P53 and retinoblastoma (Rb) oncogenes, respectively [7]. The D492 cell line can yield both luminal- and myoepithelial cells in culture and generates branching structures resembling the terminal duct lobular units of the breast (TDLUs) when cultured in 3D reconstituted basement membrane matrix (rBM). Furthermore, D492 cells show cellular plasticity and respond to changes in the microenvironment, dramatically increasing their branching potential when embedded in an rBM co-culture with endothelial cells [8]. Under co-culture conditions the endothelial cells stay as nonproliferative yet metabolically active single cells, while a subpopulation of D492 cells undergoes EMT as evidenced by formation of spindle shaped colonies in between the epithelial branching colonies (Fig. 1). It was from one of these spindle-shaped colonies that the D492M mesenchymal cell line was isolated [8]. Together D492 and D492M provide a robust model for EMT (reviewed in [9]). Moreover, D492 (epithelial) and D492M (mesenchymal-like) are isogenic nontumorigenic cell lines, whereas the D492HER2 cell line, generated by overexpression of the HER2 oncogene in D492, forms tumors in vivo and shows partial EMT [10]. In addition to being used as a model for branching morphogenesis and EMT the D492/D492M/D492HER2 cell lines have been employed to explore epithelial–endothelial interactions [11, 12]. Upon request, D492 and derivative cell lines are freely available for academic use.

Fig. 1 D492 forms branching, spindle and solid round colonies in co-culture with endothelial cells. In co-culture with breast endothelial cells in Matrigel, D492 cells form three morphologically different colony types: branching, spindle and solid round. Whereas, the branching and solid round colonies have epithelial characteristics, the spindle shaped colonies consist of mesenchymal-like cells that have undergone epithelial-to-mesenchymal transition (EMT) (Scale bar = 100 μm)

2 Materials

Use only sterilized utensils and material for breast tissue isolation and for cell culture. Utensils can be sterilized by autoclaving, whereas solutions may be sterilized by autoclaving or filtration (0.22 μm filter).

2.1 General Maintenance of D492 Cell Lines

1. Pure-Col Type I collagen solution, 3 mg/mL.
2. Cell culture flasks or dishes.
3. Phosphate buffered saline (PBS) without calcium and magnesium.
4. Trypsin–EDTA solution: 0.25% w/v trypsin and 1 mM EDTA in 1 × PBS.
5. Soybean Trypsin Inhibitor (STI), 10 mg/mL.
6. H14 medium [13] consisting of Dulbecco's Modified Eagle's Medium–Nutrient Mixture F-12 (DMEM:F12) with HEPES and L-glutamine, supplemented with; 50 IU/mL penicillin, 50 μg/mL streptomycin, and the following growth factors: 250 ng/mL insulin, 10 ng/mL EGF, 10 μg/mL transferrin, 2.6 ng/mL NaSel, 10^{-10} M estradiol, 500 ng/mL hydrocortisone, and 0.15 IU prolactin.

2.2 Isolation of Breast Endothelial Cells (BRENCs) from Reduction Mammoplasties

1. Breast tissue from reduction mammoplasty (*see* **Note 1**).
2. Polystyrene petri dishes, nontreated, 100 mm, sterile.
3. Scalpels.
4. Forceps.
5. Collagenase solution: DMEM:F12 supplemented with 900 U/mL collagenase-IA.
6. Rotary incubator shaker.
7. Anti-CD31 magnetic microbeads.
8. Magnetic cell sorter.
9. Endothelial Cell Growth Medium BulletKit™ (EGM™) (*see* **Note 2**).
10. Heat-inactivated fetal bovine serum (FBS).

2.3 Preparation of D492 3D Assays

1. Ice in a small container.
2. 24-well cell culture dishes, flat bottom.
3. Hemocytometer.
4. Corning® Matrigel® Basement Membrane Matrix (Corning, 354230) (*see* **Note 3**).
5. H14 medium for monoculture or Endothelial Cell Growth Medium BulletKit™ (EGM™) (*see* **Note 2**), for co-culture, supplemented with 5% heat-inactivated fetal bovine serum (FBS).

2.4 Isolation of 3D Structures

1. 5 mM EDTA, ice cold.
2. Small spatula.
3. Shaker.
4. Ice.

2.5 Immunostaining of 3D Structures

1. 1 × PBS.
2. 4% paraformaldehyde.
3. 10% FBS.
4. IF buffer consisting of distilled H_2O supplemented with; 0.15 M NaCl, 5 mM EDTA, 20 mM HEPES, and 0.1% Triton X-100.
5. Primary and secondary antibodies.
6. DAPI or phalloidin.
7. Distilled water.
8. Spatula.
9. Microscope glass slide.
10. Coverslips.
11. FixoGum.

3 Methods

3.1 General Maintenance of D492 Cell Lines

3.1.1 Collagen Coating of Culture Vessels

1. Add 1 mL Pure-Col Type I collagen solution, 3 mg/mL, to 44 mL 1×PBS and mix.
2. Add 3 mL collagen I solution to a 25 cm^2 culture vessel (*see* **Note 4**).
3. Incubate for 30 min at 37 °C, 1 h at room temperature or overnight at 4 °C (*see* **Note 5**).
4. Remove collagen I solution and rinse the culture vessel with 1× PBS. Remove the PBS and plate cells.

3.1.2 Retrieving D492 Cell Lines from Liquid Nitrogen

1. Thaw one vial of D492 cells in a 37 °C water bath or by the warmth of your hands.
2. When small ice crystals remain, the remaining ice is dissolved by pipetting up and down with a 1 mL pipette. Subsequently transfer cells to a 15 mL tube containing 9 mL 1× PBS (*see* **Note 6**).
3. Pellet cells at 200 × *g* for 5 min. Remove supernatant, resuspend in 4 mL of prewarmed H14 medium, and plate in a 25 cm^2 flask.

3.1.3 Maintaining D492 Cell Lines

1. D492 cells are routinely maintained in collagen I coated 25 cm^2 flasks at 37 °C, 5% CO_2 (*see* **Note 7**).
2. Cells are cultured in H14 medium prewarmed to 37 °C. Medium is changed 3 times per week.
3. D492 cells are routinely frozen in 10% DMSO, 40% FBS, and 50% DMEM:F12 when 70–80% confluent (*see* **Note 8**).

3.1.4 Passaging D492 Cell Lines

1. Passage cells when they are 90% confluent.
2. Aspirate medium and rinse the flask once with 1× PBS.
3. Add 1 mL of Trypsin-EDTA solution to each 25cm^2 flask.
4. Incubate at 37 °C until cells are detached (*see* **Note 9**).
5. Stop trypsinization by adding 20 μL of STI to the flask.
6. Add 4 mL 1× PBS and gently pipet up and down.
7. Transfer to a 15 mL tube and centrifuge at 200 × *g* for 3 min.
8. Aspirate supernatant and resuspend cell pellet in 1 mL prewarmed H14 media.
9. Cells are subcultured at 1:10 ratio for D492M and 1:20 ratio for D492 and D492HER2 (*see* **Note 10**).

3.2 Isolation of Breast Endothelial Cells (BRENCs) from Reduction Mammoplasties (Fig. 2)

Day 1

1. Place fresh pieces of breast tissue in a sterile petri dish inside the cell hood (*see* **Notes 11** and **12**).
2. The breast tissue is composed of adipose tissue as well as epithelial and stromal components (*see* **Note 13**). To facilitate digestion of the tissue, carefully chop the tissue into small pieces (~1 mm) using two opposing scalpel blades (*see* **Note 14**). Place about 30 mL of tissue in a 75 cm^2 flask (*see* **Note 15**).
3. Add 15 mL of prepared collagenase solution to each 75 cm^2 flask.
4. Digest on a rotary shaker (60 rpm) at 37 °C overnight.

Day 2

1. Transfer digested adipose tissue into 50 mL tubes (*see* **Note 16**).
2. Centrifuge at 200 × *g* for 40 s.
3. After centrifugation the liquid will have separated into layers (*see* **Note 17**). Discard the oil from the top and transfer the brown liquid phase, containing endothelial cells, into a new 50 mL tube.
4. Centrifuge at 200 × *g* for 5 min.

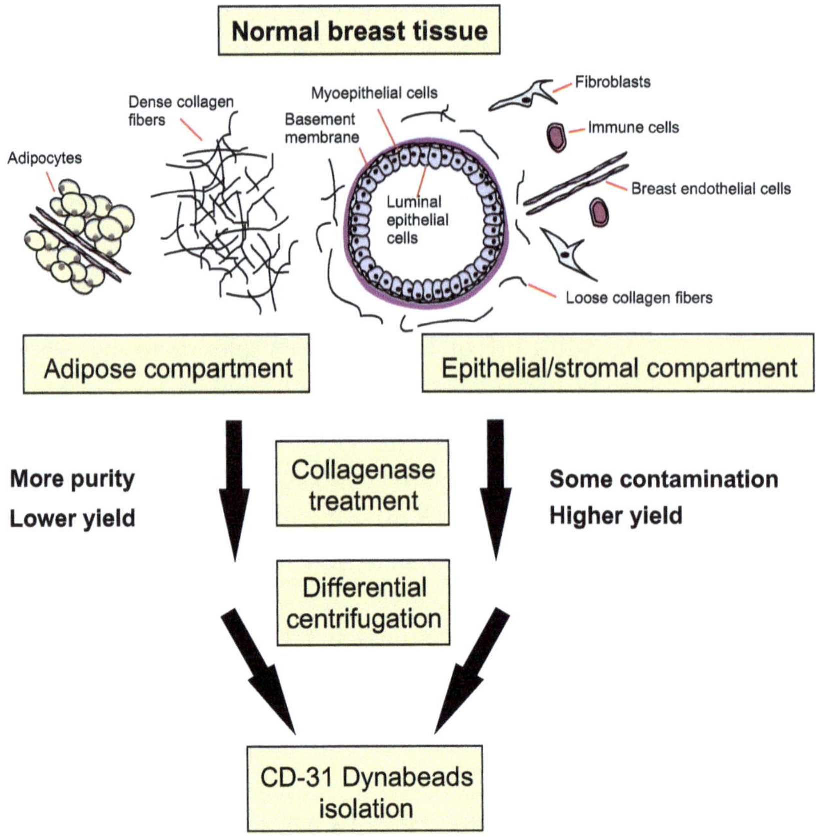

Fig. 2 Isolation of breast endothelial cells (BRENCs) from reduction mammoplasties. Human breast stroma is comprised of intralobular loose connective tissue, dense collagenous interlobular stroma and adipose tissue. Isolation of BRENCs from the adipose compartment gives more purity but lower cell yield compared to isolation from the epithelial/stromal compartment, where cellular contamination is more frequent. After treatment with collagenase and subsequent differential centrifugation, BRENCs can be isolated via magnetic sorting using CD-31 dynabeads

5. Remove supernatant. Wash once with 20 mL 1× PBS, spin down as previously and remove the supernatant.
6. Incubate pellet containing microvessel organoids with anti-CD31 magnetic microbeads as per manufacturer instructions and separate CD31 positive endothelial cells from the supernatant using a magnetic system.
7. Seed cells into 25 cm^2 flasks precoated with collagen I in EGM™ media supplemented with 30% FBS.
8. Maintain endothelial cells in EGM™ media supplemented with 30% FBS until the first passage and then reduce serum concentration to 5% (*see* **Note 18**).

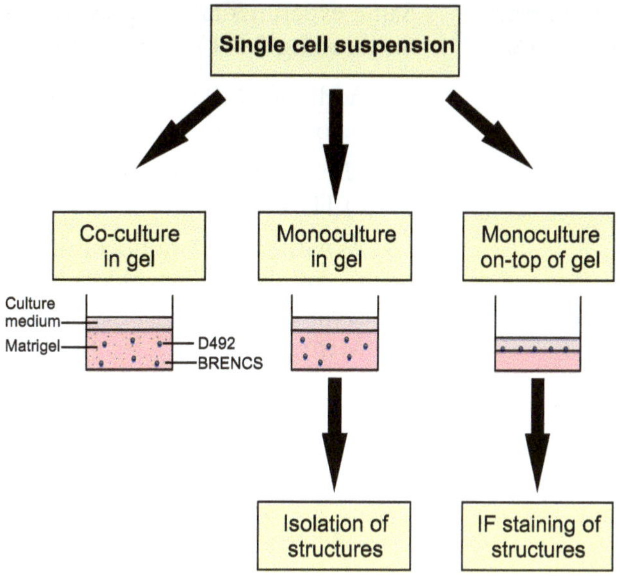

Fig. 3 Three-dimensional culture assays to study breast morphogenesis, epithelial plasticity and cellular interactions. D492 cells can be cultured in Matrigel® in monoculture or in co-culture with BRENCs. Alternatively, D492 cells can be cultured in monoculture on-top of gel, where branching colonies take less time to form and less gel is used

3.3 Preparation of D492 3D Assays (Fig. 3)

3.3.1 D492 Monoculture in Matrigel®

1. Remove an aliquot of Matrigel® from −20 °C and thaw on ice for 2–3 h or overnight (*see* **Note 19**).
2. Place appropriate number of cryovials (v-shaped bottom) on ice.
3. Detach cells (*see* Subheading 3.1.4) and resuspend in a small volume. Count cells using a hemocytometer. Add 10,000 cells (*see* **Note 20**) to a cryovial. For volume >10 μL centrifuge at 200 × *g* for 5 min and aspirate supernatant.
4. Gently resuspend cell pellet in 300 μL Matrigel®. Pipet up and down 6 times to mix (*see* **Note 21**).
5. Quickly transfer Matrigel®–cell mix to a well of a 24-well plate.
6. Let Matrigel®–cell mix set in an incubator at 37 °C for 30 min.
7. Gently add 500 μL H14 medium on top of gel (*see* **Note 22**).
8. Change media three times a week by removing 250 μL of old media and gently adding 250 μL of fresh media (*see* **Note 23**).
9. Record progress of cells by phase contrast imaging. Full branching of D492 cells is reached within 2 weeks (*see* **Note 24**).

3.3.2 D492 Co-culture with Endothelial Cells in Matrigel®

Same experimental setup as for D492 3D monoculture assay (*see* Subheading 3.3.1), with the following exceptions.

1. Add 500 D492 cells and 100,000 endothelial cells to a cryovial before spinning and removal of supernatant (*see* **Note 25**).
2. Use EGM™ medium supplemented with 5% FBS instead of H14.
3. As with the 3D monocultures, full branching will be completed in 14 days. The endothelial cells in the culture are viable throughout the 2 weeks but will not proliferate in the Matrigel®.

3.3.3 Isolation of 3D Colonies in Matrigel®

1. Remove medium.
2. Add 1 mL of ice cold 5 mM EDTA to a 15 mL tube (per sample).
3. Add 100 μL of 5 mM EDTA to each well.
4. Use a small spatula to separate the gel from the edges of the well.
5. Use a spatula to move the gel into the 15 mL tube.
6. Add 100 μL of 5 mM EDTA to the well and aspirate the remaining gel and add to the 15 mL tube.
7. Fill up to 5 mL with 5 mM EDTA in the 15 mL tube. Place the tube on ice and on a shaker.
8. After 20 min, check whether the gel is dissolved. If gel is not dissolved, continue shaking on ice for another 20 min (*see* **Note 26**).
9. Use structures either for isolation of cell populations, protein/RNA/DNA extraction or for immunofluorescent staining.

3.3.4 D492 Monoculture on-Top of Matrigel® (See Note 27)

1. Remove an aliquot of Matrigel® from −20 °C and thaw on ice for 2–3 h or overnight.
2. Place 50 μL of Matrigel® per well in a 96 well plate. Incubate at 37 °C for 30 min.
3. Detach cells (*see* Subheading 3.1.4) and resuspend in a small volume. Count cells using a hemocytometer.
4. Plate 2500 cells per well in 200 μL H14 media (*see* **Note 28**).
5. Change media three times per week by gently removing 100 μL of media and adding 100 μL of fresh media instead (*see* **Note 29**).
6. Evaluate colony morphology, size, and number, 8 days after plating (*see* **Note 30**).
7. Proceed with RNA/protein isolation and/or immunostaining (*see* **Note 31**).

3.3.5 Immunostaining of 3D Structures on-Top of Matrigel®

1. Gently remove medium from the wells.
2. Rinse wells 3 times with 1× PBS by holding the plate on its side and letting the liquid seep down the side of the well.
3. Fix cells with 4% formaldehyde in 1× PBS for 30 min at room temperature.
4. Repeat **step 2** (*see* **Note 32**).
5. Block for 10 min in IF buffer with 10% FBS.
6. Repeat **step 2**.
7. Incubate with primary antibody in IF buffer for 1 h at room temperature or overnight at 4 °C.
8. Rinse wells with IF buffer, three times, as described in **step 2**. One quick wash and subsequently twice for 5 min.
9. Incubate with fluorescent secondary antibody and DAPI or phalloidin in IF buffer for 30 min at room temperature.
10. Repeat **step 8**.
11. Rinse with distilled water.
12. Aspirate liquid and take a small spatula and scoop up the Matrigel® with colonies from each well.
13. Place Matrigel® with colonies on a glass slide (*see* **Note 33**).
14. Mount and cover with glass coverslip. Seal sides of coverslip with FixoGum and let dry at room temperature in the dark (*see* **Note 34**).
15. Proceed with imaging.

4 Notes

1. Breast tissue can be kept at 4 °C in DMEM:F12 for a few hours before the isolation procedure is carried out.
2. All growth factors are provided with the kit; however, the gentamicin–amphotericin B mixture, also provided with the kit, is not used.
3. In order to avoid repeated freeze–thaw cycles, Matrigel® should be aliquoted and kept at −20 °C.
4. For other types of culture vessels add enough collagen I solution to cover the surface.
5. Flasks can be kept at 4 °C with the collagen I solution for up to 2 weeks.
6. D492 cell lines are sensitive to the DMSO in the freezing media. Therefore, it is important to remove the DMSO before the cells are plated.
7. D492 cells are very sensitive to CO_2-levels and humidity in the incubator. Growth will halt at suboptimal growth conditions.

8. Cells from one 25 cm^2 flask can be divided into three cryovials.
9. The incubation time varies between cell types, D492 will detach after 3–5 min while D492M and D492HER2, having a more mesenchymal phenotype, need about 1 min of trypsinization. If D492 cells are overconfluent or starved, detaching the cells can be difficult. Then it is possible to facilitate detachment by either rinsing once with the trypsin–EDTA solution before 5 min trypsinization, or by trypsinizing twice and collecting all supernatant in the same tube.
10. Maintaining the cells at this ratio will require the cells to be split once a week.
11. All work with primary breast tissue should be carried out at sterile conditions.
12. The cutting of the tissue may be facilitated by adding a few mL of 1× PBS to the petri dish.
13. BRENCs are found in abundance and are most readily isolated from the adipose tissue of the breast. Isolation from the epithelial and stromal components results in higher cell yield but often has cellular contamination, mainly from fibroblasts.
14. Macroscopic examination of the breast tissue reveals two kinds of tissue: White strands contain the epithelium and stroma while yellow sections contain mostly adipose tissue.
15. Our main source of primary tissue is from breast reduction surgery. As there is plenty of tissue from each surgery we aim to process at least 60 mL of tissue in two 75 cm^2 flask.
16. All pieces of tissue should be digested at this point. If not, we recommend adding 6–12 h to the digestion time.
17. The liquid has now been separated to three layers: Top layer is adipose tissue, the pellet contains organoids and big veins and the middle phase contains microvessels and single cells. From here it is possible to isolate endothelial cells and fibroblasts.
18. Breast endothelial cells can be passaged up to 12 times. During the first passages, fibroblasts may be present in the endothelial culture. Fibroblasts take longer than endothelial cells to detach when subcultured and can thereby be gradually eliminated from the culture by stopping the trypsinization process when 50–60% of the cells have been detached.
19. Matrigel® should be kept on ice at all times until it is in the plate with the cells. Keep ice in a small container inside the hood while working.
20. The exact composition of Matrigel® can vary from lot to lot. It is essential to assess colony formation in every new batch of Matrigel®, by testing different cell numbers. If too many cells are seeded, colony growth may be prevented by steric

hindrance. To reduce the need to test the Matrigel® frequently it is possible to test 3–4 batches of Matrigel® at the same time and subsequently order multiple vials from the best lot that was tested. When Matrigel® is tested we set up 3D monocultures for D492 ranging from 5000 cells/300 μL–30,000 cells/300 μL of Matrigel®. A suitable batch will support the growth of branching 3D colonies.

21. Avoid generating bubbles when pipetting Matrigel®.
22. After cells are seeded in Matrigel®, the first 24 h are critical for their growth. If growth of 3D organoids is problematic it is possible to add EGM™ media with 5% FBS on top of the 3D cultures overnight and replace with H14 media the subsequent day.
23. The 3D assay can be scaled down to 96 well plate where 100 μL of Matrigel® are used and 100 μL of media. Also, it is possible to scale down to an 8 well chamber slide where 50 μL of gel and 200 μL of media are used. The number of cells is scaled down in line with the amount of Matrigel® used.
24. Rarely cells reach the bottom of the cell culture plate and grow in monolayer on the plastic, making it difficult to properly visualize the colonies within the gel.
25. Instead of breast endothelial cells (BRENCs), human umbilical vein endothelial cells (HUVECs) can be used.
26. The gel is dissolved when all colonies move freely in the EDTA solution.
27. In a 3D assay on-top of Matrigel®, less Matrigel® is used compared to 3D assay in gel and it takes less time for branching colonies to form. The 3D monoculture protocol on-top of Matrigel® is adapted from [14].
28. The exact number of cells may need to be adjusted.
29. Colonies growing on-top of Matrigel® can easily detach. Therefore, liquid must be carefully aspirated from the wells, both during change of media and subsequent treatments of the culture, such as during immunostaining.
30. Colonies form much faster when grown on-top of Matrigel® as compared to when they are cultured in the gel.
31. When isolating RNA or protein from 3D cultures on-top of Matrigel® please follow the same procedure as is described for isolation of 3D colonies embedded in Matrigel® (*see* Subheading 3.3.3).
32. Fixed colonies can be kept in IF buffer at 4 °C for a few days. Be careful to not let them dry out.

33. It is not necessary to transfer the Matrigel® with colonies to a glass slide if a culture vessel with high optical quality, such as μ-Slide 8 well from Ibidi, is used instead of a 96-well plate.
34. If the Matrigel® is too thick, consider adding spacers between the glass slide and the coverslip.

Acknowledgments

This work was supported by grants from Landspitali University Hospital Science Fund, University of Iceland Research Fund, and Icelandic Science and Technology Policy—Grant of Excellence: 152144051, "Göngum saman," a supporting group for breast cancer research in Iceland (www.gongumsaman.is) and the Science Fund of the Icelandic Cancer Society.

References

1. Macias H, Hinck L (2012) Mammary gland development. Wiley Interdiscip Rev Dev Biol 1(4):533–557. https://doi.org/10.1002/wdev.35
2. Visvader JE (2009) Keeping abreast of the mammary epithelial hierarchy and breast tumorigenesis. Genes Dev 23(22): 2563–2577. https://doi.org/10.1101/gad.1849509
3. Acerbi I, Cassereau L, Dean I, Shi Q, Au A, Park C, Chen YY, Liphardt J, Hwang ES, Weaver VM (2015) Human breast cancer invasion and aggression correlates with ECM stiffening and immune cell infiltration. Integr Biol (Camb) 7(10):1120–1134. https://doi.org/10.1039/c5ib00040h
4. Tomko LA, Hill RC, Barrett A, Szulczewski JM, Conklin MW, Eliceiri KW, Keely PJ, Hansen KC, Ponik SM (2018) Targeted matrisome analysis identifies thrombospondin-2 and tenascin-C in aligned collagen stroma from invasive breast carcinoma. Sci Rep 8(1): 12941. https://doi.org/10.1038/s41598-018-31126-w
5. Quail DF, Joyce JA (2013) Microenvironmental regulation of tumor progression and metastasis. Nat Med 19(11):1423–1437. https://doi.org/10.1038/nm.3394
6. Sumbal J, Budkova Z, Traustadottir GA, Koledova Z (2020) Mammary organoids and 3D cell cultures: old dogs with new tricks. J Mammary Gland Biol Neoplasia. https://doi.org/10.1007/s10911-020-09468-x
7. Gudjonsson T, Villadsen R, Nielsen HL, Ronnov-Jessen L, Bissell MJ, Petersen OW (2002) Isolation, immortalization, and characterization of a human breast epithelial cell line with stem cell properties. Genes Dev 16(6): 693–706. https://doi.org/10.1101/gad.952602
8. Sigurdsson V, Hilmarsdottir B, Sigmundsdottir H, Fridriksdottir AJ, Ringner M, Villadsen R, Borg A, Agnarsson BA, Petersen OW, Magnusson MK, Gudjonsson T (2011) Endothelial induced EMT in breast epithelial cells with stem cell properties. PLoS One 6(9):e23833. https://doi.org/10.1371/journal.pone.0023833
9. Briem E, Ingthorsson S, Traustadottir GA, Hilmarsdottir B, Gudjonsson T (2019) Application of the D492 cell lines to explore breast morphogenesis, EMT and cancer progression in 3D culture. J Mammary Gland Biol Neoplasia 24(2):139–147. https://doi.org/10.1007/s10911-018-09424-w
10. Ingthorsson S, Andersen K, Hilmarsdottir B, Maelandsmo GM, Magnusson MK, Gudjonsson T (2016) HER2 induced EMT and tumorigenicity in breast epithelial progenitor cells is inhibited by coexpression of EGFR. Oncogene 35(32):4244–4255. https://doi.org/10.1038/onc.2015.489
11. Morera E, Steinhauser SS, Budkova Z, Ingthorsson S, Kricker J, Krueger A, Traustadottir GA, Gudjonsson T (2019) YKL-40/CHI3L1 facilitates migration and invasion in HER2 overexpressing breast epithelial progenitor cells and generates a niche for capillary-like network formation. In Vitro Cell Dev Biol Anim 55(10):838–853. https://doi.org/10.1007/s11626-019-00403-x

12. Steinhaeuser SS, Morera E, Budkova Z, Schepsky A, Wang Q, Rolfsson O, Riedel A, Krueger A, Hilmarsdottir B, Maelandsmo GM, Valdimarsdottir B, Sigurdardottir AK, Agnarsson BA, Jonasson JG, Ingthorsson S, Traustadottir GA, Oskarsson T, Gudjonsson T (2020) ECM1 secreted by HER2-overexpressing breast cancer cells promotes formation of a vascular niche accelerating cancer cell migration and invasion. Lab Invest 100(7): 928–944. https://doi.org/10.1038/s41374-020-0415-6

13. Blaschke RJ, Howlett AR, Desprez PY, Petersen OW, Bissell MJ (1994) Cell differentiation by extracellular matrix components. Methods Enzymol 245:535–556. https://doi.org/10.1016/0076-6879(94)45027-7

14. Lee GY, Kenny PA, Lee EH, Bissell MJ (2007) Three-dimensional culture models of normal and malignant breast epithelial cells. Nat Methods 4(4):359–365. https://doi.org/10.1038/nmeth1015

Chapter 27

A Unified Protocol to Streamline Molecular and Cellular Analysis for Three-Dimensional Cell Cultures

Lisa M. Kim, Paul Y. Kim, and Cheuk T. Leung

Abstract

Three-dimensional (3D) cell cultures based on reconstituted basement membrane materials recapitulate features of extracellular matrix (ECM) and tissue stiffness in vivo and provide a physiologically relevant platform to study complex cellular processes, such as stem cell differentiation and tissue morphogenesis, that are otherwise difficult in animal models. The form and composition of 3D matrices in culture can interfere with and pose challenges for different experimental setups and assays, which necessitate alterations to facilitate analysis. Here, we provide a unified protocol for 3D cell cultures with modular workflows that streamline procedures for compatibility with common molecular and cellular assays such as live-cell imaging, immunofluorescence, qPCR, RNAseq, western blotting, and quantitative mass spectrometry.

Key words Three-dimensional (3D) cell culture, Extracellular matrix (ECM), Live imaging, Immunofluorescence, Protein and RNA isolation

1 Introduction

Three-dimensional (3D) cell cultures provide a system for studying cellular control in complex microenvironments reminiscent of in vivo tissues with the flexibility of modeling experimental conditions in vitro, and have emerged as an important research tool in diverse disciplines of cell biology including cancer [1–3] and development [4–6]. 3D cultures utilize reconstituted basement membrane materials to recapitulate the extracellular matrix (ECM) environment and normal tissue stiffness, offering a more physiologically relevant platform compared to standard monolayer cell cultures on rigid plastic or glass surface [7, 8]. Emerging studies have shown that molecular and mechanical cues from the microenvironment have predominant roles in governing cell growth and differentiation [9–11]. Moreover, cell–cell and cell–matrix interactions in three-dimensional space underlie tissue formation and function [12]. 3D cell culture systems have allowed the modeling

Nagarajan Kannan and Philip Beer (eds.), *Stem Cell Assays: Methods and Protocols*, Methods in Molecular Biology, vol. 2429, https://doi.org/10.1007/978-1-0716-1979-7_27,

and delineation of these complex controls that are otherwise difficult to study in vivo.

Reconstituted basement membrane materials such as Matrigel, a laminin-rich matrix derived from the Engelbreth-Holm-Swarm mouse sarcoma cell line [13], is a common scaffold used in many different 3D cell culture systems [1, 4–6, 14]. While recapitulating the molecular, physical, and spatial features of tissue-like microenvironment, the forms and compositions of these 3D matrices can pose challenges and interference to standard analysis techniques. Here, we present a unified protocol with modular workflows adapted for 3D cell culture to streamline common molecular and cellular assays including live-cell imaging, immunofluorescence, and protein and RNA isolation in small- and large-scale samples for quantitative analysis. The protocol is based on 3D mammary morphogenesis of the MCF10A nontransformed human mammary cells [15, 16], which develop growth-arrested, polarized hollow structures reminiscent of mammary acini [14], but the procedures are amenable to other 3D culture systems based on the common reconstituted basement membrane scaffold.

2 Materials

2.1 3D Culture

1. 3D culture on Matrigel bed: 8-well chamber slide with removable wells or 8-well chambered cover glass with nonremovable wells (live imaging)
2. 3D culture in suspension: ultralow-attachment 24-well plate
3. Matrigel® growth-factor reduced (GFR)
4. 1× phosphate buffered saline (PBS)
5. 0.25% trypsin
6. 3D Assay Medium (for MCF10A cells): Dulbecco's Modified Eagle Medium–Nutrient Mixture F-12 (DMEM/F-12) supplemented with 2% horse serum, 0.5 μg/mL hydrocortisone, 0.1 μg/mL cholera toxin, 10 μg/mL insulin, 50 U/mL penicillin and 50 U/mL streptomycin
7. Resuspension Buffer: DMEM/F-12 supplemented with 20% horse serum, 50 U/mL penicillin, and 50 U/mL streptomycin
8. Recombinant human epidermal growth factor (EGF)
9. Hemocytometer

2.2 Immunofluorescence (IF)

1. 16% paraformaldehyde (PFA)
2. PBS (1× and 10×)
3. Glycine Buffer: 3.75 g glycine in PBS (500 mL)
4. IF Buffer (10×): 2.5 g NaN_3, 5 g BSA, 10 mL Triton X-100, 2 mL Tween 20 in 10× PBS (500 mL)

5. Permeabilization Buffer (0.5% Triton X-100): 2.5 mL Triton X-100 in PBS (500 mL)
6. Goat serum
7. Goat anti-mouse IgG F(ab′)2 (1.3 mg/mL)
8. Primary antibody
9. Fluorescent secondary antibody
10. 4′,6-Diamidino-2-phenylindole (DAPI)
11. ProLong Diamond Antifade Mountant
12. 50 × 24 mm cover glass (thickness range 0.13 to 0.17 mm)

2.3 Protein Isolation

1. Cell Recovery Solution
2. RIPA lysis buffer with EDTA: 50 mM Tris-HCl pH 7.4, 150 mM sodium chloride, 1% NP-40, 0.5% sodium deoxycholate, 0.1% sodium dodecyl sulfate (SDS), 5 mM EDTA
3. Halt™ proteinase and phosphatase inhibitor cocktail
4. 1× PBS
5. 1-mL syringe with 27-gauge needle

2.4 Total RNA Isolation

1. TRIzol™ Reagent
2. Chloroform
3. RNeasy Mini Kit
4. 200 proof RNase-free ethanol
5. RNase-free microtubes

3 Methods

Select the experiment workflow based on applications (Fig. 1).

Use chambered cover glass for live imaging, regular chamber slides for immunofluorescence and small-scale total RNA and protein preparations, and ultralow-attachment plate with suspension 3D culture for large-scale total RNA and protein preparations (*see* **Note 1**).

3.1 3D Platforms

3.1.1 3D Culture on Gel Bed

Thaw Matrigel aliquot on ice. Chill chamber slides on ice or at −20 °C. Over-dried Matrigel bed will affect the quality of 3D culture. Timeline of gel bed and cell preparation should be arranged so that cells are ready to seed between 45 min to 1 h 15 min after laying the gel bed (*see* **Note 2**). Phenol red–free Matrigel and medium are recommended for live-cell fluorescent imaging.

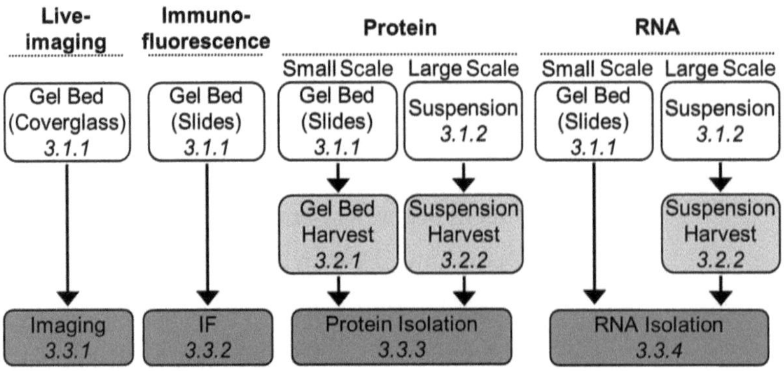

Fig. 1 Experiment workflow. Diagram illustrating choices of platform (white), cell harvest (light gray), and assays (dark gray) for different downstream applications. Numbers in italics indicate the subsections for detailed procedures

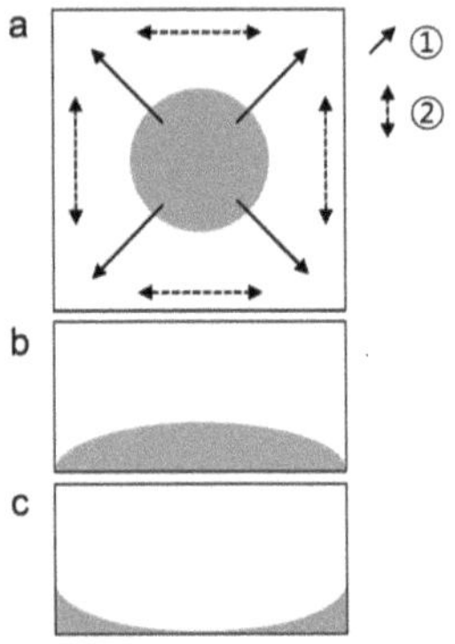

Fig. 2 Schematic diagrams of Matrigel beds. (**a**) Recommended methods for laying consistent Matrigel bed. Load Matrigel over the middle area of the well. With a fine pipette tip, spread to the four corners and then the edges. Ensure all surfaces are covered as cells will attach to uncovered glass surface and start growing and spreading as monolayers on glass under the Matrigel bed. (**b**) Convex Matrigel bed is recommended for 8-well chamber slides to allow removal of the plastic chamber without disturbing the cells and gel for mounting in immunofluorescence experiments. (**c**) Concave Matrigel bed is recommended for live imaging and for RNA and protein harvest. The thicker gel bed on the edges further reduce the chance of cells contacting the glass surface and growing in monolayer. The thinner gel bed in the middle of the well reduces the focusing distance for microscope objectives and facilitates live-cell imaging

1. Lay Matrigel to form a gel bed covering the glass surface. Work quickly and avoid bubbles to prevent inconsistency within the gel bed (*see* **Note 3** and Fig. 2).
   (a) For immunofluorescence (*see* Fig. 2b): Dispense 42 μL of Matrigel to the middle of the well of an 8-well chamber slide. Using a fine pipette tip at an angle without touching

the plastic chamber wall, spread the Matrigel to the edges to form a convex gel bed.

(b) For protein/RNA isolation (*see* Fig. 2c): Dispense 52 μL Matrigel to the middle of the well of an 8-well chamber slide. Use a fine pipette tip to spread the Matrigel to the edges and the chamber wall. Dispense an additional 42 μL Matrigel to the middle of the well.

(c) For live imaging (*see* Fig. 2c): Dispense 57 μL Matrigel to the middle of the well of an 8-well chambered cover glass. Use a fine pipette tip to spread Matrigel to the edges and chamber wall to generate a concave bed.

2. Allow the gel bed to solidify in a humidified cell culture incubator at 37 °C.
3. Wash cells with PBS and trypsinize for 25 min. Pipette to mechanically dissociate cells into single-cell suspension before inhibiting the trypsin with Resuspension Buffer.
4. Centrifuge at 200 × *g* for 3 min to pellet cells, then aspirate the Resuspension Buffer.
5. Tap loosen the cell pellet and resuspend cells in 3D Assay Medium.
6. Prepare 5,000 cells/400 μL in 3D Assay Medium with 5 ng/mL EGF and 2% Matrigel. Mix immediately (*see* **Note 4**).
7. Dispense 400 μL of the cell suspension dropwise to each well (*see* **Note 5**).
8. Gently move the chamber slides into the cell culture incubator.
9. Replace medium every 4 days with fresh 3D Assay Medium containing 5 ng/mL EGF and 2% Matrigel. From Day 12 (Seeding day is Day 0) on, refeed without Matrigel.

3.1.2 3D Culture in Suspension

1. Follow Subheading 3.1.1, **steps 3** and **4** to trypsinize cells.
2. Count cells and prepare 3D Assay Medium with 5 ng/mL EGF and 4% Matrigel.
3. Pipette thoroughly to dissociate cells into single-cell suspension. Prepare cells at 5,000 cells/500 μL in 3D Assay Medium with 5 ng/mL EGF and 4% Matrigel. Mix immediately with a serological pipet.
4. Add 500 μL of the mixture dropwise to each well of an ultralow-attachment 24-well plate (*see* **Note 6**).
5. Gently move the 24-well plates into the cell culture incubator.
6. One day after seeding, add 1 mL of 3D Assay Medium with 5 ng/mL EGF and 1% Matrigel dropwise. Do not disturb the cells.

7. On Day 4 (Seeding day is Day 0), gently remove 750 μL of the medium and add 750 μL fresh 3D Assay Medium with 5 ng/ mL EGF and 1% Matrigel (*see* **Note** 7). Add medium dropwise.
8. Refeed cells every 4 days. From Day 12, refeed without Matrigel.

3.2 Cell Harvest

3.2.1 Cell Harvest from Matrigel Bed

Overview: The yield depends on the cell growth in different experimental conditions. Start with three to four wells of 8-well chamber slides for each experimental condition and scale the protocol based on the yield.

1. Remove medium and wash cells twice with 500 μL 1× PBS per well.
2. Put the chamber slides on ice. Keep samples on ice at all times from this point on.
3. Add 200 μL of Cell Recovery Solution to each well. Scrape and pipet cells up and down to make Matrigel bed slurry. Collect the solution in a prechilled 2 mL microtube (*see* **Note 8**).
4. Rinse the wells twice with 300 μL of Cell Recovery Solution and pool the solution into the same microtube.
5. Incubate on ice for 20 min. Invert to mix during the incubation. If Matrigel has not dissolved, incubate up to 30 min.
6. Centrifuge at 3,500 × *g* for 3 min at 4 °C.
7. Remove supernatant, tap loosen the cell pellet. Add 1 mL of cold PBS and pipette up and down. Centrifuge at 3,500 × *g* for 3 min at 4 °C.
8. Repeat wash one more time.
9. Remove wash completely and tap loosen the cell pellet. Move forward to Subheading 3.3.

3.2.2 Cell Harvest from Suspension

Overview: The yield depends on the cell growth in different experimental conditions. The procedures below are for a whole 24-well plate. Scale the protocol based on the yield.

1. Prechill 15 mL conical tubes (*see* **Note 9**), 2 mL microtubes and PBS on ice.
2. Place the 24-well plate with cells on ice while harvesting. Resuspend cells by pipetting up and down and collect cells from each well into the 15 mL conical tube on ice (*see* **Note 10**). Add 1.5 mL cold PBS per well to wash and avoid drying. Pellet cells at 200 × *g* for 3 min. Remove supernatant. Do not tap loosen the cell pellet. Repeat until cells from all wells are collected.
3. Without tap loosening the cell pellet, start collecting PBS wash from each well into the 15 mL conical tube. Be careful not to

break the cell pellet while adding PBS to the 15 mL conical tube as cells do not pellet well in PBS. Centrifuge at 200 × *g* for 3 min. Remove supernatant. Repeat until the plate is empty.

4. After the last centrifugation, aspirate supernatant and tap loosen the cell pellet.
5. Add 3 mL of Cell Recovery Solution. Wash the wall while adding the solution. Do not pipette up and down.
6. Vortex gently for 10 s. Incubate for 20 min on ice with vortexing every 5 min.
7. Gently pipette up and down and transfer the cell solution into a prechilled 2 mL microtube. Centrifuge at 3,500 × *g* for 3 min at 4 °C.
8. Remove supernatant and collect the remaining cell solution into the same 2 mL microtube. Add 500 μL Cell Recovery Solution to the 15 mL conical to rinse.
9. Centrifuge the 2 mL microtube, remove supernatant, and collect the remaining Cell Recovery Solution wash from the 15 mL conical tube.
10. Centrifuge the 2 mL microtube and remove the supernatant. Tap-loosen the cell pellet.
11. Add 200 μL cold PBS to the 2 mL microtube and vortex gently for 5 s. Add an additional 500 μL cold PBS.
12. Centrifuge at 3,500 × *g* for 3 min at 4 °C.
13. Remove the supernatant completely and tap loosen the cell pellet. Move forward to Subheading 3.3.

3.3 Assays

3.3.1 Live Imaging

Overview: Live-cell reagents can be applied using standard protocols. The 3D cultures can be processed for immunofluorescence staining as **steps 1–15** in Subheading 3.3.2 for imaging without mounting. Higher reagent concentrations may be required due to possible absorption or binding by the gel bed. Inverted microscope and objectives with long working distance will be needed. For 3D reconstruction, water immersion objectives are recommended to reduce spherical aberration. Time-lapsed live-cell imaging can also be performed (Fig. 3a).

3.3.2 Immunofluorescence (IF)

1. Remove medium and wash cells once with 500 μL PBS.
2. Fix cells with 500 μL fresh 4% PFA in PBS for 20 min at 37 °C and additional 15 min at 4 °C.
3. Wash cells in 500 μL PBS at 4 °C overnight (*see* **Note 11**).
4. Equilibrate the cells to room temperature. Perform all subsequent procedures at room temperature.
5. Permeabilize cells with 500 μL Permeabilization Buffer for 10 min.

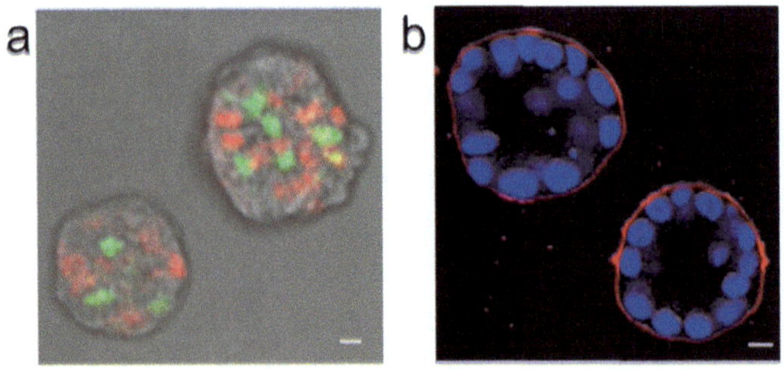

Fig. 3 Sample images of live imaging and immunofluorescence. (**a**) Acinar structures of MCF10A cells with FUCCI cell-cycle reporter [17]. The image shown is an optical section of one-time point from a multichannel time-lapsed live-cell confocal imaging experiment over 3 days. (**b**) Immunofluorescence staining of polarized MCF10A acini showing basal deposition of Laminin-332 (red) and DAPI nuclear counterstain (blue). (**a**, **b**) Scale bars, 10 μm

6. Wash with 500 μL 1× Glycine Buffer for 10 min.
7. Wash cells twice with 500 μL 1× IF Buffer for 10 min each.
8. Block with 200 μL Blocking Buffer (10% goat serum in 1× IF Buffer) for 1 h.
9. Incubate with 150 μL primary antibody (1:100–200 in Blocking Buffer with 1:100 goat anti-mouse F(ab')$_2$ fragments) in a humidified chamber overnight.
10. Wash three times with 500 μL 1× IF buffer for 20 min each.
11. Incubate with 150 μL secondary antibody (1:200 in Blocking Buffer) for 1 h. Keep slides in the dark from this step forward.
12. Wash three times with 500 μL 1× IF buffer for 10 min each. To counterstain nuclei, include 1 μg/mL DAPI in the last wash and incubate for 15 min instead.
13. Wash twice with 500 μL PBS for 10 min each.
14. Lift plastic chambers from glass slide. Remove any residual wash from the edges of the slide and mount with ProLong Diamond antifade mountant (*see* **Note 12**).
15. Dry slides overnight in the dark.
16. Store slides at 4 °C until analysis.

Figure 3b shows an example of immunofluorescence staining with this protocol.

3.3.3 Protein Isolation

1. Add 100 μL of cold RIPA buffer with 1× Halt inhibitor cocktail. Rinse cells from the wall and pipette up and down to resuspend the cells.
2. Vortex briefly and quick spin. Incubate on ice for 10 min. Vortex briefly and quick spin every 2 min.

3. Homogenize the lysate by passing through a 27-gauge needle with a 1 mL syringe several times (*see* **Note 13**).
4. Vortex briefly and quick spin. Incubate on ice for additional 15 min.
5. Centrifuge at 14,000 × *g* for 10 min at 4 °C.
6. Collect the supernatant in a fresh prechilled 1.5 mL microtubes.
7. Snap-freeze in liquid nitrogen and store samples at −80 °C for downstream applications.

3.3.4 Total RNA Isolation

1. For samples from Cell Harvest in Subheading 3.2.2: tap-loosen the cell pellet, add 1 mL TRIzol™ Reagent and mix by pipetting up and down, and incubate at room temperature for 5 min. Proceed to **step 4**.
2. Remove medium from the chamber slide.
3. Use 1 mL TRIzol™ Reagent per sample. (Add 250 μL per well for 4 wells of an 8-well chamber slide). Scrape the gel bed and homogenize by pipetting up and down. Collect into a 1.5 mL RNase-free microtube and incubate at room temperature for 5 min.
4. Centrifuge at 12,000 × *g* for 10 min at 4 °C.
5. Collect supernatant in a new 1.5 mL RNase-free microtube.
6. Add 200 μL chloroform. Shake vigorously for 20 s and incubate at room temperature for 3 min.
7. Centrifuge at 10,000 × *g* for 18 min at 4 °C.
8. Carefully transfer the top, aqueous phase solution to a new 1.5 mL RNase-free microtube.
9. Add an equal volume of 100% RNase-free ethanol and mix slowly by pipetting.
10. Load 700 μL of the lysate into a RNeasy column seated in a collection tube.
11. Centrifuge for 30 s at 8,000 × *g*. Discard the flow-through. Repeat this step if the lysate is more than 700 μL.
12. Wash the RNeasy column according to vendor protocol. Elute RNA with 30 μL of DEPC-treated water.
13. Store samples at −80 °C for downstream applications.

4 Notes

1. 3D cultures on Matrigel bed generally give more regular 3D acinar structures. Suspension 3D cultures are suitable for larger scale sample preparation. RNA isolation is not affected by the

Matrigel but extracting cells from Matrigel before RNA isolation is recommended for larger scale samples. Protein isolation is significantly affected by the Matrigel content and extracting cells from Matrigel is necessary. After extracting from Matrigel bed, cell samples can be processed for other downstream assays.

2. Matrigel has lot-to-lot variations that may show differences in the ability to maintain integrity after prolonged culture and/or fixation. Endotoxin levels lower than 1.5 units/mL are recommended. Speed of Matrigel solidification differs between lots. Test out the conditions and adjust the experiment timeline accordingly.
3. Always keep Matrigel on ice and avoid multiple freeze-thaw cycles. Matrigel solidifies in elevated temperature. Work quickly while laying the gel bed. Keeping the chamber slides on ice could slow down the solidification while spreading the gel. The suggested volumes for preparing the Matrigel bed in 8-well chamber slide format may require adjustments based on the variation in viscosity of different Matrigel lots.
4. Prepare 3D Assay Medium with 5 ng/mL EGF and 2% Matrigel first. Mix immediately and thoroughly after adding Matrigel to the medium. Dilute cells in the medium and mix immediately.
5. Mix well before dispensing cells onto the Matrigel bed. Cell clumps will grow aberrant structures.
6. If medium evaporation is a concern, use only the middle 8 wells of the ultralow-attachment 24-well plate and fill wells on the edges with 1.5 mL sterile water.
7. Tilt the plate at a 45-degree angle to the front and remove medium gently from the surface. Cells should settle on the bottom but could be washed off easily.
8. The gel slurry tends to stick to the pipette tip. Set the pipette to 300 μL when scraping and collecting the slurry.
9. Do not use 50 mL conical tubes. Cells pellet better in 15 mL conical tubes.
10. Collect the same sample condition in the same 15 mL conical tube to minimize cell loss. Repeat the spinning and discard medium when full. Collect and spin down all cells in medium from all wells first before collecting and spinning down the wash. Cells do not pellet as well in PBS.
11. Gel bed integrity varies in different Matrigel lots. Test the incubation procedures at 4 °C or room temperature for optimal conditions.
12. Add about a half drop of ProLong Diamond Antifade Mountant on each well with cells and a drop on each empty well.

Using a P200 pipette tip, spread the mountant onto the blue lines between the wells without touching the Matrigel beds. Place cover glass gently on top of the slide. Add mountant from the edges to fill any empty space between the slide and the cover glass.

13. If pellets are too large to pass through the needle, use a 25-gauge needle first and then a 27-gauge needle.

Acknowledgments

The work is supported by a grant award from NIH National Cancer Institute (#R01CA200652). Images were obtained using equipment in the University Imaging Centers at the University of Minnesota.

References

1. Huang L, Holtzinger A, Jagan I, BeGora M, Lohse I, Ngai N, Nostro C, Wang R, Muthuswamy LB, Crawford HC, Arrowsmith C, Kalloger SE, Renouf DJ, Connor AA, Cleary S, Schaeffer DF, Roehrl M, Tsao MS, Gallinger S, Keller G, Muthuswamy SK (2015) Ductal pancreatic cancer modeling and drug screening using human pluripotent stem cell- and patient-derived tumor organoids. Nat Med 21(11):1364–1371. https://doi.org/10.1038/nm.3973
2. Kenny PA, Lee GY, Myers CA, Neve RM, Semeiks JR, Spellman PT, Lorenz K, Lee EH, Barcellos-Hoff MH, Petersen OW, Gray JW, Bissell MJ (2007) The morphologies of breast cancer cell lines in three-dimensional assays correlate with their profiles of gene expression. Mol Oncol 1(1):84–96. https://doi.org/10.1016/j.molonc.2007.02.004
3. Schafer ZT, Grassian AR, Song L, Jiang Z, Gerhart-Hines Z, Irie HY, Gao S, Puigserver P, Brugge JS (2009) Antioxidant and oncogene rescue of metabolic defects caused by loss of matrix attachment. Nature 461(7260):109–113. https://doi.org/10.1038/nature08268
4. McCracken KW, Cata EM, Crawford CM, Sinagoga KL, Schumacher M, Rockich BE, Tsai YH, Mayhew CN, Spence JR, Zavros Y, Wells JM (2014) Modelling human development and disease in pluripotent stem-cell-derived gastric organoids. Nature 516(7531):400–404. https://doi.org/10.1038/nature13863
5. Leung CT, Brugge JS (2012) Outgrowth of single oncogene-expressing cells from suppressive epithelial environments. Nature 482(7385):410–413. https://doi.org/10.1038/nature10826
6. Eiraku M, Takata N, Ishibashi H, Kawada M, Sakakura E, Okuda S, Sekiguchi K, Adachi T, Sasai Y (2011) Self-organizing optic-cup morphogenesis in three-dimensional culture. Nature 472(7341):51–56. https://doi.org/10.1038/nature09941
7. Gilbert PM, Havenstrite KL, Magnusson KE, Sacco A, Leonardi NA, Kraft P, Nguyen NK, Thrun S, Lutolf MP, Blau HM (2010) Substrate elasticity regulates skeletal muscle stem cell self-renewal in culture. Science 329(5995):1078–1081. https://doi.org/10.1126/science.1191035
8. Debnath J, Brugge JS (2005) Modelling glandular epithelial cancers in three-dimensional cultures. Nat Rev Cancer 5(9):675–688. https://doi.org/10.1038/nrc1695
9. Guilak F, Cohen DM, Estes BT, Gimble JM, Liedtke W, Chen CS (2009) Control of stem cell fate by physical interactions with the extracellular matrix. Cell Stem Cell 5(1):17–26. https://doi.org/10.1016/j.stem.2009.06.016
10. Even-Ram S, Artym V, Yamada KM (2006) Matrix control of stem cell fate. Cell 126(4):645–647. https://doi.org/10.1016/j.cell.2006.08.008
11. Ng MR, Besser A, Danuser G, Brugge JS (2012) Substrate stiffness regulates cadherin-dependent collective migration through myosin-II contractility. J Cell Biol 199(3):

545–563. https://doi.org/10.1083/jcb.201207148

12. Yamada KM, Cukierman E (2007) Modeling tissue morphogenesis and cancer in 3D. Cell 130(4):601–610. https://doi.org/10.1016/j.cell.2007.08.006
13. Orkin RW, Gehron P, McGoodwin EB, Martin GR, Valentine T, Swarm R (1977) A murine tumor producing a matrix of basement membrane. J Exp Med 145(1):204–220. https://doi.org/10.1084/jem.145.1.204
14. Debnath J, Mills KR, Collins NL, Reginato MJ, Muthuswamy SK, Brugge JS (2002) The role of apoptosis in creating and maintaining luminal space within normal and oncogene-expressing mammary acini. Cell 111(1):29–40
15. Xiang B, Muthuswamy SK (2006) Using three-dimensional acinar structures for molecular and cell biological assays. Methods Enzymol 406: 692–701. https://doi.org/10.1016/S0076-6879(06)06054-X
16. Debnath J, Muthuswamy SK, Brugge JS (2003) Morphogenesis and oncogenesis of MCF-10A mammary epithelial acini grown in three-dimensional basement membrane cultures. Methods 30(3):256–268. https://doi.org/10.1016/s1046-2023(03)00032-x
17. Sakaue-Sawano A, Kurokawa H, Morimura T, Hanyu A, Hama H, Osawa H, Kashiwagi S, Fukami K, Miyata T, Miyoshi H, Imamura T, Ogawa M, Masai H, Miyawaki A (2008) Visualizing spatiotemporal dynamics of multicellular cell-cycle progression. Cell 132(3):487–498. https://doi.org/10.1016/j.cell.2007.12.033

Chapter 28

Mesenchymal Stem Cell Seeding on 3D Scaffolds

Agata Kurzyk

Abstract

Evaluation of mesenchymal stem cell seeding efficiency in three-dimensional (3D) scaffolds is a critical step for constructing a potent and useful tissue engineering product for regenerative medicine. To determine the quantity of cells seeded on a scaffold, their condition and viability, and/or to confirm cell adhesion to the scaffold surface, a number of cellular assays are used. The assays are most often based on a direct or indirect colorimetric-, fluorimetric-, bioluminescent-, or isotope-based measurement of changes reflecting the activity of cellular processes. This chapter presents a selection of assays measuring the efficiency of cell seeding on scaffolds, that is, the MTS (3-(4,5-dimethylthiazol-2-yl)-5-(3-carboxymethoxyphenyl)-2-(4-sulfophenyl)-2H-tetrazolium)) assay, the MTT (3-(4,5-dimethylthiazol-2-yl)-2,5-diphenyltetrazolium bromide) assay, the ATP (adenosine triphosphate), DAPI (4′,6-diamidino-2-phenylindole) assay, the Alamar Blue (7-hydroxy-10-oxidophenoxazin-10-ium-3-one, resazurin) assay and the Pico Green dsDNA (N′-[3-(dimethylamino)propyl]-N,N-dimethyl-N′-[4-[(E)-(3-methyl-1,3-benzothiazol-2-ylidene)methyl]-1-phenylquinolin-1-ium-2-yl]propane-1,3-diamine) assay. These assays monitor the number of viable cells, sometimes in conjunction with specifying cell membrane integrity, determine enzymatic activity associated with cell metabolism, measure cell proliferation rate, and assess the total protein or DNA content in the cell–scaffold construct. The choice of the appropriate methods and the details for testing 3D cultures are of utmost importance to properly evaluate tissue engineering products. Still, developing standards for assessment of cell–scaffold constructs remains a challenge in tissue engineering.

Key words Cell seeding, Scaffolds, Efficiency, Evaluation, 3D culture, MTS, MTT, ATP, Alamar Blue, PicoGreen

1 Introduction

Seeding mesenchymal stem cells on scaffolds is a critical step in constructing polymer-cell scaffolds and often determines the quality of tissue-engineered (TE) products. Evaluation of cell seeding efficiency on three-dimensional (3D) scaffolds and monitoring of cell growth over a desired period time in 3D cultures is a challenge [1]. Currently, the efficiency of a cell–scaffold construct is assessed by several cellular assays, commonly used to study proliferation, cytotoxicity, viability, apoptosis, or cell cycle in 2D cultures

Nagarajan Kannan and Philip Beer (eds.), *Stem Cell Assays: Methods and Protocols*, Methods in Molecular Biology, vol. 2429, https://doi.org/10.1007/978-1-0716-1979-7_28,

[2]. Those assays are classified according to the end-point detected, and include:

- Dye exclusion (e.g., trypan blue tests).
- Colorimetic assays employing tetrazolium dyes, MTT (3-(4,5-dimethylthiazol-2-yl)-2,5-diphenyltetrazolium bromide) or MTS (3-(4,5-dimethylthiazol-2-yl)-5-(3-carboxymethoxyphenyl)-2-(4-sulfophenyl)-2H-tetrazolium).
- Fluorometric assays, such as Alamar Blue (7-hydroxy-10-oxidophenoxazin-10-ium-3-one) assay.
- Luminometric assays (e.g., ATP [adenosine triphosphate] assay.

The most common and simple method to visualize cells on biomaterials is DAPI (2-(4-amidinophenyl)-1H-indole-6-carboxamidine) staining of the cell nucleus. DAPI has an absorption maximum at a wavelength of 358 nm and its emission maximum is at 461 nm (blue). This method visualizes cell adhesion to scaffolds. DAPI staining enables a quick initial assessment of the cells seeded on scaffolds, and a comparison between different cell–scaffold constructs. DAPI staining of cell nuclei allows for calculation of the percentage of viable cells on a scaffold. Furthermore, using DAPI along with other dyes (e.g., Calcein, PKH 26, PKH 67) can provide a lot of useful information on the distribution of cells on the scaffolds. A combination of DAPI staining with other techniques, such as scanning electron microscopy or micro-computer tomography, provides a complete picture of the distribution of cells in a scaffold construct [1].

In any type of cell culture, it is important to know how many cells remain viable at the end of the experiment [2]. Cell assays can destroy or interfere with cell function, some of them may be cytotoxic and require cell lysis (e.g., MTT assay), others are of minimal cytotoxicity, without the need for cell lysis (e.g., Alamar Blue assay, ATP assay) [3, 4]. In the cell viability assays based on the tetrazolium or resazurin reduction and protease activity, the typical metabolic or enzymatic activities of viable cells are assessed by spectrophotometric measurement of the colored product of a dye reduction by living cells. Resazurin-based reagents can also be used to evaluate apoptosis. Optimally selected test should serve a specific application under study, depending on the cell model, cell type, cell growth (adherent vs. nonadherent), cell density, cell line, and cell age [5].

The MTT test is applied to quantify the number of cells in small-size tissue-engineered constructs [4]. The MTT tetrazolium reduction assay was the first homogeneous cell viability assay developed for a 96-well format that was suitable for high throughput screening [3]. The MTT substrate is positively charged and it readily penetrates viable eukaryotic cells. It is prepared in a physiologically balanced solution, added to the cells in a culture, usually at a final concentration of 0.2–0.5 mg/mL, and incubated for 1 to 4 h. Formazan, the substrate reduction product, the amount of which is presumably directly proportional to the number of viable cells, is quantified by recording a changed absorbance at 570 nm in a plate-reading spectrophotometer, after dissolving the purple formazan crystals into a colored solution [6].

The MTS test is an improved version of the MTT test. Although the negatively charged MTS does not readily penetrate viable eukaryotic cells, MTS is typically used with an intermediate electron acceptor that can transfer electrons from the cytoplasm or plasma membrane. The colored reaction product of the conversion of MTS by viable cell dehydrogenase in the presence of PMS (phenazine methosulfate) is fully soluble in water. This allows fast and accurate assessment of the percentage of functional cells, that is, of the effect of the tested factor on viability of any cell line. The tests are performed by adding a 20 μL of the reagent per 100 μL of culture medium (the final concentration of MTS is 317 μg per mL of a medium in the assay wells) directly to cell cultures, 1–4 h incubation and measuring absorbance using a multifunctional plate reader at 490 nm [7].

One of the disadvantages of metabolic tests is that there is no distinction between dividing and nondividing cells which may result in an inaccurate estimation of cell numbers. Furthermore, these assays, by comparing the medium before and after the culture period (usually several hours or days), approximate the trend of changes in cell viability over a certain period of time. So, with these tests we can estimate a proportion of viable cells rather than the exact number of viable cells. In addition, at the stage of incubation of the substrate with viable cells at 37 °C, artifacts, which may interfere with the correct reading of the results, may be generated by chemical interactions between the tested compounds and cell biochemistry [6–9].

Alamar Blue assay is a useful, nontoxic alternative to the commonly used MTT cell viability test [10]. Alamar Blue is the test reagent, containing an active ingredient called resazurin. This weakly fluorescent blue indicator dye is water soluble, stable in culture medium, and passes through cell membranes [9–11]. Once permeated into living cells, resazurin is reduced to resorufin, a compound of red color and high fluorescence. The color change can be detected by absorbance (detected at 570 nm and 600 nm) or fluorescence (by excitation between 530 and 560 nm and emission at 590 nm). Alamar Blue tests provide accurate measurements over time, require no cell analysis and may be used with different cell models. In addition, Alamar Blue has a high sensitivity which can be increased by longer incubation times without impairing cell viability. Alamar Blue is suitable for the evaluation of cell count in tissue-engineered bone structures with porous surfaces [5], but may not be appropriate for studying proliferation in tendon-derived cells [12].

Among various principles of estimating proliferation of cultured cells, measuring the amount of ATP (Adenosine Triphosphate) or DNA, as with the PicoGreen dsDNA assay, is the most reliable strategy for quantifying cell proliferation in 3D cultures [13–17]. The ATP bioluminescence assay is the fastest and the most sensitive cell viability assay. This test does not need an incubation step when viable cells convert a substrate into a colored product [4]. The luminescent signal stabilizes within 10 min following the addition of 100 μL of the reagent per well of the 96-well plate, and glows with a typical half-life of over 5 h [13]. There is a linear relationship between the intensity of luminescent signal and ATP concentration or cell number [3]. The ATP assay, is one of the best methods to complement MTS/MTT assays, to assess cell seeding efficiency in a cell–scaffold construct.

The PicoGreen dsDNA assay is useful for testing cell numbers within scaffolds. PicoGreen is an ultrasensitive (with a sensitivity of +/−50 cells) fluorescent nucleic acid dye, which provides linear results over multiple orders of magnitude of a single dye concentration [15, 16]. Therefore, PicoGreen is the dye of choice over Hoechst and DAPI for this type of quantitative application [15]. The PicoGreen assay enables quantitation of as little as 25 pg/mL of dsDNA with a standard spectrofluorometer and fluorescence excitation and emission wavelengths.

An optimal method for the assessment cell seeding on scaffolds should be fast, nontoxic, accurate, and cheap. Moreover, the methods used at this stage are crucial in examining tissue-engineered structures. However, despite the large selection of tests, no single method accurately assesses proliferation, viability, and cytotoxicity of cells seeded on 3D scaffolds. Thus, to avoid mistakes and to increase the reliability of the results it is important to use adequate complementary methods.

In spite of numerous studies in the field of new biomaterials seeded with mesenchymal stem cells, requirements or standards for evaluation of the cell–scaffold constructs are lacking. Developing such guidelines would allow to effectively compare the results of different studies, as well as to optimize cell assessment in 3D structures.

2 Materials

2.1 MTS Assay

1. Commercial kit: CellTiter 96® AQueous One Solution Cell Proliferation kit (Promega Corporation Madison, WI, USA), (*see* **Note 1**).
2. ELISA plate reader, colorimetric microplate reader—equipped with a filter for OD 490 nm.

3. 24-well plate, 96-well plate: clear, flat-bottom plates for colorimetric assay.
4. Pipettes (preferably a multichannel pipette) and pipette tips.
5. Plate shaker.
6. SDS.

2.1.1 Reagent Preparation

1. MTS reagent. Store at −20 °C. Equilibrate all materials and prepared reagents to room temperature prior to use. It is recommended to assay all controls and samples in duplicate (*see* **Note 2**).

2.2 MTT Assay

1. Commercial kit: CellTiter 96® nonradioactive cell proliferation assay. Promega Corporation Cat.# G4000; Cell Growth Determination Kit, MTT based. Sigma-Aldrich Cat.# CGD1-1KT; MTT Cell Growth Assay Kit. Millipore Cat.# CT02; Thiazolyl Blue Tetrazolium Bromide (MTT Powder). Sigma-Aldrich Cat.# M2128 (*see* **Note 1**).
2. ELISA plate reader, colorimetric microplate reader—equipped with a filter for OD 490 nm.
3. 24-well plate, 96-well plate: clear, flat-bottom plates for colorimetric assay.
4. Pipettes (preferably a multichannel pipette) and pipette tips.
5. Plate shaker.
6. DMSO.

2.2.1 MTT Solution Preparation

1. Dissolve MTT in Dulbecco's phosphate buffered saline or DMEM (no phenol red) pH = 7.4 (DPBS) to 5 mg/mL or 0.5 mg/mL (*see* **Note 3**).
2. Filter-sterilize the MTT solution through a 0.2 μM filter into a sterile, light-protected container.
3. Store the MTT solution protected from light, at 4 °C, for frequent use, or at −20 °C for a long-term storage.

2.2.2 Solubilization Solution Preparation

1. Choose appropriate solvent-resistant container and work in a ventilated fume hood.
2. Prepare 40% (vol/vol) dimethylformamide (DMF) in 2% (vol/vol) glacial acetic acid.
3. Add 16% (wt/vol) sodium dodecyl sulfate (SDS) and dissolve.
4. Adjust pH with HCl/NaOH/acetic acid to 7.4.
5. Store at room temperature to avoid precipitation of SDS. If a precipitate forms, warm to 37 °C and mix to solubilize SDS.

2.3 DAPI Staining

1. DAPI (ThermoFisher Scientific, Cat.# D1306).
2. Deionized water (diH_20).

3. Glutaraldehyde or paraformaldehyde.
4. 0.1 M sodium cacodylate buffer (pH 7.3).
5. PBS (phosphate buffered saline) buffer.
6. Fluorescence microscope.

2.3.1 Preparing Solutions

1. Add 2 mL of deionized water (diH_2O) or dimethylformamide (DMF) to the entire contents of the DAPI vial to make a 14.3 mM (5 mg/mL) DAPI stock solution. DAPI stock solution may be stored at 2–6 °C for up to 6 months or at −20 °C for longer periods.
2. Add 2.1 μL of the 14.3 mM DAPI stock solution to 100 μL PBS to make a 300 μM DAPI intermediate dilution.
3. Dilute the 300 μM DAPI intermediate dilution 1:1000 in PBS if needed to make a 300 nM DAPI staining solution.

2.3.2 Fixing of Cell-Seeded Scaffolds

1. Fix the cell–scaffold constructs in 1.5% glutaraldehyde, containing a 0.1 M sodium cacodylate buffer (pH = 7.3), rinse twice with PBS, and leave to dry at room temperature. You can also fix cell–scaffold constructs with 4% paraformaldehyde, at 4 °C for 30 min on days 1, 3, and 5 after cell seeding (*see* **Note 4**).
2. Wash the cell-seeded scaffold twice with PBS.

2.4 Alamar Blue Assay

The Alamar Blue assay can be used for long-term cell proliferation assays and for repeated measurements. For these types of study, it is recommended that aliquots of the cell medium/suspension are taken at each time point for incubation with alamarBlue prior to an endpoint test [6].

1. Commercial kit: alamarBlue™ Cell Viability Reagent (cat.# DAL1025, Thermo Fisher Scientific) (*see* **Note 1**).
2. Colorimetric microplate reader—equipped with a filter for OD 570 and 600 nm.
3. Fluorescence microplate reader—530-560 nm excitation, 590 nm emission.
4. Pipettes (preferably a multichannel pipette) and pipette tips.
5. 96-well plate: clear, flat-bottom plates for colorimetric assay or fluorescence assay.

2.5 ATP Assay

1. Commercial kit: CellTiter-Glo® Luminescent Cell Viability Assay (Promega Corporation Cat.# G7570) (*see* **Note 1**).
2. 96-well plate: clear, flat-bottom plates.
3. Luminometer plate reader.

2.5.1 Reagent Preparation

1. Thaw the CellTiter-Glo® Buffer and CellTiter-Glo® Substrate and equilibrate to room temperature prior to use (*see* **Note 5**).
2. Transfer the appropriate volume (10 mL for Cat.# G7570) of CellTiter-Glo® Buffer into the amber bottle containing CellTiter-Glo® Substrate to reconstitute the lyophilized enzyme/substrate mixture. This forms the CellTiter-Glo® Reagent.
3. Mix by gently vortexing or inverting the contents to obtain a homogeneous solution. The CellTiter-Glo® Substrate should go into solution easily in less than 1 min.

2.6 PicoGreen dsDNA Assay

The PicoGreen DNA assay can be used to quantitate the cells located either on the proximal surface of scaffold, inside the scaffold, and also those still suspended in the medium around the scaffold.

1. Commercial kit: Quant-iT PicoGreen dsDNA kit (Thermo Fisher Scientific, cat. #: P7589) (*see* **Note 1**).
2. 96-well plate (*see* **Notes 6** and 7).
3. Fluorescence microplate reader.

2.6.1 Assay Buffer Preparation

Prepare 1× TE buffer (10 mM Tris-HCl, 1 mM EDTA, pH = 7.5) from 20× TE buffer included in the Assay Kit (200 mM Tris-HCl, 20 mM EDTA, pH = 7.5) using nuclease-free water (*see* **Note 8**).

2.6.2 PicoGreen Preparation

On the day of the experiment, prepare a working solution of the PicoGreen Reagent by making a 1:200 dilution of the concentrated DMSO solution in 1× TE buffer. Protect the PicoGreen Reagent from light, until adding to the plate (*see* **Note 9**).

2.7 Cell Lysis Buffer

1. 10 mM Tris pH = 8.0.
2. 1 mM EDTA.
3. 0.2% (V/V) Triton X-100.

The three ingredients of the lysis buffer may be mixed in advance and stored at room temperature.

3 Methods

3.1 MTS Standard Curve

1. Culture cells in various numbers (e.g., 20×10^3, 40×10^3, 60×10^3, 80×10^3, 100×10^3 cell/well) in a 96-well microtiter plate in a culture medium at a final volume of 200 μL/well (*see* **Note 10**).
2. Add 20 μL/well of the MTS reagent (CellTiter 96® AQueous One Solution Reagent) (*see* **Note 11**).

3. After 1–4 h at 37 °C in a humidified, 5% CO_2 atmosphere, record the absorbance at 490 nm using a colorimetric microplate reader (*see* **Note 12**).
4. For the statistical reliability, a measurement should be made in at least 3 replicates.
5. Calculate the linear correlation coefficient.

3.2 MTS Assay

1. Thaw the MTS (CellTiter 96® AQueous One Solution Reagent) [1]. It should take approximately 90 min at room temperature, or 10 min in a water bath at 37 °C, to completely thaw the 20 mL volume (*see* **Note 13**).
2. Rinse the cell-seeded scaffolds thoroughly in PBS and move into a new well plate. Rinse controls (e.g., cells in 2D culture seeded) in PBS before starting the assay (*see* **Note 14**).
3. Pipet 20 μL of MTS into each well of the 96-well assay plate containing the samples (cell-seeded scaffolds) in 100 μL of fresh culture medium (*see* **Notes 15** and **16**).
4. Incubate the plate at 37 °C for 1–4 h in a humidified, 5% CO_2 atmosphere (*see* **Notes 17** and **18**).
5. Shake the plate briefly on a shaker and measure the absorbance at 490 nm, using a 96-well plate reader (*see* **Note 19**).
6. To calculate percent difference between cell-seeded construct and control cells (total cells used for cell seeding) use the following formula [7, 11] (*see* **Note 14**):

$$\text{Cell seeding efficiency}(\%) = \frac{\text{OD cell-seeded construct}}{\text{OD of total cells used for cell seeding}} \times 100$$

3.3 MTT Assay

Prepare two experiments, for the standard curve (2D cell culture) and for the cell–scaffold construct, following each step, according to the instructions below.

1. Add cells and test compounds into 96-well plates containing a final volume of 100 μL/well (*see* **Notes 10**, **20** and **21**).
2. Incubate for the desired exposure time.
3. Add 5 mg/mL or 0.5 mg/mL MTT solution and incubate at 37 °C for 1 h or 4 h, respectively (*see* **Note 22**). You need the following amounts of the MMT solution.
   (a) 100 μL for a 96-well multiplate.
   (b) 300 μL for a 24-well multiplate.
   (c) 500 μL for a 12-well multiplate.
   (d) 1000 μL for a 6-well multiplate.

4. After incubation, add 100 μL of solubilization solution or DMSO to each well to dissolve formazan crystals (*see* **Note 23**).
5. Mix to ensure complete solubilization (*see* **Notes 24** and **25**).
6. Record absorbance at 570 nm (*see* **Note 26**).

3.4 DAPI Staining Test

1. Wash the cells 1–3 times in PBS.
2. Add 300 nM DAPI stain solution in an amount sufficient to cover the cells.
3. Incubate for 30 s to 5 min to label the cell nuclei, and rinse twice with PBS.
4. The immunofluorescence images are obtained by using a fluorescence microscope [1].
5. Use lens magnification for pixel to area calculations. Count cells, calculate, and plot vs. OD.
6. Repeat the assay using different samples to confirm the results.

3.5 Alamar Blue Assay

1. Incubate the cell–scaffold construct for 1 to 4 h in 1 mL of alamarBlue solution in culture medium, at 37 °C.
2. Measure proliferation using fluorescence spectrophotometry and read fluorescence at the excitation wavelength of 560 nm and emission wavelength of 590 nm and 600 nm after the required incubation time. As a blank, use medium only (*see* **Note 27**).
3. Wash the cell–scaffold construct in a complete medium twice.
4. Transfer the cell–scaffold construct into a fresh culture medium and return to the incubator for the remaining culture period.
5. Create a curve of the relative fluorescence units versus cell density to generate quantitative results. To calculate the percent difference in reduction between sample (cell–scaffold construct) and control use the formula [18]:

$$\text{Percentage difference between sample and control cells } [\%] = \frac{(O2 \times A1) - (O1 \times A2)}{(O2 \times P1) - (O1 \times P2)} \times 100$$

Where:

- O1 = molar extinction coefficient (E) of oxidized alamarBlue at 570 nm*.
- O2 = E of oxidized alamarBlue at 600 nm*.
- A1 = absorbance of test wells at 570 nm.
- A2 = absorbance of test wells at 600 nm.

- P1 = absorbance of positive growth control well (cells plus alamarBlue but no test agent) at 570 nm.
- P2 = absorbance of positive growth control well (cells plus alamarBlue but no test agent) at 600 nm.

* Only one appropriate substitute wavelength may be used.

3.6 ATP Assay

1. Set-up white opaque-walled microwell assay plates, containing cells in culture medium at the desired density.
2. Add the cell–scaffold construct and controls to the appropriate wells, so that the final volume is 100 μL in each well of the 96-well plate (25 μL for a 384-well plate) (*see* **Note 28**).
3. Culture cells for the desired test exposure period.
4. Equilibrate plates to ambient temperature for 30 min, to ensure uniform temperature across plate during luminescent assay.
5. Add CellTiter-Glo® Reagent in an equal volume to all wells (100 μL per well for 96-well plates or 25 μL per well for 384-well plates) [11, 13].
6. Mix contents for 2 min on an orbital shaker to induce cell lysis.
7. Allow the plate to incubate at room temperature for 10 min to stabilize luminescent signal (*see* **Note 29**).
8. Record luminescence.

3.7 PicoGreen dsDNA Assay

3.7.1 Standard Curve of 2D Cell Culture

1. To create a cell number standard curve, harvest cells from 2D cultures using an appropriate detachment method. Count cells and resuspend appropriate cell numbers in 1 mL of the buffer reagent (*see* Subheading 2.6). Always work RNase and DNase free.
2. Vortex samples for 10 s every 5 min for half an hour, keeping them on ice throughout (*see* **Note 30**).
3. Thaw on ice, if required, and homogenize samples (10–15 times, using a 21-guage needle).
4. Dilute samples 1 in 10 in 1× TE buffer by placing 90 μL of 1× TE buffer and add 10 μL of sample into each well of a black bottom 96-well plate.
5. Add 100 μL of PicoGreen reagent, mix and incubate at room temperature for 5 min, wrapped in aluminum foil (*see* **Note 31**).
6. Read fluorescence measured at the excitation and emission wavelengths of 460 nm and 540 nm, respectively, using a fluorescence microplate reader, or equivalent.

3.7.2 DNA Standard Curve (Using a Lambda DNA at the Standard-Quant-iT PicoGreen dsDNA Kit)

1. Prepare a 2 μg/mL stock solution of dsDNA in TE.
2. Determine the DNA concentration on the basis of absorbance at 260 nm (A260) in a cuvette.
3. For the Lambda DNA standard provided at 100 μg/mL in the Quant-iT™ PicoGreen® Kits, prepare a 50-fold dilution in TE to make the 2 μg/mL working solution. For example: take 30 μL of the DNA standard and mix with 1.47 mL of TE.
4. Create a five-point standard curve ranging from 1 ng/mL to 1 μg/mL. For a low-range standard curve of 25 pg/mL to 25 ng/mL, prepare a 40-fold dilution of the 2 μg/mL DNA solution to yield a 50 ng/mL DNA stock solution to be added into each cuvette as shown in Table 1. For the high-range standard curve, prepare a 2 μg/mL stock solution of dsDNA in TE. Dilute the stock solution into each cuvette as shown in Table 2.
5. Next, add 1.0 mL of the aqueous working solution of Quant-iT™ PicoGreen® reagent (*see* Subheading 2.6) to each cuvette. Mix well and incubate for 2 to 5 min at room temperature, protected from light.

Table 1
A protocol to prepare the low-range standard curve

Volume (μL) of TE	Volume (μL) of 50 ng/mL DNA stock	Volume (μL) of diluted PicoGreen reagent	Final DNA concentration in PicoGreen assay
0	1000	1000	25 ng/mL
900	100	1000	2.5 ng/mL
990	10	1000	250 ng/mL
999	1	1000	25 ng/mL
1000	0	1000	Blank

Table 2
A protocol to prepare the high-range standard curve

Volume (μL) of TE	Volume (μL) of 2 μg /mL DNA stock	Volume (μL) of diluted PicoGreen reagent	Final DNA concentration in PicoGreen assay
0	1000	1000	1000 ng/mL
900	100	1000	100 ng/mL
990	10	1000	10 ng/mL
999	1	1000	1 ng/mL
1000	0	1000	Blank

6. Read fluorescence measured at the excitation and emission wavelengths of 460 nm and 540 nm, respectively, using a fluorescence microplate reader, or equivalent (*see* **Note 32**).
7. Subtract the fluorescence value of the reagent blank from that of each of the samples. Use corrected data to generate a standard curve of fluorescence versus DNA concentration.

3.7.3 Sample Analysis

1. Dilute the experimental DNA solution in TE to a final volume of 1.0 mL in cuvettes or 2 mL Eppendorf Tube.
2. Add 1.0 mL of the PicoGreen Reagent (Quant-iT™ PicoGreen reagent) to each sample.
3. Incubate for 2 to 5 min at room temperature, protected from light.
4. Read fluorescence measured at the excitation and emission wavelengths of 460 nm and 540 nm, respectively, using a fluorescence microplate reader, or equivalent.
5. Subtract the fluorescence value of the reagent blank from that of each of the samples. Determine the DNA concentration of the sample from the DNA Standard Curve.
6. Repeat the assay using a different dilution of a sample to confirm the quantitation results.

3.7.4 Cell–Scaffold Construct

1. Wash cell–scaffold construct with PBS 3 times and transfer to 2.0 mL Eppendorf Tube using forceps (*see* **Note 33**).
2. Lyse cells with 1 mL of the Cell Lysis Buffer (*see* Subheading 2.7) and vortex samples for 10 s every 5 min for a total of 30 min, keeping on ice throughout (see **Note 34**).
3. Thaw on ice, if required, and homogenize samples (10–15 times using a 21-guage needle) (*see* **Note 35**).
4. Place cell lysate in a new 1.5 mL Eppendorf Tube.
5. Add standards and experimental samples in 100 μL volumes to a 96-well plate, along with 100 μL of the PicoGreen solution.
6. Mix and incubate at room temperature for 5 min, wrapped in aluminum foil (*see* **Note 31**).
7. Read fluorescence measured at the excitation and emission wavelengths of 460 nm and 540 nm respectively, using a fluorescence microplate reader, or equivalent. Use Lambda DNA for the standard curve to calculate the amount of DNA.
8. For the statistical reliability, a measurement should be made in at least 3 replicates.

4 Notes

1. Make sure all buffers and developing solutions are at room temperature before starting the experiment. Avoid cross contamination of samples or reagents by changing tips between sample, standard and reagent additions. Make sure you have the appropriate type of plate for the detection method of choice. Make sure the heat block/water bath and microplate reader are switched on before starting the experiment.
2. To avoid cell migration from tissue culture plate to scaffold, move the seeded-scaffolds into a new well plate after 3 days of culture. Refresh the medium every third day, until the 7-day time point is reached. At this point, perform the MTS assay to determine total cell numbers for each scaffold. To determine cell numbers on the seeded scaffolds, use an MTS assay following 1-, 3-, 6-h and/or 1-, 3-, 7-, 21-day cultures.
3. MTT should always be dissolved on the day of the experiment and warmed to 37 °C before addition to the cells.
4. Be careful; paraformaldehyde may induce autofluorescence.
5. For convenience, the CellTiter-Glo® Buffer may be thawed and stored at room temperature for up to 48 h prior to use.
6. Before starting, determine the number of samples and standards to test in 96-well plate format. Multiply by 2, if running everything in duplicate, for calculating the total number of wells.
7. Use a black-walled plate with black bottoms if possible. Black-sided wells with clear bottoms or white-sided wells will also work, but background will be higher due to reflected fluorescence in the wells. Do not use clear microtiter plates for fluorescence readings.
8. Prepare 1× TE by pipetting 2.5 mL of 20× stock TE into a sterile 50 mL centrifuge tube and fill to a 50 mL mark with molecular biology grade water. Invert tube to mix.
9. For best results, this PicoGreen reagent should be used within a few hours of starting the experiment. Prepare working solution in a plastic container. This reagent may be adsorbed into glass surfaces. A 1:200 dilution of PicoGreen reagent is prepared by adding 10 μL of PicoGreen per 2 mL of 1× TE buffer (for 20 samples—add 100 μL Quant-iT™ PicoGreen® dsDNA reagent to 20 mL TE). You will need 100 μL diluted PicoGreen per well containing 100 μL sample.
10. The cell number can be determined by performing a standard curve, or the metabolic activity should be analyzed by comparing it with the metabolic activity of a given cell numbers seeded

in the scaffold or in a 2D cell culture. Cell proliferation assays require cells to grow over a period of time. Choose an initial number of cells per well that produces an assay signal near the low end of the linear range of the assay. This helps to ensure that the signal measured at the end of the assay will not exceed the linear range of the assay. Different cell types have different levels of metabolic activity. Using high densities of cells may inhibit adhesion of cells to the scaffold, resulting in nonlinear relationship between cell number and absorbance. Factors that affect the cytoplasmic volume or physiology of the cells will affect metabolic activity. For most tumor cells and fibroblast cell lines, 5000 cells/well is optimal to initiate proliferation studies.

11. If cells are cultured in different volumes of culture medium, adjust the amount of MTS reagent accordingly.
12. Desirable incubation time will depend on the individual cell type and cell densities used. Determine the optimal incubation time for a particular experiment.
13. The reagent should always be thawed on the day of the experiment.
14. To avoid cell migration from the tissue culture plate to the scaffold, the scaffolds should be moved to a new well plate after specific number of days of culture. At this point, the MTS test should be performed to determine the total number of cells for each scaffold (*see* **Note 14e**).

    Approaches to count the number of cells seeded on a scaffold

    (a) Prepare a standard curve (with a known initial number of cells—2D cell culture) and compare it with the tested samples (cell–scaffold construct) in a period time-frame or calculate the differencc between the number of cells in the standard curve versus the number of cells seeded on the scaffold [13].
    (b) Measure the metabolic activity of a cell-seeded scaffold after 24 h (as a control in 1 day) and compare it with the metabolic activity of the cell-seeded scaffold after 48 and 72 h and so on.
    (c) Calculating cells adhered to the scaffolds. Move the cell–scaffold construct to a new plate and rinse it with PBS, to avoid cell adhesion to plastic. Perform the MTS test. Next, count the nonadherent cells in the scaffold, and calculate the difference (*see* **Note 14d**).
    (d) Counting the nonadherent cells in the scaffolds. Transfer the cell–scaffold construct to a new plate and count the cells remaining in the plate without the cell–scaffold

construct. Perform the MTS test. Calculate the difference between the cells adhering to the plate versus the cells seeded on the scaffold.

(e) Counting the total number of cells in each well and dividing by the surface area of the scaffold (cells/mm^2).

15. If different volumes of culture medium are used, adjust the volume to maintain the ratio of 20 μL MTS per 100 μL culture medium. This reagent to medium ratio results in a final concentration of 317 μg/mL of MTS in the assay wells. If the test is performed on 24- or 12-well plate, check the surface area (cm^2) of a well, calculate and add the respective number of cells per well.
16. For convenient delivery of uniform volumes of CellTiter 96® AQueous One Solution Reagent to the 96-well plate, repeating pipettes, digital pipettes, or multichannel pipettes are recommended.
17. The appropriate incubation time will depend on the cell type and cell concentration used. Therefore, it is recommended to determine the optimal incubation time for a particular experiment.
18. To measure the amount of soluble formazan produced by cellular reduction of MTS, proceed immediately to **step 4**.

    Alternatively, to measure the absorbance later, add 10–25 μL of 10% SDS to each well to stop the reaction. Store the SDS-treated plates protected from light in a humidified chamber, at room temperature for up to 18 h. Proceed to **step 4**.
19. Analyze the optical density (OD) of the formed product using a microplate reader at 490 nm with the background subtraction at 630 nm. The average medium background is then subtracted from the OD.
20. Optimal cell density:
    (a) 96-well: 2–3000 cells per well/100 μL of cell suspension;
    (b) 24-well: 12–18,000 cells per well/1000 μL of cell suspension;
    (c) 12-well: 24–36,000 cells per well/2000 μL of cell suspension;
    (d) 6-well: 60–90,000 cells per well/3000 μL of cell suspension.
21. Grow cells in the growth medium until a desired and numerically balanced population (about 90% confluency) is reached; for the given optimal cell concentrations per well, it is about 4–5 days. Remove the medium and rinse cells twice with PBS.

22. After 1 (5 mg/mL) or 4 (0.5 mg/mL) h incubation, remove MTT solution by aspiration (do not rinse with PBS).
23. Dissolve formazan using DMSO adding the following:
    (a) 100 μL DMSO for a 96-well plate;
    (b) 300 μL DMSO for a 24-well plate;
    (c) 500 μL DMSO for a 12-well plate;
    (d) 1000 μL DMSO for a 6-well plate.
24. Multiplates should be shaken for 30 min (150 RPM, at room temperature) until formazan dissolves in DMSO.
25. Before measuring, additionally stir the solution by rotating the plate placed on the table top. Check the dissolution of formazan under a microscope. 100 μL should be taken from 6-, 12-, and 24-well plates and transferred to a 96-well multiplate (in 2–4 replicates per well).
26. The reading is done at a wavelength of 550–570 nm without background, because MTT was removed before the dissolution of formazan. For cells grown in suspension, the background is subtracted by measuring the absorbance at a wavelength of 630–690 nm, relative to wells without cells (blank).
27. The signal is stable for 7 h. Excitation/Emission 560/590 nm. Fluorescence is more sensitive than absorbance and is the preferred detection method. Bottom-read is more sensitive than top-read. Correct for background fluorescence by including control wells containing only cell culture medium (no cells) on each plate.
28. If the cell–scaffold construct is too big for a 96-well plate, use 24-well plate and add the proportional, 400 μL volume to each well in a 24-well plate).
29. Uneven luminescent signal within standard plates can result from temperature gradients, uneven seeding of cells or edge effects in multiwall plates.
30. Samples can be stored at –80 °C until the assay is ready to be performed.
31. Keep plate in the dark as long as possible.
32. After incubation calibrate the fluorometer. Press standard validation button and use an option for the highest standard (1000 ng/mL). After reading the blank, insert the most fluorescent sample (1 μg/mL DNA) and press enter. After the standard is read, press enter to test samples. To minimize photobleaching effects, keep the time for fluorescence measurement equal for all samples [18].

33. Rinse the constructs with PBS three times before submerging in the lysis buffer, so that the number of dead cells on the scaffolds is insignificant relative to the live cells. Studies mainly focus on live cell counting and analysis.
34. Samples can be stored at –80 °C until the assay is ready to be performed. Sonication can be used at intervals of 1 s on/5 s off with amplitude on 50% (0.046 kJ) for a total of 1 min. Instead of Cell Lysis Buffer, 4.8 μL/tube of collagenase (Sigma C8176, 100 mg/mL) can be used for cell lysis, and incubated at 37 °C for 3 h with shaking, followed by 7.7 μL Proteinase K (Sigma P2308, 20 mg/mL) adding and subsequent incubation at 45 °C for 20 h. Spin scaffold debris at 10,000 × *g* for 5 min [16].
35. Keep the cell–scaffold construct in the 2 mL Eppendorf tube when homogenizing.

References

1. Kurzyk A, Ostrowska B, Święszkowski W, Pojda Z (2019) Characterization and optimization of seeding process of adipose stem cells on the polycaprolactone scaffolds. Stem Cells Int 2019: Article ID 1201927
2. Adan A, Kiraz Y, Baran Y (2016) Cell proliferation and cytotoxicity assays. Curr Pharm Biotechnol 17(14):1213–1221
3. Mueller H, Kassack MU, Wiese M (2004) Comparison of the usefulness of the MTT, ATP and calcein assays to predict the potency of cytotoxic agents in various human cancer cell lines. J Biomol Screen Sep 9(6):506–515
4. Ren X, Tapias LF, Jank BH et al (2015) Ex vivo non-invasive assessment of cell viability and proliferation in bio-engineered whole organ constructs. Biomaterials 52:103–112
5. Rampersad SN (2012) Multiple applications of Alamar blue as an indicator of metabolic function and cellular health in cell viability. Bioassays Sensors (Basel) 12(9):12347–12360
6. Wang P, Henning SM, Heber D (2010) Limitations of MTT and MTS-based assays for measurement of Antiproliferative activity of green tea polyphenols. PLoS One 5(4):e10202
7. Thevenot P, Nair A, Dey J et al (2008) Method to analyze three-dimensional cell distribution and infiltration in degradable scaffolds. Tissue Eng Part C Methods 14(4):319–331
8. Chemmarappally JM, Pegram HCN, Abeywickrama N et al (2020) A co-culture nanofibre scaffold model of neural cell degeneration in relevance to Parkinson's disease. Sci Rep 10: 2767
9. Dao TT, Nguyen CT, Vu NB et al (2019) Evaluation of proliferation and osteogenic differentiation of human umbilical cord-derived mesenchymal stem cells in porous scaffolds. Adv Exp Med Biol 1084:207–220
10. O'Brien J, Wilson I, Orton T, Pognan F (2000) Investigation of the Alamar blue (resazurin) fluorescent dye for the assessment of mammalian cell cytotoxicity. Eur J Biochem 267(17):5421–5426
11. Eilenberger C, Kratz RSA, Rothbauer M et al (2018) Optimized alamarBlue assay protocol for drug dose-response determination of 3D tumor spheroids. MethodsX 5:781–787
12. Mallick E, Scutt N, Scutt A, Rolf C (2009) Passage and concentration-dependent effects of indomethacin on tendon derived cells. J Orthop Surg Res 4:9
13. Crouch SP, Kozlowski R, Slater KJ, Fletcher J (1993) The use of ATP bioluminescence as a measure of cell proliferation and cytotoxicity. J Immunol Methods 160(1):81–88
14. Eriksson TM, Day RM, Fedele S, Salih MV (2016) The regulation of bone turnover in ameloblastoma using an organotypic in vitro co-culture model. J Tissue Eng 7: 2041731416669629
15. Rumiński S, Ostrowska B, Jaroszewicz J et al (2018) Three-dimensional printed polycaprolactone-based scaffolds provide an advantageous environment for osteogenic differentiation of human adipose-derived stem cells. J Tissue Eng Regen Med 12(1): e473–e485

16. Chen M, Le QS, Kjems J et al (2015) Improvement of distribution and osteogenic differentiation of human mesenchymal stem cells by hyaluronic acid and β-Tricalcium phosphate-coated polymeric scaffold in vitro. Biores Open Access 4(1):363–373
17. Kijanska M, Kelm J (2004) In vitro 3D spheroids and microtissues: ATP-based cell viability and toxicity assays assay guidance manual. Eli Lilly & Company and the National Center for Advancing Translational Sciences, Bethesda, MD
18. Protocol: Measuring Cytotoxicity or Proliferation Using alamarBlue, BioRad. https://www.bio-rad-antibodies.com/measuring-cytotoxicity-proliferation-spectrophotometry-fluorescence-alamarblue.htmL

Chapter 29

Assaying Candidate Human Skin Keratinocyte Stem Cells by Determining Their Long-Term Serial Proliferative Output in Culture

Zalitha Pieterse and Pritinder Kaur

Abstract

Stem cells are found in niches around the body, including the epidermis of the skin, and can be distinguished from their more committed progeny by their high long-term proliferative capacity in vitro. Here we describe a technique used to isolate three main epidermal cell fractions from human neonatal foreskin termed early differentiating (ED), transient amplifying (TA) and keratinocyte stem cells (KSC) based on their differential expression of two cell surface markers: CD49f and CD71. These three fractions were cultivated in parallel in a serial proliferation assay to determine their long-term proliferative output. This assay demonstrates that the KSC fraction had the highest proliferative output (total cell yield) over a long experimental timeframe of 2–3 months, as well as a higher proliferative rate compared to the other two fractions ($P > 0.05$). This assay can be utilized under similar conditions to determine the proliferative capacity of other putative stem cells using novel stem cell markers for epidermal or other stem cell populations.

Key words Keratinocyte stem cell, Long-term culture, Serial cultivation, Proliferative output, Proliferative rate

1 Introduction

Stem cells have become increasingly popular in regenerative therapies given their extensive self-renewal capacity and ability to give rise to differentiated specialized cells [1]. Stem cells are a rare population of cells recognized for their unique characteristics of being quiescent in vivo and highly proliferative in vitro [2]. There are many types of stem cells found in specific niches around the body including adipose tissue, bone marrow, and of specific interest to us, the epidermis of the skin [3].

The epidermis consists of many layers of successively differentiated keratinocytes resulting from the combined proliferative activity of keratinocyte stem cells (KSC) and transient amplifying/

Nagarajan Kannan and Philip Beer (eds.), *Stem Cell Assays: Methods and Protocols*, Methods in Molecular Biology, vol. 2429, https://doi.org/10.1007/978-1-0716-1979-7_29,

committed progenitors (TA/CP) within the basal layer. Early differentiated keratinocytes (ED) can also be found in the basal layer. The KSC fraction is the rarest population, typically only making up about 10% of the basal keratinocytes, whereas the TA/CP and ED make up the larger proportion of the basal layer [4]. These three different subsets of basal keratinocytes can be isolated from the human epidermis [4] by enzymatic digestion of skin tissue and immune-labeling the basal keratinocytes for the cell surface markers α_6 integrin/CD49f and transferrin receptor/CD71. Notably, KSCs, TA/CPs, and EDs are distinguished by the phenotypes $CD49f^{bri}CD71^{dim}$, $CD49f^{bri}CD71^{bri}$, and $CD49f^{dim}$, respectively [5]. However, any putative cell surface markers that can distinguish stem cells from their progeny can potentially be used to separate them, and their comparative long-term proliferative output determined in parallel to ascribe stem cell characteristics. Comparing the proliferative output of subpopulations of basal keratinocytes provides a simple surrogate assay to identify putative stem cell populations based on the premise that stem cells are the most potent subset of cells within any tissue and that this is reflected in their ability to regenerate the greatest number of cells (and therefore tissue) in long-term culture. In vivo, KSCs are thought to continually give rise to daughter cells during the entire life span of an organism. However, human keratinocytes have a limited life span in vitro that can most likely be attributed to our inability to replicate the stem cell niche outside the body. However, keratinocytes can be cultivated for about 3 months before senescing in culture, providing a window to assess the relative proliferative output of putative KSCs versus their more differentiated progeny. Ideally, this assay should be combined with complementary biological characterisation such as DNA cell cycle analysis, global gene expression profiling and, most importantly, skin regeneration assays in vivo as reported by us previously [4, 6].

Typically, the efficiency of culturing primary keratinocytes in vitro is poor, particularly when small numbers of cells are available for biological analysis. It has been shown that clonal keratinocyte growth can be achieved in culture using a monolayer of mouse embryonic fibroblast Swiss 3T3 feeder cells that have been irradiated or treated with Mitomycin C (MmC) to prevent proliferation—a method developed in 1975 by Rheinwald and Green [7]. We have used this technique to assess the long term proliferation potential of neonatal human keratinocyte subsets and successfully distinguish stem cells from their progeny supported by subsequent skin regeneration assays both in vitro and in vivo [4]. Notably, short-term colony-forming assays have not been able to distinguish KSCs from TA/CPs, although differentiating keratinocytes do indeed display a low colony-forming efficiency. The method for determining the comparative long-term proliferative capacity of basal keratinocytes fractionated on the basis of cell surface phenotype was reported in Li et al. 1998, which allowed us to ascribe stem cell characteristics to the $\alpha_6^{bri}CD71^{dim}$ fraction is described in detail below [4].

2 Materials

1. DMEM-10. Dulbecco's Modified Eagle Medium (Gibco) supplemented with 10% fetal bovine serum, 1% sodium pyruvate (100 mM), 1% GlutaMAX™.
2. PBS^{+++}. Phosphate buffer saline supplemented with 4% penicillin–streptomycin (10,000 U/mL), 160 μg/mL gentamicin, and 6 μg/mL fluconazole.
3. Mitomycin C. Vial of Mitomycin C (2 mg) dissolved in 4 mL of PBS to make up a 0.5 mg/mL stock concentration.
4. Coculture medium. 5× DMEM and 5× Ham's F-12 (3:1 ratio), 1.37 mg/mL $NaHCO_3$, 4 mM GlutaMAX, 0.4 μg/mL hydrocortisone, 10 ng/mL EGF, 2×10^{-9} M triiodothyronine, 5 μg/mL insulin, 20 μg/mL transferrin, 0.1 mM ethanolamine, 0.1 mM phosphorylethanolamine, 0.18 mM adenine, 0.16 mg/mL gentamicin, 0.15% FBS, 0.002 μM progesterone. For full recipe description see Gangatirkar et al. [5].
5. Soybean trypsin inhibitor. To make a stock solution at 1 mg/mL, dissolve 10 mg of soybean trypsin inhibitor (e.g., Sigma), in PBS. Can be stored at −20 °C in 1 mL aliquots. Working solution is made up at 100 μg/mL, by adding 1 mL of stock solution to 9 mL of sterile DMEM.
6. 10% BSA. Dissolve 10 g of bovine serum albumin in 100 mL of PBS.
7. Dispase. Dissolve 40 mg of Dispase II in 10 mL of PBS. Sterile filter the solution using 0.22 μm filter prior to use.
8. FACS labeling buffer. 2% BSA made up in PBS^{+++}.
9. FACS blocking buffer. 2% FBS added to FACS labeling buffer.
10. Access to fluorescence-activated cell sorting.
11. Fluorescent dye conjugated antibodies (see Table 1).

3 Methods

3.1 General Culturing and Maintenance of Swiss 3T3-J2 Cells and Establishing Feeder Cells for Long Term Epidermal Cell Culture

In order to provide epidermal cells with the appropriate support they need to be cultured in vitro, they are seeded on a monolayer of MmC treated (or irradiated) Swiss 3T3-J2 feeder cells.

1. Thaw out a vial of Swiss 3T3-J2 cells (e.g., Kerafast, Inc. #EF3003) as per routine cell culture protocol.
2. Maintain Swiss 3T3-J2 cells in DMEM-10.

Table 1
Antibody labeling for keratinocyte sorting

Tube #	Number of cells per tube	Primary antibody	Secondary antibody
1—Unstained	1×10^5 cells	Labeling buffer only	Labeling buffer only
2—7AAD viability dye (BioLegend #420404)	1×10^5 cells	Labeling buffer only	Labeling buffer only
3—Isotype control	1×10^5 cells	APC Mouse $IgG2_a$,κ Isotype control (BD Pharmingen #555576) FITC Rat $IgG2_a$,κ Isotype control (BD Pharmingen #555843)	Labeling buffer only
4—CD71 APC	1×10^5 cells	Biotin mouse anti-human CD71 (BD Pharmingen #555535)	Streptavidin-APC (BD Pharmingen #554067)
5—CD49f FITC	1×10^5 cells	FITC Rat anti-human CD49f (BD Pharmingen #555735)	Labeling buffer
6—Sample	Remaining available cells (10^7 cells/ 1000 μL)	Mouse anti-human biotinylated CD71 + FITC Rat anti-human CD49f	Streptavidin-APC

3. Split cells (3:1 ratio), every 2–3 days, or when 80% confluency is reached, using warm 0.05% trypsin–EDTA for 3–5 min, quenched with DMEM-10.
4. To establish feeder layer, seed 1.5×10^5 cells into each well of a 24-well plate for the initial keratinocyte expansion, and then 7.5×10^5 cells into each well of a 6-well plate for subsequent keratinocyte passages. Seed feeder layers 1 day before the start of the experiment or passaging of keratinocytes.
5. Treat feeder layer with 3 μg/mL MmC for 3 h at 37 °C, 5% CO_2. *See* **Note 1**.
6. After treatment, wash feeder layer with PBS for 2 min, repeating 3 times.
7. Maintain feeder layer in DMEM-10, until needed for long term epidermal culture.

3.2 Epidermal Cell Isolation

Human neonatal foreskin tissue from circumcision surgeries are routinely collected and processed in the laboratory for basal keratinocytes with informed parental consent and Institutional Ethics approval. All steps should be conducted in a sterile laminar flow hood using aseptic technique

1. Prior to each step, wash the processed skin tissue in a fresh petri dish containing PBS^{+++}.
2. Upon collection, process the skin tissue by removing excess connective tissue and fat using surgical scissors and scalpel blades.
3. Cut the skin tissue in 2 × 2 mm pieces, and incubate in 4 mg/mL Dispase (prepared in PBS^{+++}) overnight at 4 °C. Dispase is a protease that breaks down the junction between the dermis and epidermis, allowing their separation [8].
4. After overnight incubation, separate the epidermal sheets from the dermal tissue using forceps, placing them in PBS^{+++}.
5. Transfer the epidermal sheets onto a sterile 50 mL conical tube using forceps and add 5 mL of 0.05% trypsin–EDTA pre-warmed at 37 °C.
6. Isolate basal keratinocytes by triturating the epidermal sheets using a transfer pipette for 3 min precisely. *See* **Note 2**.
7. Quench the reaction by adding 5 mL soybean trypsin inhibitor to the trypsin. *See* **Note 3**.
8. Strain the cells through a 70 μm and 40 μm cell strainer sequentially.
9. Centrifuge cells at 300 ×*g* for 5 min, discard supernatant, and wash cells in PBS. Repeat centrifugation step.
10. Resuspend cells in 1 mL FACS blocking buffer. Perform a cell count using a hemocytometer and viability exclusion dye (0.4% Trypan blue solution, Sigma) to determine cell viability, cell yield and calculate antibody concentrations to be used for cell staining prior to sorting. *See* **Note 4**.

3.3 Fractionating Epidermal Cells Based on Cell Surface Phenotype

1. Incubate cells in approximately 3–4 mLs of FACS blocking buffer for 1 h on ice.
2. While cells are incubating in FACS blocking buffer, prepare cocktail of primary antibodies prepared in FACS labeling buffer as per Table 1.
3. Centrifuge cells at 300 × *g* for 5 min, discard supernatant and wash in PBS.
4. Distribute appropriate cell numbers to each tube as per Table 1. Add primary antibody cocktail to each of the relevant tubes. Incubate on ice for 1 h in the dark.
5. After primary incubation, wash cells in FACS labeling buffer and centrifuge at 300 × *g* for 5 min.
6. Add secondary antibody to each of the relevant tubes. Incubate on ice for 1 h in the dark.

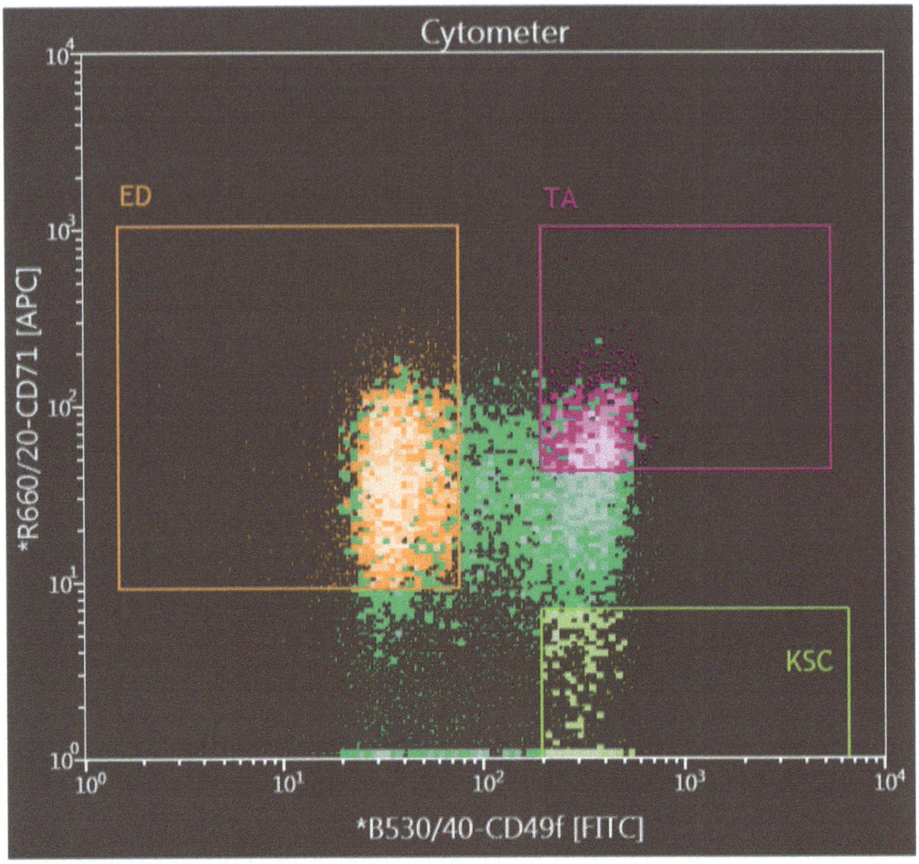

Fig. 1 Gating strategy of sorting keratinocyte fractions. A dot plot of keratinocytes costained with CD49f-FITC (*x*-axis) and CD71-APC (y-axis) reveals three phenotypically distinct cell subsets demonstrated to represent: (i) the CD49f^{bright}CD71^{dim} Keratinocyte stem cell (KSC) fraction; (ii) the CD49f^{bright}CD71^{bright} Transient amplifying (TA) fraction; and (iii) the CD49f^{dim} Early differentiating (ED) fraction. The gates used to collect the three fractions are shown by the lime green, pink, and orange boxes; and their placement as illustrated ensures minimal cross-contamination during sorting

7. Wash cells in FACS labeling buffer and centrifuge at 300 × *g* for 5 min.
8. Resuspend cells in labeling buffer and add 7AAD to tube 2 and tube 6 for cell death exclusion. Incubate on ice for 15 min.
9. Sort the basal keratinocytes into the relevant fractions, based on their cell surface expression profiles (Fig. 1):
   - Keratinocyte stem cells (KSC): CD49f^{bri} CD71^{dim}.
   - Transient amplifying/committed progenitor cells (TA): CD49f^{bri} CD71^{bri}.
   - Early differentiating cells (ED): CD49f^{dim}.
   - Unfractionated keratinocytes—collected by sorting through the same cell size and viability gates used for keratinocyte

subsets (without selection for cell surface phenotype)—as a control to demonstrate enrichment for keratinocyte stem cells.

10. Collect cells in 5 mL FACS tubes in coculture medium. Upon collection, centrifuge cells at 300 × *g* for 5 min and resuspend in coculture medium. Perform a cell count to determine cell concentration. *See* **Note 5**.

3.4 Expansion of Basal Keratinocyte Fraction to Determine Proliferation Capacity

1. Prepare a cell suspension of 5×10^3 cells/mL of each fraction of basal keratinocytes in coculture medium.
2. Remove the medium from plates containing MmC treated Swiss 3T3-J2 feeder layers (as described in Subheading 3.1). Wash with PBS.
3. Seed 1 mL of 5×10^3 cell suspension to each well of each fraction of basal keratinocytes in triplicate wells onto preprepared feeder layers.
4. Maintain coculture in coculture medium at 37 °C, 5% CO_2, changing medium every 2–3 days. *See* **Note 6**.
5. Once keratinocyte colonies start to merge to form a monolayer, they are ready to be passaged.
6. Remove medium from all wells, and wash with PBS.
7. Remove the Swiss 3T3-J2 feeder cells by adding prewarmed 0.05% EDTA for 30 s.
8. Dislodge the Swiss 3T3-J2 feeder cells by vigorously pipetting.
9. Discard the dislodged cell suspension and wash the remaining adherent keratinocytes twice with PBS.
10. Harvest the basal keratinocytes by adding prewarmed 0.05% trypsin–EDTA for 5 min at 37 °C.
11. Pipette vigorously to dislodge the keratinocytes.
12. Quench the trypsin reaction with soybean trypsin inhibitor at a 1:1 ratio.
13. Transfer cells to a centrifuge tube and perform a cell count for each replicate well to obtain cell yield. Note the cell output from 5000 cells/well and calculate mean±SEM of triplicate wells for each fraction. These cell counts will be used to determine the cell output at each passage and total cumulative yield of each fraction at the end of the experiment.
14. Pool the replicate wells for each fraction, and centrifuge at 300 × *g* for 5 min.
15. Resuspend pooled cell pellet in appropriate volume of coculture medium and perform a cell count.
16. Replate keratinocytes into 6 well plates at 5×10^3 cells per well in triplicate per cell fraction.

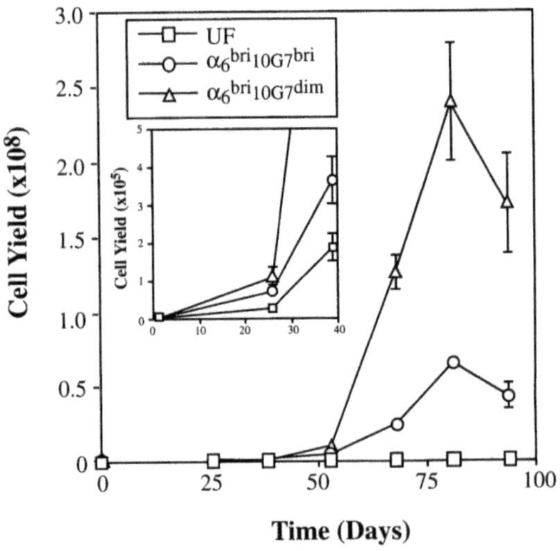

Fig. 2 Growth curve of epidermal cell fractions, representing their long-term proliferation capacity in culture. The putative keratinocyte stem cell population ($\alpha6^{bri}10G7^{dim}$ = $CD49f^{bright}CD71^{dim}$ KSC fraction) exhibited the greatest proliferative output compared to the $\alpha6^{bri}10G7^{bri}$ = $CD49f^{bright}CD71^{bright}$ TA fraction and unfractionated (UF) keratinocytes over 94 days in culture. This figure is reproduced from Li et al. PNAS-USA,1998 [4]. (Copyright (1998) National Academy of Sciences, USA)

17. Maintain in coculture medium, changing medium every 2–3 days.
18. When keratinocyte colonies have merged, repeat **steps 5–17**.
19. The proliferative capacity of each keratinocyte fraction is determined when the clonal expansion capability of each fraction has been exhausted, that is, cells reach their culture life span or replicative senescence and cannot be propagated further in vitro (typically between 2 and 3 months).
20. For analysis, accumulate all cell counts per passage and the total number of days each fraction was sustained for. Calculate mean values ± SEM and statistical significance in order to compare total cumulative cell yield per fraction (Figs. 2 and 3). *See* **Note 7**.
21. Growth curves indicate the rate of proliferation for each fraction, and in our data, we have shown that the KSC fraction had the highest proliferation rate ($P > 0.05$) compared to the other fractions (Fig. 2).
22. The total cumulative cell output is visualized as a bar graph (Fig. 3). Our studies have shown that the KSC fraction produces significantly higher ($P > 0.05$) total cell yield over the course of the experiment compared to the other fractions.

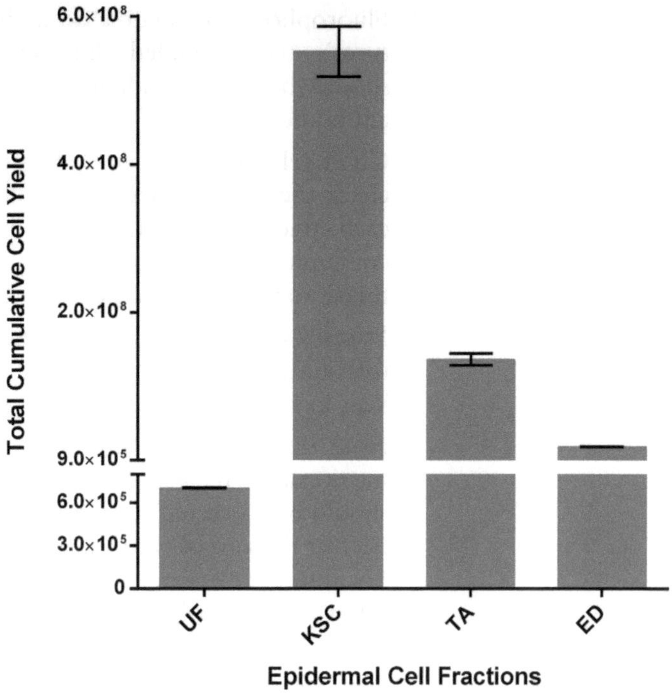

Fig. 3 Total cumulative cell yield of epidermal fractions. The total proliferative cell output from an initial seeding of 5000 cells per epidermal fraction from a representative experiment reveal that the $CD49f^{bright}CD71^{dim}$ KSC fraction displayed the greatest long-term proliferative output compared to the $CD49f^{bright}CD71^{bright}$ TA and $CD49f^{dim}$ ED fractions with clear enrichment for proliferating cells from the unfractionated (UF) keratinocytes. (Data originally published in Table 1 of Li et al., PNAS-USA, 1998 [4])

4 Notes

1. Ensure that MmC is freshly made up prior to use and is clear. Store MmC solution in the dark at 4 °C for up to 6 months. When MmC solution acquires a smoky colour, this indicates precipitation and is toxic to cells; this can affect the efficiency and viability of the feeder layers.
2. Ensure to only trypsinize the epidermal sheets for 3 min. This will allow the isolated cells to be enriched for basal keratinocytes and ensure a higher efficiency of cell yield of the three fractions during cell sorting.
3. Some studies use medium containing FBS to quench the reaction of trypsin–EDTA. We refrain from using FBS during keratinocyte isolation as it can induce differentiation of the cell fractions.

4. Fluorophore-conjugated antibodies are used at the manufacturer's recommended dilution; isotype control antibodies should be used at identical concentrations to the primary antibodies.
5. Good cell sorting practice should include a reanalysis step to check the sort purity of each collected fraction to ensure that each fraction has the selected phenotype without cross-contamination due to various reasons including machine settings, nozzle blockages, cell stream accuracy, and so on.
6. Primary keratinocytes are extremely sensitive in culture. Ensure that during culture maintenance periods, the culture plates are kept in a relatively undisturbed cell culture incubator.
7. Note that although only 5000 cells are re-plated from each fraction at each passage, the calculation for cumulative cell yield should be corrected to factor in what the cell yield would have been at the end of each passage, had all the cells been replated. For instance, if a fraction yields 15,000 cells at p1, and only 5000 cells are replated at p2 (i.e., a third), then the yield at the end of p2 should be multiplied by 3 as if the entire population had been replated. This approach eliminates the need to replate all cells obtained at each passage to determine cumulative cell yield per fraction. Some minor losses due to cell differentiation at each passage are inherent in the experimental design but can be kept to a minimum by passaging before differentiation occurs. Similarly, clonal plating of each fraction at each passage is likely to impact on the absolute cumulative cell output obtained but is used in this approach to continually demand high proliferative output from all fractions to ascertain which one is the most potent. Thus, the goal is to determine relative rather than absolute proliferative output.

References

1. Dulak J et al (2015) Adult stem cells: hopes and hypes of regenerative medicine. Acta Biochim Pol 62(3):329–337
2. Jackson CJ, Tønseth KA, Utheim TP (2017) Cultured epidermal stem cells in regenerative medicine. Stem Cell Res Ther 8(1):155–155
3. Bacakova L et al (2018) Stem cells: their source, potency and use in regenerative therapies with focus on adipose-derived stem cells—a review. Biotechnol Adv 36(4):1111–1126
4. Li A, Simmons PJ, Kaur P (1998) Identification and isolation of candidate human keratinocyte stem cells based on cell surface phenotype. Proc Natl Acad Sci U S A 95(7):3902–3907
5. Gangatirkar P et al (2007) Establishment of 3D organotypic cultures using human neonatal epidermal cells. Nat Protoc 2(1):178–186
6. Schlüter H et al (2011) Functional characterization of quiescent keratinocyte stem cells and their progeny reveals a hierarchical organization in human skin epidermis. Stem Cells 29(8): 1256–1268
7. Rheinwatd JG, Green H (1975) Seria cultivation of strains of human epidemal keratinocytes: the formation keratinizin colonies from single cell is. Cell 6(3):331–343
8. Stenn KS et al (1989) Dispase, a neutral protease from bacillus Polymyxa, is a powerful Fibronectinase and type IV collagenase. J Investig Dermatol 93(2):287–290

Chapter 30

Protocol for the Detection of Organoid-Initiating Cell Activity in Patient-Derived Single Fallopian Tube Epithelial Cells

Liang Feng, Wenmei Yang, Hui Zhao, Jamie Bakkum-Gamez, Mark E. Sherman, and Nagarajan Kannan

Abstract

Identification of serous tubal intraepithelial carcinomas (STIC) in the fallopian tubes of women who are carriers of germ line pathogenic variants in tubo-ovarian cancer predisposition genes (i.e., *BRCA1* and *BRCA2*) has led to the hypothesis that many high-grade serous carcinomas (HGSC) arise from the fimbria of the fallopian tube. However, the primitive (stem and progenitor) tubal epithelial cells that give rise to STIC and HGSC have not been defined. Further, as putative HGSC precursors are discovered at salpingectomy, the natural history of such lesions is truncated at diagnosis. Thus, living cultures of human fallopian tubes suitable for experimental studies are needed to define and characterize the cellular origin of HGSCs and thereby advance the discovery of improved methods to assess risk, develop effective early detection tests and identify novel prevention approaches. Accordingly, patient-derived tissue-organoids and isolated epithelial stem cell derived-organoids generated from average and high-risk patients are vital resources to understand the developmental biology of aging fallopian tubes and pathogenesis of HGSCs. With a vision to boost HGSC prevention research, we have established state-of-the-art protocols for the collection, processing, storage, distribution, and management of fallopian tube tissues. Here we describe the protocol for preparing these organoids, with emphasis on the key steps that require meticulous attention to achieve success.

Key words Uterine tube, Fallopian tube, Fallopian tube epithelium, Organoid, Organoid technology, High-grade serous cancer, Ovarian cancer, Epithelial stem cells, Organoid-initiating cells, Organoid-initiating cell activity

1 Introduction

Identification of serous tubal intraepithelial carcinoma (STIC) in the fimbria of the fallopian tubes of women who are carriers of pathogenic variants of *BRCA1* or *BRCA2* has led to the view that many high-grade serous carcinomas (HGSCs) arise from the tubes [1–3]. However, the cell of origin of HGSC precursors has not

Liang Feng, Wenmei Yang, Hui Zhao have contributed equally to this work.

Nagarajan Kannan and Philip Beer (eds.), *Stem Cell Assays: Methods and Protocols*, Methods in Molecular Biology, vol. 2429, https://doi.org/10.1007/978-1-0716-1979-7_30,

been definitively identified and characterized. Tubal 3D-organoids are multicellular epithelial structures derived from clones of primitive epithelial cells with the capacity for sustained self-renewal and differentiation [4]. Given that salpingectomy truncates the natural history of early lesions, experimental models represent a critical approach to defining the pathogenesis of HGSC [5]. Here, we describe a routinely used protocol to isolate tubal epithelial cells from patient specimens and grow 3D-organoids in vitro.

2 Materials

Prepare and store all reagents in −20 °C freezer (unless indicated otherwise). Milli-Q grade water (H_2O) should be used for preparation of reagents.

2.1 Growth Factors

1. B27 supplement 50×.
2. Human EGF: Add 200 μL of 10 mM acetic acid to dissolve 200 μg EGF, then bring up the volume to 10 mL by adding ~9.8 mL 0.1% BSA solution. The stock concentration will be 20 μg/mL.
3. Noggin: Dissolve 20 μg in 80 μL 0.1% BSA solution.
4. FGF10: Dissolve 25 μg in 250 μL 0.1% BSA solution.
5. Conditioned human Wnt3A medium: Culture L Wnt-3A (ATCC® CRL-26476) cell line in Dulbecco's Modified Eagle's medium (DMEM) and 10% fetal bovine serum. Let the cells grow for 4 days, collect the medium and filter-sterilize to obtain the first batch of conditioned human Wnt3A medium. Replace with fresh culture medium and culture for another 3 days. Collect the medium and filter sterilize to obtain the second batch of conditioned human Wnt3A medium. To prepare final Wnt3A condition medium, mix the first and second batch of medium 1:1. Verify the presence Wnt3A in the medium by western blot. The expected size should be approximately 40 kDa. The conditioned specimen should be kept frozen in multiple aliquots and avoid repeated freeze–thaw.
6. Conditioned human RSPO1 medium: Culture HA-R-Spondin1-Fc 293 T cells (R&D Systems, 3710–001-0) in Advanced DMEM/F12 (ADF) medium containing GlutaMAX (1%) for 7–10 days. The cells will detach from the cell culture vessel and grow in suspension after few days. Collect the supernatant and centrifuge at 3000 × *g* for 15 min at 4 °C to remove the cells and debris. Filter the supernatant through 0.22 μm filter at 4 °C. Verify the presence of HA-R-Spondin1-Fc in the medium by western blot. The expected size should be approximately 70–75 kDa. The conditioned specimen should be kept frozen in multiple aliquots and avoid repeated freeze thaw.

2.2 Chemicals, Media and Reagents

1. TGF-ß-RI kinase inhibitor IV: To prepare 5 mM stock, dissolve 2 mg in 1.04 mL of DMSO, aliquot and freeze.
2. HEPES 1 M.
3. GlutaMAX 100×: Store at room temperature.
4. ROCK inhibitor: To prepare 10 mM stock, add 1.56 mL H_2O into the vial containing 5 mg ROCK inhibitor. Resuspend and filter through 0.22 μm filter and freeze.
5. Collagenase I (EMB Millipore, SCR103, 3301502, 250 mg): Dissolve 100 mg Collagenase I in 200 mL of ADF to make a 0.5 mg/mL solution. Stir for about 30 min and filter sequentially using 0.8 μm, 0.45 μm, 0.22 μm filters. Aliquot and freeze.
6. TrypLE™.
7. Advanced DMEM/F12 (ADF).
8. DMEM/F12.
9. DMEM (with high glucose).
10. Phosphate buffered saline (PBS; pH 7.4): Store at room temperature.
11. Matrigel: Thaw on ice. Freeze aliquots and avoid repeated freeze-thaw.
12. Penicillin–streptomycin 100×.
13. 0.1% BSA solution: Dissolve in DMEM and filter-sterilize. Store at 4 °C.
14. DMSO: Store at room temperature.
15. Heat-inactivated (HI) bovine serum.
16. N2 Supplement 100×.
17. Nicotinamide.
18. Organoid and cell media: *see* Table 1 and **Note 1**.
19. Transport medium: see Table 2.
20. Cell and organoid cryomedium: see Table 3.

2.3 Other Equipment

1. 1.5 mL low-adhesion microcentrifuge tubes
2. 37 °C shaking platform
3. 24-well suspension culture plate
4. Glass slide with hemocytometer counting grid.
5. Stereomicroscope.
6. Scissors.
7. Forceps.
8. CO_2 incubator.
9. Tissue culture treated dishes 60 × 15 mm.

Table 1
Organoid and Cell Media (*see* Note 1)

Components	Organoid growth medium (adapted from [4]) (10 mL)	Cell culture medium (adapted from [4]) (10 mL)
Advanced DMEM/F12	4.5 mL	~9.2 mL
HEPES (1 M)	120 μL	120 μL
GlutaMAX (100×)	100 μL	100 μL
EGF (20 μg/mL)	5 μL	5 μL
ROCK inhibitor (10 mM)	9 μL	9 μL
Bovine serum	/	500 μL
Penicillin–streptomycin (100×)	/	100 μL
WNT3a conditioned medium	2.5 mL	/
RSPO1 conditioned medium	2.5 mL	/
B27 (50×)	200 μL	/
N2 (100×)	100 μL	/
Noggin (250 μg/mL)	4 μL	/
FGF10 (100 μg/mL)	10 μL	/
Nicotinamide (1 mM)	10 μL	/
SB431542 (5 mM)	1 μL	/

Table 2
Transport medium

Components	Total 100 mL
DMEM	94 mL
Bovine serum	5 mL
Penicillin–streptomycin	1 mL

Table 3
Cell and Organoid Cryomedium

Components	Total 100 mL
Advanced DMEM/F12	50 mL
Bovine serum	44 mL
DMSO	6 mL

10. Syringe filter Units 0.80 μm, 0.45 μm, and 0.22 μm.
11. NALGENE Mr.Frosty Cryo 1 °C freezing container.
12. EZ-Grip micropipette (CooperSurgical, 7-72-2802).

3 Methods

1. All procedures need to be carried out at room temperature, unless otherwise specified. Specimen handling should be performed in a certified laminar hood following institutionally required biosafety standards. The protocol described below works for patient specimens collected from fimbriated as well as ampullary regions of the fallopian tube.

3.1 Cell Enrichment for Tissue Sample

1. Transport the excised fresh tubal specimen in transport medium from the Operating Room or Surgical Pathology Unit within 2–3 h of removal.
2. Transfer each specimen into a 60 mm dish and record specimen weight (Fig. 1a, *see* **Note 2**).
3. Wash specimen with 1 mL PBS in the dish and remove excessive connective and vascular tissue using a sterile scalpel. It is necessary to use a new scalpel for each specimen to avoid cross-contamination of cells.
4. Using a scalpel, open tubes longitudinally to expose the epithelium.
5. Transfer tissue into a 15 or 50 mL tube and add 1–2 mL of prewarmed 0.5 mg/mL collagenase dissociation buffer and incubate at 37 °C for 45 min on a rotary shaker preset to 80 rpm (Fig. 1b, *see* **Note 3**).
6. Transfer tissue and dissociation buffer from 15 mL tube to a petri dish and use scalpel to scrape off the epithelial cells from the tissues (Fig. 1c, *see* **Note 4**).
7. Transfer tissue and cell suspension to a 15 mL tube.
8. Centrifuge at 40 × *g* for 1 min at 4 °C to separate cell suspension and partly dissociated tissue. Follow **steps 9–12** to process cell suspension and follow **steps 13–16** to process partly dissociated tissue.
9. Aspirate the supernatant to a new tube and centrifuge at 300 × *g* for 5 min at 4 °C to collect cell suspension. The cell suspension is the primary source for epithelial cells with organoid-initiating activity in 3D Matrigel.
10. Resuspend the cell pellet in 1 mL cell culture medium.
11. Count the total cell number with hemocytometer.

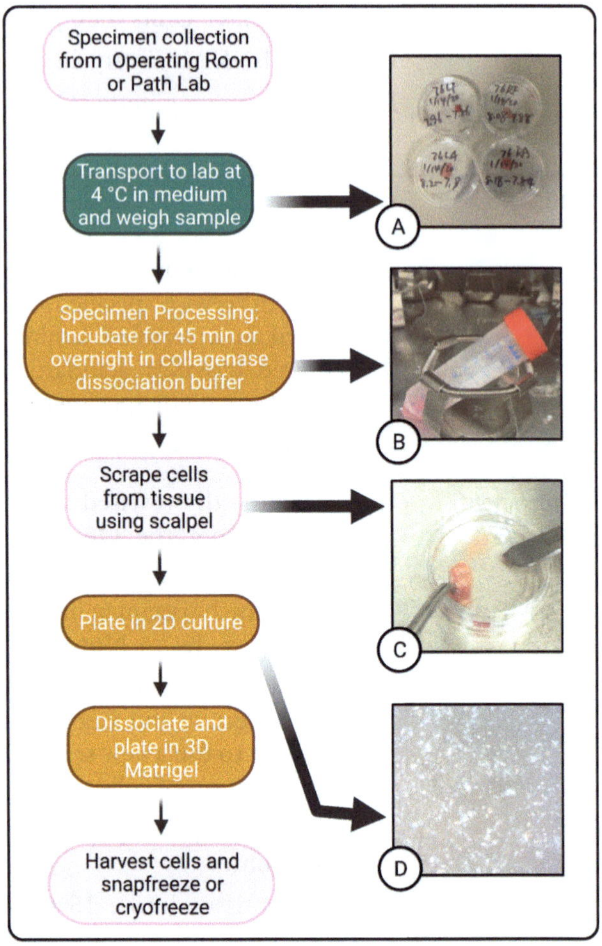

Fig. 1 Workflow involved in tubal single cell isolation and culture from patients

12. Seed up to five million viable cells in 60 mm dish with 3 mL cell culture medium, and culture in a humidified incubator at 5% CO_2 and 37 °C (Fig. 1d).
13. To release residual epithelial cells still present in the scrapped tissue, add 1 mL of prewarmed 0.5 mg/mL collagenase dissociation buffer to the scrapped tissue, resuspend it and continue dissociation at 37 °C in incubator on a rotary shaker overnight.
14. Next morning, centrifuge the dissociated tissues at 40 × *g* for 1 min at 4 °C.
15. Aspirate the supernatant to a new tube and centrifuge at 300 × *g* for 5 min at 4 °C to collect cell suspension. Resuspend the cell pellets in 1 mL cell culture medium. The residual tissue can be used for genotyping if necessary.
16. Count the cell number with hemocytometer.
17. Seed the cells in 60 mm dish with 3 mL Cell culture medium, and keep at 37 °C, 5% CO_2 in a humidified incubator (Fig. 1d).

3.2 Assay to Detect Organoid-Initiating Cells from Uterine Tubal Derived Single Cell Suspension

1. Once the cells reach 70% confluence (*see* **Note 5**), remove the cell culture medium. Wash with 1 mL PBS and then add 1 mL prewarmed TrypLE (*see* **Note 6**).
2. Collect the cells and centrifuge at 300 × *g* for 5 min at 4 °C.
3. Count the cell number with hemocytometer.
4. Seed 25,000–50,000 cells in 50 μL Matrigel in the middle of a well in 24-well plate (*see* **Notes 7** and **8**).
5. Place the plate into the humidified incubator at 37 °C for 20 min to allow the Matrigel to solidify.
6. Gently pipet 500 μL of prewarmed (37 °C) organoid culture medium into each well without disturbing the Matrigel.
7. Refresh medium carefully every 3 days for 2 weeks or more depending on experimental needs.
8. Single layers of cells forming lumens in mostly spherical organoids should appear around 1 week (Fig. 2) and stable structures at 2 weeks (Fig. 3) that can sustain in culture for several weeks and in many cases can be further passaged for several months as previously shown [4].

3.3 Storage Procedure for Cells

1. Centrifuge cells in 1 mL microfuge tube at 300 × *g* for 5 min.
2. Discard supernatant and add 1 mL of cryomedium to cell pellet. Triturate cells until homogeneous. Screw each vial closed.
3. Transfer vials into NALGENE Mr.Frosty Cryo 1 °C freezing container and then transfer and store the container overnight at −80 °C deep freezer.

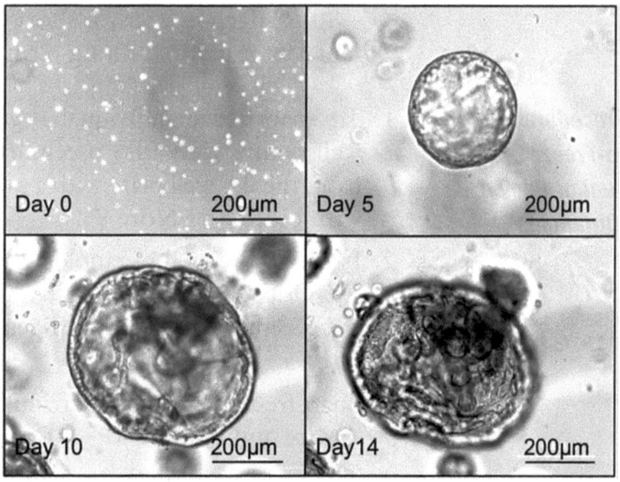

Fig. 2 Tubal organoid-initiating cell activity in 3D-Matrigel. Representative images of patient-derived tubal single cell culture in 3D-Matrigel on day 0, 5, 10, and 14 depicting growth of organoid-initiating cells

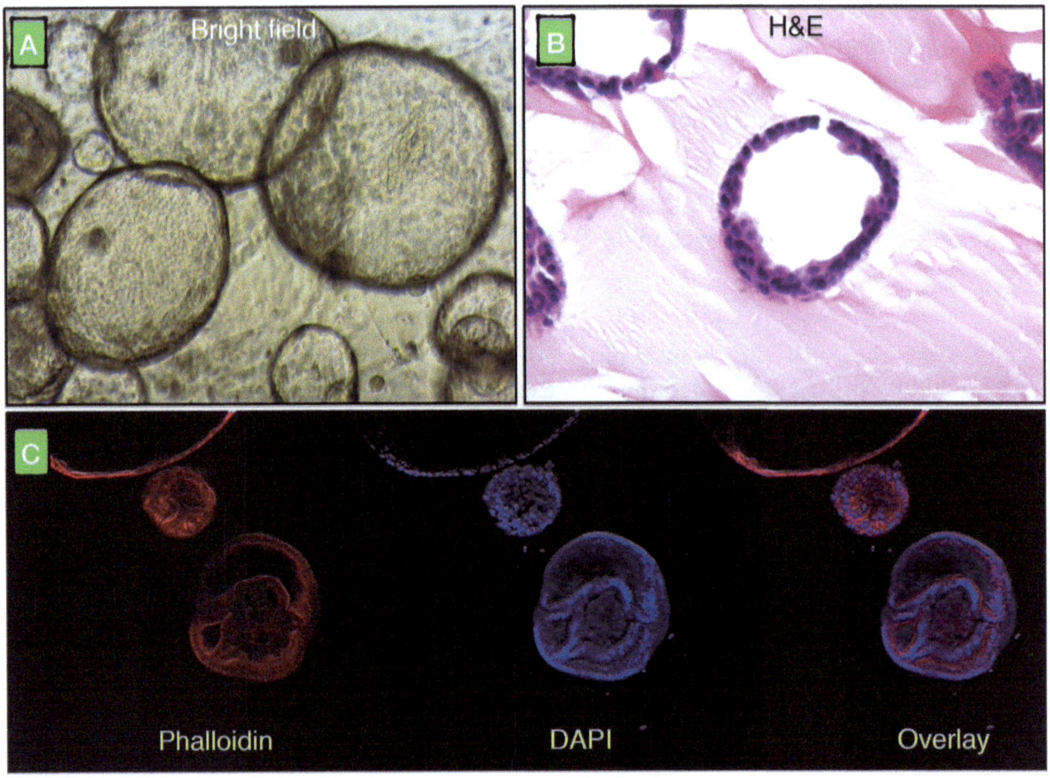

Fig. 3 Microscopy of 3D-Matrigel of patient derived tubal organoids. (**a**) Bright field images of 3D-Matrigel culture of tubal single cells. (**b**) Matrigel culture with tube organoids fixed in formalin and embedded in paraffin for histology. A representative H&E stained section of Matrigel culture is shown. (**c**) Confocal fluorescent images of tubal organoids stained for filamentous actin with phalloidin-TRITC and nucleus with DAPI

4. Next day, transfer vials into storage box and store them in −80 °C deep freezer for short term (weeks) or liquid nitrogen tank vapor phase for long term storage.

3.4 Procedure to Snap-Freeze Individual or Pooled Tubal Organoids

1. To snap-freeze individual organoids, wash organoid culture well gently with PBS prechilled at 4 °C, scrape the Matrigel from the 24-well plate with a pipette and transfer to a petri dish with buffer (*see* **Note 9**).
2. Isolate individual organoids under the stereomicroscope using EZ-Grip micropipette and transfer to 1.5 mL low-adhesion microcentrifuge tubes after gently washing the single organoid in PBS.
3. Snap-freeze vials with individual organoids by placing them on dry ice and transfer to −80 °C deep freezer.
4. To snap-freeze pooled organoids, wash organoid culture well gently with buffer prechilled at 4 °C, and scrape the Matrigel from the 24-well plate with a pipette and transfer to 1.5 mL low-adhesion microcentrifuge tubes (*see* **Note 9**).

5. Centrifuge at 300 × *g* for 5 min and collect pooled organoids.
6. Follow **step 3** to snap-freeze pooled organoids and store at −80 °C for long-term storage.

3.5 Procedure to Cryofreeze and Thaw Tubal Organoids

1. To cryofreeze organoids, wash organoid culture well gently with buffer prechilled at 4 °C, scrape the Matrigel from the 24-well plate with a pipette and transfer to a 15 mL Falcon tube in cold buffer (*see* **Note 9**).
2. Spin down the organoids at 300 × *g* and 4 °C for 5 min and discard supernatant.
3. Resuspend organoids in cryomedium and transfer vials into NALGENE Mr.Frosty Cryo 1 °C freezing container and then transfer and store the container overnight at −80 °C deep freezer.
4. Next day, transfer vials into storage box and store them in −80 °C deep freezer for short term (weeks) or liquid nitrogen tank vapor phase for long term storage.
5. To revive organoids, rapidly thaw the cyrofrozen organoid vial in 37 °C water bath, wash the organoids in PBS, plate them in 50 μL Matrigel and add organoid growth medium as above to regrow/expand 3D organoids (*see* **Note 10**).

4 Notes

1. The organoid growth medium and cell culture medium should be prepared fresh for best results. The media should not be stored longer than 2 weeks at 4 °C. Conditioned medium batches should be tested in organoid culture system for consistent results and the best batch should be used. Alternatively, purified recombinant RSPO1 and Wnt3A can be used. Please note that the organoid growth medium supports growth of tubal organoid from primary cells isolated from patients and may not support organoid growth from established fallopian tube epithelial cell lines.
2. Enter weight of the tissues on the petri dish and take a photograph of the specimen for record.
3. Use specified speed, that is, 80 rpm on shaker platform in standard cell culture incubator for best results.
4. To improve viable epithelial cell yield, try to scrape off the epithelial cells very gently instead of chopping tissue into pieces.
5. Cells from ~200 mg tubal tissue reach 70% confluence in 4 to 8 days in 2D culture.

6. TrypLE should be previously warmed at 37 °C water bath. After adding TrypLE, place the petri dish in the incubator, and quickly check every 2–3 min if cells have detached from the bottom of the dish. The time to detach may vary but, on average, can take 5–10 min.
7. The Matrigel stock vials should be thawed on ice and not room temperature. Pay extra care to ensure that Matrigel does not solidify before plating and that the Matrigel does not contain air bubbles. Tissue culture plates should be prewarmed (1 h at 37 °C). 200 μL micropipette tip boxes used for Matrigel handling are typically kept frozen at −20 °C and placed on ice during liquid handling to avoid Matrigel solidifying inside micropipette tips.
8. The cell seeding density should be adjusted based on prior results. 25,000 cells per 50 μL Matrigel is a good seeding density to begin with.
9. Gentle handling is recommended at this step to maintain organoid structural integrity.
10. The method described here can be adapted to study various influences on organoid-initiating cells following lab specific genetic or pharmacological manipulations, or various molecular analyses. An example of application of organoids for discovery research using this protocol can be found in Mun et al. [6].

Acknowledgments

This work was supported partly through Developmental Research Program (DRP) award grant to NK from Mayo-NCI SPORE in Ovarian Cancers.

References

1. Carlson JW, Miron A, Jarboe EA et al (2008) Serous tubal intraepithelial carcinoma: its potential role in primary peritoneal serous carcinoma and serous cancer prevention. J Clin Oncol 26(25):4160–4165
2. Labidi-Galy SI, Papp E, Hallberg D et al (2017) High grade serous ovarian carcinomas originate in the fallopian tube. Nat Commun 8(1):1093
3. Ducie J, Dao F, Considine M et al (2017) Molecular analysis of high-grade serous ovarian carcinoma with and without associated serous tubal intra-epithelial carcinoma. Nat Commun 8(1):990
4. Kessler M, Hoffmann K, Brinkmann V et al (2015) The notch and Wnt pathways regulate stemness and differentiation in human fallopian tube organoids. Nat Commun 6:8989
5. Hill SJ, Decker B, Roberts EA et al (2018) Prediction of DNA repair inhibitor response in short-term patient-derived ovarian cancer organoids. Cancer Discov 8(11):1404–1421
6. Mun DG, Renuse S, Saraswat M et al (2020) PASS-DIA: a data-independent acquisition approach for discovery studies. Anal Chem 92(21):14466–14475

Chapter 31

Quantification of Muscle Stem Cell Differentiation Using Live-Cell Imaging and Eccentricity Measures

Paige C. Arneson-Wissink and Jason D. Doles

Abstract

Culturing primary muscle stem cells ex vivo is a useful method for studying this cell population in controlled environments. Primary muscle stem cells respond to external stimuli differently than immortalized myoblasts (C2C12 cells), making ex vivo culture of muscle stem cells an important tool in understanding cell responses to stimuli. Primary muscle stem cells cultured ex vivo retain a majority of the characteristics they possess in vivo such as the abilities to differentiate into multinucleated structures, and self-renew a stem cell-like population. In this chapter, we describe methods for isolating primary muscle stem cells, controlled differentiation into myotubes, and quantification of differentiation using IncuCyte live cell imaging and analysis software.

Key words Muscle stem cells, Satellite cells, Myotubes, Differentiation, Eccentricity, IncuCyte ZOOM

1 Introduction

Skeletal muscle is a highly dynamic tissue that provides bodily structural support, facilitates locomotion, and supplies/stores critical metabolites and small molecules such as those involved in energy metabolism and endocrine function. In addition to continuous protein anabolism and catabolism necessary to support these functions within mature muscle myofibers, skeletal muscle is highly regenerative and can efficiently repair or replace lost/compromised tissue mass. For example, approximately 30 days after major experimental chemical injuries, murine skeletal muscle not only recovers, but often hypertrophies to a greater mass than contralateral (uninjured) control limbs [1]. The major cell type responsible for this remarkable recovery is the muscle stem cell, also termed satellite cell. Muscle stem cells are an adult stem cell population that com-

Nagarajan Kannan and Philip Beer (eds.), *Stem Cell Assays: Methods and Protocols*, Methods in Molecular Biology, vol. 2429, https://doi.org/10.1007/978-1-0716-1979-7_31,

prise approximately 2% of the total nuclei in skeletal muscle [2]. Muscle stem cells lie mitotically quiescent within the basal lamina, but outside the sarcolemma, poised for activation stimuli to push them into the cell cycle [3]. Once activated, muscle stem cells proliferate and then either commit to differentiation or self-renewal. The process of differentiation, or myogenesis, is orchestrated on a molecular level by myogenic regulatory factors (MRFs), but also on a larger scale by the microenvironment within the muscle [2]. Other mononuclear cell types, such as endothelial cells, fibro/adipogenic-progenitors, and immune cells, are present in muscle and are critical contributors to muscle regeneration [1, 2, 4]. The mature myofibers, which comprise the bulk of muscle tissue, are also critical in regulating myogenesis as they provide signals and a physical niche for the muscle stem cells [5, 6]. The natural regenerative potential of skeletal muscle makes it an obvious target for regenerative medicine-based therapies, but to move from bench to bedside, a critical understanding of myogenesis and how it changes with disease is required.

Although some complexities of myogenesis remain unknown, models for characterizing myogenesis, both in vitro and in vivo, have allowed scientists to better understand the repercussions of defective myogenesis and the potential benefits of boosting muscle regeneration. Models for studying myogenesis in vivo include: lineage tracing using transgenic mice that express fluorescent protein under a muscle stem cell-specific driver, such as Pax7; and monitoring MRF expression using transgenic reporters [7, 8]. While these methods preserve the muscle stem cell niche, including complex interactions with other cell types, it can be difficult to pinpoint stage-specific defects in myogenesis (i.e., proliferation, commitment, differentiation, or fusion). To achieve this degree of mechanistic granularity, an in vitro approach is more appropriate. In vitro models range from more basic, immortalized cell lines such as C2C12, MM14, and L6, to complex cocultures of primary myoblasts on isolated myofibers [9–12]. Immortalized cell lines are typically easy and economical to culture, retain MRF expression, and they exhibit a robust ability to differentiate into multinucleated myotubes. However, concerns have been raised about the ability of immortalized cells to model the specific functions of muscle stem cells, which are tightly tied to changes in the cell cycle [13, 14]. Primary muscle stem cells with or without associated myofiber cultures, are a step closer to the behavior of muscle stem cells in vivo, and retain properties like self-renewal more robustly than immortalized cells [12, 15]. For these reasons, it is important to use primary muscle stem cells to validate and extend observations made using cell line-based approaches. Additionally, primary muscle stem cell cultures are an excellent tool to study potential myogenesis defects in experimental mouse disease models [16, 17].

To date, the largest drawback of using primary muscle stem cells is that there is extensive animal-to-animal variation and a lack of standardized protocols across laboratories, which results in data that is often difficult to compare. Therefore, it is critical that we continue to build upon existing literature and standardize protocols. Additionally, quantification of differentiation (be it from immortalized or primary cell cultures) can be done in myriad ways (nuclei per myotube, myotube diameter/length, expression of the terminal marker myosin heavy chain), leading to variability in how labs "score" myotube differentiation. Most of these methods rely on immunostaining, imaging, and hand quantification in software such as ImageJ or Adobe Photoshop, which introduces multiple places for human error and bias [18, 19]. Hand quantification also severely limits the number of samples that can be reasonably quantified. While methods have been established (MyoVision, Open-CSAM) to limit human bias and increase efficiency when quantifying myofiber diameters/cross-sectional area (a critical parameter for studying muscle atrophy), such methods are underdeveloped for assessments of muscle stem cell differentiation [20–22]. The methods described in this chapter provide the opportunity to increase cell isolation consistency, differentiation quantification efficiency, and data reproducibility.

In this chapter we outline methods for successfully isolating, culturing, and quantifying differentiation of primary muscle stem cells. First, we describe methods for isolating primary muscle stem cells from murine hind limb muscles, using negative selection columns and magnetic beads to yield a high-purity culture [18]. We highly recommend utilizing transgenic mice with endogenously fluorescent muscle stem cells because this further increases the efficiency and accuracy of quantification. In our figures we use muscle stem cells from the lineage-tracing transgenic model $\mathrm{ROSA26}^{\mathrm{lsl\text{-}TdTomato}}/\mathrm{Pax7}^{\mathrm{Cre/ERT2}}$, but many other models, including $\mathrm{Pax7}^{\mathrm{zsgreen}}$ mice, would also suffice [1, 8]. Second, we describe how to differentiate primary muscle stem cells into myotubes in a controlled manner, leveraging continuous, longitudinal cell monitoring using the IncuCyte ZOOM platform [8]. Live imaging during primary muscle stem cell differentiation is particularly advantageous because it captures all stages of in vitro myogenesis, from early expansion to fusion of multinucleated myotubes. Finally, we describe how to quantify differentiated myotubes using IncuCyte software. Our quantification strategy eliminates a large portion of human bias and works well on all stages of differentiated cultures. Ultimately, pairing these methods together creates a reproducible system for monitoring and quantifying muscle stem cell differentiation.

2 Materials

2.1 Muscle Stem Cell Isolation

1. 2% collagenase II w/v in Hanks's Balanced Salt Solution (HBSS)
2. Ham's F-12 Nutrient Mixture.
3. Muscle Stem Cell Isolation Medium (IM): Ham's F-12 medium, 15% horse serum (HS), 1% penicillin–streptomycin (10,000 U/mL stock).
4. Muscle Stem Cell Growth Medium (GM): Dulbecco's Modified Eagle Medium–Nutrient Mixture F-12 (DMEM/F-12 GlutaMAX™), 20% fetal bovine serum (FBS), 10% HS, 1% chick embryo extract, 1% penicillin–streptomycin (10,000 U/mL stock), 0.1% amphotericin B, 2.5 ng/mL fibroblast growth factor 2 (FGF-2).
5. Gelatin-coated cell culture plates (*see* **Note 1**).
6. Cell Isolation Buffer: 1× Dulbecco's phosphate-buffered saline (DPBS), 2 mM ethylenediaminetetraacetic acid (EDTA), 0.5% bovine serum albumin (BSA).
7. Dulbecco's phosphate-buffered saline.
8. 70% ethanol: 30% deionized water, 70% absolute ethanol
9. Dissection tools: fine forceps (such as #5), blunt forceps, scissors.
10. 14 mL round bottom culture tubes
11. Cell strainers: 100 μm, 70 μm, and 40 μm (*see* **Note 2**).
12. 18G needles with syringes
13. 50 mL conical tubes
14. 1.5 mL microcentrifuge tubes
15. Incubator with rotator.
16. Satellite cell isolation kit (e.g., Miltenyi).
17. Separation columns (e.g. LS columns, Miltenyi).
18. Magnetic separator (e.g. QuadroMACS, Miltenyi).

2.2 Muscle Stem Cell Differentiation

1. Muscle stem cells isolated in Subheading 3.1.
2. Gelatin-coated cell culture plates.
3. Muscle stem cell differentiation medium (DM): DMEM high glucose pyruvate, 2% horse serum, 1% penicillin–streptomycin (10,000 U/mL stock).
4. IncuCyte Zoom, dual color model filter module (4459), Nikon 10× objective.

2.3 Myotube Quantification

1. IncuCyte Zoom Software (*see* **Note 3**).

3 Methods

3.1 Muscle Stem Cell Isolation

1. Prepare dissection tools by cleaning with 70% ethanol.
2. Begin thawing collagenase II on ice.
3. Euthanize a mouse by cervical dislocation or other approved methods of euthanasia. We recommend using muscle stem cells that are endogenously fluorescent, because this will make live-cell imaging and downstream analysis in the IncuCyte more accurate. Examples of useful transgenic strains were mentioned in the introduction.
4. Wet hindlimb with 70% ethanol to prevent fur shedding. Remove skin from hindlimb to expose muscle. Remove fascia and fat from hind limb muscles as much as possible.
5. Dissect muscles from hind limb from both legs (tibialis anterior, extensor muscles, gastrocnemius and soleus, quadriceps, hamstring group). Place dissected groups in small amount, approximately 750 μL, F12 medium as others are being dissected (*see* **Note 4**).
6. Determine weight of muscles.
7. Mince muscle tissue using scissors until it looks like a paste (at least 5 min).
8. Transfer minced muscle tissue into a 14 mL round-bottom tube containing 4.5 mL F12 medium (no serum) (*see* **Note 5**).
9. To the tube, add 0.5 mL of 2% w/v collagenase II and shake vigorously for 5 s. Seal tube well using tape and parafilm.
10. Incubate with rotation (5 rpm) at 37 °C for 60 min.
11. Dissociate digested muscle tissue by syringing through 18G needle. Return to 14 mL round bottom tube.
12. Incubate with rotation (5 rpm) at 37 °C for 15 min.
13. Add 25 mL of IM to a 50 mL tube.
14. Dissociate digested muscle tissue by syringing through 18G needle and depositing in the IM prepared in **step 13**. Digested muscles in mass prep medium will be denoted as "muscle suspension" from here on.
15. Gently pour muscle suspension into a 100 μm cell strainer set on a 50 mL tube. Allow the muscle suspension to filter by gravity. Do not force filtering by pipetting. If strainer is clogged, change to a fresh strainer for remainder of muscle suspension.

16. Repeat **step 15** using a 70 μm cell strainer.
17. Repeat **step 15** using a 40 μm cell strainer.
18. Centrifuge the filtered muscle suspension at 400 × *g* × 5 min at 4 °C, and remove supernatant (*see* **Note 6**).
19. Wash the pellet by adding 1 mL PBS and resuspend well with gentle pipetting. Split sample into two 1.5 mL Eppendorf microfuge tubes.
20. Centrifuge muscle suspension at 700 × *g* × 5 min at 4 °C, and remove supernatant.
21. Add 80 μL of cell isolation buffer to each tube.
22. Add 20 μL of satellite cell isolation cocktail (e.g., Miltenyi, 130-104-268) to each tube, and mix by pipetting gently.
23. Incubate the muscle suspension and isolation beads for 15 min in refrigerator (NOT ICE).
24. Add 350 μL of cell isolation buffer to each tube of the muscle suspension and mix well by pipetting gently.
25. Prepare the magnetic columns. Place two LS column in the magnetic MACS separator. Flow 3 mL of cell isolation buffer through each column, and discard flow-through (*see* **Note** 7).
26. Place a flow tube below each column. Apply cells from each tube (resuspended in cell isolation buffer) onto each LS column and let them drop by gravity.
27. When the reservoir of LS column is empty, rinse column with 1 mL of cell isolation buffer.
28. Repeat **step 27**.
29. Transfer flow-through to 1.5 mL microfuge tubes (dividing as needed to accommodate final volume) and spin down 700 × *g* × 5 min, 4 °C.
30. Resuspend pellet (at this point the pellet will be very small) in 1 mL GM. To a collagen-coated plate add 9 mL GM. Gently pipette the 1 mL cell suspension onto the culture plate.
31. Culture in a humidified incubator at 37 °C 5% Oxygen, 5% CO_2 for 48–72 h (*see* **Notes 8–10**).

3.2 Muscle Stem Cell Differentiation

1. Begin by splitting cells (*see* **Note 9**), and plating them at a much higher than normal density on a collagen-coated cell culture dish in GM. For a 96 well plate, 1000 cells per well is sufficient. This is **Day 0** of differentiation.
2. Move culture dish to IncuCyte Zoom to begin live cell imaging (*see* **Note 11**). All images taken prior to Day 1 will be considered baseline images.
3. Set up IncuCyte Zoom scan. Open IncuCyte Zoom software and navigate to the "Schedule Scans" tab. Click "Add a Vessel"

then, select the cell culture plate that matches what you are using and click "OK." In the right-side menu, "Scan Setup" tab, to select both phase and red as the "Channel Selection". In the "Properties" tab you can add details for your experiment, if desired (*see* **Note 12**). Set the scan frequency (*see* **Note 13**). Click "Apply" to save the changes. Scanning will begin as shown in the scan schedule at the top of the software window.

4. **Day 1**: Allow the cells 12–24 h to adhere to plate in GM, then gently aspirate off GM and replace with appropriate amount of DM. Be sure to only remove cells from the IncuCyte when it is not in the process of scanning.
5. **Day 2**: Add 50% of the total volume in the culture as fresh DM (e.g., if you have a 6-well dish with 1.5 mL DM per well, on day 2 you will add 750 μL fresh DM).
6. **Day 3**: Gently aspirate off medium and replace with fresh DM.
7. **Day 4**: Repeat **step 2**.
8. **Day 5**: End of differentiation. If downstream assessment (i.e., qPCR, immunofluorescence) will be done, process cells as you normally would for those applications. All images are automatically saved in the IncuCyte. To end IncuCyte run, in "Schedule Scans" tab, click on the plate you would like to end, then click "Remove Vessel." Click "Apply" to finalize changes. *See* Fig. 1 for example of differentiation time course.

3.3 Myotube Quantification Using IncuCyte Zoom

1. **Create Image Collection**. To do this, navigate to the "Search Scans," "Scanned Vessels" tab of the IncuCyte Software, and double click on your differentiation assay to open the main scan window. Your Image Collection needs to be representative of the whole experiment, take into consideration changes in confluence, cell health/morphology, and differentiation that occurred over the course of your experiment. Keeping this mind, select an image by expanding the calendar menu on the left site of the window and selecting a date/time; then, in the right-side menu click "Create or Add to Image Collection." In the pop-up window, select "New" and add a name for your Image Collection, then select the appropriate channels (phase and red, in this case), and click "add" (*see* **Note 14**). To add another image to the image set, click "Create or Add to Image Collection," then the pop-up window should default to be selected "existing" with your newly created Image Collection listed as the name. Click "Add." Add more images for a total of 5–10 images in the Image Collection.
2. **Create a Processing Definition**. In the main window for your differentiation assay, click the "New Processing Definition" button on the right-hand menu (*see* **Note 15**). In the pop-up window, select your newly made image collection from the

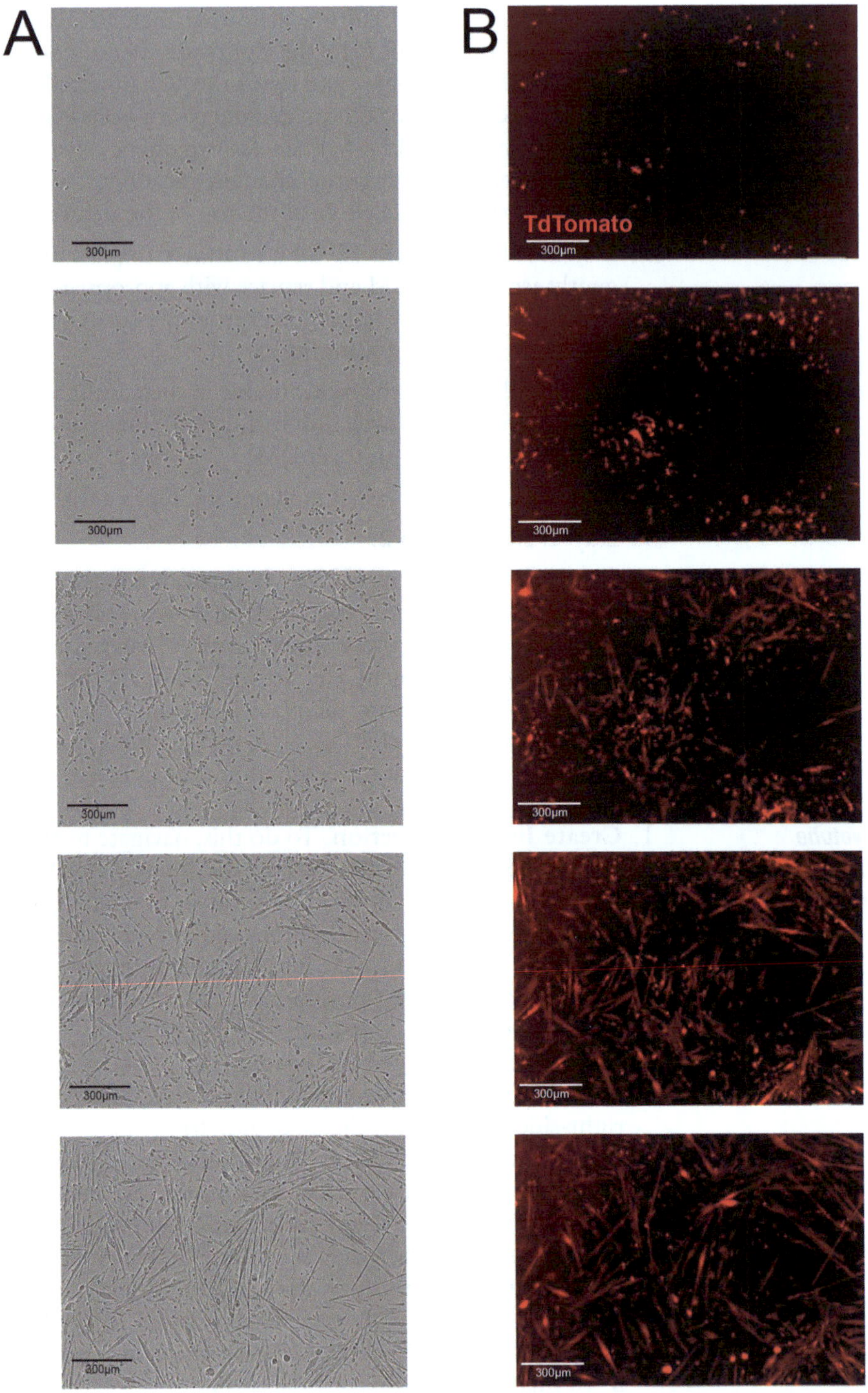

Fig. 1 Differentiation of muscle stem cells. (**a**) Phase contrast images of muscle stem cells differentiating over the course of 5 days. (**b**) Red fluorescence images of same cells. Day 1 on top to day 5 on bottom. All images were acquired on IncuCyte Zoom 10× objective. Scale bars represent 300 μm

drop-down menu. Click "continue". In the new window you will have a series of controls to set a mask to mark cell area in each of your channels (in this case phase and red fluorescence). To start, click the small down arrow next to "Preview Current" in the upper right corner and click "Preview All." This will apply the masks for each channel to your images. To visualize the masks, check the box next to each mask in the "Analysis Mask" menu on the left side of the image. Once checked, a colored mask to apply to the image you are viewing (you can navigate through all the images in your Image Collection using the arrows at the top center of the window, under "Preview Image Stack"). Start in the "Phase (Analyze)" menu on the left side of the window and use the "Segmentation Adjustment" slider to put more mask on cells (slide toward "cells") or remove mask and create more background (slide toward "background"). To view the adjustments to the mask, click "Preview Current" or "Preview All" in the upper right corner. You can toggle between the image and image + mask using the" Image Channels/Analysis Mask" window and checking the boxes next to the items you want to view. *See* **Note 16** for recommendations for advanced masking. Repeat this methodology until the masks for each channel are as accurate as possible, only masking objects in that channel for all images in the Image Collection (i.e., in the red fluorescent channel, only red cells are masked and nonred cells and debris are unmasked). Once the mask is sufficient, go to file, save, and name the Processing Definition. You may now close the Processing Definition window. (*See* Fig. 2 for example of processing definition applied to differentiating myotubes.

3. **Launch New Analysis Job**. In the main window for your differentiation assay, click the "Launch New Analysis Job" button on the right-hand menu. In the pop-up window, select your newly made processing definition from the drop-down menu (*see* **Note 17**). Add a name for your analysis job. Select the time range you would like to use (select "open-ended" if you are still scanning your plate (*see* **Note 17**)). Select the wells of the plate you are assessing. Click "Launch" in the lower right corner. A pop-up window will tell you that the Analysis Job has been created and will provide updates on the progress of this job. You may close all windows (for the differentiation assay), except the main Incucyte home window, now.
4. **Confirm mask accuracy**. Once the Analysis Job is complete, the status window will show a date and time under "Date Completed." To view your analysis, navigate to the "Search" tab of the main IncuCyte window, and select the "Analysis Jobs" tab at the top. Identify your scan/analysis job and double click to open the analysis window. This window will look nearly

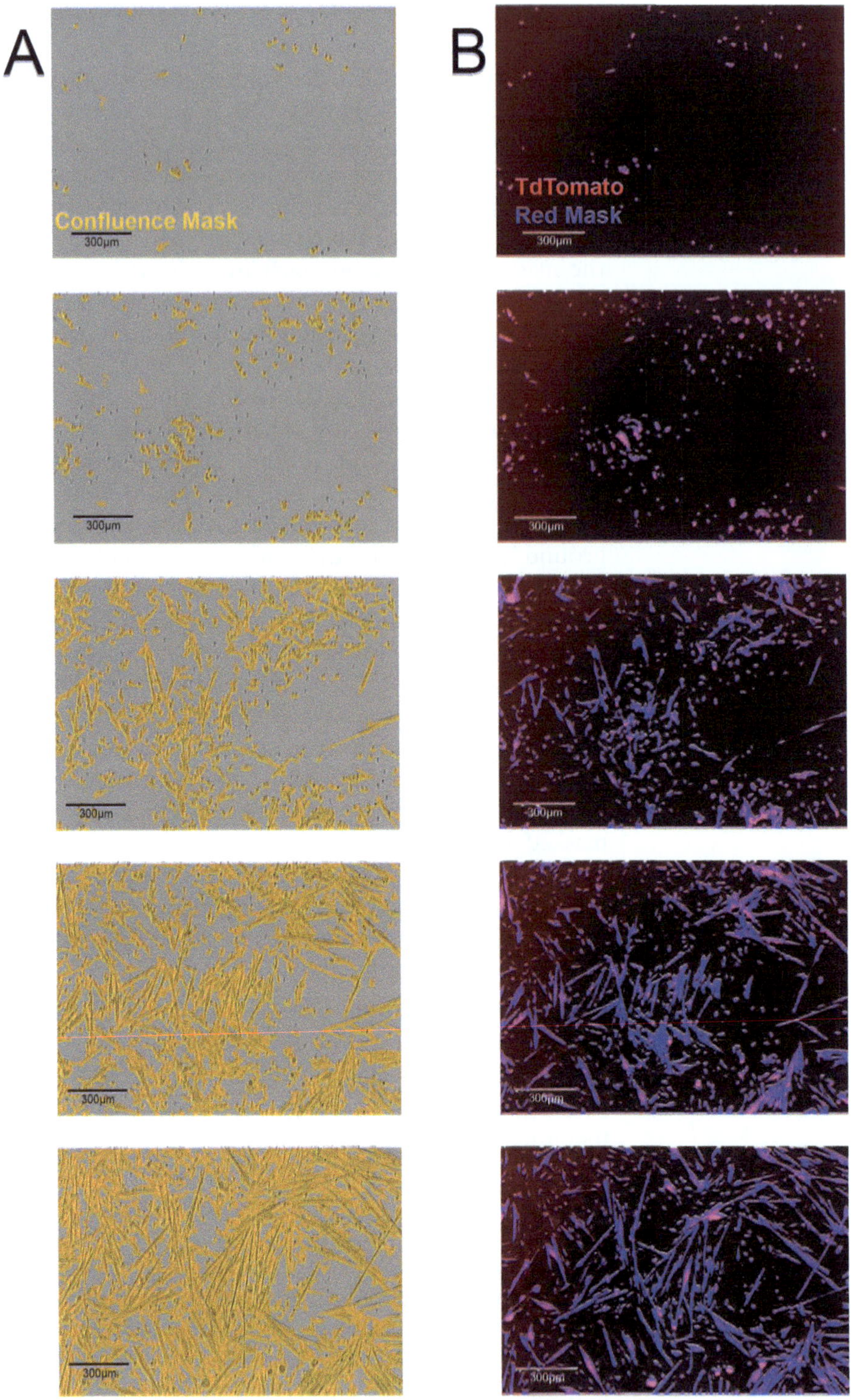

Fig. 2 Applying mask to differentiated cells. (**a**) Phase contrast images of muscle stem cells differentiating over the course of 5 days with the "Confluence Mask" overlaid (yellow). (**b**) Red fluorescence images of same cells with "Red Mask" overlaid (purple). The "Red Mask" is better for distinguishing individual myotubes than the "Confluence Mask" is, which makes downstream applications easier (Fig. 3). Day 1 on top to day 5 on bottom. All images were acquired on IncuCyte Zoom 10× objective, masks were applied in IncuCyte Zoom2016B software. Scale bars represent 300 μm

identical to the main window of your differentiation assay, except that there will be an added menu to display the "Analysis Masks" on the left-hand side of the screen (to check the fit of your mask *see* **Note 17**).

5. **Assess differentiation**. Begin analysis by navigating to the "Metrics" tab on the left-hand menu. You will see a set of metrics for each channel you analyzed and also for the scan (scan metrics can be ignored for the purposes of this protocol). You can measure many relevant parameters using the same basic protocol. The analysis protocol is listed in steps here, and an explanation of useful parameters can be found in **Note 18**. *See* Fig. 3 for myotube quantification using IncuCyte Zoom.
   (a) Click the + next to "Red Metrics" and select the parameter of interest from the list. Click "Graph/Export," which will open a new window.
   (b) Set the time range you are interested in (likely the whole duration of the experiment, from the first scan to the last). In the "Region" menu, select custom region, then select the wells you would like to assess. In the "Group" menu, select none to ensure that data is not autoaveraged by the IncuCyte Software (*see* **Note 19** for previewing instructions).
   (c) Click "Data Export," which opens a window. Select the layout you prefer for your exported data. Under the "Destination" header, select "Clipboard." Check the box next to "include experiment details in header," and click "Export." After clicking this you may open Microsoft Excel (or other data processing software) and paste the data. The data will paste as a table formatted by scan time and well number.
   (d) Plot data as a time course or use only the endpoint data to make comparisons in differentiation capacity of muscle stem cells.

4 Notes

1. To coat cell culture plates with gelatin: in a biosafety hood, cover the bottom of the cell culture surface with 0.1% gelatin. Allow to sit for 10–30 min at room temperature. Pour off gelatin (save gelatin, as it can be reused) and allow plates to dry until there is no moisture left on plate. Store coated plates in cool dry place. Plates can be made in advance. Gelatin and collagen-coated plates both will work well for the purposes of this protocol and have been used interchangeably. Gelatin is recommended because it is less expensive.

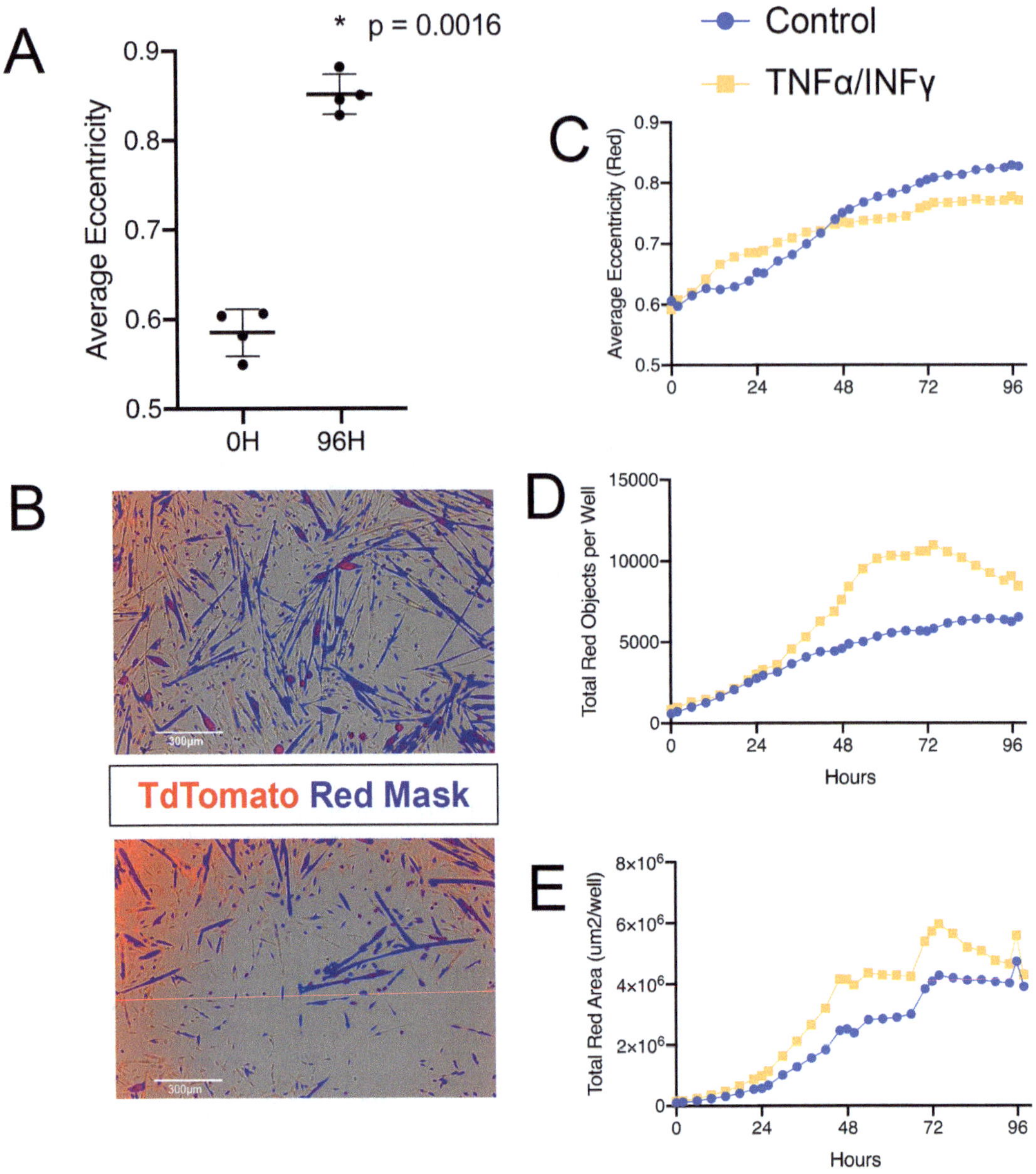

Fig. 3 Assessing differentiation using eccentricity and other parameters. (**a**) Average eccentricity of cells at 0 h and 96 h of differentiation from 4 independent experiments. Data points each represent one biological replicate that is the average of 4–8 technical replicates. These data show that undifferentiated cultures are easily distinguished from differentiated cultures using this method. $p = 0.0016$ by student's t test. (**b**) Phase contrast images overlaid with red fluorescence channel and "Red Mask" from day 5. Top: control differentiation conditions. Bottom: cells cultured in the presence of 40 ng/mL TNFα and 40 ng/mL IFNγ to impair differentiation [23, 24]. All images were acquired on IncuCyte Zoom 10× objective. Scale bars represent 300 μm. (**c–e**) Comparison of well-differentiated (control) and poorly differentiated (TNFα/IFNγ) muscle stem cells using different metrics measured in the IncuCyte Zoom2016B software. One representative biological replicate is shown. Data points are the average of 8 technical replicates. (**c**) Average eccentricity shows that poorly differentiated cultures had lower average eccentricity after 4 days of differentiation. (**d**) Total red

2. Fisherbrand cell strainers are recommended because other brands of cell strainers have been found to clog easier and prevent muscle stem cells from flowing through, leaving yield very low.
3. The protocol and data presented in this chapter were generated using IncuCyte Zoom software 2016B; however, more recent versions of the software (2018A, most recent at time of publication) are essentially the same and can be used with this protocol.
4. Muscles may be collected from one or both hindlimbs, depending on the yield of MuSCs desired. If collecting from both hindlimbs, pool tissues together for all downstream steps.
5. Typical 15 mL conical tube may also be used, although digestion may be hindered by tissue accumulating and getting stuck in the conical end of the tube.
6. The supernatant at this step will be cloudy and the pellet will be quite large, but loose. Be careful not to aspirate the pellet, leaving a small amount of supernatant on pellet is preferred over aspirating pellet.
7. The number of columns used depends on the amount of tissue isolated and the desired purity of the end isolate. For optimal purity, use 1 column per limb of tissue collected. If optimal purity is not needed, you may use 1 column per 2 limbs of tissue collected.
8. After isolation, cells take 2–3 days to adhere and begin proliferation. It is advisable to check cells once per day, but do not be alarmed if there are few (or no) visible colonies of cells prior to 3 days post isolation. Cells will grow in clonal clusters, but may begin spontaneous differentiation if clusters become larger than 8–16 cells. It is important to split cells before they reach this point.
9. Cells can be lifted off of plate or split using 1 of 3 methods:
    (a) TrypLE reagent for approximately 4 min (highest purity and viability).
    (b) Trypsin EDTA reagent for approximately 4 min.

Fig. 3 (continued) objects per well shows an increase in both conditions as noncommitted cells continue to proliferate. Poorly differentiated cultures may exhibit an increase in total red objects if there are defects in fusing to form myotubes. (**e**) Total red area per well (μm^2) is the most consistent parameter across cultures of varying differentiation capacity as this parameter does not distinguish between differentiated/undifferentiated and fused/unfused cells

(c) Mitotic shake: while cells are in GM gently tap sides of plate to loosen any cells that may be mid division. Remove the medium and add to a new plate. This medium can be supplemented with fresh medium.

10. If fibroblast colonies start to appear on the culture dish, mitotic shake to split satellite cells and then preplate cells for 1 h. Preplating will remove fibroblasts. Preplating can be done by adding the cell suspension from the mitotic shake to an UNCOATED plate and leaving the plate at room temperature (in hood) for 1 h. After 1 h, remove the medium (and muscle stem cells) and transfer to a coated plate and return to incubator.

11. Models of IncuCyte newer than Zoom will also work for the methods described here, keeping in mind that setting up the scan and analysis will be slightly different. The general methodology holds true across models of IncuCyte.

12. One useful property to add is a Plate Map. In the properties tab, click the small arrow next to "Plate Map" and click "Add." A new window appears displaying the wells in your plate. Click "Growth Condition" from the left-side menu, click "New," and in the popup menu, and your desired information. To assign this condition to a well, select the growth condition you have added, highlight the wells in the plate you would like to assign to that condition, and click "Add Growth Condition to Plate." Repeat as needed. Click the save button on the top menu to save for reuse or click "OK" to apply to scan.

13. For muscle stem cell differentiation, scan frequency of every 4–8 h is sufficient. Less frequent scan frequency can also be used, keeping in mind that you will have fewer intermediate data points.

14. If you are repeating an experiment, you may use a previous image set, or add images on to an existing image set. Keep in mind that if you add images to an existing collection, you will want to form a new processing definition (*see* Subheading 3.3, **step 2** and **Note 15**).

15. You can create multiple processing definitions from the same image collection. This may be advantageous if you want to test different processing definitions in their accuracy to established methods.

16. Sometimes the "Segmentation Adjustment" slider will not create a perfect mask. You can use other tools in the menu to make smaller adjustments to the mask. Of particular use is the area filter, which can be used to filter small debris or large clumps of floating cells out of the mask. For fluorescent channels with high background the top-hat parameter is useful for

eliminating background. The adaptive parameter is used to set a fluorescent threshold to eliminate objects with low fluorescence. The area filter may also be applied to the fluorescent channels, usually to eliminate autofluorescent debris. It is important to remember that no mask is perfect in all scenarios, you should strive for a mask that works best in most images. In the context of differentiating primary muscle cells, it is important that the mask works well on undifferentiated cells (day 1) and differentiated cells (day 6 or 7). Ensuring this will allow you to assess the degree of differentiation under conditions that may impair or accelerate complete differentiation.

17. Once you have a processing definition that works well for your studies and encompasses all stages of differentiation, you may apply it to other experiments. If you already have a processing definition you may create an "open-ended" analysis job that will work as your plate scans. If you do this, it is best to double check that the processing definition is compatible with the new experiment. After the experiment has completed and the analysis job has applied to all scans, open the Analysis Job window and on the left-hand side menu, "Analysis Masks," check the box next to the mask you are interested in. The mask will appear on the image you are viewing and you can check the fit of the mask on all your images from different areas of the plate, different time points, and different conditions. If you feel that the mask does not fit properly, return to Subheading 3.3, **step 2** and create a new processing definition.
18. Useful parameters.
    (a) **Object Confluence (Percent)—Phase**: The percentage of the image area occupied by objects masked as phase objects. This parameter is useful to determine the overall health of your culture (i.e., whether the cells between all conditions proliferate to the same degree). If different conditions have vast differences in percent confluence (phase), we recommend looking for fibroblast contamination (*see* **Note 10**) or assessing apoptosis rates under your experimental conditions. This is also an effective way to observe inconsistencies in your initial plating.
    (b) **Object Confluence (Percent)—Red**: The percentage of the image area occupied by objects masked as red objects. In the experiments outlined in this chapter, this would be an indication of cumulative muscle stem cell proliferation and differentiation. This can help you make the distinction between fibroblast contamination and differing rates of apoptosis/proliferation as mentioned in **Note 16(a)**. This is also an effective way to observe inconsistencies in your initial plating.

(c) **Average Object Eccentricity—Red**: The average of how round or compact the objects are, which ranges from 0 (perfect circle) to 1 (perfect line). Average eccentricity is the most useful parameter in defining the degree of differentiation in your experiment. As myotubes differentiate and elongate they will transition from eccentricity scores near 0 to scores closer to 1 (although do not expect that myotubes reach a score of 1).

(d) **Object Count (per well)—Red**: The number of red-masked objects per well. This is another way to assess proliferation and plating differences between wells and conditions. The red object count per well should not change significantly from day 1 to day 7 of differentiation (Fig. 3).

(e) **Total Object Area (μm^2/well)—Red:** The total area of red-masked objects per well. This is another way to assess proliferation and plating differences between wells and conditions. The red object area per well should may change slightly from day 1 to day 7 of differentiation if myotubes become wider as they differentiate (Fig. 3).

19. You can preview the analysis metrics by two options. (1) After setting your time range and wells, click "Microplate Graph" in the lower right corner. This will open a window where a graph of the selected metric is displayed for each well separately. This option is particularly useful for determining if you have outlier wells (e.g., contamination or acute toxicity in one well) and to get a general idea of differentiation in different conditions. (2) After setting your time range and wells, in the "Group" menu, select "by condition" (this will only work if you created a plate map, *see* **Note 12**). Click "Graph" in the lower right corner. This will open a new window showing a graph representation of the selected metric where the average for each condition is plotted. This option is good for getting a general idea of how your assay worked, but does not provide information for individual replicates.

References

1. Murphy MM, Lawson JA, Mathew SJ, Hutcheson DA, Kardon G (2011) Satellite cells, connective tissue fibroblasts and their interactions are crucial for muscle regeneration. Development 138(17):3625–3637. https://doi.org/10.1242/dev.064162
2. Yin H, Price F, Rudnicki MA (2013) Satellite cells and the muscle stem cell niche. Physiol Rev 93(1):23–67
3. Montarras D, L'honoré A, Buckingham M (2013) Lying low but ready for action: the quiescent muscle satellite cell. FEBS J. https://doi.org/10.1111/febs.12372
4. Tidball JG (2017) Regulation of muscle growth and regeneration by the immune system. Nat Rev Immunol 17(3):165–178. https://doi.org/10.1038/nri.2016.150

5. Kästner S, Elias MC, Rivera AJ, Yablonka-Reuveni Z (2000) Gene expression patterns of the fibroblast growth factors and their receptors during myogenesis of rat satellite cells. J Histochem Cytochem 48(8):1079–1096
6. Bischoff R (1990) Interaction between satellite cells and skeletal muscle fibers. Development 109(4):943–952
7. Pawlikowski B, Pulliam C, Betta ND, Kardon G, Olwin BB (2015) Pervasive satellite cell contribution to uninjured adult muscle fibers. Skelet Muscle 5:42. https://doi.org/10.1186/s13395-015-0067-1
8. Bosnakovski D, Xu Z, Li W, Thet S, Cleaver O, Perlingeiro RCR, Kyba M (2008) Prospective isolation of skeletal muscle stem cells with a Pax7 reporter. Stem Cells 26(12):3194–3204. https://doi.org/10.1634/stemcells.2007-1017
9. Yaffe D, Saxel O (1977) Serial passaging and differentiation of myogenic cells isolated from dystrophic mouse muscle. Nature 270(5639):725–727
10. Richler C, Yaffe D (1970) The in vitro cultivation and differentiation capacities of myogenic cell lines. Dev Biol 23(1):1–22
11. Linkhart TA, Clegg CH, Hauschka SD (1981) Myogenic differentiation in permanent clonal mouse myoblast cell lines: regulation by macromolecular growth factors in the culture medium. Dev Biol 86(1):19–30
12. Keire P, Shearer A, Shefer G, Yablonka-Reuveni Z (2013) Isolation and culture of skeletal muscle myofibers as a means to analyze satellite cells. Methods Mol Biol 946:431–468. https://doi.org/10.1007/978-1-62,703-128-8_28
13. Cornelison DD, Wold BJ (1997) Single-cell analysis of regulatory gene expression in quiescent and activated mouse skeletal muscle satellite cells. Dev Biol 191(2):270–283. https://doi.org/10.1006/dbio.1997.8721
14. Olguin HC, Olwin BB (2004) Pax-7 up-regulation inhibits myogenesis and cell cycle progression in satellite cells: a potential mechanism for self-renewal. Dev Biol 275(2):375–388. https://doi.org/10.1016/j.ydbio.2004.08.015
15. Montarras D, Morgan J, Collins C, Relaix F, Zaffran S, Cumano A, Partridge T, Buckingham M (2005) Direct isolation of satellite cells for skeletal muscle regeneration. Science 309(5743):2064–2067. https://doi.org/10.1126/science.1114758
16. Loro E, Rinaldi F, Malena A, Masiero E, Novelli G, Angelini C, Romeo V, Sandri M, Botta A, Vergani L (2010) Normal myogenesis and increased apoptosis in myotonic dystrophy type-1 muscle cells. Cell Death Diff 17(8):1315–1324
17. Yablonka-Reuveni Z, Anderson JE (2006) Satellite cells from dystrophic (mdx) mice display accelerated differentiation in primary cultures and in isolated myofibers. Dev Dynam 235(1):203–212
18. Hogan KA, Cho DS, Arneson PC, Samani A, Palines P, Yang Y, Doles JD (2017) Tumor-derived cytokines impair myogenesis and alter the skeletal muscle immune microenvironment. Cytokine. https://doi.org/10.1016/j.cyto.2017.11.006
19. Agley CC, Velloso CP, Lazarus NR, Harridge SD (2012) An image analysis method for the precise selection and quantitation of fluorescently labeled cellular constituents: application to the measurement of human muscle cells in culture. J Histochem Cytochem 60(6):428–438
20. Desgeorges T, Liot S, Lyon S, Bouviere J, Kemmel A, Trignol A, Rousseau D, Chapuis B, Gondin J, Mounier R (2019) Open-CSAM, a new tool for semi-automated analysis of myofiber cross-sectional area in regenerating adult skeletal muscle. Skeletal Muscle 9(1):2
21. Wen Y, Murach KA, Vechetti IJ Jr, Fry CS, Vickery C, Peterson CA, McCarthy JJ, Campbell KS (2018) MyoVision: software for automated high-content analysis of skeletal muscle immunohistochemistry. J Appl Physiol (1985) 124(1):40–51. https://doi.org/10.1152/japplphysiol.00762.2017
22. Murphy DP, Nicholson T, Jones SW, O'Leary MF (2019) MyoCount: a software tool for the automated quantification of myotube surface area and nuclear fusion index. Wellcome Open Res 4
23. Langen RC, Van Der Velden JL, Schols AM, Kelders MC, Wouters EF, Janssen-Heininger YM (2004) Tumor necrosis factor-alpha inhibits myogenic differentiation through MyoD protein destabilization. FASEB J 18(2):227–237
24. Miller S, Ito H, Blau H, Torti F (1988) Tumor necrosis factor inhibits human myogenesis in vitro. Mol Cell Biol 8(6):2295–2301

Part IV

Malignancy

Chapter 32

The Enrichment of Breast Cancer Stem Cells from MCF7 Breast Cancer Cell Line Using Spheroid Culture Technique

Anan A. Ishtiah and Badrul Hisham Yahaya

Abstract

Breast cancer is the most common malignancy worldwide in females, representing 29% of all cancer new cases and 14% of cancer deaths in the world. Amongst the reasons for the high mortality rate is resistance to chemotherapy resulting in therapeutic failure. Various studies have shown that the presence of cancer stem cells (CSCs) in breast tumors is responsible for chemotherapy resistance and tumor recurrence. This CSC population possesses the characteristics of normal stem cells, including their ability to self-renewal and give rise to other epithelial cells. One thing that unique to the CSC population is their ability to escape from chemotherapy drugs; this can make them resistant to therapy and able to repopulate the cancer. Isolation and enrichment of breast CSCs (BCSCs) is required in order to study their characteristics and the behavior that enables them to drive breast tumor development, in order to develop better therapies. This chapter describes a method for the isolation and enrichment of BCSCs from the MCF7 breast cancer cell line, which consists of a heterogeneous breast cancer cell population. This method depends on cancer stem cell behavior, specifically an ability to self-renew and form spheroids in harsh conditions that allow only cancer cells with stem cell characteristics to survive and form spheroids.

Key words MCF7, Breast cancer stem cells, Spheroid, 3D culture

1 Introduction

Cancer stem cell (CSCs) are a unique subpopulation of tumor cells first identified in breast cancer by Al-hajj et al. [1] based on cell surface marker expression of $CD44^{+}CD24^{low/-}$; these markers are now being used extensively to isolate and characterize breast CSCs. Following that report, many studies have used different approaches to identify breast CSCs [2–6]. CSCs possess similar characteristics to normal stem cells, including the ability to undergo self-renewal and to give rise to other heterogeneous population of tumor cells. CSCs are a potential target for therapy as they may be responsible for drug resistance, tumor recurrence, and metastases

Nagarajan Kannan and Philip Beer (eds.), *Stem Cell Assays: Methods and Protocols*, Methods in Molecular Biology, vol. 2429, https://doi.org/10.1007/978-1-0716-1979-7_32,

[7–11]. Although the use of cell surface markers to isolate and identify CSCs is relatively straightforward as compared to other methods, there are some limitations regarding the purity of the CSCs obtained. Since this method for the isolation of CSCs is based on the expression of certain cell surface markers, it will not detect CSCs that may be present in other cellular fractions. This argument has been proven in many studies where cells that are negative for putative CSC markers were also be able to form tumors when injected into immunocompromised recipients. These studies demonstrate the presence of CSCs in cell fractions that do not express traditional CSC markers. On the other hand, some cells that do express CSC cell surface markers may not behave as a CSC [12]. Therefore, in order to increase the likelihood of isolating pure CSCs, it's advisable to use multiple cell surface markers and/or a combination of markers [6]. Another limitation of using cell surface marker expression to isolate CSCs is the low number of BCSCs within a primary tumor, which represent only about 0.1–1% of tumor cells [13]. This means that isolating these cells from cultured cell lines, primary tumors or primary cultures required a large amount of starting cells.

To overcome these disadvantages of isolating and enriching CSCs, the spheroid technique is now accepted and proven as a common tool to isolate and enrich CSCs [4, 5, 14]. This assay depends on the ability of CSCs to self-renewal [15–17] in a non-supporting matrix environment [18] supplemented with serum-free culture medium [19]. This chapter describes a detailed method for BCSCs isolation and enrichment from Michigan Cancer Foundation-7 (MCF7) breast cancer cell line. The MCF7 cell line was derived from breast adenocarcinoma, and has characteristics of differentiated estrogen receptors (ER)-positive progesterone receptor (PR)-positive mammary epithelium. This method will enable the researcher to isolate and enrich BCSCs from either breast cancer cell lines and/or breast cancer primary cultures, and generate a large amount of BCSCs for downstream applications.

2 Materials

Make sure all the reagents and disposables are cell culture-grade and sterile.

2.1 Cell Line

MCF-7 breast cancer cell line.

2.2 Cell Culture Medium

1. MCF7 medium: supplement Dulbecco's Modified Eagle's Medium (DMEM) with 10% fetal bovine serum (FBS), and 1% penicillin–streptomycin mix.

2. Spheroid complete medium: DMEM-F12 with 1% penicillin–streptomycin mix, 1× N21-MAX medium supplement, 10 ng/mL basic fibroblast growth factor (bFGF) and 20 ng/mL epidermal growth factor (EGF).

   Note: add growth factors (bFGF and EGF) to the working medium only (*see* **Note 1**).

2.3 Spheroid Culture Medium and Plates

1. 1.2% Poly (2-hydroxyethyl methacrylate) (pHEMA): prepare 1.2% pHEMA by adding 12 g of pHEMA to 1 L of 95% ethanol and stirring constantly on a heated plate. Once it dissolved completely, cool it down to room temperature, and pass it through 0.2 μm filter, and store at room temperature.
2. Ultralow-attachment 6-well plates and T25 flasks: pipet 100 μL/cm^2 of 1.2% pHEMA solution into a 6-well plate (1 mL/well) and T25 flasks (2.5 mL/T25 flask). Place in an oven at 40 °C and leave for 48 h (*see* **Note 2**).

3 Methods

3.1 MCF7 Culture

1. Culture MCF7 breast cancer cell line in T25 cell culture flask in MCF7 medium (Subheading 2.2, **item 1**).
2. Incubate the cells at 37 °C supplied in humidified incubator with 5% carbon dioxide.
3. Change medium on the next day, and then every 2 days until the cells reach 70–80% confluence (*see* **Note 3**).

3.2 MCF7 Cells Harvest and Preparation of Single-Cell Suspension

1. Discard the culture medium and add 4 mL of prewarmed 1× PBS; cover the monolayer with PBS and swirl the flask gently to wash out any remaining serum; discard PBS.
2. Add prewarmed trypsin–EDTA (0.05%) and incubate at 37 °C for 4 min.
3. Tap the edges of the flask gently for several times, and check under microscope for cell detachment.
4. Pipet the cells for several time using a 5 mL serological pipette to reduce cells clumping.
5. Add 4 mL of MCF7 medium (Subheading 2.2, **item 1**) to neutralize trypsin.
6. Transfer all the contents of T25 flask to 15 mL falcon tube, and centrifuge at 300 × *g* for 5 min.
7. Discard supernatant and add 2 mL of serum-free DMEM medium to the pelleted cells, then mix gently by pipetting.
8. Pass the cells suspension through 0.40 mm cell strainer to get a single cell suspension.

9. Take 50 μL of the single-cell suspension add 50 μL Trypan blue and perform viable cells count using hemocytometer. Viable cells appear bright under light microscope, while nonviable (dead) cells appear dark. Determine the cells concentration/mL (*see* **Note 4**).

3.3 Preparing Single-Cell Suspension of 500 Cell/mL Concentration and Culture

1. Prepare a single cell suspension of 500 cell/mL concentration by diluting the main cell suspension with spheroid medium (Subheading 2.3, **item 1**).
2. Do not prepare the 500 cell/mL cell suspension in a single step, instead make several dilutions (*see* **Note 5**).
3. Transfer 2 mL (1000 cells) of 500 cell/mL suspension to each well of ultralow-attachment 6-wells plate (1000 cells/well).
4. Incubate at 37 °C, 5% CO_2 incubator for 3 days. Take care not to move the plates, in order to prevent cell aggregation.

3.4 Spheroid Incubation Time, Medium Refresh, and Counting

1. On day 3, count the number of spheroids, measure their size, and add 1 mL spheroid complete medium to each well with slight mixing by pipetting. MCF7 spheroids originating from single cells are smooth, perfectly rounded spheres without projections (Fig. 1a), while cells aggregates appear as rough nonrounded like structures (Fig. 1b).
2. Repeat **step 1** every 3 days; on day 10 most of the spheroids will be >50 μm in size (Fig. 2).
3. Count the spheroids on Day 10, count only spheroids of greater than 50 μm.
4. Do not count any viable cells aggregate; if all cells form aggregates (Fig. 3), discard the plate and repeat the experiment.
5. In case of few cells aggregate, sometimes a CSC will be present within the aggregate and it will form a spheroid (Fig. 4), while other non-CSCs within the aggregate will die; if observed, count spheroids originating from aggregates of dead cells if (Fig. 4) (*see* **Notes 6–9**).

3.5 Spheroid Dissociation, and Passaging

1. On day 10 mix each well by pipetting and transfer the content of all wells to a 15 mL Falcon tube.
2. Add 1 mL serum-free DMEM medium to each well, wash the well thoroughly, and transfer the contents to same 15 mL Falcon tube as in **step 1**.
3. Leave the 15 mL Falcon tubes standing at room temperature for 20 min; this will allow the spheroids to settle to the bottom, while cell debris will remain in suspension.
4. Discard the supernatant carefully by pipetting using a 10 mL serological pipette.

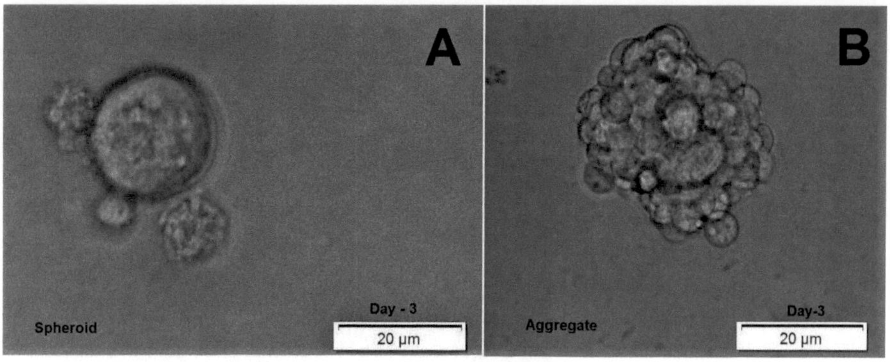

Fig. 1 MCF7 spheroid development, Day-3. (**a**) Typical spheroid structure and shape. (**b**) Cells aggregated

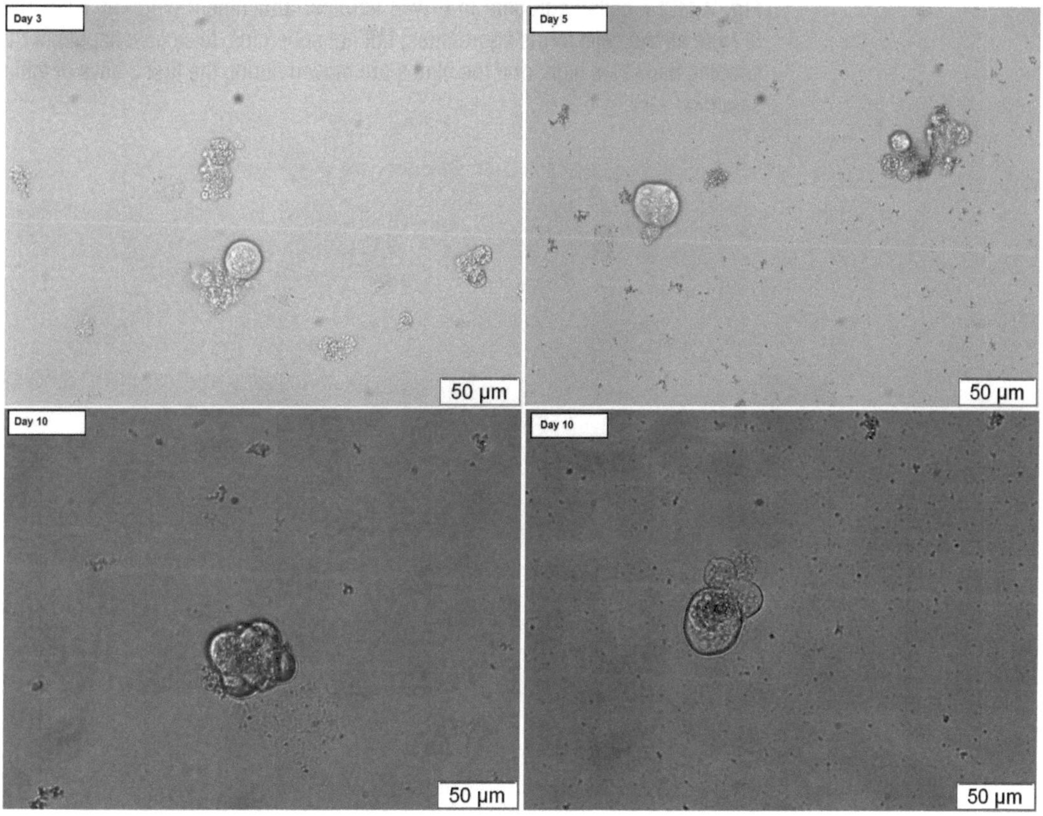

Fig. 2 MCF7 spheroid development, in 10 days, in 6-wells ultralow-attachment plate

5. Add 0.5–1 mL of 0.25% trypsin (according to pellet size), mix by pipetting, and incubate for 5 min at 37 °C.
6. Mix by pipetting several times (10–20 times).
7. Pass the solution sequentially through 21, 23, and 25 gauge needles 5–6 times per needle.

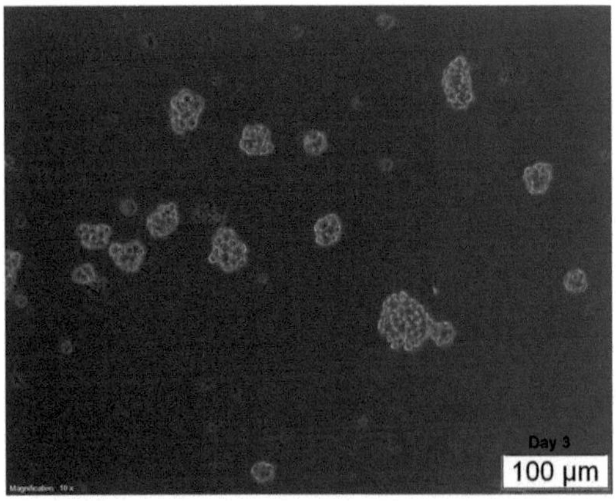

Fig. 3 MCF7 cells aggregate in 6-well ultralow-attachment plate, at 10×, Day 3. Note all the cells forms aggregates, but not spheroids, this could happen if the seeding density is high, or if the plates are moved during the first 3 days of initial seeding

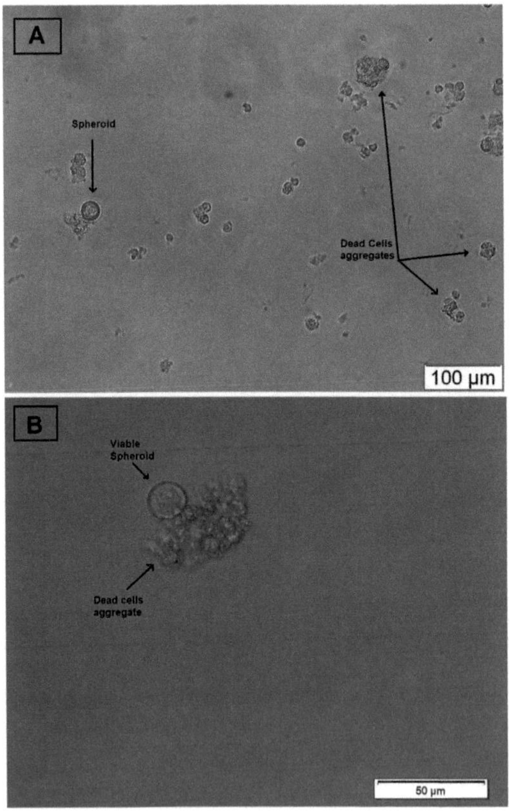

Fig. 4 MCF7 spheroids development, initial seeding after 3 days in culture. (**a**) MCF7 spheroid come out from a lift over dead cells aggregate, (**b**) Died cells, died cells aggregates, and viable spheroids

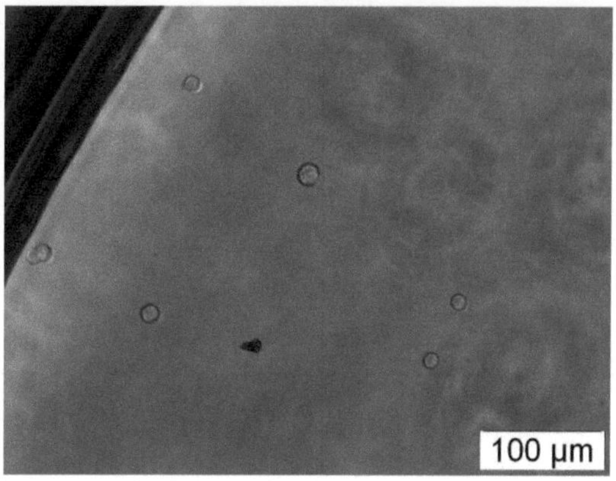

Fig. 5 MCF7 spheroids(20uM), Day-3, Passage-3, in ultralow-attachment 6-well plate, most of the cells (70–80%) are able to form spheroids in passage-3, compared to only (1–2%) in initial seeding, this confirms an enrichment of MCF7 cells with spheroids developments abilities (BCSCs). Note there is no dead cells compared to initial seeding in Fig. 4a

8. Add 0.5–1 mL MCF7 medium to neutralize trypsin, centrifuge, discard the supernatant, and add 1 mL serum-free DMEM.
9. Take 50 μL of the dissociated spheroid suspension, add 50 μL Trypan blue, and check for viable cells count and single cell formation (*see* **Note 10**).
10. According to the cells count, add spheroid complete medium (Subheading 2.2, **item 2**) and seed the first passage of dissociated cells in a new ultralow-attachment 6-well plate, making sure that each well contains 2 mL of spheroid medium containing a maximum of 1000 cells per well.
11. Repeat all steps of Subheadings 3.4 and 3.5 for second and third passages, each passage with 10 days, seeding all the dissociated cells each time (*see* **Note 11**).
12. Finally, count the spheroids at day 10, passage 3; this is the final step, and spheroid efficiency is expected to be as high as 80% (Fig. 5).

3.6 Spheroid Efficiency Calculation

1. At each passage, count the number of viable cells before seeding and then count the number of developed spheroids at day 10.
2. The spheroid efficiency ratio (%) = (number of spheroids/cell count) × 100 (*see* **Note 12**).

3.7 Storing Enriched BCSCs

1. After passage 3, count the viable dissociated single cells from passage 3 spheroids.

2. Prepare a cell suspension of 0.5–1 × 10^6 cell/ml in spheroid complete medium (Subheading 2.2, **item 2**).
3. Add DMSO (10% of the total cells suspension volume) 100 μL DMSO with 900 μL cells suspension.
4. Transfer 1 mL of DMSO-Cell suspension to 1.5 mL cryogenic tubes.
5. Store at −80 °C or liquid nitrogen.

4 Notes

1. For spheroid culture medium, it is recommended to add the growth factors (bFGF and EGF) to the working medium only because these factors have a short stability time after dilution in culture medium.
2. Make sure that pHEMA solution is covering all the surface area of each well or the flask by swirling the plate or the flask thoroughly.
3. Make sure the MCF7 monolayer cells are in the log phase of growth before harvest; this is accomplished by changing the medium every 48 h and harvesting before 80% confluence.
4. For manual viable cell count using hematocytometer: take 50 μL of the cell suspension, and mix it with 50 μL trypan blue, load the chamber and count 4 big squares (the 4 corner squares). Then multiply the cell count with the dilution factor (which is 2) and with the volume correction factor (which is 2.5); thus the formula for viable cells count will be (counted cells × 2 × 2.5), and the result will be viable cells per μL.
5. For preparing a 500 cell/mL cell suspension from the harvested cells, perform a serial dilution procedure, as this will reduce the level of error. The serial dilution should be performed using spheroid complete medium (Subheading 2.2, **item 2**).
6. During spheroid culture, you may see some cell aggregation (not spheroids). But those aggregates will start dying at a later time point (Fig. 4a), and if they contain a CSC, you will notice a spheroid growing out from the dead cells aggregates on later days (after day 3) (see Fig. 4a, b).
7. If many cells aggregates developed, then repeat the experiment, and make sure not to move or shake the spheroid culture plates during the first 3 days of culture (Fig. 3).
8. During spheroid culture, some spheroids will die after day 3 or day 6; these are spheroids originating from progenitor cancer cells, which have a limited renewal capacity.

9. On day 10, spheroids of <50 μm are not considered to have originated from a BCSC.
10. If you still can see cell duplicates or triplicates after spheroid dissociation, then pass the cells through 0.4 μm strainer
11. If the dissociated spheroid cells count is high, you can use more than one ultralow-attachment 6-well plate, or you can use a T25 ultra low attachment flask. For T25 ultralow-attachment flasks, seed a maximum of 6000 cell per flask in 4 mL complete spheroid medium, and every 3 days add 2 mL complete spheroid medium.
12. Spheroid efficiency should increase after each passage, until you reach a high spheroid efficiency rate at passage 3; first spheroid culture will only yield 2–4% spheroids, while at passage 3 most of the seeded cells will develop spheroids, with an efficiency of 70–80% (Fig. 5).

References

1. Al-Hajj M, Wicha MS, Benito-Hernandez A, Morrison SJ, Clarke MF (2003) Prospective identification of tumorigenic breast cancer cells. Proc Natl Acad Sci U S A 100(7): 3983–3988
2. Ghanei Z, Jamshidizad A, Joupari MD, Shamsara M (2020) Isolation and characterization of breast cancer stem cell-like phenotype by Oct4 promoter-mediumted activity. J Cell Physiol. [cited 2020 Apr 8]. http://www.ncbi.nlm.nih.gov/pubmed/31904128
3. Wang H, Zhang Y, Du Y (2013) Ovarian and breast cancer spheres are similar in transcriptomic features and sensitive to fenretinide. Biomed Res Int 2013:510905
4. Saadin K, White IM (2013) Breast cancer stem cell enrichment and isolation by mammosphere culture and its potential diagnostic applications. Expert Rev Mol Diagn 13:49–60
5. Harrison H, Shaw FL, Harrison H, Spence K, Ablett MP, Simões BM et al (2012) A detailed mammosphere assay protocol for the quantification of breast stem cell activity. J Mammary Gland Biol Neoplasia 17(2):111–117
6. Sin WC, Lim CL (2017) Breast cancer stem cells—from origins to targeted therapy. Stem Cell Investig 6
7. Velasco-Velázquez MA, Popov VM, Lisanti MP, Pestell RG (2011) The role of breast cancer stem cells in metastasis and therapeutic implications. Am J Pathol Am Soc Investig Pathol 179:2–11
8. Geng SQ, Alexandrou AT, Li JJ (2014) Breast cancer stem cells: multiple capacities in tumor metastasis. Cancer Lett 349:1–7. [cited 2020 Apr 8]. http://www.ncbi.nlm.nih.gov/pubmed/24727284
9. Fico F, Bousquenaud M, Rüegg C, Santamaria-Martínez A (2019) Breast cancer stem cells with tumor- versus metastasis-initiating capacities are modulated by TGFBR1 inhibition. Stem Cell Rep 13(1):1–9
10. Ayob AZ, Ramasamy TS (2018) Cancer stem cells as key drivers of tumour progression. J Biomed Sci 25
11. Park SY, Choi JH, Nam JS (2019) Targeting cancer stem cells in triple-negative breast cancer. Cancers 11
12. Pastrana E (2013) Eyes wide open: a critical review of sphere-formation as an assay for stem cells. Cell Stem Cell 8(5):486–498
13. Bao B, Ahmad A, Azmi AS, Ali S, Sarkar FH (2013) Overview of cancer stem cells (CSCs) and mechanisms of their regulation: implications for cancer therapy. Curr Protoc Pharmacol 61(1):14.25.1–14.25.14. [cited 2020 May 2]
14. Zhang J, Zhang Y, Cheng L, Li C, Dai L, Zhang H et al (2018) Enrichment and characterization of cancer stem-like cells in ultra-low concentration of serum and non-adhesive culture system. Am J Transl Res 10(5): 1552–1561
15. Lee CH, Yu CC, Wang BY, Chang WW (2016) Tumorsphere as an effective in vitro platform for screening anticancer stem cell drugs. Oncotarget 7(2):1215–1226

16. Herheliuk T, Perepelytsina O, Ugnivenko A, Ostapchenko L, Sydorenko M (2019) Investigation of multicellular tumor spheroids enriched for a cancer stem cell phenotype. Stem Cell Investig 6
17. Ishiguro T, Ohata H, Sato A, Yamawaki K, Enomoto T, Okamoto K (2017) Tumor-derived spheroids: relevance to cancer stem cells and clinical applications. Cancer Sci 108: 283–289
18. Zhong Y, Guan K, Guo S, Zhou C, Wang D, Ma W et al (2010) Spheres derived from the human SK-RC-42 renal cell carcinoma cell line are enriched in cancer stem cells. Cancer Lett 299(2):150–160
19. Lee JW (2018) Isolation of spheroid-forming single cells from gastric cancer cell lines: enrichment of cancer stem-like cells. BioTechniques 65(4):197–203

Chapter 33

Assessment of Breast Cancer Stem Cell Activity Using a Spheroid Formation Assay

Ajeya Nandi and Rumela Chakrabarti

Abstract

Evidence is emerging that cancer cells are arranged as a hierarchy that spans from stem cells to lineage-restricted progenitor cells. The recent development of spheroid cultures with several tissue type has provided new opportunities to assess cancer stem cell (CSC) activity by allowing them to propagate under conditions that resemble the microenvironment for growth of tumors. One tissue type widely used for stem cell investigations is mammary tissue, and the sphere formation assay has been used in both normal mammary tissue and in breast cancer. Here, we describe detailed experimental methodology for generating and propagating spheres from normal mammary tissue and primary breast tumors of mice, patient derived xenografts (PDXs) and breast cancer cell lines. We further describe how these sphere cultures can be employed for coculture assays to assess the effect of tumor microenvironment (TME) on self-renewal ability of CSCs in breast cancer.

Key words Cancer stem cells, Spheroid formation assay, Breast cancer, Tumorsphere, Matrigel, Patient derived xenografts

1 Introduction

The development of innovative ex vivo culture approaches has enabled the study of putative cancer stem cells (CSCs), a subpopulation of cancer cells, which have garnered substantial interest in the scientific community over the past few years. Prevailing CSC hypothesis posits that cancer cells are hierarchically organized and that CSCs with the characteristics of self-renewal, multipotent differentiation and unlimited propagation are responsible for tumor initiation, progression, resistance to clinical therapies and subsequent tumor recurrence [1–5]. However, the existence of CSCs in solid tumors remains elusive due to a lack of cell-surface markers and the plasticity of CSC phenotypes. As a consequence, the identification of CSCs relies on functional assays which can

Nagarajan Kannan and Philip Beer (eds.), *Stem Cell Assays: Methods and Protocols*, Methods in Molecular Biology, vol. 2429, https://doi.org/10.1007/978-1-0716-1979-7_33,

determine the capability of CSCs to generate tumors through self-renewal and differentiation. The existence of CSCs has been demonstrated in variety of tumors including hematopoietic cancer [6], breast cancer [7], colon cancer [8], and glioma [9]. These findings prompted researchers to establish advanced ex vivo methods to propagate CSCs from solid tumors. The mammosphere assay was developed as a method to propagate mammary epithelial stem cells (MaSCs) in nonadherent and undifferentiated culture conditions by Dontu and colleagues [10], as a modification of the neurosphere assay developed by Reynolds et al. [9]. Ponti and coworkers showed that this method was also efficient in maintaining breast CSCs from breast cancer cell lines [11]. Therefore, in breast cancer biology, this assay has been considered as a convenient surrogate to evaluate the functionality of CSCs from breast cancer cell lines or primary breast tumors [12–15]. In addition, Matrigel matrix, one of the most popular extracellular matrix (ECM) components used for spheroid assay, is enriched in basement membrane proteins and provides essential ligands needed for adhesion during cancer cell invasion [16, 17]. Therefore, the Matrigel matrix sphere culture system provides a useful method for simulating the microenvironment of CSCs in vitro to study cell morphology and enable high throughput screening for the study of growth kinetics and the development of drug screening [18–22]. In addition, coculture assay finds myriad applications in breast cancer for studying complex interactions between breast epithelium and stromal cells which can help to elucidate the role of stromal variation in tumor phenotypes [23]. Here, we report the detailed methodology used in our laboratory for generating and propagating mammospheres and tumorspheres in cell suspension and Matrigel matrix culture system from murine mammary tissue, breast tumors, patient-derived xenograft (PDX) tumors, and breast cancer cell lines. We then briefly highlight methodology for coculture assay of tumor cells with stromal cells such as fibroblasts. We also discuss how to analyze and interpret these data.

2 Materials

1. Mammary epithelial cell growth medium (MEGM): Dulbecco's Modified Eagle Medium (DMEM)–F12, 1 mM glutamine, 5 μg/mL insulin, 500 ng/mL hydrocortisone, 10 ng/mL epidermal growth factor (EGF), 1 mg/mL cholera toxin, 5% bovine calf serum. Filter and store at 4 °C for a month.
2. Mammary gland dissociation medium (MGDM) (MGDM is the same medium as MEGM with addition of collagenase and hyaluronidase): Dulbecco's Modified Eagle Medium (DMEM)–F12, 1 mM glutamine, 5 μg/mL insulin, 500 ng/

mL hydrocortisone, 10 ng/mL EGF, 1 mg/mL cholera toxin, 5% bovine calf serum, 300 U/mL collagenase (type 1A), 100 U/mL hyaluronidase, 1× pen/strep. MGDM needs to be filtered and sterilized through 0.45 μm and 0.22 μm filters sequentially. It should always be freshly prepared on the day of experiment.

3. Dispase–DNase I solution: 200 mg Dispase and 4 mg DNase I dissolved in 40 mL 1× DPBS. It needs to be filtered and sterilized through 0.45 μm. It can be stored at 4 °C up to 1 month without loss of activity.
4. Red blood cell lysis buffer (0.64% NH_4Cl): 6.4 g NH_4Cl, dissolved in 1 L sterile water. It needs to be filtered and sterilized through 0.22 μm. Store at 4 °C.
5. FCS medium: 5% newborn calf serum in 1× DPBS with pen/-strep. Filter and store at 4 °C.
6. Mammosphere/tumorsphere medium: Dulbecco's Modified Eagle Medium (DMEM)/ F12, 1× B27 supplements, 20 ng/mL EGF, 20 ng/mL basic fibroblast growth factor (FGF). It should always be freshly prepared on the day of experiment (*see* **Note 1**).
7. HBEC medium: Dulbecco's Modified Eagle Medium (DMEM) –F12, 10 mM HEPES, 1 μg/mL insulin, 500 ng/mL hydrocortisone, 1 mg/mL bovine serum albumin (BSA), 5% FCS, 2.5 μg/mL fungizone, 50 μg/mL gentamicin. Filter and store at 4 °C for a month.
8. Patient-Derived Xenograft (PDX)—HBEC digestion medium: Dulbecco's Modified Eagle Medium (DMEM)-F12, 10 mM HEPES, 10% FBS (fetal bovine serum), 1 μg/mL insulin, 50 μg/mL gentamicin, 10 mg/mL collagenase (type 3 or IV), 100 U/mL hyaluronidase, 10 ng/mL EGF. It needs to be filtered and sterilized through 0.45 μm and 0.22 μm filters sequentially. It should always be freshly prepared on the day of experiment.

3 Methods

The protocols below describe how to generate spheres from normal mammary tissue, primary murine breast tumors and patient derived xenograft (PDX) tumors as well as breast cancer cell lines (*see* Fig. 1). All following steps should be carried out in a cell culture laboratory under class II biosafety cabinet to maintain sterility of the sample.

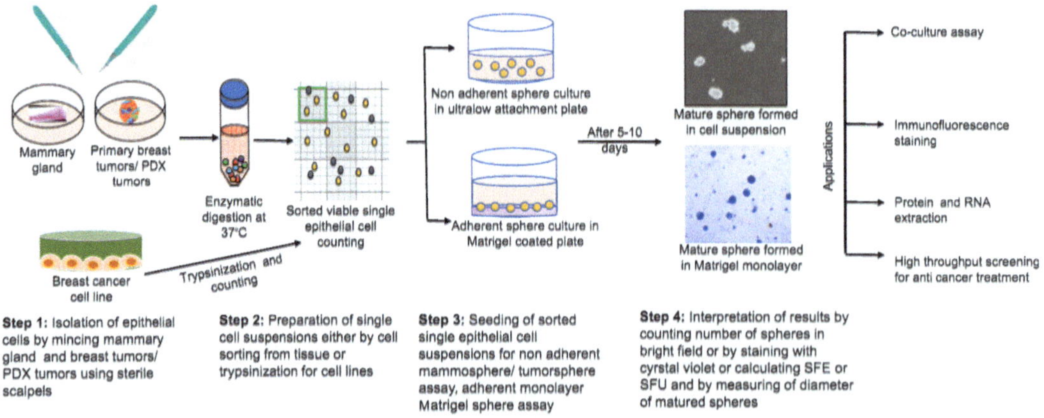

Fig. 1 A schematic representation of sphere formation assay. Samples from different sources are enzymatically digested, then validated as viable, single Lin$^-$ epithelial tumor cell suspensions after cell sorting, and plated either on nonadherent plates in serum-free sphere medium with growth factors for cell suspension sphere culture or on growth factor reduced Matrigel-coated plates in serum-free sphere medium for adherent sphere culture assay

3.1 Isolation of Epithelial Cells from Murine Mammary Tissue, Primary Breast Tumors and PDX Tumors

1. Harvest mammary tissues and tumors in sterile environment. Mechanically separate tissue or tumor from connective and fat tissues using sterile scalpels and forceps and transfer into DMEM-F12 (1:1) medium. For PDX tumors, use HBEC medium. Subsequently, mince the tissues into small fine pieces of around 0.3–0.5 mm in size with two scalpels inside the petri dish under sterile conditions until uniform in appearance.
2. Transfer the minced tissue to a 50 mL falcon tube containing 10–15 mL digestion buffer MGDM (for murine mammary tissue and breast tumors) or PDX- HBEC medium (for PDX tumors). The volume of digestion medium should depend on size of the tissue. For 2 pairs of mammary glands at 6 weeks and/or a volume of 300 mm^3 tumors, 10–15 mL digestion buffer will be required. Place the falcon tube vertically on a water bath at 37 °C for 45–50 min for normal mammary tissues and for 1 h for tumors, till the color of digestion medium changes to pink from red (*see* **Note 2**). Shake the tube vigorously every 15 min to ensure thorough mixing.
3. On completion of digestion, centrifuge the mixture at 500 × *g* for 3–5 min at 4 °C. Discard the floating layer of fat and supernatant. Resuspend the pellet in 2 mL of prewarmed 0.25% trypsin–EDTA solution and incubate at 37 °C for 90 s for normal mammary tissues and 2 min for tumors. This step helps break the organoid clumps. The volume of trypsin–EDTA solution should depend on size of the tissue. For 2 pairs of mammary glands at 6 weeks and/or a volume of 300 mm^3 tumors, 2–3 mL Trypsin-EDTA solution will be

required. After incubation, neutralize the trypsin with triple the volume of serum containing medium, that is, 6–10 mL of FCS medium and pellet at 500 × *g* for 3 min at 4 °C.

4. Incubate the pellet with 2–3 mL 5 mg/mL Dispase with 0.1 mg/mL DNase I for 5 min at 37 °C. Centrifuge the suspension at 500 × *g* for 3 min. Volume may need to be adjusted depending on the size of the pellet. This step further helps to break the cell clumps from basement membrane and surrounding matrix.

5. Resuspend the pellet with 2–3 mL 0.64% NH_4Cl to lyse red blood cells for 5 min at 37 °C. Spin the suspension at 500 × *g* for 3 min. Resuspend the resultant cells in DMEM/F12 medium and/or HBEC medium for PDX tumors and pass through a 40 μm strainer to enrich for single cells in a 50 mL tube.

6. In order to isolate lineage negative (Lin^-) epithelial cells and/or lineage negative basal cells ($Lin^-CD24^+CD29^{high}$) from mammary tissue and tumors, treat the single cell suspension with a combination of biotinylated lineage (Lin) primary antibodies (CD45, CD31, and Ter119 [24] to remove hematopoietic cells, endothelial cells and red blood cells respectively) and anti-CD24 and anti-CD29 antibodies at 1:50 dilution for 30 min at room temperature in the dark. Different primary antibody cocktails will be required depending on whether mouse or human cells are being processed. After the primary antibody staining, wash in 5 mL FCS medium and further incubate with secondary antibody, conjugated with PE-cy7 fluorochrome at 1:50 dilution for another 30 min at room temperature in the dark. After secondary staining, wash the cells with 5 mL FCS medium and resuspend the pellet with DAPI solution.

7. Sort DAPI negative live cells and lineage negative (Lin^-) mammary epithelial cells from normal mammary gland or tumors by cell sorting and collect into sterile MEGM medium. Magnetic beads may be used if available with respective lineage antibodies.

8. Count the sorted cells using a hemocytometer and trypan blue (to differentiate live cells from dead ones) (*see* **Note 3**). Avoid large cell clumps which could be cell debris. At this stage, make sure cells are properly dissociated as single cells and of appropriate morphology with round shiny properties.

9. After counting, spin down the cells at 500 × *g* for 3 min and add the appropriate volume of tumorsphere medium. This cell suspension should be kept on ice and mixed well for sphere formation and coculture assays (*see* **Note 4**).

3.2 Isolation of Epithelial Cells from Breast Cancer Cell Lines

1. Breast cancer cell lines are grown as monolayers according to the supplier's recommendations on tissue culture treated 10 cm dishes. Ensure the cells cultured as monolayers are healthy and 80–90% confluent. Overgrown cultures will have high levels of cell death.
2. Aspirate culture medium from a 10 cm dish of adherent monolayer cells and rinse with ice cold 1× DPBS to remove residual culture medium from the plate. This step needs to be repeated twice.
3. Add 3 mL of prewarmed 0.25% trypsin–EDTA solution to detach the adherent cells and incubate in a humidified incubator at 37 °C for at least 3 min.
4. Observe the cells under the microscope to ensure all cells have dissociated from the plate before proceeding to the next step. If less than 90% of cells are detached, increase the incubation time by few more minutes (*see* **Note 5**). When ≥90% cells are detached, neutralize the trypsin with 10 mL of appropriate culture medium containing 5–10% FBS.
5. Transfer cells and medium into a 15 mL conical tube and spin down the cells at 500 × *g* for 3 min. Resuspend the cell pellet in 3 mL of culture medium without serum (*see* **Note 6**).
6. Count the viable cells using a hemocytometer and trypan blue. After counting, the single cell suspension can be kept on ice for up to 1 h before use.

3.3 Seeding Cells for Sphere Formation Assay

3.3.1 Cell Suspension Sphere Culture

1. A nonadherent ultralow-attachment culture surface is required for cell suspension sphere cultivation, which can be purchased from commercial companies or homemade by coating tissue culture plates with poly hydroxyl ethyl methacrylate polymer (pHEMA) as 12% (w/v) [25] (*see* **Note 7**).
2. Add 2 mL of freshly prepared sphere medium (*see* Subheading 3) to each well in a 6-well plate.
3. Before plating, wash cells with serum-free culture medium. Then, plate out single cell suspension of freshly isolated lineage negative (Lin^-) tumor cells at appropriate density in triplicate. To generate spheres from primary breast tumor cells, plate 5000–10,000 cells/well in a 6-well plate, whereas from breast cancer cell lines, plate 100–500 cells/well in a 6-well plate. However, the cell number differs depending on breast cancer subtypes. The number of cells seeded per well must be optimized for each primary breast tumor subtypes and for each breast cancer cell line through serial dilution of cells across different wells (*see* **Note 8**).
4. Incubate the plates at 37 °C in a humidified incubator for 5–10 without disturbing the plates, particularly during the growth phase of the first 5 days. Depending on subtype of breast tumors and different breast cancer cell lines this time may vary.

3.3.2 Adherent Sphere Culture

1. For adherent sphere culture establishment, growth factor reduced Matrigel matrix is used as a basement membrane matrix which mimics in vivo extracellular matrix and promotes sphere formation from single cell suspensions. Once formed, spheres are less likely to migrate and aggregate, which contrasts with spheres growing in nonadherent suspension. Other semi-solid matrices can also be used, depending on the requirements of the used cell lines, including collagen-based matrices or hydrogel-based matrices (*see* **Note 9**).
2. Always keep the Matrigel on ice in order to prevent the solidification of Matrigel at room temperature. Slowly pipet 150 μL ice chilled Matrigel around the bottom rim of a well in a 12-well plate uniformly by going around in a circular manner (*see* **Note 10**). Double the amount of Matrigel is required for every well of a 6-well plate (i.e., 300 μL Matrigel), and half the amount for a 24-well plate (i.e., 80 μL Matrigel). Perform gentle pipetting to avoid forming any bubbles in the Matrigel matrix during the process (*see* **Note 11**).
3. Allow the Matrigel to solidify in a humidified incubator at 37 °C for 20–30 min. Avoid tilting the plate while plating and transferring the plate to the incubator.
4. After solidification of Matrigel, freshly isolated lineage negative (Lin^-) epithelial cells and/or Lin^- $CD24^+CD29^{high}$ mammary basal cells from either mammary tissues or Lin^- tumor cells are seeded at desired density into Matrigel-coated wells of a 24-well plate in triplicate. Make sure to pipet up and down to keep the cells uniformly suspended before plating.
5. After seeding the cells, slowly add 1 mL of warm culture medium in the middle of each well and incubate the plate in a humidified incubator at 37 °C for 7–10 days. It is very important to use warm culture medium as cold medium will lead to Matrigel dissociation. Of note, culture periods may differ with different types of breast cancer cells.
6. Replenish cells with warm serum-free culture medium every 2–3 days. It is very important to aspirate the culture medium from the center of the well very gently as to not disturb the Matrigel matrix.

3.4 Counting of Spheres from Cell Suspensions and Matrigel Matrix

1. Growth and formation of spheres from both non adherent cell suspensions and Matrigel matrix must be followed up daily after 4–5 days. In general, spheres will start to appear by 3–4 days. Spheres should be solid and round in structure. However, based on the doubling time, each cell line exhibits a distinct pattern in the process of sphere formation and growth, which must be monitored to optimize the follow-up

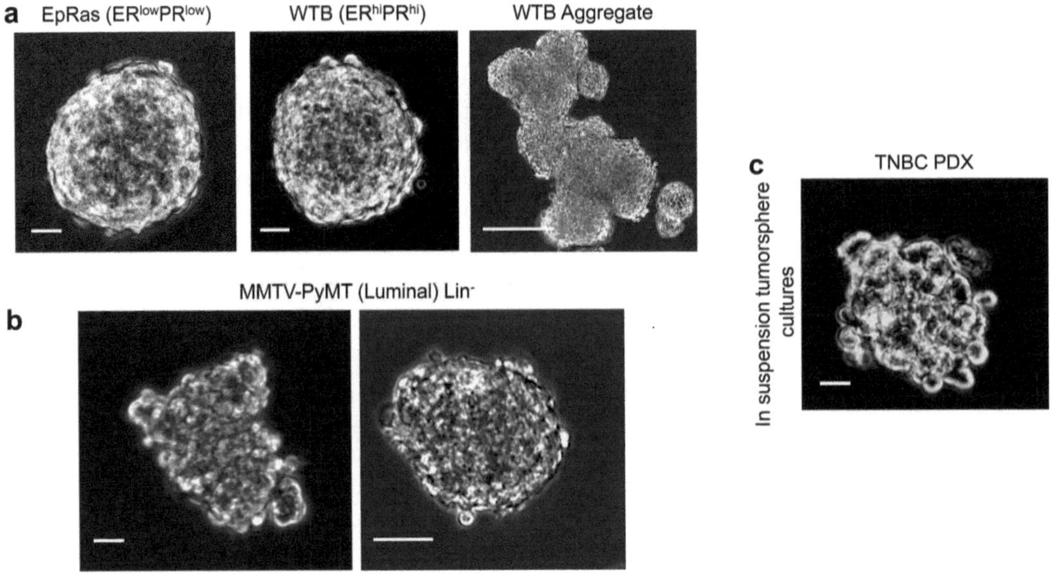

Fig. 2 In suspension sphere formation with breast cancer cell lines, murine primary breast tumors, and patient-derived xenograft (PDX) tumors. (**a–c**) Representative bright field microscopic images of tumorsphere of aggressive EpRas (**a**, left panel) and WTB (**a**, middle panel) breast cancer cell lines. Representative bright field microscopic image showing aggregation from WTB breast cancer cell line in cell suspension medium (**a**, right panel). (**b**) Primary PyMT luminal breast cancer cells derived tumorspheres. (**c**) Tumorsphere derived from triple negative breast cancer (TNBC) patient derived xenograft (PDX) breast tumors. Scale bars in **a** (left and middle panel), **b** (left panel) and **c** are 20 μm, in **b** (right panel) is 40 μm and **a** (right panel) is 100 μm

schedule. It is very important to periodically acquire bright field images, using an inverted light microscope, of developing spheres to assess any changes in morphology.

2. After 5 days, count the number of spheres using a microscope at 40× magnification, noting the number of spheres which are greater than 50 μm diameter or more than 4–6 cells (*see* **Note 12**). Bright-field images showing morphology of mature spheres (in suspension) are illustrated in Fig. 2; from different breast cancer cell lines (*see* Fig. 2a left and middle panel) [26], murine primary breast tumors (*see* Fig. 2b) and PDX-tumors (*see* Fig. 2c) [27]. Some cell lines may not form typical sphere structures as shown in Fig. 2 and thus it may be difficult to count or classify definitively as spheres. Aggregates (*see* Fig. 2a right panel) should be avoided as they may form from many spheres and the quantification will be incorrect. Figure 3 bright-field images showing the morphology of mature Matrigel culture spheres from normal mammary gland (MG) (*see* Fig. 3a left panel) [28, 29] and from primary breast tumor (*see* Fig. 3b left panel). Spheres can also be stained with antibodies for immunofluorescence (*see* Fig. 3a right panel).

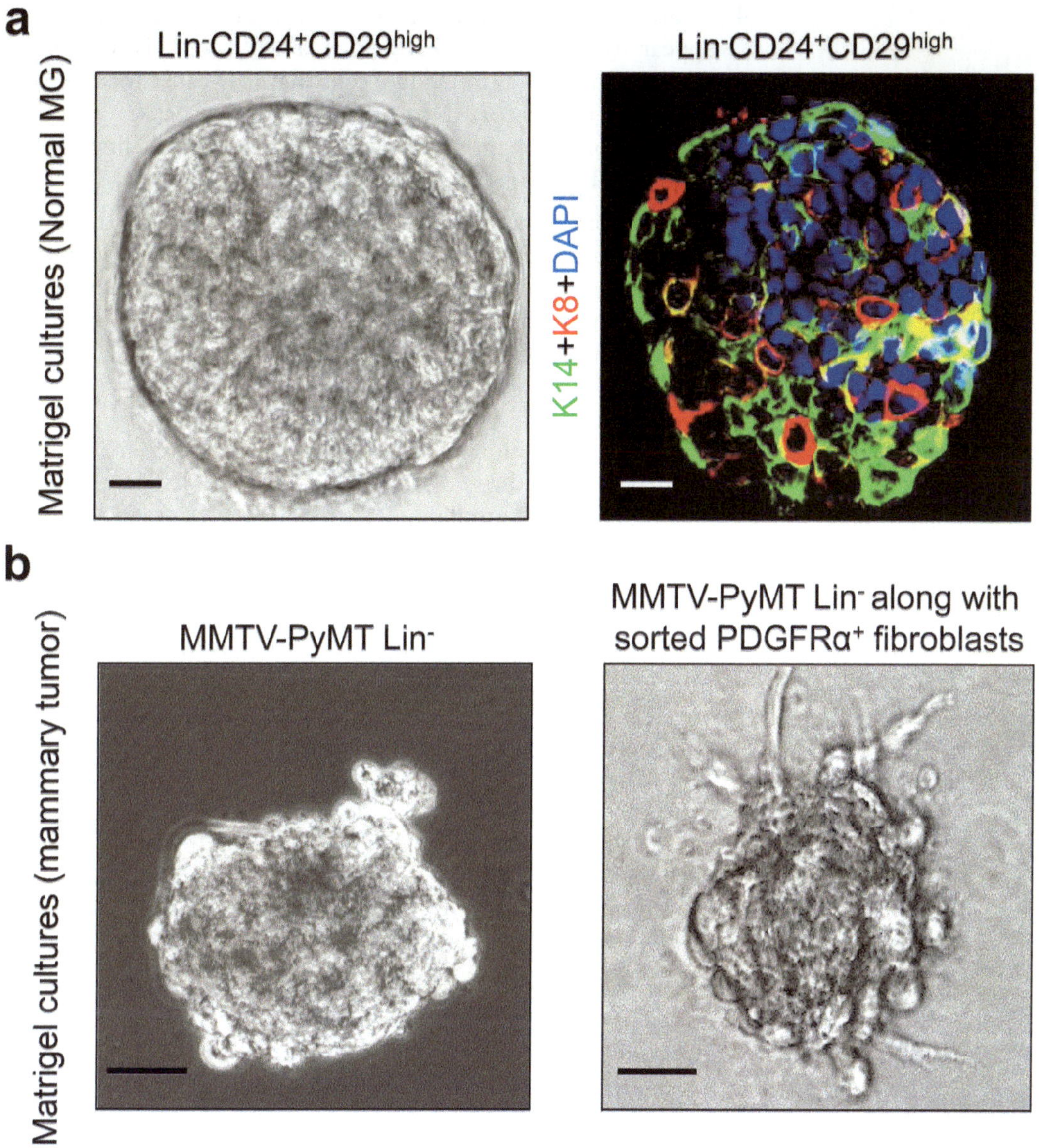

Fig. 3 Sphere formation in Matrigel from murine normal mammary gland and primary mammary tumor cells. (**a**, **b**) Representative bright field images of mammospheres from $Lin^{-}CD24^{+}CD29^{high}$ basal cells (**a**, left panel) from normal mammary gland (MG). (**a**, right panel) Representative confocal images of $Lin^{-}CD24^{+}CD29^{high}$ basal cell derived mammosphere stained with K14 (green), K8 (red) and DAPI (blue) in Matrigel. (**b**, left panel) Spheres from sorted Lin^{-} epithelial tumor cells alone from MMTV-PyMT luminal mammary tumors in Matrigel. (**b**, right panel) Representative bright field image of tumorsphere when sorted $PDGFR\alpha^{+}$ fibroblasts are cocultured with Lin^{-} tumor cells from MMTV-PyMT in 1:4 ratio in Matrigel. Scale bars in (**a**) are 20 μm and (**b**) are 40 μm respectively

3. The results of sphere formation assay are expressed as size or total number of formed spheres. This is crucial as the number of spheres can be employed to characterize the population of cancer stem cells and/or progenitor cells within a population of in vitro cultured cancer cells and within in vivo tumors.

4. Aggregation in sphere forming assays can occur where cells appear to become fused together. Therefore, it is difficult to distinguish them as individual cells that may cause misinterpretation of results (*see* **Note 13**). A representative image of aggregation formation from a breast cancer cell line is illustrated in Fig. 2a, right panel.
5. In addition, sphere forming efficiency (SFE) (%) or sphere forming unit (SFU) can also be calculated based on the following simple formula: SFE or SFU = (number of spheres counted per well/number of cells seeded per well) × 100. For example, the formation of 10 spheres at day 10 after plating 1000 cells indicates an SFE of 1%.

3.5 Passaging of Spheres for Assessment of Self-Renewal (Second Generation)

1. From cell suspension sphere culture, once spheres are properly formed (10–15 days) and scored for SFE or SFU, pipet the entire medium, containing spheres from each well, into a 15 mL centrifuge tube. Wash each well with ice cold 1 mL 1× DPBS and transfer into the same falcon tube. Spin the cells at 200 × *g* for 5 min to ensure only spheres are pelleted. Supernatant is discarded and the pellet is ready to use for second generation.

   For second generation assay from adherent sphere culture, aspirate the culture medium from the center of each Matrigel-coated well and add 500 μL of ice-cold 1 mg/mL Dispase solution, dissolved in 1× DPBS and kept at 37 °C (*see* **Note 14**). Use a pipette tip to gently agitate the mixture every 15 min. Collect the medium with the released spheres from each well and transfer into a 15 mL centrifuge tube with more ice cold 1× DPBS (*see* **Note 15**).
2. Centrifuge the cells at 200 × *g* for 5 mins and gently remove the supernatant as much as possible without disturbing the pellet at the bottom of tube (*see* **Note 16**). Wash twice with ice cold 1× DPBS to remove leftover Matrigel.
3. Resuspend the cell pellet in 300 μL of 0.25% trypsin- EDTA and incubate at 37 °C for 2–3 min. After the incubation period, disaggregate the spheres by pipetting vigorously up and down a few times until a single cell suspension is produced. Neutralize trypsin with double the volume of serum-containing medium and centrifuge at 600 × *g* for 5 min.
4. Resuspend the cell pellet in a small volume around 100–200 μL of ice-cold 1× DPBS and count the cells with a hemocytometer.
5. Plate out the entire single cell suspension into mammosphere or tumorsphere medium or Matrigel, whichever was used in the primary generation.

6. Incubate the plate in a humidified incubator at 37 °C for the same number of days as the primary culture without replenishing the medium (*see* **Note 17**). If need be, medium can be added on top to prevent drying of plate.

3.6 Coculture Assay

A coculture assay is where two or more different populations of cells are grown with some degree of contact between them. This technique is of great importance as it can imitate cellular interplay observed in vivo, and allow for high-throughput testing and in-depth monitoring of drug effects on cell-cell interactions [30, 31]. In the context of cancer, this method allows for direct cell-cell interaction facilitating study of tumor-associated stroma, as stromal populations play an important role in regulating the emergence and the progression of epithelial malignancy [32] (*see* **Note 18**). Fibroblasts, one of the predominant stromal cells, are prominent modifiers of cancer progression. In breast cancer, there is compelling evidence that fibroblasts surrounding cancer epithelial cells can influence epithelial transformation by producing paracrine factors [33] (*see* **Note 19**).

1. Isolate fibroblasts from breast tumors by cell sorting or magnetic bead separation using CD140 (PDGFRα) and plate with Lin^- breast epithelial cells in 1:4 ratio on Matrigel-coated plates for 3–7 days. Monitor spheroid formation daily under bright field microscope. Spheroids are generally mature after 5 days. Coculture spheroids show close existence of both tumor cells and fibroblast (long thread like structures). Take the sphere images with a bright field microscope (*see* Fig. 3b right panel). Spheroid size and projection length can be measured under light microscopy.
2. Isolate cells from spheroid coculture system as described before in Subheading 3.6. Analyze cells by flow cytometry for population purity using appropriate cell surface markers and further separate different cell populations by cell sorting, if necessary. This approach can be used to assess the anticancer effects of drugs on tumor cells and cytotoxicity on epithelial cells.

3.7 Propagation of Spheres for Assessment of Different Functional Studies

Each sphere is presumably descended from a single stem/progenitor cell and therefore, the differentiation potential within spheres should be tested. These methods can be used to further assess the effects of different drug treatments on differentiation pathways of targeted tumors/cells. To do so, spheres need be collected at any generation (preferably at each generation or first generation) and be subjected to different analysis as follows.

1. *Immunofluorescence analysis.*

   Take Lin^- and/or Lin^- $CD24^+CD29^{high}$ epithelial mammosphere/tumorsphere from any generation of cell suspension culture or Matrigel matrix culture in a polylysine-coated slide as a small drop containing spheres. Once the medium evaporates, fix cells with 4% paraformaldehyde on ice for 15 min and permeabilize with 0.5% Triton X-100 for another 15 min. Block the cells with 20% goat serum for 1 h at room temperature [28, 29] (*see* **Note 20**). After blocking, incubate cells overnight at 4 °C with primary antibodies. Here, we stain spheres with cytokeratin14 (K14), basal cell marker and cytokeratin8 (K8), luminal cell marker antibodies. On the next day, following three washes with 1× DPBS, add Alexa Fluor conjugated 488 and 568 secondary antibodies and incubate for 1 h at room temperature. After secondary antibody staining, following three washes with 1× DPBS, mount slides in mounting medium with DAPI. Take the fluorescence images using a confocal microscope (*see* Fig. 3a right panel).

2. *Protein and RNA extraction.*

   Total protein and RNA can be extracted from treated and nontreated spheres from any generation and subjected to Western blot and qRT-PCR analyses for stemness and differentiation markers.

4 Notes

1. The addition of B27 has been shown to increase sphere forming efficiency and sustain propagation of spheres for several passages of sphere cultures. Morphology of spheres cultured in B27 free medium shows rough margins due to adherence of cells to the spherical edges during serial passaging.
2. The shaking should be at such a speed to cause thorough mixing but should not be too fast to cause any physical damage to the cells. In addition, enzymatic digestion time of around 1 h usually gives rise to enough single cells with the highest cell viability. Note that digestion time may need to adjust according to the pathological characteristics of the tumor subtype to avoid cell death at the end of the procedure. The degree of digestion should be checked at every 15 min.
3. Only the bright cells within the four corner quadrants of the hemocytometer counting chamber should be considered as viable cell count.
4. The isolation of primary breast cancer cells from murine breast tumors should be done as fast as possible, to preserve survival of these cells. Cells can be kept on ice for up to 1 h before use.
5. This incubation time is highly cell type-dependent. Gently tap the plate to expedite cell detachment.

6. More than 90% single cells suspension from cell line is usually obtained after trypsinization. Also, try to avoid any clumps at the time of counting the cells.
7. For coating tissue culture plates with hydroxyl ethyl methacrylate polymer (pHEMA), dissolve 12 g pHEMA in 1 L of 95% ethanol by stirring constantly on a heated plate. pHEMA can be stored as a flammable liquid at room temperature for several months. Adding natural polymers to the culture system prevents sedimentation of the sphere, but does not disturb their inherent properties or ability to proliferate. Once the solution has cooled, pipet 1 mL/well into a 6-well plate. Incubate the plate at 40 °C for 48 h, ensuring that the plates remain sterile.
8. The optimal density of cells to be seeded should be one that is high enough to generate a sufficiently good number of countable spheres, but it should not be so high that aggregation occurs. It is observed that a decrease in sphere formation happens with increasing seeding density.
9. A bottle of Matrigel containing 10 mL should be thawed on ice 12 h before performing the experiment. Aliquots of 1 mL should be prepared to prevent repeated freeze/thaw cycles of Matrigel.
10. There are several advantages for plating the Matrigel matrix around the rim of the well. First, it allows feeding and washing from the center of the empty well without disturbing the Matrigel matrix on the rim. Second, the rim provides edge support ensuring physical stability for the period of sphere formation which may last for 2 weeks. Lastly, it also confines growth of the sphere to a defined area in the well that makes more precise counting of the spheres. In addition, it ensures the distribution of cells in a single cell format, avoiding clumping and overgrowth.
11. During coating with Matrigel matrix, bubble formations should be avoided as bubbles will create gaps in the Matrigel matrix that will lead to its breakage. To get rid of a bubble, carefully prick it with the needle of a 27-gauge syringe.
12. A floating cell clump with a diameter less than 50 μm should not be considered as a sphere.
13. This step highlights the importance of minimizing such aggregation through optimization of seeding densities and corrects cell manipulation.
14. Note that spheres that are formed in the primary sphere-formation assay, referred to as first generation spheres, which do not account for stem cells, as progenitor cells can form short-lived spheres too. Therefore, multiple serial generation of the spheres are essential to truly identify stem cell

characteristics like self-renewal ability. Thus, after counting the number of spheres and calculating the SFE, spheres propagation must follow.

15. Dispase helps to break down the Matrigel and allows releasing the spheres into the culture medium. Incubate the plate for a minimum of 30 min in a humidified incubator at 37 °C.
16. At this stage, pellet may not be visible. Therefore, remove carefully the supernatant, so as not to dislodge the pellet.
17. The SFE or SFU must be calculated through manual counting of the spheres, as stated in Subheading 3.5. Both SFE and/or SFU may change from one generation to another and from one cell line to another throughout the five generations starting from the same number of cells at the start point. This step is important to assess the self-renewal ability of the cancer stem cells.
18. This system recreates a realistic scenario for study of tumors, in which cancer cells receive mechanical and physical signals from their environment that are needed by them to undergo their biological functions. Stromal cells will contribute by providing biological signals (such as growth factors and cytokines) to tumor cells that could modulate tumorigenesis.
19. Various stromal populations as desired can be isolated by fluorescence sorting from normal mammary tissue and tumors and cocultured with epithelial cells for 3–5 days. This coculture assay can also be done in nonadherent cell suspension context.
20. For immunofluorescence staining, after fixation, wash the spheres with 1 mL 1× DPBS. This washing step is required to repeat three times to remove any excess PFA. Fixed spheres can be kept at −20 °C up to 1 month before further proceeding to immunofluorescence staining.

Acknowledgments

We thank Dr. Snahlata Singh for critical reading of the manuscript and helpful discussions. This work was supported by grants from American Cancer Society and NCI-R01 (R01CA237243) grant to R.C.

References

1. Ishiguro T, Ohata H, Sato A et al (2017) Tumor-derived spheroids: relevance to cancer stem cells and clinical applications. Cancer Sci 108:283–289
2. Anderson K, Lutz C, van Delft FW et al (2011) Genetic variegation of clonal architecture and propagating cells in leukaemia. Nature 469: 356–361
3. Kitamura H, Okudela K, Yazawa T et al (2009) Cancer stem cell: implications in cancer biology and therapy with special reference to lung cancer. Lung Cancer 66:275–281

4. Schulenburg A, Ulrich-Pur H, Thurnher D et al (2006) Neoplastic stem cells: a novel therapeutic target in clinical oncology. Cancer 107: 2512–2520
5. Jordan CT, Guzman ML, Noble M (2006) Cancer stem cells. N Engl J Med 355: 1253–1261
6. Bonnet D, Dick JE (1997) Human acute myeloid leukemia is organized as a hierarchy that originates from a primitive hematopoietic cell. Nat Med 3:730–737
7. Al-Hajj M, Wicha MS, Benito-Hernandez A et al (2003) Prospective identification of tumorigenic breast cancer cells. Proc Natl Acad Sci U S A 100:3983–3988
8. Ricci-Vitiani L, Lombardi DG, Pilozzi E et al (2007) Identification and expansion of human colon-cancer-initiating cells. Nature 445: 111–115
9. Reynolds BA, Weiss S (1992) Generation of neurons and astrocytes from isolated cells of the adult mammalian central nervous system. Science 255:1707–1710
10. Dontu G, Abdallah WM, Foley JM et al (2003) In vitro propagation and transcriptional profiling of human mammary stem/progenitor cells. Genes Dev 17:1253–1270
11. Ponti D, Costa A, Zaffaroni N et al (2005) Isolation and in vitro propagation of tumorigenic breast cancer cells with stem/progenitor cell properties. Cancer Res 65:5506–5511
12. Rosenbluth JM, Schackmann RCJ, Gray GK et al (2020) Organoid cultures from normal and cancer-prone human breast tissues preserve complex epithelial lineages. Nat Commun 11:1711
13. Grimshaw MJ, Cooper L, Papazisis K et al (2008) Mammosphere culture of metastatic breast cancer cells enriches for tumorigenic breast cancer cells. Breast Cancer Res 10:R52
14. Liu JC, Deng T, Lehal RS et al (2007) Identification of tumorsphere- and tumor-initiating cells in HER2/Neu-induced mammary tumors. Cancer Res 67:8671–8681
15. Farnie G, Clarke RB, Spence K et al (2007) Novel cell culture technique for primary ductal carcinoma in situ: role of Notch and epidermal growth factor receptor signaling pathways. J Natl Cancer Inst 99:616–627
16. Hughes CS, Postovit LM, Lajoie GA (2010) Matrigel: a complex protein mixture required for optimal growth of cell culture. Proteomics 10:1886–1890
17. Shaw LM, Chao C, Wewer UM, Mercurio AM (1996) Function of the integrin alpha 6 beta 1 in metastatic breast carcinoma cells assessed by expression of a dominant-negative receptor. Cancer Res 56:959–963
18. Chaicharoenaudomrung N, Kunhorm P, Noisa P (2019) Three-dimensional cell culture systems as an in vitro platform for cancer and stem cell modeling. World J Stem Cells 11: 1065–1083
19. Goldhammer N, Kim J, Timmermans-Wielenga V, Petersen OW (2019) Characterization of organoid cultured human breast cancer. Breast Cancer Res 21:141
20. Dekkers JF, Alieva M, Wellens LM et al (2019) High-resolution 3D imaging of fixed and cleared organoids. Nat Protoc 14:1756–1771
21. Roelofs C, Hollande F, Redvers R et al (2019) Breast tumour organoids: promising models for the genomic and functional characterisation of breast cancer. Biochem Soc Trans 47: 109–117
22. Smart CE, Morrison BJ, Saunus JM et al (2013) In vitro analysis of breast cancer cell line tumourspheres and primary human breast epithelia mammospheres demonstrates inter- and intrasphere heterogeneity. PLoS One 8: e64388
23. Casbas-Hernandez P, Fleming JM, Troester MA (2011) Gene expression analysis of in vitro cocultures to study interactions between breast epithelium and stroma. J Biomed Biotechnol 2011:520987
24. Shackleton M, Vaillant F, Simpson KJ et al (2006) Generation of a functional mammary gland from a single stem cell. Nature 439: 84–88
25. Shaw FL, Harrison H, Spence K et al (2012) A detailed mammosphere assay protocol for the quantification of breast stem cell activity. J Mammary Gland Biol Neoplasia 17:111–117
26. Kumar S, Srivastav RK, Wilkes DW et al (2019) Estrogen-dependent DLL1-mediated Notch signaling promotes luminal breast cancer. Oncogene 38:2092–2107
27. DeRose YS, Wang G, Lin YC et al (2011) Tumor grafts derived from women with breast cancer authentically reflect tumor pathology, growth, metastasis and disease outcomes. Nat Med 17:1514–1520
28. Chakrabarti R, Wei Y, Romano RA et al (2012) Elf5 regulates mammary gland stem/progenitor cell fate by influencing notch signaling. Stem Cells 30:1496–1508
29. Chakrabarti R, Wei Y, Hwang J et al (2014) DeltaNp63 promotes stem cell activity in mammary gland development and basal-like breast

cancer by enhancing Fzd7 expression and Wnt signalling. Nat Cell Biol 16:1004–1015. 1001–1013

30. Wu MH, Huang SB, Lee GB (2010) Microfluidic cell culture systems for drug research. Lab Chip 10:939–956

31. Chakrabarti R, Celia-Terrassa T, Kumar S et al (2018) Notch ligand Dll1 mediates cross-talk between mammary stem cells and the macrophageal niche. Science 360

32. Kumar S, Wilkes DW, Samuel N et al (2018) DeltaNp63-driven recruitment of myeloid-derived suppressor cells promotes metastasis in triple-negative breast cancer. J Clin Invest 128:5095–5109

33. Kuen J, Darowski D, Kluge T, Majety M (2017) Pancreatic cancer cell/fibroblast co-culture induces M2 like macrophages that influence therapeutic response in a 3D model. PLoS One 12:e0182039

Chapter 34

Enrichment of Cancer Stem Cells in a Tumorsphere Assay

Abhijeet P. Deshmukh, Petra den Hollander, Nick A. Kuburich, Suhas Vasaikar, Robiya Joseph, and Sendurai A. Mani

Abstract

Cancer stem cells (CSCs) are a small subpopulation of self-renewing cancer cells that are present within tumors. CSCs possess tumor initiation potential as well as the ability to resist toxic compounds and chemotherapeutic agents through the upregulation of drug efflux transporters, DNA repair pathways, and survival cascades. Accumulating evidence suggests that CSCs are responsible for tumor relapse and resistance to chemotherapeutic agents and that targeting CSCs is critical to inhibition of cancer progression. Therefore, isolation and characterization of CSCs is important in studying tumor initiation and progression. In this chapter, we provide a detailed method for the identification and isolation of CSCs.

Key words Cancer stem cells, CSC Enrichment, Primary Tissue, Tumorspheres, Serum free conditions

1 Introduction

Cancer stem cells (CSCs) with self-renewing ability [1, 2] possess unique cellular characteristics [3–7] that are implicated in driving tumor relapse and resistance to chemotherapeutic agents [8]. Three methods are commonly used to identify and isolate CSCs. One is isolation based on the side population phenotype: stem cells, including CSCs, efflux Hoechst 33342 dye [9], and this characteristic can be used to select these cells using fluorescence-activated cell sorting (FACS). A second method relies on the use of relevant surface markers during FACS or magnetic-activated cell sorting (MACS) [4]. A number of markers have been shown to be indicative of CSCs in various cancers (Table 1). In pancreatic cancer, CSCs express surface markers such as CD133, CXCR4, CD24, CD44, ESA, c-Met, and ALDH1 [10]. In breast cancers, CSCs have been characterized by low or no expression of CD24 and expression of CD44 [4]. CD44 expression is also characteristic of CSCs in colon, gastric, head and neck, and ovarian cancers. In leukemia, CSCs are identified as cells that express CD34 but not

Nagarajan Kannan and Philip Beer (eds.), *Stem Cell Assays: Methods and Protocols*, Methods in Molecular Biology, vol. 2429, https://doi.org/10.1007/978-1-0716-1979-7_34,

Table 1
CSC surface markers expressed in various tissues (table adapted from Lathia et al., *Oncologist* [15])

CSC Markers / Tissues	CD133	ALDH1	CD44	CD24	CD90	ACBD3	ALDH1A1	ALDH1A3	Bmi1	Gli1	LGR5	β-catenin	Oct4	PLAG2G16	PROCR	STMN1	CD44V2	CD44V6	ACBDG2	CXCR4	Nanog	SOX4	SOX9
Colon	+	+	+	+									+				+	+	+	+	+	+	
Breast	+	+	+	-	+	+	+	+	+	+	+	+	+	+	+	+							
Liver	+	+	+	+	+							+											+
Ovary	+	+	+	+			+											+			+		
Pancreas	+	+	+	+			+													+			+
Melanoma	+	+																					
Glioma	+	+	+		+																		
Prostate	+	+																					
Lung	+	+	+		+		+			+		+	+								+		
Oral squamous cell carcinoma			+																				
Thyroid																				+			+
Head and neck		+																					
Esophageal		+	+				+			+													+
Gastric	+		+						+		+												
Small intestine																		+					
Cervical and vulvar		+											+										
Kidney and bladder	+	+	+	+					+														

CD38 [11]. Other markers characteristic of CSCs are CD13, CD45, and CD90 in liver cancer, CD117, CD90, and EpCAM in lung cancer; CD20, CD166, and Nestin in melanoma; and CD133 and ABCG2 in glioblastoma multiforme [11]. A third method involves culture of sphere-like cellular aggregates of CSCs in serum-free medium (SFM) containing growth factors EGF, bFGF, and b-27 [12]. This method was derived from the culturing of neural stem/progenitor cells [13].

Each of these methods has pros and cons. Hoechst 33342 can decrease clonogenicity [14], limiting reliability of subsequent functional experiments on isolated cells. The cell surface markers for CSCs for various cancer tissues are still not well characterized, and there does not appear to be a "universal" set of markers characteristic of CSCs from all cancers. Finally, the sphere cells formed in serum-free medium may develop a necrotic center due to diffusion limits of nutrients and oxygen, and the spheroids may also contain differentiated cells and dead cells. Nevertheless, we found that under serum-free, stem cell-selective culture conditions, the cells in nonadherent spheres formed from cancer specimens display remarkable stem cell properties, drug resistance, and propagation of their original tumor phenotype—the properties expected for CSCs. Here, we describe CSC isolation from primary cancer tissues and cancer cell lines using these techniques.

2 Materials

2.1 Tissue and Cells

1. Mouse tumor tissues (*see* **Notes 1** and **2**).
2. Tumor cell lines.

2.2 Digestion Medium

1. DMEM F-12 Medium.
2. 0.5 mg/mL hyaluronidase.
3. 3 mg/mL collagenase Type 1A.

2.3 Reagents and Equipment for Separation (See Note 3)

1. 40-μm cell strainer
2. Phosphate-buffered saline (PBS, pH 7.4).
3. Collagenase Type A medium.
4. 15-ml and 50-ml centrifuge tubes.

2.4 Cell Culture Reagents (See Note 3)

1. TrypLE Express Enzyme (Thermo-Fisher; *see* **Note 4**).
2. DMEM 1× complete medium (CM): DMEM 1× medium, 10% fetal bovine serum (FBS), 1% penicillin–streptomycin solution; store at 4 °C.
3. Serum-free medium (SFM): DMEM 1× medium + DMEM-HG (HyClone, SH30081.02) supplemented with the growth

factors 20 ng/mL bFGF and 20 ng/mL EGF, B27 supplement, and 1% penicillin–streptomycin solution (*see* **Note 5**).

4. Dimethyl sulfoxide (DMSO), tissue culture grade (*see* **Note 6**).
5. Cryopreservation reagent: 10% DMSO, 30% FBS, 60% DMEM 1× CM.
6. Ultralow-attachment culture plates and dishes: 6-well plate, 100 mm, and 60 mm.

3 Methods (*See* Note 7)

3.1 Enrichment from Primary Solid Tumors

3.1.1 Isolation of Primary Tumor Cells from Mouse Solid Tumors

1. Rinse the primary solid tumor with PBS at least three times (*see* **Note 8**).
2. Transfer the tissue to a 50-mL centrifuge tube.
3. Manually homogenize the tumor using a blade.
4. Transfer the homogenized tumor to a 15-mL tube.
5. Add 5 mL of Digestion medium to the tube and submerge the tissue (*see* **Note 9**).
6. Incubate at 37 °C for 30 min with rocking (*see* **Note 10**).
7. Add 5 mL of CM to neutralize collagenase (*see* **Note 11**).
8. Filter through a 40-μm cell strainer to remove clumps and transfer the filtrate to a new 15-mL tube (*see* **Note 12**).
9. Centrifuge at 1000 × *g* for 5 min at 4 °C.
10. Remove the supernatant and resuspend the cells in 10 mL. PBS.
11. Repeat **steps 9** and **10** twice.

3.1.2 Culture of Primary Mouse Tumor Cells

1. Culture cells in 100-mm culture plates at 37 °C in a humidified incubator overnight (*see* **Note 13**).
2. Replace the medium with fresh CM after the cells are attached then change the medium every 3 days depending on the cell growth rate (*see* **Notes 14** and **15**).
3. When the cells reach 70–80% confluence, passage the cells with TrypLE express enzyme. Once cells become rounded and detach, add CM. Centrifuge cells at 150 × *g* for 5 min. Remove supernatant and resuspend cells in CM and distribute them in two or three 100-mm culture plates (*see* **Note 16**). To use fresh cells, skip to **step 6**.
4. To freeze cells, after centrifugation, resuspend the cells in cryopreservation reagent, and distribute cells into several cryogenic vials. Store the vials at −80 °C.

5. To revive frozen cells, thaw and transfer the cells to a 15-ml tube containing CM. Centrifuge at 150 × *g* for 5 min. Discard supernatant. Resuspend the cells in CM.
6. Passage primary cancer cells until a sufficient number of cells have been obtained to do sphere culturing. Grow cells to 70–80% confluence, harvest using TrypLE, centrifuge cells at 150 × *g* for 5 min, and resuspend in 1 mL SFM.

3.1.3 Culture of Tumorspheres from Primary Mouse Tumor Cells

1. Count the number of cells using a hemocytometer or automated cell counter.
2. Transfer the cells at 25,000–50,000 cells/cm^2 to an ultralow-attachment plate containing SFM. Culture at 37 °C, 5% CO_2 in a humidified incubator.
3. Observe the cells each day until nonadherent spheroids form (*see* **Note 17**).

3.2 Tumorspheres of Cancer Cells

1. Thaw cells and follow the cell culture procedures (Subheading 3.1.2, from **step 5**).
2. Culture the cells in SFM in ultralow-attachment plates at 10,000 cells/cm^2.
3. Observe the cells each day until nonadherent tumorspheres form (*see* **Note 17**).

4 Notes

1. Appropriate biohazard authorization is required for experimentation with mouse tissues.
2. Tumor samples should be collected at the time of surgery under sterile conditions. The amount of tissue obtained will depend on the mouse strains. Store the samples on ice and transfer them to the laboratory as soon as possible.
3. Prepare all solutions using DNase- and RNase-free distilled water. Prepare and store all reagents at 4 °C unless otherwise noted.
4. The TrypLE express enzyme should be stored at room temperature.
5. Fresh preparation of SFM is recommended as growth factors may degrade during storage. If you use SFM frequently, you can store it at 4 °C for no more than a week.
6. DMSO stock solution should be stored at room temperature.
7. All procedures should be performed inside a sterilized culture hood. All reagents and apparatus must be sterilized.

8. Wash the solid tumor at least three times. More washes may be required as the tissue should be free of blood and dead tissue clumps.
9. Mix during the digestion to digest completely to obtain more cells.
10. If the tumor is large, the digestion time may need to be extended.
11. Tumor digestion using digestion medium can be stopped using culture medium containing 10% FBS. The amount of culture medium should be equivalent to the volume of the collagenase medium.
12. Prevent the tissue from leaking outside the tube opening.
13. Avoid plating too many cells in one culture dish. Over-confluence prevents cell growth in the exponential phase and could result in detachment from the surface.
14. The tumor primary cells should attach to the surface within 24 h. If bacterial contamination occurs in the cell culture, discard the cells immediately.
15. The SFM should be changed every 3 days unless the color of the media changes (indicative of a pH change). If this occurs, centrifuge at 150 g for 5 min, resuspend the cells in SFM, and inoculate them at the center of the plate.
16. Trypsin is not recommended for use with primary cells. Trypsin is known to cleave epitopes from cell-surface receptors.
17. The cells should begin to form spheres in 3–4 days and can be kept in the medium for 1 week. We often start experiments after 4 days of culturing.

Acknowledgments

This work was supported by NIH/ NCI grants (2R01CA155243;1R01CA200970); and NSF grant (PHY-1605817) awarded to S.A.M. and the NIH T32 grant (5T32CA186892) awarded to N.A.K.

References

1. Clevers H (2011) The cancer stem cell: premises, promises and challenges. Nat Med 17: 313–319
2. Reya T, Morrison SJ et al (2001) Stem cells, cancer, and cancer stem cells. Nature 414: 105–111
3. Lathia JD, Heddleston JM et al (2011) Deadly teamwork: neural cancer stem cells and the tumor microenvironment. Cell Stem Cell 8: 482–485
4. Al-Hajj M, Wicha MS et al (2003) Prospective identification of tumorigenic breast cancer cells. Proc Natl Acad Sci U S A 100: 3983–3988
5. Bao S, Wu Q et al (2006) Glioma stem cells promote radioresistance by preferential

activation of the DNA damage response. Nature 444:756–760
6. Dean M, Fojo T et al (2005) Tumour stem cells and drug resistance. Nat Rev Cancer 5: 275–284
7. Deshmukh A, Kumar S et al (2017) Secreted frizzled-related protein 4 (sFRP4) chemo-sensitizes cancer stem cells derived from human breast, prostate, and ovary tumor cell lines. Sci Rep 7:2256
8. Vidal SJ, Rodriguez-Bravo V et al (2014) Targeting cancer stem cells to suppress acquired chemotherapy resistance. Oncogene 33: 4451–4463
9. Szotek PP, Pieretti-Vanmarcke R et al (2006) Ovarian cancer side population defines cells with stem cell-like characteristics and Mullerian inhibiting substance responsiveness. Proc Natl Acad Sci U S A 103:11154–11159
10. Matsuda Y, Ishiwata T et al (2016) Systemic administration of small interfering RNA targeting human nestin inhibits pancreatic cancer cell proliferation and metastasis. Pancreas 45: 93–100
11. Tomao F, Papa A et al (2013) Emerging role of cancer stem cells in the biology and treatment of ovarian cancer: basic knowledge and therapeutic possibilities for an innovative approach. J Exp Clin Cancer Res 32:48
12. Zhang S, Balch C et al (2008) Identification and characterization of ovarian cancer-initiating cells from primary human tumors. Cancer Res 68:4311–4320
13. Uchida N, Buck DW et al (2000) Direct isolation of human central nervous system stem cells. Proc Natl Acad Sci U S A 97:14720
14. Zhong Y, Zhou C et al (2007) Most MCF7 and SK-OV3 cells were deprived of their stem nature by Hoechst 33342. Biochem Biophys Res Commun 364:338–343
15. Lathia J, Liu H et al (2020) The clinial impact of cancer stem cells. Oncologist 25(2): 123–131

Chapter 35

In Vitro Quantification of Cancer Stem Cells Using a Mammosphere Formation Assay

Nick A. Kuburich, Petra den Hollander, Abhijeet P. Deshmukh, Suhas Vasaikar, Robiya Joseph, Max S. Wicha, and Sendurai A. Mani

Abstract

Cancer stem cells (CSCs) are a small subpopulation of self-renewing cancer cells that are present within tumors. In this chapter, we provide a detailed method for the quantification of CSCs in vitro through mammosphere formation.

Key words Cancer stem cells (CSC), Mammosphere, Stemness, Suspension, Tumorsphere

1 Introduction

The method for cultivating mammary stem and progenitor cells in vitro was initially described nearly 20 years ago by Max Wicha's lab [1]. This method was a derivative of methods for culture of undifferentiated neural cells that grow as neurospheres in suspension [2, 3]. Under the described conditions, which include various growth factors, the cells that grow in suspension are undifferentiated; differentiated cells undergo anoikis due to lack of attachment. The cells that grow in solution are able to undergo differentiation upon implantation in a mammary fat pad where they are able to regenerate an entire mammary gland [4]. Mani and colleagues demonstrated that the number of stem cells can be increased by inducing the epithelial–mesenchymal transition (EMT) program, which results in cells with enhanced tumorigenesis [5]. Here we describe the quantification of cancer stem cells (CSCs) in vitro through mammosphere formation after culturing cancer cells in serum-free media in nonadherent conditions.

Nagarajan Kannan and Philip Beer (eds.), *Stem Cell Assays: Methods and Protocols*, Methods in Molecular Biology, vol. 2429, https://doi.org/10.1007/978-1-0716-1979-7_35,

2 Materials

2.1 Mammosphere Medium

1. Aliquot ~150 mL of Mammary Epithelial Cell Growth Medium (MEGM; Lonza, CC-3150) into a 1-L glass bottle containing a sterile magnetic stir bar. Warm to 37 °C. Keep a second ~350 mL aliquot at 4 °C (*see* **Note 1**).
2. Add 1% w/v (5 g; calculated for 500 mL of final volume) of methylcellulose to the bottle with the 150 mL aliquot of MEGM at 37 °C. Manually stir with a sterile glass rod to equally distribute the methylcellulose. Transfer the solution to a magnetic stir plate and stir overnight. The resulting solution will appear to be a gel-like slurry (*see* **Note 2**).
3. Add the ~350 mL aliquot of cold MEGM. Place the bottle back on a stir plate and stir overnight in a cold room at 4 °C as cooler temperatures increase the ability for methycellulose to go into the solution.
4. Once the methylcellulose is fully in solution, filter sterilize the 1% methylcellulose MEGM with a vacuum polyethersulfone filtration apparatus with 0.45-μm pore size (*see* **Note 3**).
5. Add the human epidermal growth factor (hEGF), insulin, and hydrocortisone from the BulletKit provided with the MEGM kit and store at 4 °C.

2.2 Complete Mammosphere Medium

1. To Mammosphere Medium, add the following: 20 ng/mL basic fibroblast growth factor (bFGF), 10 ng/mL epidermal growth factor (hEGF), and 4 μg/mL heparin (*see* **Notes 4** and **5**).
2. Allow to warm to 37 °C before use.

2.3 Additional Materials

1. Clear, flat bottom, ultralow-attachment 96-well microplates.
2. Micropipettes.
3. Light microscope.
4. Image analysis software for measuring sphere size.

3 Methods

3.1 Cell Preparation

1. Wash cells that are ~70% confluent with PBS.
2. Detach cells with trypsin or TrypLE (Thermo-Fisher).
3. Neutralize with serum-containing media and spin down cells at 150 g for 5 min.
4. Resuspend cell pellet in base media that does not have serum or added growth factors (*see* **Note 6**).
5. Perform cell count with trypan blue to determine the number of viable cells.

3.2 Cell Plating

1. Calculate the cell number needed for 5,000 cells per ml and dilute the calculated volume of suspended cells into the calculated volume of complete Mammosphere Medium.
2. Plate cells at least in triplicate for each condition at 500 cells per well in a volume of 100 μL in the ultralow-attachment plates (*see* **Note** 7).
3. Incubate at 37 °C, 5% CO_2.

3.3 Media Addition and Replacement

1. Media for the mammosphere cultures should be added or replaced every 3 days. On day 3 after seeding, warm Complete Mammosphere Medium to 37 °C. Add 100 μL of Complete Mammosphere Medium to each well. Add slowly to avoid bubbles.
2. Return plates to the incubator for 3 more days.
3. On days 6 and 9, carefully pipet 100 μL from the top of the well (Fig. 1).
4. Pipette 100 μL of warmed Complete Mammosphere Medium into the well slowly to avoid addition of bubbles.
5. Return plates to the incubator.

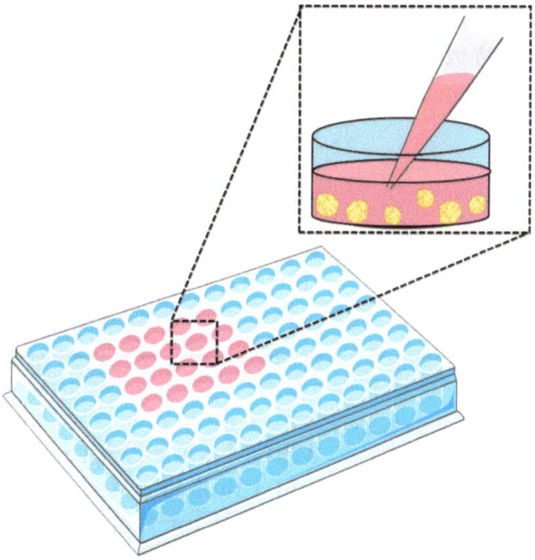

Fig. 1 Schematic of removing media with a pipette. Handle plates carefully to ensure that spheres remain at the bottom of the well. Slowly aspirate 100 μL medium from the top of the well. Do not pipette from the bottom of the well in order to avoid sphere loss. Carefully add 100 μL fresh medium to each well. (This figure was created by modifying Servier Medical Art templates, which are licensed under a Creative Commons Attribution 3.0 Unported License; https:/ smart.servier.com)

Fig. 2 Quantification of mammospheres. HMLER-FOXC2 cells were grown for 12 days as per the mammosphere protocol. Spheres were photographed with a ZEISS ZEN Microscope, and diameters measured with instrument software. Mammospheres larger than 70 μm in diameter were counted. Green checkmarks indicate examples of accepted spheres based on size and shape. Examples of nonspheres are indicated with red X's. The bottom X (44.6 μm) is not a sphere due to small size, and the top X (circled) is not a sphere based on lack of spherical shape

3.4 Mammosphere Quantification

1. On day 12, remove the ultralow-attachment plate from the incubator and view the wells under a light microscope.
2. Count the number of spheres that are over 70 μm in diameter (Fig. 2). Record as the number of mammospheres per 500 cells seeded [5]. This reflects the number of CSCs plated.
3. Determine the average and standard deviation of mammosphere quantity for the replicate samples.

4 Notes

1. MEGM comes with additives of bovine pituitary extract (BPE) and a BulletKit with additional additives. The BPE additive is not used for this protocol. The BulletKit contents should be added when specified in the protocol.
2. Mammosphere medium should be prepared well in advance due to the length of time required to dissolve and filter methylcellulose [6, 7]. The methylcellulose will not go into the 150 mL solution completely, but ensure there are no large clumps of powder prior to adding the remaining 350 mL of media.

3. The methylcellulose can clog the filter during this process so filter 100-mL aliquots at a time. As necessary, replace the filter apparatus and collect into the same collection bottle until the entire batch of medium is filter sterilized.
4. Prepare 10 mL of Complete Mammosphere Medium; this medium can be stored for up to 2 weeks.
5. These additives are not supplied by the MEGM kit and must be prepared separately.
6. When suspending for counting prior to dilution in mammosphere media, use a low volume such that the methylcellulose and growth factors are not diluted, but cells can still be counted accurately.
7. Pipette up and down slowly to mix the cells before plating and plate slowly to ensure that bubbles are not present and that cells are evenly distributed.

Acknowledgments

This work was supported by the NIH/NCI grants 2R01CA155243; 1R01CA200970 awarded to S.A.M, and the NIH T32 grant 5T32CA186892 to N.A.K.

References

1. Dontu G, Abdallah WM et al (2003) In vitro propagation and transcriptional profiling of human mammary stem/progenitor cells. Genes Dev 17:1253–1270
2. Reynolds BA, Weiss S (1996) Clonal and population analyses demonstrate that an EGF-responsive mammalian embryonic CNS precursor is a stem cell. Dev Biol 175:1–13
3. Weiss S, Reynolds BA et al (1996) Is there a neural stem cell in the mammalian forebrain? Trends Neurosci 19:387–393
4. Liao MJ, Zhang CC et al (2007) Enrichment of a population of mammary gland cells that form mammospheres and have in vivo repopulating activity. Cancer Res 67:8131–8138
5. Mani SA, Guo W et al (2008) The epithelial-mesenchymal transition generates cells with properties of stem cells. Cell 133:704–715
6. Ergun R, Guo J et al (2015) Cellulose. Encycl Food Heal:694–702
7. Kapoor D, Maheshwari R (2020) Coating technologies in pharmaceutical product development. In: Drug delivery systems. Elsevier Inc, p 665

Chapter 36

Designing Genetically Engineered Mouse Models (GEMMs) Using CRISPR Mediated Genome Editing

Jade Desjardins, Mitra Cowan, and Yojiro Yamanaka

Abstract

Genetically engineered mouse models (GEMMs) are very powerful tools to study lineage hierarchy and cellular dynamics of stem cells in vivo. Stem cell behavior in various contexts such as development, normal homeostasis and diseases have been investigated using GEMMs. The strategies to generate GEMMs have drastically changed in the last decade with the development of the CRISPR/Cas9 system for manipulation of the mammalian genome. The advantages of the CRISPR/Cas9 are its simplicity and efficiency. The bioinformatics tools available now allow us to quickly identify appropriate guide RNAs and design experimental conditions to generate the targeted mutation. In addition, the genome can be manipulated directly in the zygote which reduces the time to modify target genes compared to other technologies such as Embryonic Stem (ES) cells. Equally important is that we can manipulate the genome of any mouse background with the CRISPR/Cas9 system which omits time-consuming backcrossing processes, accelerates research and increases flexibility. Here, we will summarize basic allelic types and our standard strategies of how to generate them.

Key words CRISPR/Cas9, Genetically engineered mouse models, GEMM, Genotyping, Breeding, Mosaicism

1 Introduction

Genetically engineered mouse models (GEMMs) have been a powerful tool to study in vivo behavior of various types of stem cells. GEMMs have tremendously contributed to the understanding and emergence of specific stem cells in development, cell lineages derived from single stem cell types and their roles in homeostasis and regeneration of tissues and organs. The studies using GEMMs have revealed that deficiency in stem cell function is tightly linked with aging and other diseases such as cancer. Various types of GEMMs such as conventional knock-out (KO), conditional KO (e.g., Cre/loxP system) and Cre-reporter lines have been generated to study gene functions and cell lineages. Until 10 years ago, the majority of GEMMs were generated by targeting embryonic stem

Nagarajan Kannan and Philip Beer (eds.), *Stem Cell Assays: Methods and Protocols*, Methods in Molecular Biology, vol. 2429, https://doi.org/10.1007/978-1-0716-1979-7_36,

(ES) cells with a donor template with insertion of the mutation of interest by homologous recombination and drug selection. The subsequent targeted ES cells are then injected into a blastocyst for chimera production. These chimeric mice are bred to wild-type (WT) mice to assess germline transmission of the targeted mutation. When germline transmission of the targeted allele is confirmed in the heterozygote F1 animals, often backcrossing on another strain to generate the correct experimental mouse strain was necessary. The standard timeline to generate a mutant mouse model using ES cell technology is usually between 18 and 24 months.

In the last decade, the CRISPR/Cas9 system has become the main technological tool to generate GEMMs because of its simplicity and efficiency. This relatively simple system consists of two components (Fig. 1a): a guide RNA (gRNA) and a CRISPR-associated protein (Cas) endonuclease. Each gRNA consists of a spacer and scaffold sequence. A spacer sequence is a 17-20 nt RNA sequence recognizing a target DNA sequence. A scaffold sequence binds to a Cas protein to form a ribonucleoprotein (RNP) complex. The Cas protein most frequently used at this time is the WT Cas9 endonuclease from *Streptococcus pyogenes*. This Cas9 endonuclease recognizes a 5′-NGG-3′ protospacer adjacent motif (PAM) and cleaves the DNA 3–4 nucleotides upstream of the PAM sequence if the adjacent sequence is complementary to the gRNA spacer sequence [1]. A DNA double stranded break (DSB) induced by a gRNA/Cas9 RNP complex will be repaired by two intrinsic cellular DNA repair mechanisms: nonhomologous end joining (NHEJ) or homology directed repair (HDR). These two DNA repair mechanisms can be used to create specific genetic alterations for various research purposes. Another important advantage of the CRISPR/Cas9 system is that it can be used in any background mouse strain (*see* **Note 1**), therefore the time-consuming backcrossing process can be completely omitted. Here, we describe different types of genetically modified alleles and experimental strategies to generate GEMMs using CRISPR/Cas9 mediated genome editing. In this chapter we will share our standard operating procedures on how to select allelic types and design genotyping strategies.

2 Materials

1. DNA extraction buffer: 10 mM Tris, pH 8, 100 mM NaCl, 10 mM EDTA, pH 8, 0.5% sodium dodecyl sulfate (SDS). Make up to 500 mL using DNase/RNase-free water. Store at room temperature.
2. Microinjection buffer: 10 mM Tris-HCl, pH 7.4, 0.1 mM EDTA. Make up to 500 mL using DNase/RNase-free water.

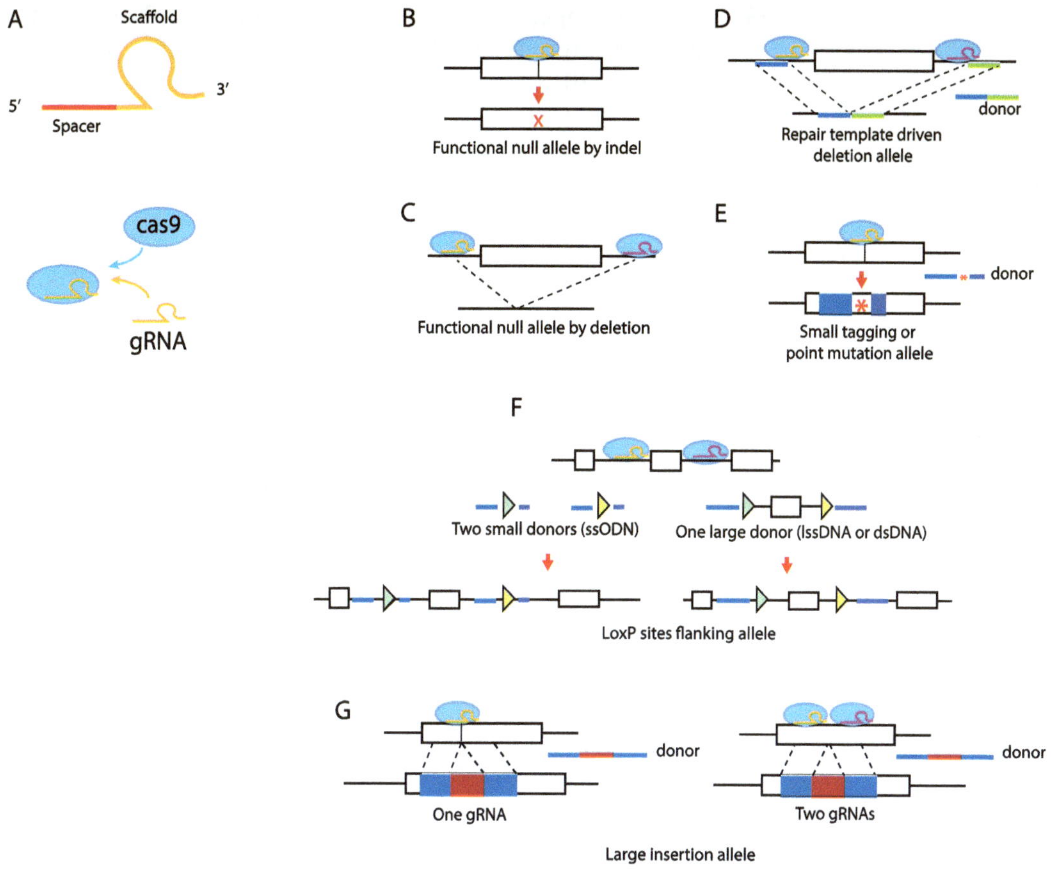

Fig. 1 Allelic types and CRISPR/Cas9 strategies for their generation. (**a**) A gRNA consists of a 17–20 nt spacer sequence specific to the target site and a scaffold sequence that anneals to Cas9 protein to form a ribonucleoprotein (RNP) complex. (**b**) A functional null allele can be created by a NHEJ-mediated indel mutation. If a frame-shift mutation is created, it could lead to an early stop codon. (**c**) A functional null allele can be created by a large deletion. Two gRNAs are selected at both ends of the region of interest. A deletion can happen if two DSBs are joined together by the NHEJ repair mechanism. (**d**) A functional null allele can be created by HDR-mediated deletion allele. Two gRNAs targeting both ends of the region of interest with a donor ssODN carrying 100 nt homology arms facilitate the precise deletion. (**e**) A small tagging or point mutation allele can be created using one gRNA at the target site and one donor ssODN with homology arms of 91 nt (5′) and 36 nt (3′) containing a tag sequence or nucleotide changes. (**f**) *LoxP* sites can be introduced with two gRNAs in the introns at both ends the region of interest and two small donor ssODNs with asymmetric arms of 91 nt (5′) and 36 nt (3′) or with one large donor lssDNA or dsDNA with long (>100 nt or bp) and symmetric arms. (**g**) Large insertion alleles can be generated using one or two gRNAs and a donor lssDNA or dsDNA with long (>100 nt or bp) containing the insert and symmetric arms

Pass through 0.22 μm syringe filter. Store at room temperature. dsDNA donor templates are diluted in microinjection buffer and filtered using a 0.22 μm membrane (Corning® Costar® Spin-X® centrifuge tube filters, Sigma-Aldrich). Large donor templates are stored at 4 °C.

3. All other reagents (primers, gRNAs and lyophilized donor templates) are resuspended in DNase/RNase free water and stored at −20 °C.

3 Methods

3.1 Model Strategy

The NHEJ pathway is active in all cell cycle phases and is the main repair mechanism happening after a DSB caused by the RNP complex. NHEJ leads to ligation of both ends of the break or a larger deletion can occasionally be observed [2].

1. Functional null allele by indel mutation (Fig. 1b) can be generated by using one or multiple gRNAs spaced by 10–200 bp within the coding region [3]. Generally, targeting early exons or coding functional domains is recommended. However, if alternative transcriptional initiation or alternative splicing happens, the gene function may not be compromised. Truncated products might also function as a dominant negative or hypomorphic.
2. Functional null allele by deletion (Fig. 1c) can be created by deleting a portion of gene between two gRNAs. Deleting crucial exon(s), enhancers or promoter regions can result in a nonfunctional allele.

The HDR pathway is a high-fidelity process that uses homology regions from sister chromatids or exogenous donor template to repair DSBs [2]. HDR occurs only in the late S/G2 phase of the cell cycle. Therefore, the HDR pathway is less frequent than the NHEJ pathway [4]. The efficiency of HDR is highly dependent on locus and size of insert.

1. Repair template driven deletion alleles (Fig. 1d) can be achieved using a pair of gRNAs flanking the region to delete and a repair template to ligate both ends by HDR. With this approach, a small or relatively large deletion can be efficiently generated, and the resulting genotype can be precisely predicted.
2. Point mutation alleles (Fig. 1e) are alleles where one or a few nucleotides are changed without changing the sequence length. This approach can be used for introduction of single nucleotide polymorphisms (SNPs), amino acid substitutions to modulate posttranslational modifications or insertion of an early stop codon. For this strategy, one gRNA is chosen close to a target site and nucleotide alterations are introduced using a small donor template.

3. Small tagging alleles (Fig. 1e) are alleles where short sequences (15–100 bp) like V5 and His tags are inserted. For this strategy, one gRNA is chosen close to a target site and a tag is introduced using a small donor template.
4. Conditional alleles using *loxP* sequences (Fig. 1f) are created by flanking one or more critical coding exon(s) with *loxP* sites. gRNAs are designed in the intronic regions at least 100 bp away from the exon boundaries and *loxP* sites can be introduced by two small or one large donor template(s).
5. Large insertion alleles (Fig. 1g) such as fluorescent protein genes and Cre recombinase gene are inserted at a specific locus for lineage tracing and conditional gene deletion, respectively. The *Rosa26* locus is the most common site-directed insertion locus in mice for ubiquitous or conditional expression of gene-of-interest [5]. Depending on the goals, one or two gRNAs are used, and the insertion is introduced using a large single-stranded or double-stranded DNA donor template [6–8].

3.2 Design and Choice of gRNAs

1. The genomic structure of a gene-of-interest is analyzed through the Ensembl database (http://uswest.ensembl.org/index.html). The region to target is chosen based on project goals, structure of coding sequence and transcript variants.
2. gRNAs are selected by entering the sequence of interest, typically between 50–200 bp, into the online tool CRISPOR (http://crispor.tefor.net/). The genome and PAM motif are selected based on project goals (*see* **Notes 2–7**).
3. We suggest staying within 10 bp between the gRNA's cut site and the target site. The cutting efficiency of gRNAs can vary even if there is only a difference of a few base pairs in the sequence. Therefore, three candidates per target site are selected for testing.
4. Synthetic single gRNAs can be purchased from either Synthego or Integrated DNA Technologies (IDT).

3.3 PCR Strategy Design

1. A genomic PCR protocol for screening gRNA cutting efficiency and further genotyping should be optimized before ordering gRNAs. Some regions of the genome might be difficult to amplify by PCR. If a region is too difficult for genotyping, redesigning the gRNAs is suggested.
2. Primers are selected using the NBCI primer tool. (https://www.ncbi.nlm.nih.gov/tools/primer-blast/).
3. PCR products between 0.4 and 1 kb encompassing the gRNA cut site are ideal.

4. There must be at least 100 bp between the primers and the cut site for readable Sanger sequencing around the cut site.
5. The GC content of the PCR product is ideally between 35 and 60% and repeated sequences should be avoided.
6. PCR products can be sent for Sanger sequencing to test feasibility. It is important to BLAST the sequencing product against the database to confirm sequence and identify possible SNPs.

3.4 Validation of gRNAs in Mouse Zygotes

On-target cutting efficiency of the gRNA is the first key element for a successful CRISPR/Cas9 project. A gRNA that has a low cutting efficiency has a small chance to generate the desired genome modification by either NHEJ or HDR. We test gRNAs in mouse zygotes before attempting to make any animal model. Using the most efficient gRNA(s) will generate founder mice quicker and require less resources to produce the founder mice.

1. Mouse zygotes (E0.5) are harvested from 3-week-old superovulated females as previously described [9]. We suggest validating gRNAs in the same background strain that is chosen to make the final model since different strains and substrains occasionally have single nucleotide polymorphisms (SNPs) that can affect their cutting efficiency. All zygotes are cultured in EmbryoMax® Advanced KSOM Embryo Medium (Millipore-Sigma) under mineral oil (Millipore-Sigma) in a 35 mm dish at 37 °C and 5.5% CO_2.
2. Zygotes are either microinjected or electroporated with each gRNA/Cas9 RNP complex (*see* Subheading 3.8). All gRNAs are tested separately (*see* **Note 8**).
3. After microinjection or electroporation, the embryos are cultured and monitored daily until they develop to the blastocyst stage (E3.5). Lysed embryos are removed from the culture dish.

3.5 DNA Isolation and Genotyping of Blastocysts

1. Embryos are cultured up to blastocyst stage (E3.5) with conditions mentioned above and genomic DNA is isolated as previously described [10].
2. Blastocysts are collected individually using a mouth pipette and placed in a PCR tube containing 0.4 μL of extraction solution + 1.1 μL of tissue preparation solution (REDExtract-N-Amp™ Tissue PCR kit, Millipore-Sigma).
3. Tubes are placed in PCR machine with one cycle of 25 °C for 20 min, 56 °C for 30 min, 95 °C for 5 min and held at 4 °C.
4. 4.4 μL neutralization buffer is added to each tube, mixed by vortexing and samples are stored at −20 °C.
5. The PCR is conducted using Q5® Hot Start High-Fidelity 2× Master Mix (NEB) following a protocol from the supplier

(*see* **Note 9**). 2 μL of DNA extraction is used for a 25 μL PCR reaction. WT (e.g., nonmodified DNA) and negative (e.g., H20) controls should be included.

6. The annealing temperature of the primers is calculated using the NEB TM calculator (https://tmcalculator.neb.com/).
7. Because of the low amount of genomic DNA from a single blastocyst, second or nested PCR amplification might be necessary. 1 μL of the PCR product from the first reaction is used for the second PCR reaction.
8. 5 μL of the final reaction is loaded on a 1–2% agarose gel to confirm correct amplification.
9. The remaining nonpurified PCR products (20 μL) are sent for Sanger sequencing with one of the primers. WT control (i.e., nonmodified DNA) is sequenced along with the samples.
10. PCR products of blastocysts with visible shifts on gel are a sign of detectable deletions or insertions and can be considered as positive results without sequencing.

3.6 Sequence Analysis and Interpretation

1. Chromatograms are downloaded as Ab1 format.
2. It is crucial to look at chromatogram peaks instead of text sequences when analysing sequencing results (*see* **Note 10**).
3. Chromatograms are aligned using the Benchling website (https://benchling.com/).
4. The WT control is compared to the database sequence to confirm the presence of SNPs, accuracy of the sequence and absence of contamination.
5. Chromatograms from each blastocyst are aligned individually at the gRNA spacer sequence against the WT sequence to locate the cut site.
6. Indels and point mutations around the cut site are evidence of misrepair and are considered positive results. Indel mutations can result in unreadable sequences downstream from the cut site.
7. About 10 blastocysts are used to validate the efficiency of each gRNA. The efficiency is evaluated by the ratio of DNA modified and nonmodified embryos (i.e., an efficiency score (%) = (# of embryos with modifications)/(total numbers used in validation) × 100). We use this efficiency score to compare gRNAs.
8. The best gRNA is selected based on its cutting efficiency as well as the proximity of its cut site to the target site. An efficiency score closer to 100% is ideal. Avoid gRNAs with efficiency scores lower than 60–75% if other candidates are available. If the nearest gRNA does not have the best score, this gRNA could still be chosen if the efficiency score is acceptable.

3.7 Designing a Donor Template for HDR (Fig. 1)

1. Whenever possible, inserts (tags, *loxP* sites etc.) are inserted at gRNA's cut site so that the gRNA sequence is disrupted, and donor template will not be recognized by gRNA/Cas9 RNP complex. Otherwise, the PAM or at least 2 or 3 nucleotides of the gRNA's seed sequence are mutated. These mutations can delete or create a unique restriction site for restriction fragment length polymorphism (RFLP) genotyping.
2. Donor template is complementary to PAM strand when using one gRNA or 2 gRNAs on the same strand. If gRNAs are on different strands, donor sequence can be either strand.
3. Homology arms for small donors (< 200 nt) are 91 nt for the 5′ arm and 36 nt for the 3′ arm except for deletion donors where both arms are 100 nt. Homology arms for large donors are 100–500 nt or bp depending on insert size, goal of the project, region specificities and genotyping strategy.
4. Small donors (< 200 nt) are ordered as single-stranded Ultramers® from IDT. Large donors (> 200 nt) are ordered as long single-stranded DNA (lssDNA) Megamers® from IDT for double-stranded plasmids from Genescript (*see* **Note 11**).

3.8 CRISPR/Cas9 Delivery Methods (Fig. 2)

Mouse zygotes (E0.5) on the background of interest are harvested from 3-week-old superovulated females as previously described [9]. Zygotes are cultured in EmbryoMax® Advanced KSOM Embryo Medium (Millipore-Sigma) under mineral oil (Millipore-Sigma) in a 35 mm dish at 37 °C and 5.5% CO_2.

1. For microinjection of mouse zygotes or two-cell stage embryos, a 20 μL microinjection mix is prepared in a 0.6 mL Eppendorf tube. Add 1 μL of Cas9 protein (1 mg/mL stock) and 6.67 μL of gRNA (150 ng/μL stock) and incubate at room temperature for 5 min to form the RNP complex before adding other components. The final concentration is 50 ng/μL of Cas9 and 50 ng/μL of gRNA. If multiple gRNAs are used, each gRNA is complexed with Cas9 separately before being combined to the same final concentration. For HDR projects, add 0.6 μL of ssODN (1000 ng/μL stock) or 4 μL of ssDNA or dsDNA (100 ng/μL stock) for a final concentration of 30 ng/μL or 20 ng/μL respectively. Add 2 μL of 1 M KCl for a final concentration of 100 mM. Add DNase/RNase free water up to 20 μL Spin the tube at 20,000 × *g* for 10 min to precipitate debris. Carefully transfer 15 μL of the supernatant without disturbing the pellet to a new tube for microinjection. Microinject the mix into either the pronucleus of zygotes or the nucleus of two-cell stage embryos as previously described [6, 11]. For two-cell injection, one nucleus or both nuclei are injected based on project goals.

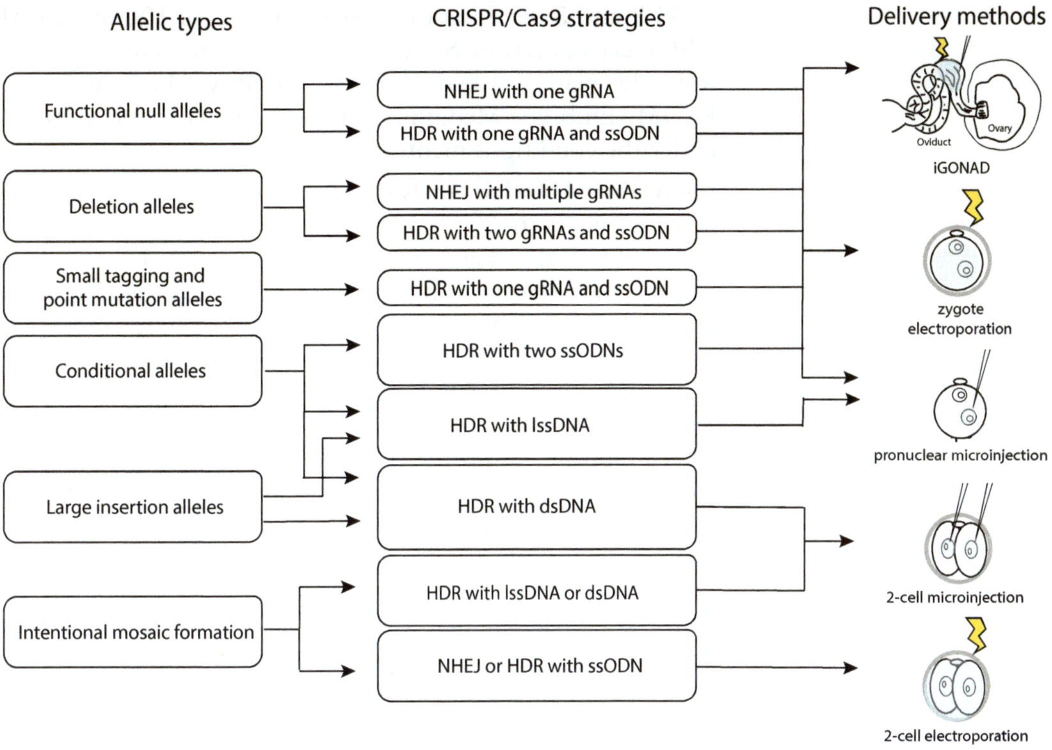

Fig. 2 Flowchart for allelic types, CRISPR/Cas9 strategies, and delivery methods. Functional null alleles, deletion alleles, small tagging alleles, and point mutation alleles made with either one or two gRNAs and with or without an ssODN can be introduced in the mouse embryos by iGONAD, zygotes electroporation or by pronuclear microinjection. Conditional alleles made with two ssODNs can be introduced using the same three methods. For *loxP* flanking alleles or large insertion alleles using a lssDNA, pronuclear microinjection is used and for dsDNA donors, two-cell microinjection is used. Intentional mosaic alleles are made using two-cell microinjection for long donors whereas electroporation at two-cell stage can be used when using a short or no donor

2. For electroporation of mouse zygotes or two-cell stage embryos, a 10 μL electroporation mix is prepared in a 0.6 mL Eppendorf tube. Add 1 μL of Cas9 protein (62 μM or 10 mg/mL stock) and 1.44 μL gRNA (approx. 42 μM or 1500 ng/μL stock) and incubate at room temperature for 5 min to form the RNP complex before adding other components. The final concentration is 6 μM, approx. or 1 μg/μL of Cas9 and 6 μM or approx. 216 ng/μL of gRNA. If multiple gRNAs are used, each gRNA is complexed with Cas9 separately before being combined to the same final concentration. For HDR projects, add 2.4 μL of ssODN (1000 ng/μL or approx. 25 μM stock) for a final concentration of 6 μM. Add Opti-MEM media with L-glutamine (Fisher) up to 10 μL. Embryos are washed 3 times in Opti-MEM and in a final 5 μL drop of the mix. Embryos are transferred to an electroporation 1 mm gap slide chamber containing 5 μL of the mix and electroporated using 10 pulses

of 30 V and 3 ms duration with an interval of 100 ms (ECM 830 Electroporation System, BTX). Zygotes are recovered from the plate, washed 3 times in EmbryoMax® Advanced KSOM Embryo Medium (Millipore-Sigma) and cultured under the conditions mentioned above.

3. For oviductal Nucleic Acids Delivery (iGONAD) [12], sexually mature females, 6–8-week-old depending on strain, are mated with fertile males. Females are not superovulated but oestrus can be synchronized using the Whitten effect by adding male used-bedding in female cages 3 days prior to mating [13]. Copulation plugs are observed the next morning (E0.5). A 10 μL iGONAD mix is prepared in a 0.6 mL Eppendorf tube. Add 1 μL of Cas9 protein (62 μM or 10 mg/mL stock) and 1.44 μL gRNA (approx. 42 μM or 1500 ng/μL stock) and incubate at room temperature for 5 min to form the RNP complex before adding other components (final concentration of 6 μM, approx. 1 μg/μL Cas9 and 6 μM, approx. 216 ng/μL gRNA). If multiple gRNAs are used, each gRNA is complexed with Cas9 separately before being combined to the same final concentration. For HDR projects, add 2.4 μL of ssODN (1000 ng/μL or approx. 25 μM stock) for a final concentration of 6 μM. Add 1 μL of 1 M KCl for a final concentration of 100 mM. Add DNase/RNase free water up to 10 μL. The electroporation procedure is performed at E0.4, E0.7 or E1.5 depending on project goals. Oviducts are electroporated using the CUY21 EDIT II electroporator (BEX) and parameters as previously described [12].

3.9 Embryo Implantation and Monitoring of Pregnancies

1. After microinjection or electroporation, the embryos are transferred into the oviducts of pseudopregnant CD-1 females (E0.5) [9].
2. Females are carefully monitored on the due date of birth (E19.5). If there are only 3 pups or less in a litter, pups from other mating pairs with a different coat color or marked by a tail snip are added to the litter. This helps to increase milk production and reduce mortality [14]. Caesarean delivery is performed if the pregnant females do not deliver by E20.5. The pups are fostered by other mating pairs with a litter of similar age. They should have a different coat color or the receiving litter should be marked by a tail snip.
3. Pups are monitored weekly to identify unusual phenotypes or need for special treatments.

3.10 DNA Isolation from Ear Punch or Tail Snips

1. Ear punches are collected for genotyping at 2–3 weeks of age. A metal tag with a unique number is applied on the right ear and a punch is collected from the left ear. A tail snip is additionally collected upon request. Pups are weaned at 3–4 weeks of age.

2. Tissues are collected in 1.5 mL Eppendorf tubes and kept at −20 °C (*see* **Note 12**).
3. Lysis buffer is prepared by diluting Proteinase K to 500 μg/μL in DNA extraction buffer.
4. Add 500 μL lysis buffer to each sample.
5. Incubate the samples overnight at 37 °C.
6. Add 250 μL saturated (6 M) NaCl to each tube and vortex for 2–5 min.
7. Place the tubes on ice for 10 min.
8. Spin the tubes for 10 min at low speed (9000 × *g*).
9. Transfer 500 μL of the supernatant to a fresh tube with 1 mL 100% EtOH and mix by inversion.
10. Centrifuge at high speed for 10 min (20,000 × *g*).
11. Discard supernatant by vacuum and wash the pellet with 500 μL 70% EtOH to remove salts.
12. Repeat washing twice with 500 μL 70% EtOH.
13. Dry the pellet with the lid opened at room temperature or on a heat block. Do not over dry the pellet.
14. Resuspend the DNA pellet in 40 μL (ear punch) or 100 μL (tail snip) DNase/RNase free water.

3.11 Genotyping of Animals (Fig. 3)

Although the gRNA/Cas9 RNP complex is delivered into zygotes, the F0 mice that are born often carry more than two distinct allelic types due to mosaicism (Fig. 4). DNA cutting and repairing events take place independently on each allele and it is impossible to control their precise timing which will then generate a variety of mutations. DNA editing events can happen at the zygote, two-cell, or later developmental stages of the preimplantation embryo. We can use the ability to generate edit in different developmental stages to intentionally create mosaic animals from a few genetically distinct cell populations each carrying a unique set of two targeted alleles. Generally, precise genotyping of genetically mosaic mice is challenging (*see* **Note 13**). We recommend final confirmation of the precise genotype in the F1 generation.

1. To detect indel and point mutation modifications, a pair of primers external to the donor template is selected. The size of the PCR product should be between 0.4 and 1 kb. Always include WT and negative controls in PCR reactions. The samples are loaded on a 1–2% agarose gel. It is rare to observe a visible size difference between small mutations and the WT alleles on the gel unless an occasional large deletion occurs. If there are multiple bands in a PCR reaction, the band at the

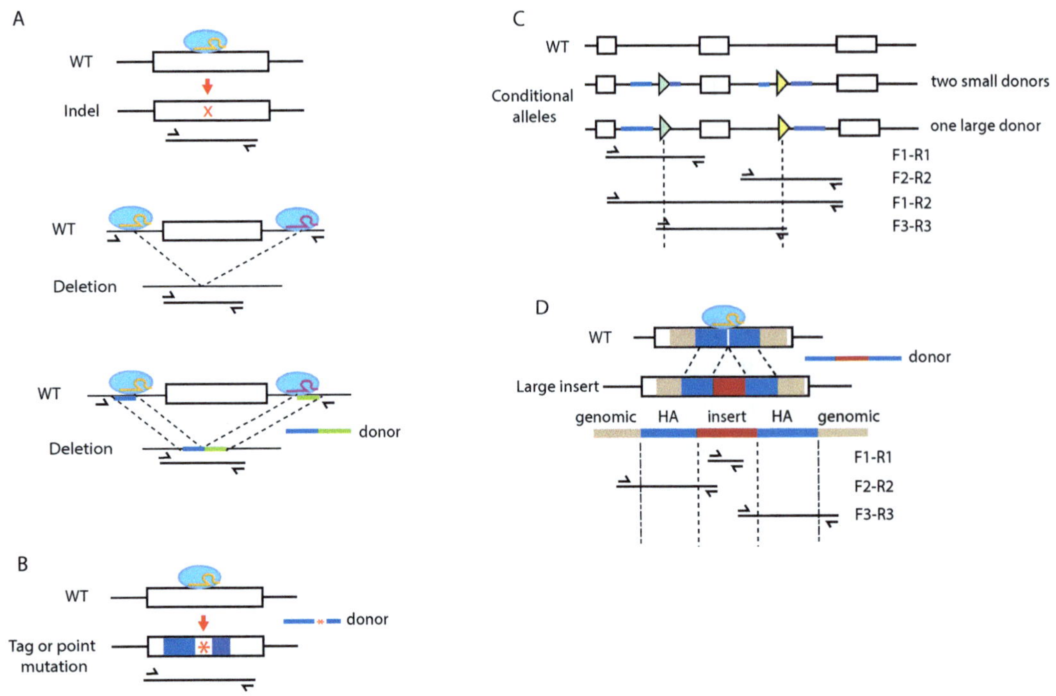

Fig. 3 Genotyping strategies to detect desired genome modifications. (**a**) To genotype deletion alleles, primers external to the donor template or at least 100 bp from the gRNA(s) cut site are selected. (**b**) Small tagging and point mutation alleles are detected by primers outside the homology arms (HA). (**c**) Conditional alleles are first genotyped by detecting each *loxP* site individually using primers sets F1-R1 and F2-R2. Cis-integration of *loxP* sites is evaluated by primers encompassing both *loxP* sites using primers F3-R3 or F1-R2. (**d**) Large insertion alleles are detected by first identifying the insertion with primers F1-R1 specific to the insert sequence. Both junctions are assessed using primers F2-R2 and F3-R3 in genomic DNA to confirm targeted insertion

expected size is isolated from the gel and reamplified by PCR to get a single product. Bands of the expected size are sent for Sanger sequencing with both primers.

2. To detect small tagging or *loxP* alleles, a small PCR product (0.4–0.7 kb) is designed and loaded on a 2–3% thick agarose gel to permit the separation of two ore more PCR products. If there are multiple bands in a PCR reaction, the band at the expected size is isolated from the gel and reamplified by PCR to get a single product. Bands of the expected size are sent for Sanger sequencing to make sure of the insert and surrounding sequence integrity.
3. RFLP genotyping is used when a restriction site has been added or removed by the introduced mutation. The size of the PCR product should be between 0.4 and 1 kb. After PCR amplification, the PCR product is digested with a restriction enzyme according to the supplier's protocol. The digested product is loaded on a 1–3% agarose gel depending on product sizes.

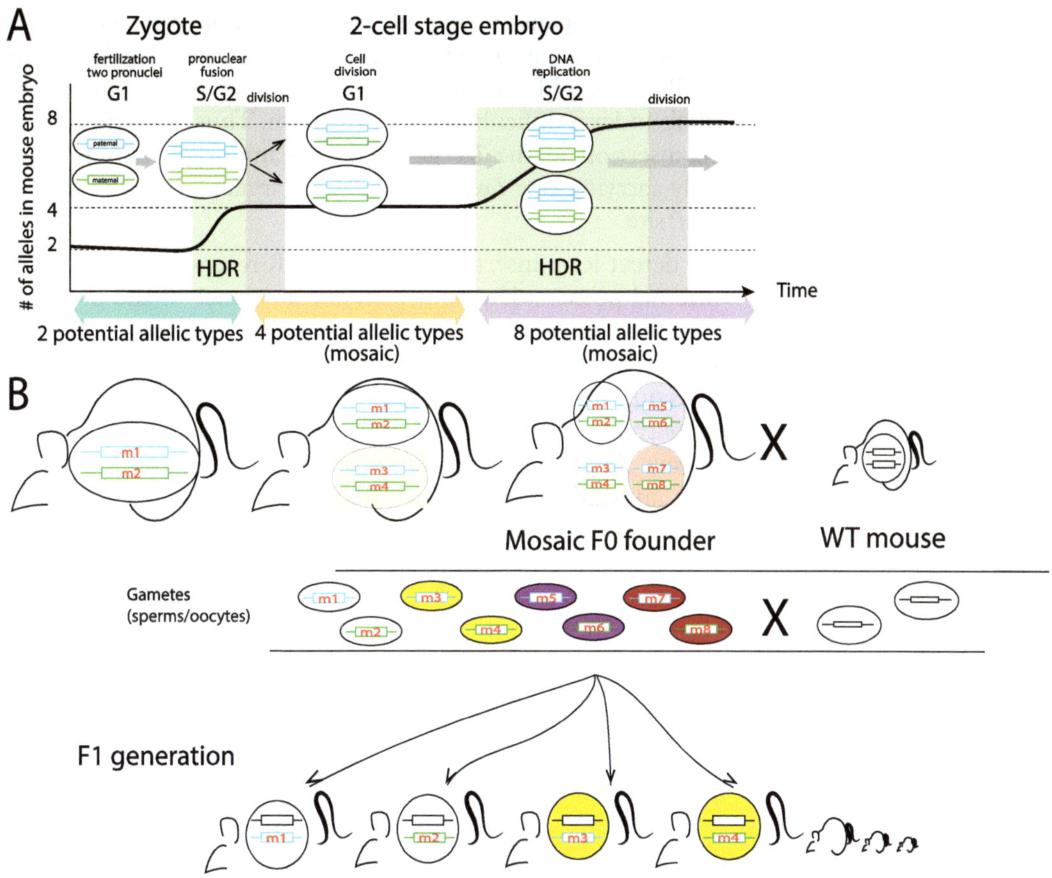

Fig. 4 Mechanism of mosaicism during CRISPR/Cas9-mediated genome editing. (**a**) The number of alleles in an embryo increases as the embryo develops. Because CRISPR/Cas9 mediated genome editing events take place in each allele independently and the precise timing of editing cannot be controlled, the developed animals will have a chance to become genetically mosaic if the editing occurs after the first cell division. (**b**) Due to meiosis during gamete (sperm/oocyte) formation, each gamete carries only one allele. Therefore, by crossing the F0 mosaic founders with WT mice, the next F1 generation will be heterozygous. Typically, because a mosaic animal consists of several genotypic cells, the inheritance of each allele does not follow the Mendelian ratio. Sometimes, an allele detected in the F0 generation is not recovered in the F1 generation. Contrarily, an allele barely detected in the F0 generation can be found in the F1 generation. It also happens that the correct allele can be inherited only after several litters are produced. It is very important to genotype several litters to identify all transmitted alleles

4. When introducing *loxP* sites, it is crucial to determine their cis-insertion. Each *loxP* insertion is detected as described in Subheading 3.10, **step 2**. If possible, a long-range genomic PCR encompassing both *loxP* sites can confirm cis-insertion. Although short PCR genotyping detecting individual *loxP* insertions cannot usually rule out cis- or trans-insertion, if one of the *loxP* sites appears homozygous (i.e., no WT allele) and the second *loxP* site is inserted, one allele can be cis-insertion. One exception is if a large NHEJ-mediated

deletion happens in one allele, the deletion can eliminate a primer annealing site resulting in no detection of the WT allele by PCR. In addition, in some cases, precise genotyping in the F0 founder is difficult due to mosaicism. Thus, confirmation of cis-insertion in the F1 generation is important. Those mice only carrying one *loxP* site can be retargeted with the second *loxP* site.

5. To detect long insertion alleles, a PCR reaction with a pair of internal primers detecting a unique sequence of the insertion (i.e., not detecting the host genome) is designed. This PCR reaction can detect both random integration and targeted insertion. Then, two genomic PCR reactions are designed to amplify both ends of the donor template using an external primer detecting a genome sequence outside the homology arm and an internal primer detecting a unique sequence of the insert. Each PCR reaction encompasses the 5′ or 3′ end of a homology arm of the donor template. Mosaicism in F0 founders may cause difficulty to confirm locus specific insertion. It is easier and important to confirm proper insertion in the F1 generation. The mice carrying the correct insertion should be validated by Sanger sequencing. The integrity of the insertion as well as the junctions should be assessed by Sanger sequencing.

3.12 Breeding F0 Founder Animals

We suggest breeding 3 founders (F0) or more to establish a new mutant line. The goal is to ensure having the correct genotype transmitted to the next generation and that the phenotype is not a result of a secondary insertion of the donor template, off-targets or random mutations generated by the CRISPR/Cas9 system. Two backcrosses to WT mice are suggested to segregate alleles (*see* **Note 14**). Mate each founder line individually when mice are between 6 and 8 weeks of age to produce the next generation. F1 mice must be genotyped carefully as described in Subheading 3.10. Each F1 offspring inherits a single allele from the F0 parent founder (*see* **Note 15**, Fig. 4). This allele segregation permits accurate genotyping and confirms germline transmission. Internal crossing can be performed with F2 mice to generate homozygous mice. F2 mice can be mated to other mouse lines to create experimental models (e.g., multiple KOs or Cre-lox models).

3.13 Long Term Genotyping Strategy

After F0 and F1 animals are thoroughly analyzed and the mutations are properly sequenced, a simple long-term genotyping strategy can be developed. A primer specific to a unique mutation (point and indel mutations, etc.) can be designed for PCR genotyping without sequencing. A primer pair specific to the equivalent WT allele can also be designed to differentiate heterozygous and homozygous animals.

3.14 Cryo-preservation

We strongly encourage cryopreserving embryos or sperm at the F2 generation (*see* **Note 16**). It reduces actively breeding all different founder lines. Cryopreservation of mouse lines also protects from breeding issues, diseases, human errors, or disasters.

4 Notes

1. Any mouse strain can be used for CRISPR/Cas9 mediated genome editing. It is also possible to introduce a second genome modification in preexisting mouse lines. This is useful to create multiple KO lines, make specific modifications in complex genetic models or remove a selection gene cassette from mouse lines originally generated with ES cells.
2. gRNA candidates are selected based on the proximity to target sites, their predicted efficiency scores and potential off-targets.
3. In absence of an 5′-NGG-3′ PAM motif close to the target site, other CRISPR-associated proteins recognizing different PAM motifs like Cas12a/Cpf1 can be used.
4. The efficiency of HDR mediated modifications is strongly linked to the distance between a cut site and a target modification site. Both sites should be as close as possible and within 10 bp is preferred [15]. We have successfully created modified alleles with longer distances when no optimal gRNA was available, but this can create potential pitfalls.
5. gRNAs having off-targets with 3 mismatches or less should be avoided as they have a higher risk to induce other mutations in the genome. They should be avoided especially if the potential off-target sites are located in a coding region as mutations might cause unexpected phenotypes.
6. The scores we look at are the MIT specificity score for off-targets and the Doench'16 and Mor.-Mateos for on-target efficiency. These scores can be used as reference but their predictive value is limited. Validation in zygotes is crucial [16].
7. gRNAs can also be designed manually by looking for a PAM motif close to the target site and taking the upstream 17–20 nt sequence with a GC content percentage of 40–80%.
8. For multiplexing projects, multiple gRNAs can be tested in the same mix to reduce the number of animals used for screening. It is important that each gRNA can be evaluated separately using independent PCR products.
9. Optimization of the PCR protocol using WT genomic DNA is critical before attempting to genotype blastocysts.

10. It is critical to look at the chromatogram peaks and not the text formats as the text will indicate only the predominant peak at each position. Embryos targeted with CRISPR/Cas9 may have more than 2 alleles due to mosaicism and minor mutated alleles could be masked by the dominant WT sequence.
11. Large donor templates can be synthetized using the targeted integration with linearized dsDNA-CRISPR (TILD) [8].
12. When collecting tissue samples, we suggest cutting each sample into two pieces and freeze them separately. These backup tissue samples can be used when doing a final confirmation of the genotype of founder animals or to optimize the PCR protocol without having to take another biopsy from the animals.
13. Having an optimized PCR genotyping protocol ahead of time is crucial to avoid delay for breeding the next generation. Older mice might not breed efficiently, and lineage could be lost.
14. Always mate the founders with WT purchased mice of the same background and supplier that was used to generate the line. Do not mate founders with WT mice from an existing colony. The goal is to start the new line on a clean genetic background exempt from genetic drift, random mutations etc.
15. Modifications on autosomal genes should follow the mendelian inheritance model if they do not cause early lethality.
16. If F1 generation is cryopreserved, the second backcross needs to be done upon reviving the line.

References

1. Cong L, Ran FA, Cox D, Lin S, Barretto R, Habib N, Hsu PD, Wu X, Jiang W, Marraffini LA, Zhang F (2013) Multiplex genome engineering using CRISPR/Cas systems. Science 339:819–823. https://doi.org/10.1126/science.1231143
2. Hilton IB, Gersbach CA (2015) Enabling functional genomics with genome engineering. Genome Res 25:1442–1455. https://doi.org/10.1101/gr.190124.115
3. Zuo E, Cai Y-J, Li K, Wei Y, Wang B-A, Sun Y, Liu Z, Liu J, Hu X, Wei W, Huo X, Shi L, Tang C, Liang D, Wang Y, Nie Y-H, Zhang C-C, Yao X, Wang X, Zhou C, Ying W, Wang Q, Chen R-C, Shen Q, Xu G-L, Li J, Sun Q, Xiong Z-Q, Yang H (2017) One-step generation of complete gene knockout mice and monkeys by CRISPR/Cas9-mediated gene editing with multiple sgRNAs. Cell Res 27:933–945. https://doi.org/10.1038/cr.201
4. Lieber MR (2010) The mechanism of double-strand DNA break repair by the nonhomologous DNA end-joining pathway. Annu Rev Biochem 79:181–211. https://doi.org/10.1146/annurev.biochem.052308.093131
5. Casola S (2010) Mouse models for miRNA expression: the ROSA26 locus. In: Monticelli S (ed) MicroRNAs and the immune system. Humana Press, Totowa, NJ, pp 145–163
6. Gu B, Posfai E, Rossant J (2018) Efficient generation of targeted large insertions by microinjection into two-cell-stage mouse embryos. Nat Biotechnol 36:632–637. https://doi.org/10.1038/nbt.4166
7. Miura H, Quadros RM, Gurumurthy CB, Ohtsuka M (2018) Easi-CRISPR for creating knock-in and conditional knockout mouse models using long ssDNA donors. Nat Protoc 13:195–215. https://doi.org/10.1038/nprot.2017.153
8. Yao X, Zhang M, Wang X, Ying W, Hu X, Dai P, Meng F, Shi L, Sun Y, Yao N, Zhong W, Li Y, Wu K, Li W, Chen Z, Yang H (2018) Tild-CRISPR allows for efficient and

precise gene knockin in mouse and human cells. Dev Cell 45:526–536.e5. https://doi.org/10.1016/j.devcel.2018.04.021

9. Behringer R, Gertsenstein M, Nagy KV, Nagy A (2014) Manipulating the mouse embryo: a laboratory manual, 4th edn. Cold Spring Harbor, New York

10. Yamanaka Y (2016) CRISPR/Cas9 genome editing as a strategy to study the tumor microenvironment in transgenic mice. In: Ursini-Siegel J, Beauchemin N (eds) The tumor microenvironment. Springer, New York, pp 261–271

11. Posfai E, Petropoulos S, de Barros FRO, Schell JP, Jurisica I, Sandberg R, Lanner F, Rossant J (2017) Position- and Hippo signaling-dependent plasticity during lineage segregation in the early mouse embryo. elife 6:e22906. https://doi.org/10.7554/eLife.22906

12. Ohtsuka M, Sato M, Miura H, Takabayashi S, Matsuyama M, Koyano T, Arifin N, Nakamura S, Wada K, Gurumurthy CB (2018) i-GONAD: a robust method for in situ germline genome engineering using CRISPR nucleases. Genome Biol 19:25. https://doi.org/10.1186/s13059-018-1400-x

13. Whitten WK (1956) Modification of the oestrous cycle of the mouse by external stimuli associated with the male. J Endocrinol 13: 399–404. https://doi.org/10.1677/joe.0.0130399

14. Pollock JD (1996) Mouse genetics: concepts and applications. In: Silver LM (ed) The quarterly review of biology, vol 71, pp 123–123. https://doi.org/10.1086/419294

15. Liang X, Potter J, Kumar S, Ravinder N, Chesnut JD (2017) Enhanced CRISPR/Cas9-mediated precise genome editing by improved design and delivery of gRNA, Cas9 nuclease, and donor DNA. J Biotechnol 241:136–146. https://doi.org/10.1016/j.jbiotec.2016.11.011

16. Haeussler M, Schönig K, Eckert H, Eschstruth A, Mianné J, Renaud J-B, Schneider-Maunoury S, Shkumatava A, Teboul L, Kent J, Joly J-S, Concordet J-P (2016) Evaluation of off-target and on-target scoring algorithms and integration into the guide RNA selection tool CRISPOR. Genome Biol 17: 148. https://doi.org/10.1186/s13059-016-1012-2

Chapter 37

Assays for the Spectrum of Circulating Tumor Cells

Xuanmao Jiao, Chandan Upadhyaya, Zhao Zhang, Jun Zhao, Zhiping Li, Vivek I. Patel, and Richard G. Pestell

Abstract

Cancer cells sharing stem cell properties are called "cancer stem cells" (CSCs). CSCs have distinct metabolic properties, are intrinsically drug resistant evading chemotherapies, are regulated by miRNA networks and participate in tumor relapse and metastases. During metastatic dissemination, circulating tumor cells (CTCs) invade distant organs and settle in supportive niches. In this process, the stem cell–like properties within CTCs contribute to CTC survival and eventually seed the growth of a secondary tumor. We herein describe methodologies for the analysis of CTCs as they reside in distinct functional pools with distinct characteristics.

Key words Circulating tumor cells, Genomics, Metastasis, Patient-derived xenografts (PDX), Self-seeding

1 Introduction

Cancer cells sharing certain properties with stem cells are often referred to as called "cancer stem cells" (CSCs). CSCs have distinct metabolic properties [1], are intrinsically drug resistant evading the cell killing by chemotherapy and radiation therapy, are regulated by miRNA networks [2, 3], and participate in tumor relapses and metastases [4].

During metastatic dissemination, circulating tumor cells (CTCs) invade distant organs and settle in supportive niches. In this process, subsets of CTCs with stem cell-like properties contribute to CTC survival and eventually seed the growth of a secondary tumor (Fig. 1).

CTCs may reside in the blood stream (i) in isolation, (ii) in clusters, and (iii) associated with immune cells. CTC clusters arise from oligoclonal tumor cell groupings and not from intravascular aggregation events. Both single CTCs and CTC clusters undergo dynamic expression of epithelial and mesenchymal (EMT) markers.

Nagarajan Kannan and Philip Beer (eds.), *Stem Cell Assays: Methods and Protocols*, Methods in Molecular Biology, vol. 2429, https://doi.org/10.1007/978-1-0716-1979-7_37,

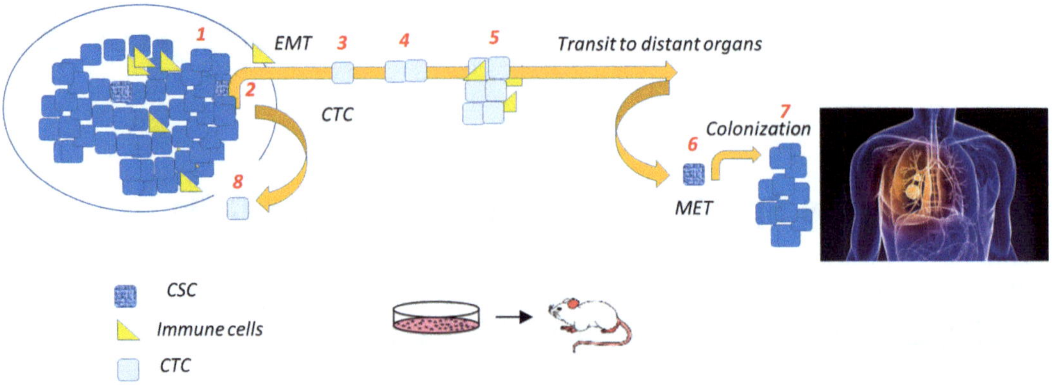

Fig. 1 Schematic representation of the circulating tumor cell spectrum. During metastatic dissemination, circulating tumor cells (CTCs) invade distant organs and settle in supportive niches. In this process, the stem cell–like properties within CTCs contribute to CTC survival and eventually seed the growth of a secondary tumor. A subset of CTCs give rise to metastases. CTC may reside in the blood stream in (i). isolation, in (ii) clusters, and (iii) associated with immune cells. (iv) CTC in turn may give rise to autometastasis

Although rare in the circulation compared with single CTCs, CTC clusters have up to 50-fold increased metastatic potential [5, 6].

In this chapter, we describe assays to determine CTC number and function at each stage of the CTC spectrum.

2 Materials

2.1 Reagents and Culture Dishes

1. Calcium chloride ($CaCl_2$).
2. Sodium chloride (NaCl).
3. Magnesium chloride ($MgCl_2$).
4. Ammonium chloride (NH_4Cl).
5. Disodium phosphate (Na_2HPO_4).
6. Monosodium phosphate (NaH_2PO_4).
7. Sodium carbonate (Na_2CO_3).
8. Monopotassium carbonate ($KHCO_3$).
9. Disodium ethylenediaminetetraacetate (Na_2EDTA).
10. Ethanol.
11. Paraformaldehyde.
12. Propidium iodide (PI).
13. Hoechst 33342.
14. Bovine serum albumin (BSA).
15. Epidermal growth factor human (EGF-h).
16. Fibroblast growth factor-basic human (bFGF-h).

17. Insulin, human recombinant.
18. Cholera toxin A subunit from *Vibrio cholerae*.
19. Hydrocortisone (suitable for cell culture).
20. Heparin.
21. Dulbecco's Modified Eagle's Medium.
22. DMEM/F12 (50/50) medium.
23. RPMI 1640 medium.
24. Fetal bovine serum (FBS).
25. Penicillin–streptomycin solution (100×).
26. 0.25% trypsin-2.21 mM EDTA solution
27. B-27™ Plus Supplement (50×).
28. Trypan Blue solution (0.4%, liquid, sterile-filtered, suitable for cell culture).
29. Corning Matrigel Matrix.
30. Rat tail collagen I.
31. Normal Rat IgG.
32. Alexa Fluor® 488–conjugated rat anti-mouse EpCAM antibody.
33. Biotinylated mouse anti-human EpCAM antibody.
34. Biotinylated rat anti-mouse EpCAM antibody.
35. EpCAM (VU1D9) Mouse mAb (Alexa Fluor® 488 Conjugate).
36. ErbB2/Her2 Antibody (191924) [Alexa Fluor® 488].
37. Texas Red–conjugated antibodies against CD45.
38. Anti-EGFR antibody [ICR10] (FITC).
39. CELLection™ Biotin Binder Kit from Invitrogen.
40. DynaMag™ 2 magnet.
41. Arcturus® PicoPure® RNA Isolation Kit.
42. Parsortix GEN3D6.5 Cell Separation Cassette (Angle Europe).
43. Eppendorf tube, 0.5 mL/1.5 mL.
44. Conical tube, 15 mL/50 mL.
45. Corning® Transwell® 24-well plate.
46. Corning® ultralow-attachment culture dishes (60 mm petri dish, low attachment).
47. Corning® ultralow-attachment 6-well culture plate.
48. 96-well cell culture plate.

2.2 Recipes of Solutions

Note: All solutions should be passed through a 0.45 μm filter prior to use.

1. Phosphate buffer saline (PBS): 137 mM NaCl, 10 mM phosphate, 2.7 mM KCl; pH 7.4.
2. PBS-BSA solution: PBS supplement with 0.1% BSA.
3. PBS-BSA-EDTA solution: PBS supplement with 0.1% BSA and 2 mM EDTA.
4. Dulbecco's Modified Eagle's Medium (DMEM): DMEM supplement with 10% fetal bovine serum (FBS) and 1× penicillin–streptomycin stock solution.
5. Hoechst 33342 counter stain solution 50×: 50 μg of Hoechst 33342 in 1× PBS.
6. RPMI 1640 medium supplement with 1% FBS, 1 mM $CaCl_2$ and 5 mM $MgCl_2$.
7. CTC culturing medium: DMEM/F12 (50/50) medium supplement with insulin (5 μg/mL), hydrocortisone (500 ng/mL), recombinant human Epidermal Growth Factor (20 ng/mL), recombinant human Fibroblast Growth Factor (20 ng/mL), B27 supplement (1×), penicillin–streptomycin solution (1×).
8. Hemolysis Buffer: Add 8.26 g NH_4Cl, 1 g $KHCO_3$, and 0.037 g Na_2EDTA into 1 L ddH_2O.

3 Methods

3.1 CTCs Detection and Isolation (See Notes 1 and 2)

3.1.1 CTC Detection in Mouse Blood Samples by Fluorescence Microscope (See Note 3)

1. Collect 200 μL of mouse blood into a EDTA precoated 1.5 mL Eppendorf tube by tail-vein bleeding and diluted with 400 μL of PBS-BSA-EDTA.
2. Centrifuge at 600 × *g* for 10 min. Discard the supernatant.
3. Resuspend the cell pellet in 1 mL of Hemolysis buffer and incubate for 5 min.
4. Centrifuge at 600 × *g* for 5 min. Discard the supernatant.
5. Wash the cell pellet twice with PBS-BSA-EDTA buffer.
6. Resuspended the cells with 50 μL DMEM cell culture medium containing 10% FBS and 100 μg/mL penicillin–streptomycin (1/100) and normal rat IgG 1/50 and incubate on ice for 15–20 min. Normal rat IgG is used to block nonspecific binding sites of rat IgG antibody on the cells.
7. Add 1 μL of Alexa Fluor 488 conjugated rat anti mouse EpCAM antibody. Incubate on ice for 20–30 min.
8. Wash the cells twice with PBS containing 1% BSA, resuspend in 20 μL of Hoechst 33342 counter stain solution (1 μg/mL) and incubate on ice for 20 min.

9. Drop all of the 20 μL of cell suspension on a slide and covered with a coverslip. Count the green fluorescent cells under a fluorescent microscope. The green fluorescent cells are supposed to be CTCs.

*3.1.2 CTC Isolation from Mouse or Human Blood Samples by CELLction™ Biotin Binder Kit and Biotinylated EpCAM Antibody (See **Notes 4** and **5**)*

1. Resuspend Cellection Biotin Binder Dynabeads in the vial by vortex for 30 s. For 1 mL of whole blood samples, Transfer 25 μL of the beads to a 1.5 mL Eppendorf tube.
2. Add 1 mL of PBS supplemented with 0.1% bovine serum albumin (PBS-BSA) to the tube and vortex for 2 s.
3. Place the tube in a DynaMag™-2 magnet for 1 min.
4. Discard the supernatant and remove the tube from the magnet. Resuspend the beads in 25 μL PBS-BSA buffer.
5. Transfer 300 μL Releasing Buffer Component II to the tube of Releasing Buffer Component I (DNase I). Dissolve DNase I gently.
6. Aliquot the reconstituted release buffer and store at −20 °C.
7. Thaw immediately before use and keep on ice once thawed. Thawed release buffer can be refrozen once.
8. Dilute 1 mL of whole blood in 2 mL PBS supplement with 0.1% BSA and 2 mM EDTA (PBS-BSA-EDTA). Centrifuge at 600 × *g* for 10 min at room temperature.
9. Discard the supernatant and resuspend the cells to 1 mL of PBS-BSA-EDTA at 2–8 °C.
10. Add 10 μg biotinylated EpCAM antibody to 1 mL cell suspension and mix. Incubate for 10 min at 2–8 °C.
11. Wash the cells by adding 2 mL of PBS-BSA-EDTA and centrifuge at 350 × *g* for 8 min. Discard the supernatant.
12. Resuspend the cells in 1 mL PBS-BSA-EDTA buffer. Add the cell suspension to the Eppendorf tube containing 25 μL of the prewashed CELLection Biotin Bonder Dynabeads.
13. Incubate at 2–8 °C with gentle tilting and rotation for 20 min.
14. Place the tube in a DynaMag™-2 magnet for 2 min.
15. While the tube is still in the magnet, carefully remove and discard the supernatant by tipping.
16. Remove the tube from the magnet and add 1 mL PBS-BSA buffer, pipet 2–3 times and place the tube back to the magnet for 2 min.
17. Repeat **steps 15** and **16** at least twice to wash the cells.
18. Resuspend the bead-bound cells in 200 μL of RPMI 1640 medium containing 1% fetal bovine serum (FBS), 1 mM $CaCl_2$ and 5 mM $MgCl_2$ which was prewarmed at 37 °C.

19. Add 4 μL reconstituted Release Buffer (DNase I).
20. Incubate at room temperature with gentle tilting and rotation for 15 min.
21. Pipet thoroughly with a 100–200 μL pipette at least 5–10 times to maximize cell release.
22. Place in a magnet for 2 min and transfer the supernatant with released cells into a new tube which precoated for 5 min with RPMI 1640 medium containing 1% FBS, 1 mM $CaCl_2$, and 5 mM $MgCl_2$.
23. Resuspend the bead fraction in 200 μL RPMI 1640 medium and repeat **steps 21** and **22** once to collect residual cells.

3.1.3 CTC Isolation from Mouse or Human Blood Samples by the Parsortix™ System

1. 7.5 mL of blood is retrieved from deidentified patient samples. Up to 1 mL is retrieved from mice. The blood samples are drawn into EDTA washed tubes.
2. Processed blood samples through a Parsortix™ GEN3D6.5 Cell Separation Cassette (Angel Europe) in the Parsortix™ System according to the manufacturer's protocol within 1 h of collection.
3. As desired, stain the captured CTCs within the cassette with antibodies directed to the targets of interest, including but not limited to EpCAM-AF488, Her2-AF488, EGFR-FITC, and TexasRed-CD45.
4. Elute the captured CTCs by reversing the liquid flow through the systems and directing the cells into an external tube. The volume of the cells harvested from the cassette should be 100–200 μL [7].

3.2 CTC Culture (See Notes 6 and 7)

1. CTCs are cultured in Corning ultra-low attachment (ULA) 6-well plates with 2 mL/well of serum-free DMEM/F12 (50/50) medium supplement with insulin (5 μg/mL), hydrocortisone (500 ng/mL), recombinant human Epidermal Growth Factor (20 ng/mL), recombinant human fibroblast growth factor (20 ng/mL), B27 supplement (1×), penicillin–streptomycin solution (1×) at 37 °C, 5% CO_2 incubator.
2. Change CTC culture medium every 3 days.
3. To passage cells (every 3 days), transfer CTCs to a 15 mL Conical tube and centrifuge at 600 × *g* for 5 min. Aspirate supernatant, wash cells once with PBS and dissociate using 0.25% Trypsin.
4. Resuspend dissociated CTCs in 2 mL/well CTC medium and plate in 6-well ULA plates for next passage culture.

3.3 CTC Tumorsphere Formation Assay [8, 9] (See Notes 7 and 8)

1. Mix 10 μL of CTCs with an equal amount of 0.4% Trypan Blue solution, count cells with a hemocytometer and adjust the concentration of the harvested cells to 1×10^5 cell/mL.
2. Seed 4000–10,000 cells per mL in tumorsphere-forming medium using 6-well Corning ultra-low attachment culture plates. 2 mL of the medium is used for each well.
3. Culture cells in CTC culturing medium (DMEM/F-12, 5 μg/mL of insulin, 500 ng/mL of hydrocortisone, 20 ng/mL of EGF-h, 20 ng/mL of FGF-h, 1× B27, 1% penicillin–streptomycin) in hypoxic conditions (5% oxygen, incubator) at 37 °C for 7–14 days to form the mammospheres.
4. Transfer tumorspheres to a 15 mL conical tube and centrifuged at 300 × *g* for 5 min. Resuspend pellet in 400 μL of PBS.
5. Add 100 μL of suspended tumorspheres into the grid wells of the 96-well plate. Count the number of tumorspheres under a 10× objective lens on an inverted microscope.
6. Photograph and measure mammosphere diameter using Zeiss Axiovert software.
7. For second or more generation of tumorsphere formation, digest the tumorspheres from **step 6** with 500 μL of 0.25% trypsin–EDTA for 10 min at 37 °C and wash once with full cell culture medium.
8. Further digest the trypsinized tumorspheres with Dispase–DNase solution for 5 min at 37 °C, then wash sequentially with full cell culture medium and PBS.
9. To gain a single-cell suspension, filter the cells through a 40 μm mesh.
10. Count the cells and seed 1000–5000 of cells per mL in an ultra-low attachment culture dish as **step 4**.

3.4 3-Dimensional Invasion Assay (See Note 7)

1. Take one Corning Transwell 24-well plate with 8 μm pore size.
2. Make up collagen I gel mix with prechilled collagen I solution, 5× DMEM, and double distilled water.
3. The final concentration of collagen I is about 1.67 mg/mL in 1× DMEM.
4. Adjust the pH of the gel mix to neutral with prechilled 2 M sodium carbonate solution.
5. Carefully pipet 100 μL of the gel mix to each transwell insert and avoid air bubbles.
6. Solidify the gel by incubating the plate with the transwell insert at 37 °C over 30 min in a cell culture incubator.
7. Invert the 24-well plate with the transwell containing the collagen gel.

8. Plate 3 × 10^4 CTC cells in 100 μL of serum-free medium on the bottom of the transwell insert (now membrane side up).
9. Cover the inverted transwell insert with the plate so that the medium containing the cells touches the plastic surface of the well. Keep the membrane side of the transwell insert up.
10. Incubate the plate in the incubator for more than 4 h (or overnight) to allow cells to attach to the membrane of the transwell insert.
11. Reinvert the transwell plate and add 1 mL serum-free medium to the bottom well and 200 μL medium with 2% FBS or other chemoattractant to the transwell insert with the collagen gel.
12. Incubate the plate for more than 1 day (depending on the type of the cells).
13. Fix the collagen gel in the transwell insert with 4% paraformaldehyde in PBS for 15 min.
14. Wash the collagen gel in the transwell insert with PBS 1–2 times.
15. Permeabilize the cells within the collagen gel in 0.2% Triton-X in PBS for 30 min.
16. Wash the collagen gel in the transwell again with PBS for 2–3 times.
17. Stain the collagen gel in the transwell insert with 40 μg/mL propidium iodide in PBS for 2 h at 37 °C.
18. Wash the collagen gel in the transwell inserts with PBS and store at 4 °C.
19. Invert the transwell (filter side up). Using the 10× lens of an inverted confocal microscope, take 50 confocal Z-sections 4 μm apart starting from above the cells that are on the filter (−28 μm, with the filter as 0). Repeat for a total of 4 areas on the transwell.
20. Use Fiji (Image J) software to determine levels of fluorescence of each section. The sum of fluorescence from 20 μm below the filter is divide by the total fluorescence. The numbers for the 4 spots on the transwells are then averaged.
21. Alternatively, (instead of **step** of **20**), use Fiji (image J) software to determine the distance of invasion based on the number of sections in which the invading cells are located.
22. Alternatively, (instead of **steps** of **20**), count the number of cells at different depths.

3.5 Animal Model Experimental CTC Metastatic Variants

1. Wash CTCs twice in PBS.
2. Count cells with a hemocytometer and assess with Trypan blue dye to exclude dead cells.

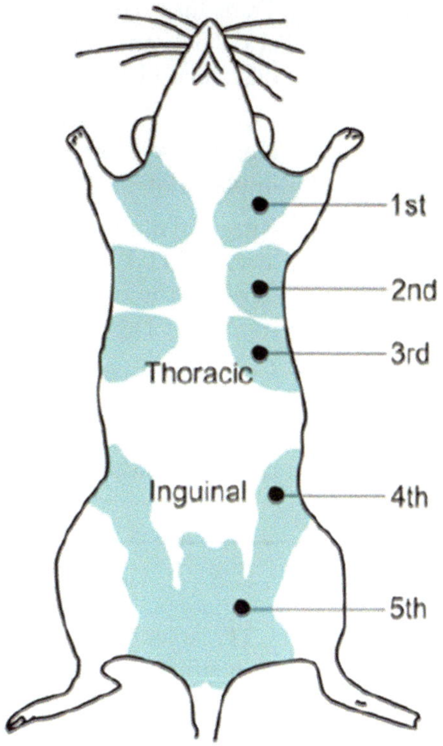

Fig. 2 The location of fourth pair of mammary fat pads in a mouse

3. Adjust cell concentration to 2×10^5 cell/mL in PBS. Mix cell suspension with an equal volume of 40% Matrigel solution in PBS.
4. Disinfect the mouse injection region is disinfected with 70% ethanol, and then inject 100 μL of the cell mixture into one side of the fourth pair of the mammary fat pad (Fig. 2).
5. Measure the tumor size with calipers every week.

3.6 Tumor Cell Self-Seeding [10]

1. Culture MDA-MB-231 cells in Dulbecco's modified Eagle's, high glucose supplemented with 10% FBS.
2. Transduce the lung metastatic derivative line MDA-MB-231-LM2 [11] with a GFP-luciferase fusion vector and transplant into one mammary gland in mice.
3. Transplant unlabeled MDA-MB-231-LM2 cells into the contra-lateral mammary gland to form a "recipient" mass.
4. After 60 days of transplantation, examine the recipient tumors for the presence of seeding cells by means of an in vivo imaging system.

3.7 CTC Clustering Using PDX (See Note 9)

1. Transduce cells with eGFP, tdTomato, luciferase 2-eGFP (LG2G), and luciferase 2-tdTomato (L2T) as previously described [12].
2. Compare the frequencies of dual-color, polyclonal lung colonies between mixed-color implants and the separate -color implants of TN1 PDX.
3. Intravital multiphoton microscopic imaging can be deployed to assess TN1 PDX breast tumors, human MDA-MB-231 cell-derived tumor models or mouse PyMT transgenic tumor models [12].
4. Collect invasive tumor cells in vivo from the models using TN PDXs that display individual cell migration patterns [12] to examine cellular patterns upon invasion into chemoattractant-containing Matrigel.

4 Notes

1. CTC isolation approaches can be based on magnetic separation, affinity chromatography, size- and/or deformability, and dielectrophoretic approaches.
2. All human blood samples should be collected in compliance with NIH guidelines for human subject studies and approved by the employing institution [12].
3. This method also can be used for detecting human blood cells to use Alexa Fluor 488 conjugated mouse anti-human EpCAM antibody. In this case, the cells should be blocked with normal mouse antibody first.
4. Magnetic-based CTC isolation systems use antibody-coated magnetic beads to bond a cell via its surface antigens through antibody–antigen interactions. Then the magnetized CTCs can be separated from the nonmagnetized cells of the ambient blood sample by applying a magnetic field. The antibody used to capture CTCs is often directed Epithelial Cell Adhesion Molecule (EpCAM), in some cases with a second antibody directed to CD45 to exclude other hemopoietic cells.
5. CELLection Biotin Binder Kit (Invitrogen, Fisher Scientific) and biotinylated human and mouse EpCAM antibody (BioLegend) are used in this protocol based on the manufacture's manual. Alternatively, Dynabeads FlowComp Flexi Kit from Invitrogen or human EpCAM MicroBeads from Miltenyi Biotech also can be used. For FlowComp Flexi Kit, the EpCAM antibody need to be DSB-X™ biotinylated first. DSB-X™ biotin has less affinity to bind with streptavidin, so the capture cells can be eluted by the buffer containing normal form biotin. For EpCAM MicroBeads, the capture cells cannot be eluted, but can be cultured with the Beads.

6. The size of CTCs are significantly larger than other blood cells [13, 14]. The Parsortix™ Cell Separation System use the single-use separation cassettes which contain a continuous precision molded separation structure laid out in a "step" configuration yielding a final "critical gap" to capture the larger CTCs [7]. Another popular and widely used technique is deterministic lateral displacement (DLD) fluidics [15] which is a robust passive microfluidic particle separation technique to sort particles based on their size with pillar arrays and was validated for CTC isolation [16, 17].
7. CTCs are collected either by Parsortix System (Subheading 3.1.3) or by positive selection of EpCAM antibody with CELLective Biotin Binder Dynabeads (Subheading 3.1.2).
8. Sphere formation assays originally developed for neural stem cells [18] were later modified to assess the activity of breast cancer stem cells (BCSC) [19] (Fig. 3). Both BCSC and some progenitor cells can form spheres [20], potentially leading to an overestimation of stem cell percentages. Moreover, assessing self-renewal of the tumorsphere (i.e., secondary generation) requires reseeding of all cells derived from the primary tumorsphere. Reseeding a proportion or lower density, and aggregation in sphere formation assays, can lead to distorted results [21].
9. To maximize the BLI detection sensitivity of breast cancer stem cells (BCSCs), use a modified, codon-optimized version of Luciferase: Luc2. In addition to the widely used enhanced GFP (eGFP), consider using the red fluorescent protein td-Tomato (Tom). Establish optimal parameters to transduce primary or passaged BCSCs with lentiviral vectors encoding Luc2-eGFP (L2G) or Luc2-Tom (L2T) fusion genes in order to shorten view time.

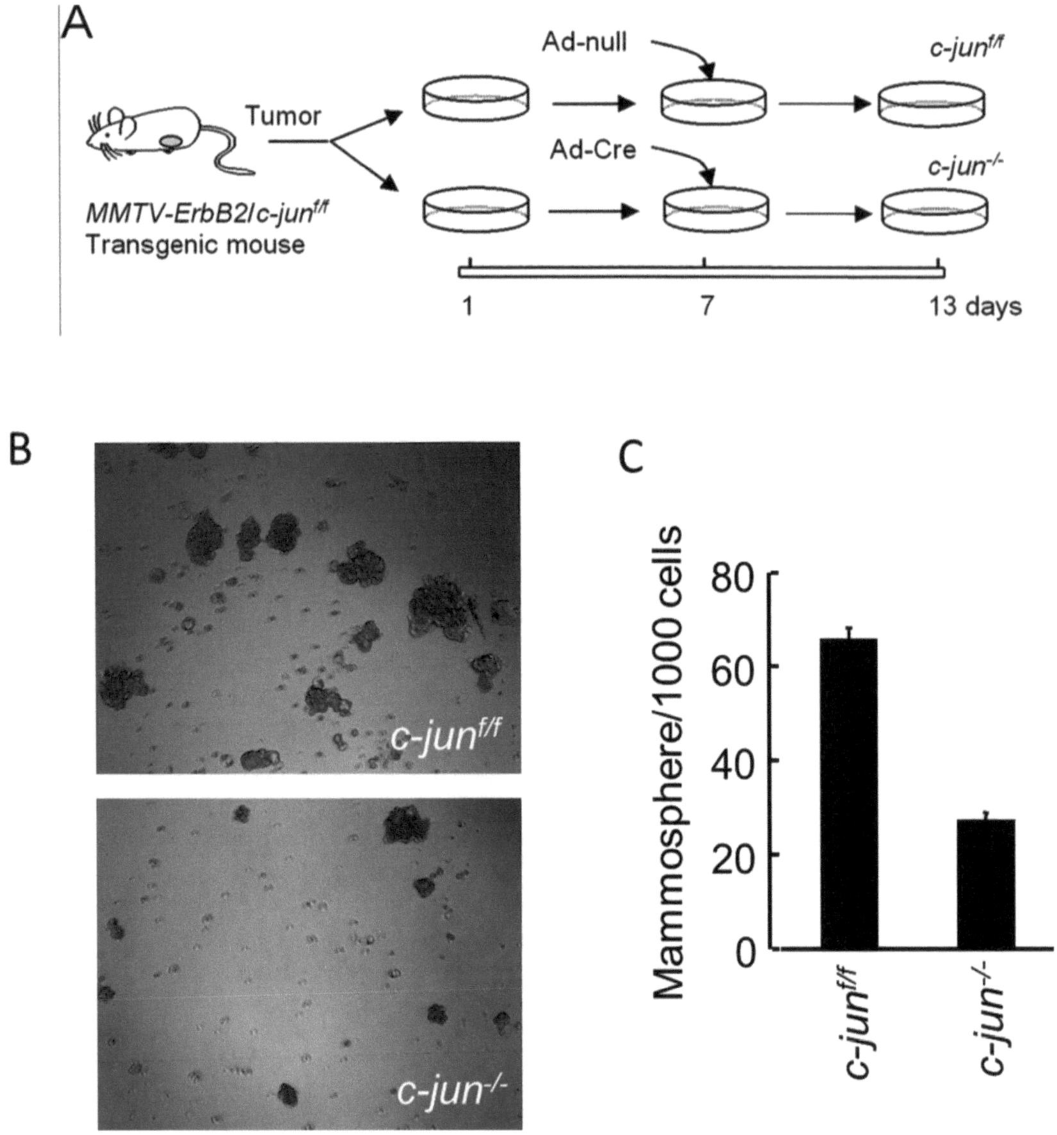

Fig. 3 Tumorsphere formation. (**a**). Schematic representation of experimental protocol in which MMTV-ErbB2-c-*jun*$^{f/f}$ double transgenic mice tumors were analyzed. (**b**). Comparison of tumorsphere formation between *c-jun*$^{+/+}$ and *c-jun*$^{-/-}$ murine ErbB2 mammary tumor cells with representative phase contrast image and (**b**) the number of tumorspheres formed from these two cell lines shown as mean $\pm$ SEM. (Adapted from [9])

Acknowledgments

Support was provided in part by R01CA132115, R21CA235139-01 and the Breast Cancer Research Program (W81XWH1810605, Breakthrough Award) (R.G.P), and the Wistar Cancer Center Support Grant (P30 CA10815) (R.G.P), and R01CA188575 (H.R.).

Conflicts of Interest *RGP is founder and CEO of ProstaGene, and LightSeed and owns several issued and pending patents. There are no conflicts of interest associated with this manuscript.*

References

1. Peiris-Pages M, Martinez-Outschoorn UE, Pestell RG, Sotgia F, Lisanti MP (2016) Cancer stem cell metabolism. Breast Cancer Res 18:55
2. Sun X, Jiao X, Pestell TG, Fan C, Qin S, Mirabelli E et al (2014) MicroRNAs and cancer stem cells: the sword and the shield. Oncogene 33:4967–4977
3. Kranjc MK, Novak M, Pestell RG, Lah TT (2019) Cytokine CCL5 and receptor CCR5 axis in glioblastoma multiforme. Radiol Oncol 53:397–406
4. Malanchi I, Santamaria-Martinez A, Susanto E, Peng H, Lehr HA, Delaloye JF et al (2011) Interactions between cancer stem cells and their niche govern metastatic colonization. Nature 481:85–89
5. Cheung KJ, Padmanaban V, Silvestri V, Schipper K, Cohen JD, Fairchild AN et al (2016) Polyclonal breast cancer metastases arise from collective dissemination of keratin 14-expressing tumor cell clusters. Proc Natl Acad Sci U S A 113:E854–E863
6. Aceto N, Toner M, Maheswaran S, Haber DA (2015) En route to metastasis: circulating tumor cell clusters and epithelial-to-mesenchymal transition. Trends Cancer 1:44–52
7. Miller MC, Robinson PS, Wagner C, O'Shannessy DJ (2018) The parsortix cell separation system—a versatile liquid biopsy platform. Cytometry A 93:1234–1239
8. Jiao X, Li Z, Wang M, Katiyar S, Di Sante G, Farshchian M et al (2019) Dachshund depletion disrupts mammary gland development and diverts the composition of the mammary gland progenitor pool. Stem Cell Rep 12: 135–151
9. Jiao X, Katiyar S, Willmarth NE, Liu M, Ma X, Flomenberg N et al (2010) c-Jun induces mammary epithelial cellular invasion and breast cancer stem cell expansion. J Biol Chem 285: 8218–8226
10. Kim MY, Oskarsson T, Acharyya S, Nguyen DX, Zhang XH, Norton L et al (2009) Tumor self-seeding by circulating cancer cells. Cell 139:1315–1326
11. Minn AJ, Gupta GP, Siegel PM, Bos PD, Shu W, Giri DD et al (2005) Genes that mediate breast cancer metastasis to lung. Nature 436:518–524
12. Liu X, Taftaf R, Kawaguchi M, Chang YF, Chen W, Entenberg D et al (2019) Homophilic CD44 interactions mediate tumor cell aggregation and polyclonal metastasis in patient-derived breast cancer models. Cancer Discov 9:96–113
13. Vona G, Sabile A, Louha M, Sitruk V, Romana S, Schutze K et al (2000) Isolation by size of epithelial tumor cells: a new method for the immunomorphological and molecular characterization of circulatingtumor cells. Am J Pathol 156:57–63
14. Low WS, Wan Abas WA (2015) Benchtop technologies for circulating tumor cells separation based on biophysical properties. Biomed Res Int 2015:239362
15. Salafi T, Zhang Y, Zhang Y (2019) A review on deterministic lateral displacement for particle separation and detection. Nano-Micro Lett 11:77
16. Loutherback K, D'Silva J, Liu L, Wu A, Austin RH, Sturm JC (2012) Deterministic separation of cancer cells from blood at 10 mL/min. AIP Adv 2:42107
17. Zeng Y, Gao L, Luo X, Chen Y, Kabeer MH, Chen X et al (2018) Microfluidic enrichment of plasma cells improves treatment of multiple myeloma. Mol Oncol 12:1004–1011
18. Reynolds BA, Weiss S (1992) Generation of neurons and astrocytes from isolated cells of the adult mammalian central nervous system. Science 255:1707–1710
19. Dontu G, Al-Hajj M, Abdallah WM, Clarke MF, Wicha MS (2003) Stem cells in normal breast development and breast cancer. Cell Prolif 36(Suppl 1):59–72
20. Stingl J (2009) Detection and analysis of mammary gland stem cells. J Pathol 217:229–241
21. Shaw FL, Harrison H, Spence K, Ablett MP, Simoes BM, Farnie G et al (2012) A detailed mammosphere assay protocol for the quantification of breast stem cell activity. J Mammary Gland Biol Neoplasia 17:111–117

Chapter 38

Limiting Dilution Tumor Initiation Assay: An In Vivo Approach for the Study of Cancer Stem Cells

Petra den Hollander, Robiya Joseph, Suhas Vasaikar, Nick A. Kuburich, Abhijeet P. Deshmukh, and Sendurai A. Mani

Abstract

Cancer stem cells (CSCs) are a small subpopulation of self-renewing cancer cells that are present within tumors. Calculating the frequency of tumor-initiating cells is important in the assessment of the number of CSCs present in a cell population. In this chapter, we present a protocol developed for quantification of CSCs from breast cancer tumors that can be adapted to CSCs from other types of tumors.

Key words Cancer stem cells, Limiting dilution tumor initiation, Tumor-initiating cell frequency, In vivo

1 Introduction

Calculating the frequency of tumor-initiating cells is important in assessment of the number of cancer stem cells (CSCs) present in a cell population. There are different approaches for characterization of CSCs [1]. The mammosphere assay is widely utilized to determine CSC frequency in vitro [2]. The most commonly used in vivo approach for calculation of CSC frequency is the limiting dilution tumor initiation (LDTI) assay described in this chapter. The LDTI assay is used to quantify the number of tumor-initiating cells (TICs) in a cell line or tumor tissue by evaluation of the self-renewal capacity of cancer cells in a mouse model (Fig. 1). The cancer cells that can initiate tumors are considered to be TICs. This technique can be applied to cell lines as well as primary cells from tumors isolated from mouse models or human samples. Through serial dilution of the number of cells injected into a mouse, the TICs are diluted until there is less than one cell per injection. It is important to serially dilute the cells until no tumors are formed in order to calculate the TIC frequency. To estimate the frequency of tumor-initiating cells (TICs) in a given cell line or tumor, an

Nagarajan Kannan and Philip Beer (eds.), *Stem Cell Assays: Methods and Protocols*, Methods in Molecular Biology, vol. 2429, https://doi.org/10.1007/978-1-0716-1979-7_38,

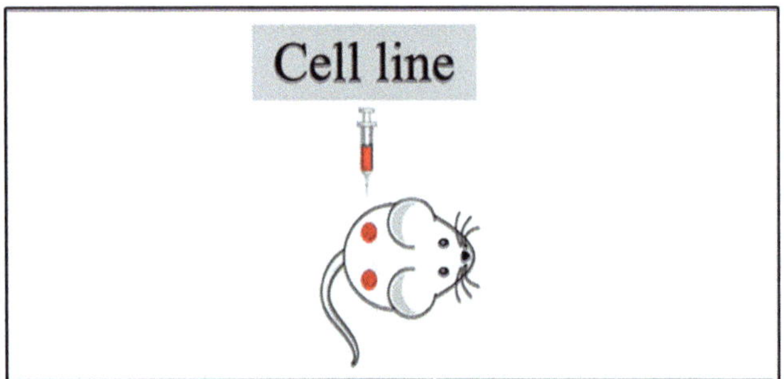

Dilutions of cells per injection	Number of study points
10,000,000	10
1,000,000	10
100,000	10
10,000	10
1000	10
100	10
10	10

Fig. 1 Plan for LDRI assay

extreme limiting dilution assay (ELDA) can be performed. Interpretation of data from this model assumes that the number of biological active particles in each culture follows a Poisson distribution and that a single biologically active cell is sufficient for a positive response from a culture [3], although ELDA results might deviate from the Poisson distribution [4]. The protocol detailed here was developed for quantification of CSCs from breast cancer tumors but can be adapted to CSCs from other types of tumors.

2 Materials

2.1 Cells for Injection

1. Cancer cell lines or cells isolated from tumors.

2.2 Cell Preparation

1. Growth medium: The medium used to grow the cells depends on the cell line used.
2. Trypsin or TrypLE (Thermo-Fisher) or another reagent to detach the cells from the tissue culture vessel. The choice and concentration of reagent depends on the cells used.
3. Phosphate buffered saline (PBS).
4. Trypan blue.
5. Matrigel (*see* **Note 1**).

2.3 Mice

1. Female mice (*see* **Note 2**). If using mouse cells, the same background strain should be used for the assay. If using human cells, immunodeficient mice should be used.

2.4 Cell Injection

1. Syringes with 27-gauge needles.
2. Hair clippers.
3. Isoflurane vaporizer.
4. 70% EtOH.

2.5 Tumor Monitoring

1. Caliper, preferably digital.
2. Electronic balance.

2.6 Tumor Harvest

1. Isoflurane.
2. Dissection kit.

2.7 Additional Equipment

1. 37 °C incubator.
2. Cell counter.
3. Pipette aid and pipettes.
4. 1.5-mL microtubes.
5. 15-mL tubes.
6. Cell culture flasks or dishes.
7. Cell counter.
8. Anesthesia induction chamber and isoflurane vaporizer.

2.8 Software

1. The ELDA software available at http://bioinf.wehi.edu.au/software/elda/.
2. R (https://www.r-project.org/) and library "statmod" [5].

2.9 Input Data

1. Data should be in a table that consists of columns headed *Cells*, *Tested*, *Responders*, and *Groups* (optional). Examples are shown in Table 1. The "Cells" column contains the number of cells injected into the mouse. The "Tested" column consists of the number of mice tested. The "Responders" column consists of the number of tumors with volumes exceeding the cutoff. The "Groups" column consists of optional annotation that describes treatment.

2.10 Parameters

1. The confidence interval is used to compute TIC frequency in each population group.
2. By default, ELDA computes a 95% confidence for the active cell frequency in each population group. The confidence interval can be varied.

Table 1
Example datasets from limiting dilution analyses

Cells	Tested	Responders	Group
(a) Single group data			
100,000	3	3	hGBM
10,000	3	3	hGBM
1000	3	3	hGBM
100	3	1	hGBM
(b) Multiple group data			
2,000,000	8	6	TGF
1,000,000	8	4	TGF
100,000	8	1	TGF
10,000	8	0	TGF
1000	8	0	TGF
2,000,000	8	2	Inhibitor
1,000,000	8	7	Inhibitor
100,000	8	4	Inhibitor
10,000	8	1	Inhibitor
1000	8	0	Inhibitor

(a) Data from an analysis in mice of tumor cells from a freshly dissociated human glioblastoma multiforme tumor from Richichi et al. [7] and (b) data from an analysis of breast tumor cell lines analyzed in a mouse mammary fat pad model. The breast tumor cell lines were treated with TGFβ1 or a TGFβ1 inhibitor.

3. Additional output can be requested including a graphical display, a comparison between groups, and a report on the adequacy of the single-hit model.

3 Methods

3.1 Cell Preparation

1. Grow the cells of interest in an incubator at 37 °C with 5% CO_2 until they are approximately 70% confluent (*see* **Note 3**).
2. Aspirate the medium from the culture flask containing the cells and wash the cells with PBS.
3. Detach cells with trypsin or TrypLE (Thermo-Fisher).
4. Add serum-containing medium and centrifuge at 150 × *g* for 5 min.
5. Resuspend cells in complete medium.
6. Using a cell counter, count an aliquot of cells stained with trypan blue to determine the concentration of viable cells.

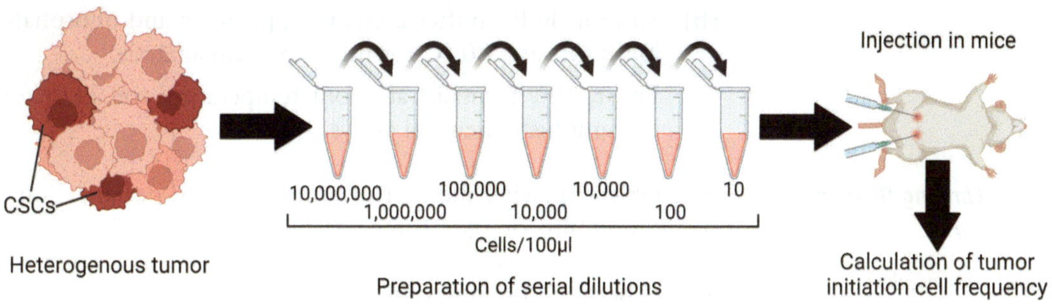

Fig. 2 The in vivo approach for the calculation of tumor-initiating cell frequency. Created with BioRender.com

3.2 Serial Dilution of Cells

1. Prepare the different dilutions of cells required for injection in mice (Fig. 1) in complete medium. The desired number of cells should be in suspended in 100 μL of complete medium for mouse cells or 50 μL of 50% Matrigel/50% medium for human cells (*see* **Notes 4** and **5**). Prepare extra volume so that there is enough for the number of injections plus one extra.
2. Serially dilute the cells to generate ten-fold dilutions (Fig. 2).
3. Keep the cell preparations on ice until injection.

3.3 Injection of Cells

1. Anesthetize the mice using 2% isoflurane in oxygen using an anesthesia induction chamber (*see* **Notes 6** and 7).
2. Trim the abdominal fur of the mouse and wipe with 70% EtOH to sterilize the area and to expose the nipples.
3. Inject 100 μL of cells into the fourth pair of mammary fat pads (*see* **Note 8**).

3.4 Tumor Monitoring

1. Monitor tumor growth every week until termination of the experiment (*see* **Note 9**).
2. Determine the length (l) and width (w) of the developing tumor using a caliper.
3. Calculate tumor volume using the formula $V = (W(2) \times L)/2$ in mm^3, with the largest number being the length and the smaller number being the width [6].

3.5 Tumor Harvest (See Note 10)

1. Anesthetize the mouse and sacrifice by cervical dislocation.
2. Spray 70% EtOH on the abdomen to sterilize the area.
3. Use scissors and forceps to dissect out the tumor.
4. Measure the tumor dimensions using calipers.
5. Store the tumor for future analysis:
   (a) Store in RNAlater (Qiagen, Cat #76104) at −80 °C for future RNA extraction,

(b) Store in RIPA buffer containing protease and phosphate inhibitors at −80 °C for future protein analysis,

(c) Fix in 10% formalin at room temperature for 24 h for immunohistochemical assays.

3.6 Limiting Dilution Analysis

3.6.1 Visualization

Create a table displaying number of tumors formed per injection/transplantation site or plot tumor frequency on the *Y*-axis and number of cells of the cell injected on the *X*-axis to visually compare differences between the tested groups.

3.6.2 Multiple Group Comparison

(a) Use the Fisher's exact test to compare take rate.

3.7 Calculation of Tumor-Initiating Cell Frequency

3.7.1 Calculation of Tumor-Initiating Cell Frequency

1. Create an input data table. An example for a single population group is shown in Table 1a.
2. Click on link http://bioinf.wehi.edu.au/software/elda/.
3. Paste data to be analyzed into the data entry module.
4. Change the confidence interval if desired.
5. Click the box to output results as a plot.
6. Click the "Run" button to calculate TIC.
7. The output will be given in table format where "TIC Frequency^{-1}" is the estimated TIC frequency with lower and upper bounds based on the confidence interval.
8. Also output will be a log-fraction plot of the limiting dilution model fitted to the data with the number of cells injected on the x-axis vs. nonresponders. The trend line indicates the estimated active cell frequency.

3.7.2 Analysis of Multiple Population Group Data

1. Create an input data table. An example for a single population group is shown in Table 1b.
2. Click on link http://bioinf.wehi.edu.au/software/elda/.
3. Paste data to be analyzed into the data entry module.
4. Change the confidence interval if desired.
5. Click the box to output results as a plot.
6. Click the "Run" button to calculate TIC.
7. The output will be given in table format where "TIC Frequency^{-1}" is the estimated TIC frequency with lower and upper bounds based on the confidence interval.
8. Perform an overall test for differences in stem cell frequencies between any of the groups.
9. Perform tests for pair-wise differences in active cell frequency between groups.
10. Calculate goodness of fit for all the groups at once.

3.7.3 Analysis in R for a Large Number of Datasets

1. Download R (https://www.r-project.org/) and library "statmod" [5].
2. Create an input data table. Examples are shown in Table 1.
3. Use function *elda* with input parameters such as,

```
out <- elda(dose=c(100000,10000,1000,100), tested=c
(3,3,3,3), response=c(3,3,3,1), group=c(1,1,1,1), test.unit.
slope=TRUE)
print(out$CI)
plot(out)
```

4. The output is TIC Frequency^{-1} with lower and upper bounds are based on the confidence interval in "out" parameter and corresponding plot obtained from "out" as mentioned above.

4 Notes

1. Matrigel is preferred when assaying human cells but may influence the absolute values in the ELDA.
2. Female mice that are 4–6 weeks old are used here, since this protocol is for analysis of breast cancer cells. Use disease-appropriate mouse strains and gender for your cancer type of interest.
3. All the media and equipment used in cell maintenance, preparation, and injection into mice should be maintained in sterile conditions to minimize chance of contamination of cells before injection and to prevent infection at the injection site.
4. Prepare extra volume at each concentration to ensure that the desired numbers of cells are injected; prepare sufficient volume for the number of injections plus one extra.
5. It is important that the Matrigel, if used, is kept at 4 °C to prevent it from solidifying. Tips and syringes used to resuspend the Matrigel containing cells should ideally also be maintained at 4 °C. This will maintain cell viability and will ensure uniform suspension and smooth injection into mice.
6. After injection, place the mice on a pad or in a chamber at 37 °C until they recover from the effects of anesthesia.
7. Ensure a continuous flow of oxygen while isoflurane is in the vaporizer.
8. A bubble will be observed upon injection of the cells.
9. The experiment should be terminated when the primary tumor diameter is 2 mm.

10. This step is not essential for the calculation of TICs but harvesting and storing the tumors could be valuable for further analysis.

Acknowledgments

This work was supported by the NIH/NCI grants (2R01CA155243; 1R01CA200970), the CPRIT grant (RP170172) awarded to S.A.M, and the NIH T32 grant (5T32CA18682) awarded to N.A.K.

References

1. Rycaj K, Tang DG (2015) Cell-of-origin of cancer versus cancer stem cells: assays and interpretations. Cancer Res 75:4003–4011
2. Dontu G, Wicha MS (2005) Survival of mammary stem cells in suspension culture: implications for stem cell biology and neoplasia. J Mammary Gland Biol Neoplasia 10:75–86
3. Hu Y, Smyth GK (2009) ELDA: extreme limiting dilution analysis for comparing depleted and enriched populations in stem cell and other assays. J Immunol Methods 347:70–78
4. Bonnefoix T, Callanan M (2010) Radioresistance mechanisms of mammary tumor initiating cells: unreliable biological conclusions based on limiting dilution assays. Proc Natl Acad Sci U S A 107(21):E83–E84
5. Giner G, Smyth GK (2016) Statmod: probability calculations for the inverse Gaussian distribution. R J 8(1):339–351
6. Faustino-Rocha A, Oliveira PA et al (2013) Estimation of rat mammary tumor volume using caliper and ultrasonography measurements. Lab Anim (NY) 42:217–224
7. Richichi C, Osti D et al (2016) Tumor-initiating cell frequency is relevant for glioblastoma aggressiveness. Oncotarget 7(44): 71491–71503

Chapter 39

Orthotopic Xenografts of Colorectal Cancer Stem Cells

Maria Laura De Angelis, Federica Francescangeli, Ann Zeuner, and Marta Baiocchi

Abstract

Cancer stem cells (CSCs) are responsible for the initiation of primary tumors and for metastasis seeding at distant organs. Therefore, they represent crucial targets for the study and preclinical testing of new antimetastatic approaches. We recently generated a molecularly characterized biobank of colorectal CSCs, isolated from individual patients and cultured in serum-free medium as multicellular spheroids. Here, we describe in detail the generation of a metastatic model of colorectal cancer based on the orthotopic injection of CSCs into the cecum serosa of immunodeficient mice. Such a model represents an excellent experimental system to investigate the cellular and molecular mechanisms involved in colorectal cancer metastasis, to analyze rare premetastatic elements such as circulating and disseminated tumor cells, and for the preclinical testing of new agents with potential antimetastatic activity.

Key words Metastatic models, Cancer stem cells, Orthotopic xenografts, Cancer stem cell-based xenografts, NSG mice, In vivo preclinical models

1 Introduction

Despite the recent advancements in prevention and therapy, colorectal cancer (CRC) still represents the third cancer-related cause of death worldwide [1], primarily due to the development of liver metastases at the advanced stages of the disease. The metastatic process thus constitute a key point for therapeutic intervention, requiring faithful animal models of metastasis as essential tools for drug design and preclinical testing. Animal models that reproduce step-by-step the spontaneous migration of CRC cells to the liver, the subsequent homing, and the formation of secondary tumors, provide an invaluable tool to unravel the molecular mechanisms involved in the whole metastatic process and to identify new targeted therapies. Colorectal cancer stem cells (CSCs) are a small subpopulation of self-renewing cells that drive tumor growth and metastasis to distant organs [2]. Recently, we described the workflow for the isolation, in vitro expansion, and banking of colorectal

Nagarajan Kannan and Philip Beer (eds.), *Stem Cell Assays: Methods and Protocols*, Methods in Molecular Biology, vol. 2429, https://doi.org/10.1007/978-1-0716-1979-7_39,

CSCs from individual patients' tumor tissues [3]. CSCs isolated from surgical specimens grow in culture as multicellular spheroids and retain the long-term capability to generate xenografts with the same histology of the original patient's tumor. CSC lines isolated in our laboratory are routinely characterized for mutational status, stem cell marker expression and phosphoprotein activation state, before being validated in vivo and subsequently stored in the CSC biobank [3]. We have previously employed this platform to analyze the differential effects of the EGFR inhibitor cetuximab on CSCs carrying different mutations, through in vitro screening followed by subcutaneous xenografting [3, 4]. Furthermore, we have used CSC-based strategies to identify factors involved in the stemness and dormancy of CRC cells [5, 6].

Recently, we have established CSC-based murine models of CRC liver metastasis, suitable for the investigation of metastatic mechanisms and for the preclinical testing of stem cell-directed antimetastatic agents [7]. Since CSCs are injected as dissociated single cells, the transplantation of cells derived from multicellular spheroid cultures (MSCs) is technically easier than grafting of organoids [8–10] or tumor fragments [11–13]. Importantly, MSCs are easier, faster, and less expensive to expand as compared to organoid cultures or patient-derived xenografts (PDX). Furthermore, CSCs can easily be genetically modified to express reporter genes such as firefly luciferase, thus allowing monitoring of tumor development by bioimaging systems.

Here we describe in detail the method for orthotopic xenografting of CSCs from multicellular spheroids by injection into the cecal wall serosa of NSG mice (Fig. 1). This procedure reproducibly generates primary intestinal tumors detectable from 4 to 6 weeks postgrafting and liver metastases detectable from 8 to 10 weeks postgrafting. Lung metastases are also observed occasionally, depending on the CSCs strain used.

2 Materials

2.1 Cell Processing

1. PBS–BSA 1%: 1% BSA w/v, PBS, sterile.
2. TrypLE Express Enzyme (1×), phenol red–free.
3. Matrigel, Growth Factor Reduced (GFR) Basement Membrane Matrix.
4. Trypan Blue–PBS: 0.4% w/v Trypan Blue, PBS.

2.2 Surgery

1. Mice: NSG (NOD.Cg-Prkdc^{scid} Il2rg^{tm1Wjl}/SzJ.
2. Buprenorphine hydrochloride solution, 0.3 mg/mL, veterinary.
3. Ketamine hydrochloride solution, 100 mg/mL, veterinary.

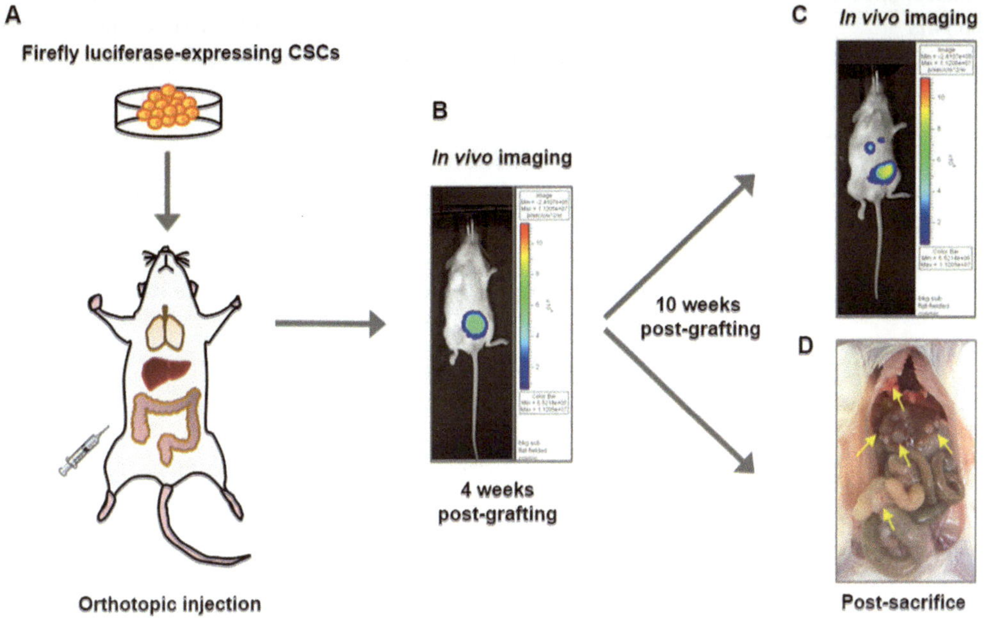

Fig. 1 Schematic rapresentation of CSCs orthotopic injection and detection. (**a**) CSCs, labeled with firefly luciferase or other bioimaging markers, are inoculated into immunodeficient mice cecal serosa; (**b**) Primary tumor is detected by bioimaging after 4–6 weeks; (**c**) Liver and lung metastases are detected by bioimaging at week 8–10 postgrafting, and (**d**) confirmed by postsacrifice examen. Images from one representative experiment are shown

4. Xylazine hydrochloride solution, 25 mg/mL, veterinary.
5. Ophthalmic ointment.
6. Ethanol/water 70% v/v.
7. Sterile cotton gauze.
8. Cotton-tipped applicators.
9. Sodium chloride solution, 0.9% w/v, sterile.
10. Hair trimmer.
11. Microsurgery scissors, straight.
12. Microsurgery forceps, straight, large (tip approximately 4 mm) and thin (tip approximately 2 mm).
13. Olsen-Hegar scissors with suture cutter.
14. Micro-Fine Syringes, 1 mL, with 29G needle.
15. Sterile syringes with needles, single use, 1 mL.
16. Synthetic absorbable suture with needle, size 5-0 (Ethicon Vicryl).
17. Heated pad.

2.3 Miscellaneous Equipments

Sterile ventilated cabinet, light microscope, centrifuge, thermostated water bath, cell count chamber, pipet-aid, conical Falcon tubes (50 mL, 15 mL), Eppendorf microtubes, disposable serological pipettes (2 mL, 10 mL, 25 mL), Gilson pipettes and tips (1 mL, 200 μL).

3 Methods

Handle cells and reagents under a ventilated sterile cabinet to avoid microbial contamination.

Use cell culture–grade sterile solutions and reagents.

Use cell culture–grade sterile, disposable plasticware at all the steps.

3.1 Cell Processing

3.1.1 Cell Culture

CSC medium formulation and procedures for colorectal CSC culture have been described in detail in a previous protocol [14]. Most relevant points to consider while preparing cultures for xenografting are the following:

1. Standard CSC cultures are routinely passaged weekly by dissociating spheroids and reseeding single cells at a density of about 5×10^4/mL in 20 mL of fresh medium, in 75 cm^2 ultralow-attachment flasks. Doubling time is about 2–3 days for most of the strains, therefore after 6 days of culture CSCs density should be about 2–4 $\times 10^5$/mL.
2. We routinely inject about 2×10^5 cells/graft. For a safe workflow, prepare cultures and start the dissociation calculating a 10–30% cell loss (depending on the CSC strain) during the whole procedure of dissociation/washing.

3.1.2 Preliminary Assessment of Spheroid Culture Quality

The quality of the culture at the time of dissociation is critical for xenografting result. Specifically, the spheroid culture used for xenografting should be in logarithmic growth phase, and clusters should be viable and healthy, as indicated by a proper morphology at the time of harvesting (as described below). The day before the procedure, check the quality of the culture and evaluate the gross amount of cells in the flask:

1. Carefully observe the general appearance of the culture under a light microscope. Good quality spheroids have a regular shape, a translucent appearance, and contain few or no dark, dead cells at the center or at the borders of the clusters.
2. Harvest, dissociate, and count a small aliquot of cells (see below). Verify log growth and estimate if enough cells will be available for the procedure the next day (*see* **Note 1**).

3.1.3 Spheroid Dissociation

1. Transfer spheroids from culture flask into a 15 mL conical tube.
2. Centrifuge 5 min at 150 × *g* to remove medium.
3. Wash spheroids: resuspend cell pellet in about 12 mL of PBS, then centrifuge 5 min at 150 × *g*. Remove the supernatant by pipetting.
4. Repeat **step 3**.
5. After the second wash, carefully remove PBS to avoid diluting TrypLE solution at the next step: aspirate most of the volume with a 10 mL pipette and then remove residual PBS with a Gilson micropipette.
6. Add TrypLE solution to the cell pellet, adjusting its volume to the expected number of cells in the pellet: TrypLE volume should be approximately 0.25 mL/10^6 cells. Mix by gentle pipetting.
7. Put the tube into a thermostated water bath and incubate the tube at 37 °C for 3 min.
8. At the end of the incubation, observe the tube upon gentle shaking, to verify whether large clusters are still visible. In this case, incubate for an additional 3 min. Up to three rounds of 3 min incubations are acceptable in our experience; usually at this point no residual clumps are visible. If, however, clusters are still present, proceed anyway to the next step, since further incubation with TrypLE will damage the cells.
9. After the enzymatic treatment, gently pipet the cell suspension through a 200 μL Gilson tip mounted on a 2 mL serological pipette. Depending on the CSC strain, 5–20 passages will be sufficient to dissociate any residual visible spheroids, and to obtain a single-cell suspension.
10. Add 12 mL of PBS–BSA 1% to the cells: this will inactivate TrypLE.
11. Centrifuge at 150 × *g* for 5 min. Discard the supernatant and suspend the cell pellet in PBS–BSA 1%, at a density suitable for counting (i.e., approximately 10^6 cells/mL), based on the grossly expected total cell number.
12. Harvest 25 μL of suspension from the tube and mix with an equal volume of Trypan Blue solution.
13. Count cells using a cell count chamber under a microscope. The procedure should have produced a single-cell suspension, containing no more than 15% of dead cells, ready for use (*see* **Notes 2** and **3** for troubleshooting).
14. Centrifuge at 150 × *g* for 5 min. Discard the supernatant and suspend the cell pellet in PBS–BSA 1%, at a density of 8 × 10^6/mL. For each mouse to inoculate, transfer 25 μL (containing

2×10^5 cells) of the suspension into a 1.5 mL sterile Eppendorf tube. Place tube(s) into an icebox; wait 1 or 2 min to allow tubes to cool down.

15. Without removing tubes from ice, add 25 μL of cold Matrigel to each Eppendorf tube. At this step, also place the Micro Fine syringes needed for grafting, wrapped into sterile aluminum foil, into the icebox. Transfer the icebox to the animal working station (*see* **Note 4**).

3.2 Orthotopic Xenografting

Carry out all the procedures under a ventilated sterile cabinet.

Autoclave all surgical instruments prior to use.

Use sterile solutions at all the steps.

3.2.1 Analgesia and Anesthesia

1. Dilute buprenorphine stock solution 1:10 with sodium chloride solution, to obtain a 30 μg/mL working solution. Inject 100 μL into the mouse, subcutaneously; the dose administered corresponds approximately to 0.15 μg/g body weight.
2. Prepare the anesthetic cocktail by mixing: 200 μL ketamine (100 mg/mL), 100 μL xylazine (25 mg/mL), and 700 μL sterile sodium chloride solution, in order to obtain a working solution containing 20 mg/mL ketamine and 2.5 mg/mL xylazine.
3. Weight the mouse. Inject the ketamine–xylazine cocktail intraperitoneally into mice with a 1 mL insulin syringe. Inject 5 μL/g of body weight: the dose administered will be approximately 100 μg/g and 12.5 μg/g body weight for ketamine and xylazine, respectively.
4. Verify hind limb reflex.
5. Apply ophthalmic ointment to both eyes of the mouse to prevent eye dryness.

3.2.2 Cell Grafting

The main steps of the surgical procedure are shown in Fig. 2.

1. Remove mouse ventral hair with a trimmer.
2. Put the mouse in recumbent position on the heating pad, covered by a sterile towel.
3. Wipe skin repeatedly with 70% ethanol.
4. Fill a 1 mL Micro-Fine syringe (cold) with 50 μL Matrigel–CSC suspension (**step 15** in Subheading 3.1.3), then place back the syringe in aluminum foil, on ice.
5. Grab the mouse skin by thinner forceps, and make a 1.5 cm vertical incision from the lower toward the middle abdomen by scissor.
6. Cut the underlying peritoneum by scissors.

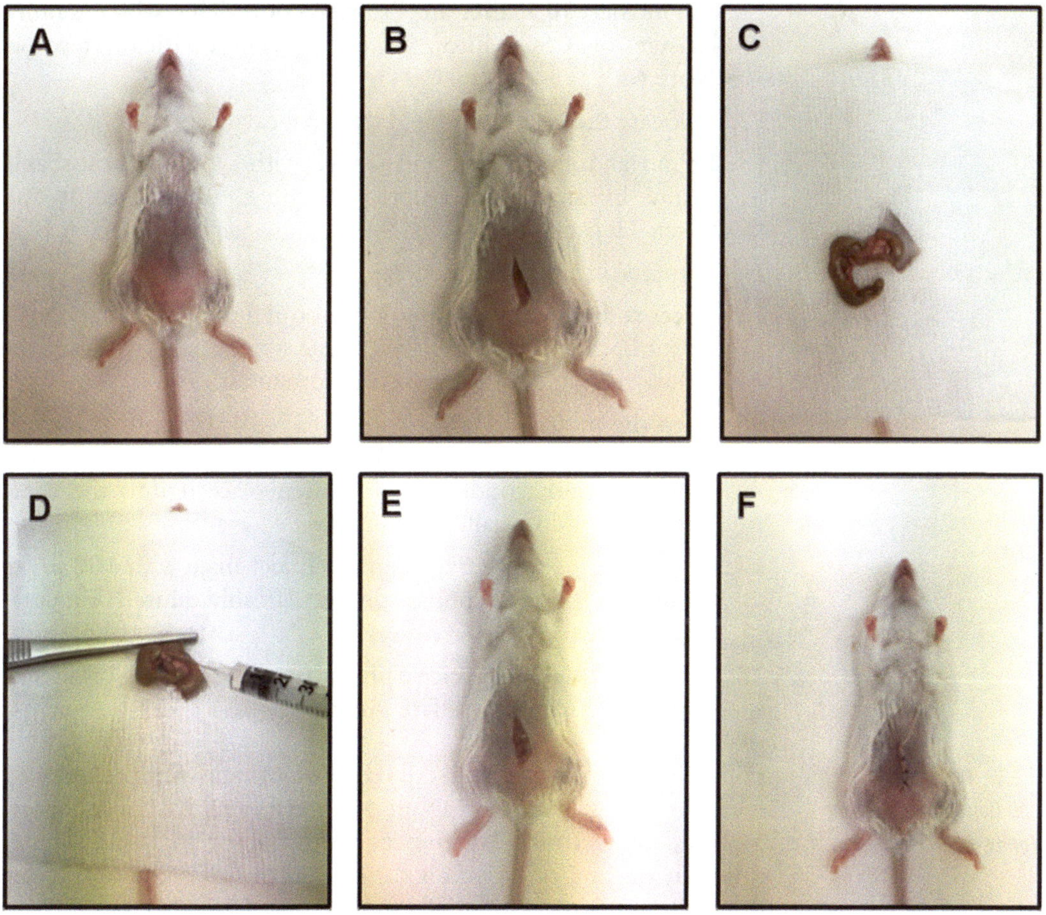

Fig. 2 Surgical procedure for CSC orthotopic grafting. Abdomen trimming (**a**); Skin and abdominal wall incision (**b**); Exposure of the cecum (**c**); Cell injection (**d**); Peritoneum suture (**e**); Skin suture (**f**)

7. Place a sterile gauze on the abdomen of the mouse, to avoid contact with exposed cecum.
8. Exteriorize cecum and ascending colon by larger forceps.
9. During the following procedure, keep cecum humidified by dropping sterile sodium chloride solution with a pipette when needed.
10. Keeping the cecum still with the larger forceps, carefully insert the needle, and then slowly inject the Matrigel–cell suspension into the serosa: Matrigel will immediately solidify within tissue, and a small bubble will be visible at the site of the injection (**critical step**: *see* **Note 5**).
11. After the injection, slowly retract the needle from the cecum, while gently pressing a cotton-tipped applicator on the needle insertion point. Keep the applicator pressed on injection site

for further 30 s after needle retraction. This will minimize leakage of CSCs into the peritoneum (**critical step**: *see* **Note 6**).

12. Relocate the cecum in the abdominal cavity.
13. Rejoin abdominal wall and suture with synthetic absorbable suture by simple interrupted stitching, using forceps and Olsen-Hegar scissors.
14. Suture skin as above (*see* **Note 7**).
15. Inject sodium chloride solution (about 1 mL) subcutaneously on the back of the mouse, by a sterile syringe. This will keep the mouse hydrated until recovered from anesthesia.
16. Keep the mouse on a heated pad (37–40 °C) until awakened from anesthesia: this will take about 30–60 min. Verify the mouse's health conditions, and then place it into the cage. Cage mice individually postsurgery.
17. Administer postsurgery analgesia: Inject mice with 100 μL of 30 μg/mL buprenorphine solution (freshly diluted) subcutaneously every 12 h, for 2 days.
18. Monitor postsurgery mice daily for signs of suffering or distress according to animal welfare guidelines [15].

3.2.3 Development and Detection of Primary Tumor and Metastases

Following the procedure described above primary tumors and liver metastases usually develop about 4–6 weeks and 8–10 weeks postgrafting, respectively. Differences in tumor or metastases onset primarily depend on the individual strain of CSCs employed, due to specific replication and grafting capability. Individual CSC strains may also give raise to lung metastases, developing concurrently or at later times as compared to liver metastases. Following mice sacrifice, tumors can be harvested and validated by histological staining. CSCs can be easily genetically modified to express common bioluminescent markers by different standard methods [16], thus allowing the monitoring of tumor progression through in vivo bioluminescence-detecting instruments. We routinely graft firefly luciferase-expressing spheroids, and then analyze mice upon D-luciferin administration by an in vivo imaging detection apparatus (*see* **Note 8**). We analyze mice weekly, starting at week 2 postgrafting. Tumor development can also be monitored by intravital imaging methods such as x-ray computed tomography (CT) or magnetic resonance imaging (MRI).

4 Notes

1. If ongoing CSC cultures look poor, consider postponing the grafting and restart the culture by dissociating the spheroids and seeding a new flask.

2. If residual clusters are still visible after the dissociation, two actions can be taken, alternatively or sequentially.
   (a) Filter the suspension through a 100-µm sterile nylon mesh filter. Count again cell suspension.
   (b) Leave the tube standing in vertical position for few minutes, until residual clusters sediment on the bottom by gravity. Carefully retrieve supernatant with a serological pipette. Count again cell suspension.
3. If too many dead cells are present in the suspension at this point, this is likely due to poor cell conditions in the original culture flask, and/or to excessive cell damage produced by the dissociation. In this case, consider delaying xenografting until a good quality culture is available.
4. Store Matrigel in aliquots of about 200 µL, in sterile cryovials, frozen at −20 °C. Before the use, get the number of aliquots needed and place them on ice. It is very important to keep Matrigel, cells, tubes, and syringes refrigerated on ice at all the times until the inoculum, since Matrigel quickly solidifies at room temperature: this would make it impossible to retrieve cells from tubes or from syringes.
5. The cecum serosa is an extremely thin layer, and correct needle insertion requires technical skill and experience. The most frequent undesired event is perforation of the serosa by the needle: this results in the loss of the cells in the intestinal lumen, with no tumor grafting. The use of an appropriate syringe is of utmost importance for a correct injection. We currently use Micro-Fine Syringes BD cat. #320924, which have needles of adequate thinness and sharpness, and no dead volume, altogether allowing an acceptable control of the actual volume injected.
6. A frequent undesired event is leakage of cells outside the injection site, externally to the cecum, resulting in intraperitoneal cell spreading. In this case, intraperitoneal tumors may develop, alone or together with the cecal graft. Intraperitoneal CSC tumors do not give raise to metastases, furthermore they usually grow faster than the orthotopic xenograft, interfering with its detection, and/or imposing early mouse euthanasia. Measures to avoid this event include the use of Matrigel, which solidifies immediately after the injection; the use of thin needles, as described in the previous Note; the injection of a maximum volume of 50 µL; the use of a cotton-tipped applicator during and after needle retraction, as described.
7. Skin may be closed with wound clips by a wound clip applier, instead of suturing.

8. Firefly luciferase is an enzyme that emits light in the yellow-green region, through oxidation of D-Luciferin in the presence of ATP, magnesium, and oxygen. Bioluminescence can be detected and quantified by bioimaging apparatuses. We routinely perform bioimaging testing by an IVIS Lumina III In Vivo Imaging System (PerkinElmer) as follows:
   (a) Initialize the imaging system and verify that the charge coupled device (CCD) camera has cooled down to −90 °C.
   (b) Shortly before the analysis, inject mice intraperitoneally with 150 mg/kg of D-luciferin potassium salt (the working solution is prepared by dissolving D-Luciferin in sterile PBS. The solution is filtered and then stored in aliquots frozen at −80 °C until use).
   (c) Wait 5 min to reach tumor saturation plateau.
   (d) Anesthetize mice with Isoflurane (2% in 100% oxygen) by an anesthesia chamber.
   (e) Place the mouse in the IVIS black light tight chamber, in supine position.
   (f) Exposure time should be appropriately set, in order not to saturate the image. The settings we routinely use for the detection primary tumors and metastases, from week 2 to 10 postgrafting, are the following: Field of view C; Exposure time 1 min; Medium binning; 1.8 f/stop; open filter.
   (g) Analyze images using Living Image Software from PerkinElmer as described in the software manual.

References

1. Bray F, Ferlay J, Soerjomataram I, Siegel RL, Torre LA, Jemal A (2018) Global cancer statistics 2018: GLOBOCAN estimates of incidence and mortality worldwide for 36 cancers in 185 countries. CA Cancer J Clin 68(6): 394–424
2. Zeuner A, Todaro M, Stassi G, De Maria R (2014) Colorectal cancer stem cells: from the crypt to the clinic. Cell Stem Cell 15(6): 692–705
3. De Angelis ML, Zeuner A, Policicchio E, Russo G, Bruselles A, Signore M, Vitale S, De Luca G, Pilozzi E, Boe A, Stassi G, Ricci-Vitiani L, Amoreo CA, Pagliuca A, Francescangeli F, Tartaglia M, De Maria R, Baiocchi M (2016) Cancer stem cell-based models of colorectal cancer reveal molecular determinants of therapy resistance. Stem Cells Transl Med 5(4):511–523
4. De Angelis ML, Bruselles A, Francescangeli F, Pucilli F, Vitale S, Zeuner A, Tartaglia M, Baiocchi M (2018) Colorectal cancer spheroid biobanks: multi-level approaches to drug sensitivity studies. Cell Biol Toxicol 34(6):459–469
5. Francescangeli F, Contavalli P, De Angelis ML, Baiocchi M, Gambara G, Pagliuca A, Fiorenzano A, Prezioso C, Boe A, Todaro M, Stassi G, Castro NP, Watanabe K, Salomon DS, De Maria R, Minchiotti G, Zeuner A (2015) Dynamic regulation of the cancer stem cell compartment by Cripto-1 in colorectal cancer. Cell Death Differ 22(10):1700–1713. https://doi.org/10.1038/cdd.2015.19
6. Francescangeli F, Contavalli P, De Angelis ML, Careccia S, Signore M, Haas TL, Salaris F, Baiocchi M, Boe A, Giuliani A, Tcheremenskaia O, Pagliuca A, Guardiola O, Minchiotti G, Colace L, Ciardi A, D'Andrea V,

La Torre F, Medema J, De Maria R, Zeuner A (2020) A pre-existing population of ZEB2(+) quiescent cells with stemness and mesenchymal features dictate chemoresistance in colorectal cancer. J Exp Clin Cancer Res 39(1):2. https://doi.org/10.1186/s13046-019-1505-4

7. Baiocchi M, Biffoni M, Ricci-Vitiani L, Pilozzi E, De Maria R (2010) New models for cancer research: human cancer stem cell xenografts. Curr Opin Pharmacol 10(4):380–384
8. Fumagalli A, Drost J, Suijkerbuijk SJ, van Boxtel R, de Ligt J, Offerhaus GJ, Begthel H, Beerling E, Tan EH, Sansom OJ, Cuppen E, Clevers H, van Rheenen J (2017) Genetic dissection of colorectal cancer progression by orthotopic transplantation of engineered cancer organoids. Proc Natl Acad Sci U S A 114(12):E2357–E2364
9. O'Rourke KP, Loizou E, Livshits G, Schatoff EM, Baslan T, Manchado E, Simon J, Romesser PB, Leach B, Han T, Pauli C, Beltran H, Rubin MA, Dow LE, Lowe SW (2017) Transplantation of engineered organoids enables rapid generation of metastatic mouse models of colorectal cancer. Nat Biotechnol 35(6):577–582
10. Roper J, Tammela T, Cetinbas NM, Akkad A, Roghanian A, Rickelt S, Almeqdadi M, Wu K, Oberli MA, Sanchez-Rivera FJ, Park YK, Liang X, Eng G, Taylor MS, Azimi R, Kedrin D, Neupane R, Beyaz S, Sicinska ET, Suarez Y, Yoo J, Chen L, Zukerberg L, Katajisto P, Deshpande V, Bass AJ, Tsichlis PN, Lees J, Langer R, Hynes RO, Chen J, Bhutkar A, Jacks T, Yilmaz OH (2017) In vivo genome editing and organoid transplantation models of colorectal cancer and metastasis. Nat Biotechnol 35(6):569–576
11. Julien S, Merino-Trigo A, Lacroix L, Pocard M, Goere D, Mariani P, Landron S, Bigot L, Nemati F, Dartigues P, Weiswald LB, Lantuas D, Morgand L, Pham E, Gonin P, Dangles-Marie V, Job B, Dessen P, Bruno A, Pierre A, De The H, Soliman H, Nunes M, Lardier G, Calvet L, Demers B, Prevost G, Vrignaud P, Roman-Roman S, Duchamp O, Berthet C (2012) Characterization of a large panel of patient-derived tumor xenografts representing the clinical heterogeneity of human colorectal cancer. Clin Cancer Res 18(19):5314–5328
12. Puig I, Chicote I, Tenbaum SP, Arques O, Herance JR, Gispert JD, Jimenez J, Landolfi S, Caci K, Allende H, Mendizabal L, Moreno D, Charco R, Espin E, Prat A, Elez ME, Argiles G, Vivancos A, Tabernero J, Rojas S, Palmer HG (2013) A personalized preclinical model to evaluate the metastatic potential of patient-derived colon cancer initiating cells. Clin Cancer Res 19(24):6787–6801
13. Roque-Lima B, Roque C, Begnami MD, Peresi P, Lima ENP, Mello CAL, Coimbra FJ, Chojniak R, Goss Santos T (2019) Development of patient-derived orthotopic xenografts from metastatic colorectal cancer in nude mice. J Drug Target 27(9):943–949
14. De Angelis ML, De Maria R, Baiocchi M (2018) How to assess drug resistance in cancer stem cells. Methods Mol Biol 1692:107–115
15. Workman P, Aboagye EO, Balkwill F, Balmain A, Bruder G, Chaplin DJ, Double JA, Everitt J, Farningham DA, Glennie MJ, Kelland LR, Robinson V, Stratford IJ, Tozer GM, Watson S, Wedge SR, Eccles SA (2010) Guidelines for the welfare and use of animals in cancer research. Br J Cancer 102(11):1555–1577. https://doi.org/10.1038/sj.bjc.6605642
16. Mezzanotte L, van't Root M, Karatas H, Goun EA, Lowik C (2017) In vivo molecular bioluminescence imaging: new tools and applications. Trends Biotechnol 35(7):640–652. https://doi.org/10.1016/j.tibtech.2017.03.012

Index

Nagarajan Kannan and Philip Beer (eds.), *Stem Cell Assays: Methods and Protocols*, Methods in Molecular Biology, vol. 2429, https://doi.org/10.1007/978-1-0716-1979-7, © Springer Science+Business Media, LLC, part of Springer Nature 2022

Q

R

S

T

U

V

Z